Selected Titles in This Subseries

43 **A. Yu. Morozov and M. A. Olshanetsky, Editors,** Moscow Seminar in Mathematical Physics (TRANS2/191)

42 **S. Tabachnikov, Editor,** Differential and Symplectic Topology of Knots and Curves (TRANS2/190)

41 **V. Buslaev, M. Solomyak, and D. Yafaev, Editors,** Differential Operators and Spectral Theory (M. Sh. Birman's 70th anniversary collection) (TRANS2/189)

40 **M. V. Karasev, Editor,** Coherent Transform, Quantization, and Poisson Geometry (TRANS2/187)

39 **A. Khovanskiĭ, A. Varchenko, and V. Vassiliev, Editors,** Geometry of Differential Equations (TRANS2/186)

38 **B. Feigin and V. Vassiliev, Editors,** Topics in Quantum Groups and Finite-Type Invariants (Mathematics at the Independent University of Moscow) (TRANS2/185)

37 **Peter Kuchment and Vladimir Lin, Editors,** Voronezh Winter Mathematical Schools (Dedicated to Selim Krein) (TRANS2/184)

36 **V. E. Zakharov, Editor,** Nonlinear Waves and Weak Turbulence (TRANS2/182)

35 **G. I. Olshanski, Editor,** Kirillov's Seminar on Representation Theory (TRANS2/181)

34 **A. Khovanskiĭ, A. Varchenko, and V. Vassiliev, Editors,** Topics in Singularity Theory (TRANS2/180)

33 **V. M. Buchstaber and S. P. Novikov, Editors,** Solitons, Geometry, and Topology: On the Crossroad (TRANS2/179)

32 **R. L. Dobrushin, R. A. Minlos, M. A. Shubin, and A. M. Vershik, Editors,** Topics in Statistical and Theoretical Physics (F. A. Berezin Memorial Volume) (TRANS2/177)

31 **R. L. Dobrushin, R. A. Minlos, M. A. Shubin, and A. M. Vershik, Editors,** Contemporary Mathematical Physics (F. A. Berezin Memorial Volume) (TRANS2/175)

30 **A. A. Bolibruch, A. S. Merkur′ev, and N. Yu. Netsvetaev, Editors,** Mathematics in St. Petersburg (TRANS2/174)

29 **V. Kharlamov, A. Korchagin, G. Polotovskiĭ, and O. Viro, Editors,** Topology of Real Algebraic Varieties and Related Topics (TRANS2/173)

28 **L. A. Bunimovich, B. M. Gurevich, and Ya. B. Pesin, Editors,** Sinai's Moscow Seminar on Dynamical Systems (TRANS2/171)

27 **S. P. Novikov, Editor,** Topics in Topology and Mathematical Physics (TRANS2/170)

26 **S. G. Gindikin and E. B. Vinberg, Editors,** Lie Groups and Lie Algebras: E. B. Dynkin's Seminar (TRANS2/169)

25 **V. V. Kozlov, Editor,** Dynamical Systems in Classical Mechanics (TRANS2/168)

24 **V. V. Lychagin, Editor,** The Interplay between Differential Geometry and Differential Equations (TRANS2/167)

23 **Yu. Ilyashenko and S. Yakovenko, Editors,** Concerning the Hilbert 16th Problem (TRANS2/165)

22 **N. N. Uraltseva, Editor,** Nonlinear Evolution Equations (TRANS2/164)

Published Earlier as Advances in Soviet Mathematics

21 **V. I. Arnold, Editor,** Singularities and bifurcations, 1994

20 **R. L. Dobrushin, Editor,** Probability contributions to statistical mechanics, 1994

19 **V. A. Marchenko, Editor,** Spectral operator theory and related topics, 1994

18 **Oleg Viro, Editor,** Topology of manifolds and varieties, 1994

17 **Dmitry Fuchs, Editor,** Unconventional Lie algebras, 1993

16 **Sergei Gelfand and Simon Gindikin, Editors,** I. M. Gelfand seminar, Parts 1 and 2, 1993

(Continued in the back of this publication)

Moscow Seminar in Mathematical Physics

American Mathematical Society
TRANSLATIONS
Series 2 • Volume 191

Advances in the Mathematical Sciences — 43

(*Formerly Advances in Soviet Mathematics*)

Moscow Seminar in Mathematical Physics

A. Yu. Morozov
M. A. Olshanetsky
Editors

American Mathematical Society
Providence, Rhode Island

Translation edited by A. B. Sossinski.

1991 *Mathematics Subject Classification.* Primary 58Fxx; Secondary 81Txx.

ABSTRACT. The volume contains articles resulting from talks given at the seminar in mathematical physics at Moscow Institute of Theoretical and Experimental Physics. The articles are mainly devoted to various aspects of Knizhnik-Zamolodchikov-Bernard connections and integrable models in two-dimensional quantum field theory.

The book is useful for researchers and graduate students working in various areas of mathematical and theoretical physics.

Library of Congress Card Number 91-640741
ISBN 0-8218-1388-9
ISSN 0065-9290

∞ The paper used in this book is acid-free and falls within the guidelines
established to ensure permanence and durability.
Visit the AMS home page at URL: `http://www.ams.org/`

10 9 8 7 6 5 4 3 2 1   04 03 02 01 00 99

Contents

Preface

ITEP: Proceedings of the Mathematical Physics Seminar

The Theory Department of the Institute of Theoretical and Experimental Physics was founded by I. Pomeranchuk in 1946 and is internationally recognized for achievements in various branches of theoretical physics. For many years, ITEP and especially its seminars have been among the main centers of scientific life in Moscow. The most remarkable feature of the best Moscow seminars is that these are places where science is done, not just reported.

This volume represents some results from the seminar on mathematical physics, which has been conducted at ITEP since 1983. This is not the only ITEP theoretical seminar. For this reason many of the famous results of the ITEP school are not reflected in this book. Another restriction on its content is the emphasis on the current activities in one particular field: integrable systems and their role in modern quantum theory. This has been one of the favorite topics of research at ITEP for many years. A lot of old results due to the ITEP theory group in integrability are beyond the scope of this volume. Still, it represents to a good extent the general style and one particular direction of our work. At the present time it attracts new attention because of the role played by integrability in describing nonperturbative phenomena in the framework of contemporary string theory.

Generally speaking, research topics included in the volume originally appeared in modern physics, and their transformation into well-defined mathematical problems is due to the intensive cooperation between physicists and mathematicians. The current mainstream of theoretical physics is becoming more and more geometrical. Thus, collaboration between physicists and mathematicians (with ITEP as one of the meeting points in Moscow) will undoubtedly produce new results in the same spirit in the future.

The seminar, although affiliated to ITEP, is in fact a meeting place which attracts people from different institutes. We have also been happy to see many mathematicians coming to the seminar. We are grateful to all those who have shared their time and inspiration with us during all these years. Moscow science could not have achieved its success without such scientific seminars, and it is important that they continue to work. We are glad that so many people feel the same way and are working toward this. Their support means a lot.

Most of the papers in this volume ([**1, 3, 4, 6, 8, 9**]) describe a certain universal structure in 2d quantum field theory, namely the Knizhnik–Zamolodchikov–Bernard connection and its far-reaching generalizations. The remaining papers are related to other aspects of 2d quantum field theory and integrable models. In [**7**] quantum Lax operators with elliptic dependence on a spectral parameter are analyzed by methods of algebraic geometry. A family of current algebras associated with complex curves is investigated in [**2**]. In [**5**] the relationship between matrix models and integrable systems is described.

We appreciate the kind suggestion of Simon Gindikin that we should publish this book. We are also indebted to Sergey Kharchev and Alesha Rosly for their help in the work on the volume.

A. Morozov, M. Olshanetsky

References

[1] G. Arutyunov, L. Chekhov and S. Frolov, *Quantum dynamical R-matrices*, this volume, pp. 1–32.

[2] B. Enriques and V. Rubtsov, *Some examples of quantum groups in higher genus*, this volume, pp. 33–65.

[3] V. Fock and A. Rosly, *Poisson structure on moduli of flat connections on Riemann surfaces and the r-matrix*, this volume, pp. 67–86.

[4] D. Ivanov and A. Losev, *KZB equations as a flat connection with spectral parameter*, this volume, pp. 87–117.

[5] S. Kharchev, *Kadomtsev–Petviashvili hierarchy and generalized Kontsevich model*, this volume, pp. 119–162.

[6] S. Khoroshkin, D. Lebedev, and S. Pakuliak, *Yangian algebras and classical Riemann problems*, this volume, pp. 163–198.

[7] I. Krichever and A. Zabrodin, *Vacuum curves of elliptic L-operators and representations of the Sklyanin algebra*, this volume, pp. 199–221.

[8] A. Levin and M. Olshanetsky, *Hierarchies of isomonodromic deformations and Hitchin systems*, this volume, pp. 223–262.

[9] N. Nekrasov, *Infinite-dimensional algebras, many-body systems and gauge theories*, this volume, pp. 263–299.

Amer. Math. Soc. Transl.
(2) Vol. **191**, 1999

Quantum Dynamical R-Matrices

G. E. Arutyunov, L. O. Chekhov, and S. A. Frolov

ABSTRACT. It is shown that the classical L-operator algebra of the elliptic Ruijsenaars–Schneider model can be realized as a subalgebra of the algebra of functions on the cotangent bundle over the centrally extended current group in two dimensions. It is governed by two dynamical r- and $\bar{r}$-matrices satisfying a closed system of equations. The corresponding quantum R- and $\overline{R}$-matrices are found as solutions to quantum analogs of these equations. We present the quantum L-operator algebra and show that the system of equations for R and $\overline{R}$ arises as the compatibility condition for this algebra. The simplest representation of the quantum L-operator algebra corresponding to the elliptic Ruijsenaars–Schneider model is obtained. The connection of the quantum L-operator algebra to the fundamental relation $R^B LL = LLR^B$ with the Belavin elliptic R-matrix is established. As a by-product of our construction, we find a new N-parameter elliptic solution of the classical Yang–Baxter equation.

1. Introduction

The appearance of classical dynamical r-matrices [**1, 2**] in the theory of integrable many-body systems gives rise to the interesting problem of their quantization. In this way, one may hope to separate the variables explicitly.

At present, the classical dynamical r-matrices are known for the rational, trigonometric [**2, 3**], and elliptic [**4, 5**] Calogero–Moser (CM) systems, as well as for their relativistic generalizations—the rational, trigonometric [**6, 7**], and elliptic [**8, 9**] Ruijsenaars–Schneider (RS) systems [**10**]. It is recognized that Calogero-type dynamical systems can be understood in the framework of the Hamiltonian reduction procedure [**11, 12**]. Moreover, this reduction procedure provides an effective scheme for computing the corresponding dynamical r-matrices [**3, 13**].

Depending explicitly on the phase variables, the dynamical r-matrices do not satisfy a single closed Yang–Baxter-type equation, which makes the problem of

---

1991 *Mathematics Subject Classification*. Primary 81R50; Secondary 58F07.

This work is supported in part by RFBR Grants 95-01-01106 and 96-01-00551 and by ISF Grant No. a96-1516.

their quantization rather nontrivial. In [**14**] the spin generalization of the Calogero–Sutherland system was quantized using the particular solution [**15**] of the Gervais–Neveu–Felder equation [**16, 15**], and in [**17**] it was interpreted in terms of quasi-Hopf algebras. Though this system is not integrable, the zero-weight representations of the quantum L-operator algebra admit a proper number of commuting integrals of motion. However, it is important to find a quantum L-operator algebra for Calogero-type systems that possesses a sufficiently large Abelian subalgebra.

Recently, an algebraic scheme for quantizing the rational RS model in the R-matrix formalism was proposed [**18**], where a special parameterization of the cotangent bundle over $GL(N, \mathbb{C})$ was introduced. In the new variables, the standard symplectic structure was described by a classical (Frobenius) r-matrix and by a new dynamical $\bar{r}$-matrix. The classical L-operator was introduced as a special matrix function on the cotangent bundle. The Poisson algebra of L, inherited from the cotangent bundle, coincides with the L-operator algebra of the rational RS model. For this reason we call L the L-operator. Quantizing the Poisson structure for L, two of the present authors found the quantum L-operator algebra and constructed its particular representation corresponding to the rational RS system. This quantum algebra has a remarkable property, namely, it possesses a family of N mutually commuting operators directly on the algebra level.

It is well known that the elliptic RS model is the most general among the CM- and RS-type systems. In this paper, we include this model in our scheme. Recall that the classical L-operator algebra for the elliptic RS model can be obtained by means of the Poisson [**19**] or the Hamiltonian [**20**] reduction scheme. In the first scheme, the affine Heisenberg double is used as the initial phase space. In the second, the cotangent bundle over the centrally extended current group in two dimensions is considered. Thus, the appropriate phase space that we choose for dealing with the elliptic RS model is the cotangent bundle $T^\star\widehat{GL}(N)(z,\bar{z})$ over the centrally extended group $\widehat{GL}(N)(z,\bar{z})$ of double loops. The application of our approach [**18**] is not straightforward, since one must work with infinite-dimensional phase space and, therefore, the correct description of the Poisson structure on $T^\star\widehat{GL}(N)(z,\bar{z})$ in the desired parameterization requires an intermediate regularization.

Let us briefly describe the content of the paper and the results obtained. In Section 2, we begin by giving a description of the Poisson structure on $T^\star\widehat{GL}(N)(z,\bar{z})$ that depends on the two complex parameters k and α; after that we parametrize $T^\star\widehat{GL}(N)(z,\bar{z})$ in a special way. The Poisson structure in the new variables is ill defined due to the presence of singularities. To overcome this problem, we introduce an intermediate regularization. Removing the regularization, we find that the resulting Poisson structure is well defined only for $\alpha = 1/N$. The value $\alpha = 1/N$ corresponds, in fact, to the case where only the subgroup $SL(N)(z,\bar{z})$ is centrally extended. The corresponding Poisson structure is described by two matrices $\mathbf{r}$ and $\bar{\mathbf{r}}$, which depend on N dynamical variables q_i. The Jacobi identity implies (Theorem 1) that $\mathbf{r}$ is an N-parameter elliptic solution to the classical Yang–Baxter equation (CYBE). It is worthwhile mentioning that the main elliptic identities (see Appendix A) follow the fact that the CYBE is fulfilled for $\mathbf{r}$. We expect the matrix $\mathbf{r}$ to be related to a special Frobenius subgroup in $GL(N)(z,\bar{z})$ as it was in the finite-dimensional case [**18**]. The Jacobi identity also implies a closed system of equations for $\mathbf{r}$ and $\bar{\mathbf{r}}$.

Further we define a special matrix function L on $T^\star\widehat{GL}(N)(z,\bar{z})$. We call this function the "L-operator" since, as we show, the Poisson algebra of L inherited from $T^\star\widehat{GL}(N)(z,\bar{z})$ literally coincides with the one for the elliptic RS model [**9, 20**]. Thus, the classical L-operator algebra can be realized as a subalgebra of the algebra of functions on $T^\star\widehat{GL}(N)(z,\bar{z})$. It turns out that the L-operator, as a function on $T^\star\widehat{GL}(N)(z,\bar{z})$, admits the factorization $L = WP$, where $p_i = \log P_i$ are the variables canonically conjugate to q_i and W belongs to a certain special subgroup in $GL(N)(z,\bar{z})$. The Poisson bracket for the entries of W is given by the matrix $\mathbf{r}$ and coincides with the Sklyanin bracket defining the structure of the Poisson–Lie group.

Although the quantum analogs of equations for $\mathbf{r}$ and $\bar{\mathbf{r}}$ can be easily established, it is rather difficult to find the corresponding quantum $\mathbf{R}$- and $\overline{\mathbf{R}}$-matrices. The problem is that the matrices $\mathbf{r}$ and $\bar{\mathbf{r}}$ have a complicated structure, $\mathbf{r} = r - s\otimes I + I\otimes s$ and $\bar{\mathbf{r}} = \bar{r} - s\otimes I$, due to the presence of the s-matrix. However, we observe that the classical L-operator algebra does not depend on s and, moreover, that the matrices r and $\bar{r}$ also obey a closed system of equations. We show that this system arises as the compatibility condition for a new Poisson algebra. This algebra contains both the Poisson algebra of $T^\star\widehat{GL}(N)(z,\bar{z})$ and the classical L-operator algebra as its subalgebras. In Section 3, using this key observation, we pass to the quantization.

We find the corresponding quantum R- and $\overline{R}$-matrices as solutions to quantum analogs of the equations for r and $\bar{r}$. In particular, the R-matrix satisfies a new triangle relation, which differs from the standard quantum Yang–Baxter equation by a special shift of the spectral parameters. The Felder elliptic R^F-matrix naturally arises in our construction. It turns out that the R-matrix is twist-equivalent to the R^F-matrix, with the $\overline{R}$-matrix playing the role of the twist.

Next, we derive a new quadratic algebra which is satisfied by the "quantum" L-operator. This algebra is described by the quantum dynamical R-matrices, namely, R, R^F, and $\overline{R}$:

$$R_{12}L_1\overline{R}_{21}L_2 = L_2\overline{R}_{12}L_1R^F_{12}.$$

We show that the system of equations for the R-, R^F-, and $\overline{R}$-matrices arises as the compatibility condition for this algebra. We present the simplest representation of the quantum L-operator algebra corresponding to the elliptic RS model. After a simple canonical transformation is performed, the quantum L-operator coincides, in essence, with the classical L-operator found in [**10**].

The quantum integrals of motion for the elliptic RS model were obtained in [**10**]. In [**21**], it was shown that any operator from the Ruijsenaars commuting family can be realized as the trace of a proper transfer matrix for the special $\widehat{L}$-operator obeying the relation $R^B\widehat{L}\widehat{L} = \widehat{L}\widehat{L}R^B$ with the Belavin elliptic R-matrix [**26**]. We note that our L-operator is gauge-equivalent to $\widehat{L}$. It follows from this observation that the determinant formula for the commuting family [**21**] is also valid for L. We show that any representation of our L-operator algebra is gauge equivalent to a representation of the relations $R^B\widehat{L}\widehat{L} = \widehat{L}\widehat{L}R^B$.

In conclusion, we discuss some unsolved problems.

2. Classical L-operator algebra

2.1. Poisson structure of $T^*\widehat{GL}(N)(z,\bar{z})$. Let T_τ be a torus endowed with the standard complex structure and periods 1 and τ. Denote by G the group of smooth mappings from T_τ into the group $GL(N,\mathbb{C})$; then $g \in G$ is a doubly-periodic matrix function $g(z,\bar{z})$. The space dual to the Lie algebra of G is spanned by the double-periodic functions $A(z,\bar{z})$ with values in $\mathrm{Mat}(N,\mathbb{C})$. In what follows, we often use the concise notation $g(z,\bar{z}) = g(z)$ and $A(z,\bar{z}) = A(z)$. The group G admits central extensions $\widehat{G}$ [**22**]. The Poisson structure on $T^*\widehat{G}$ with fixed central charges reads

$$\{A_1(z), A_2(w)\} = \frac{1}{2}[C, A_1(z) - A_2(w)]\delta(z-w) - k(C-\alpha I)\frac{\partial}{\partial \bar{z}}\delta(z-w), \tag{2.1}$$

$$\{g_1(z), g_2(w)\} = 0, \tag{2.2}$$

$$\{A_1(z), g_2(w)\} = g_2(w)C\delta(z-w), \tag{2.3}$$

where k and α are central charges and $\delta(z)$ is the two-dimensional δ-function

$$\int_{T_\tau} d^2z f(z)\,\delta(z) = f(0), \qquad d^2z = d\bar{z}\wedge dz.$$

Here we use the standard tensor notation; C is the permutation operator.

Consider the following Hamiltonian action of G on $T^*\widehat{G}$:

$$\begin{aligned} A(z) &\to T^{-1}(z)A(z)T(z) + kT^{-1}(z)\bar{\partial}T(z), \\ g(z) &\to T^{-1}(z)g(z)T(z). \end{aligned} \tag{2.4}$$

We restrict our consideration to the case of smooth elements $A(z)$.

LEMMA 2.1 ([**23**]). *The generic element $A(z)$ can be diagonalized by the transformation* (2.4),

$$A(z) = T(z)DT^{-1}(z) - k\,\bar{\partial}T(z)T^{-1}(z), \tag{2.5}$$

where D is a constant diagonal matrix with elements D_i, $D_i \neq D_j$, and $T(z)$ is a double-periodic matrix.

The matrix D is defined up to the action of the elliptic Weyl group. One can fix D by choosing the fundamental Weyl chamber.

The matrix $T(z)$ in Eq. (2.5) is not uniquely defined. Any element $\tilde{T}(z,\bar{z}) = T(z,\bar{z})h(z)$, where the diagonal matrix $h(z)$ is an entire function of z, also satisfies (2.5). Stipulating that $\tilde{T}(z,\bar{z})$ is double-periodic, we find that $h(z)$ is a constant matrix. We can remove the remaining ambiguity by imposing the condition

$$T(\varepsilon)e = e, \tag{2.6}$$

where e is a vector such that $e_i = 1$ for all i and ε is an arbitrary point on T_τ. In what follows, we denote the matrix $T(z)$ that solves Eq. (2.5) and satisfies Eq. (2.6) by $T^\varepsilon(z)$. Such matrices obviously form a group.

Now we rewrite the Poisson structure (2.1) in terms of variables T and D. Since D_i are G-invariant functions, they belong to the center of (2.1) and, therefore, it suffices to calculate the bracket $\{T^\varepsilon(z), T^\varepsilon(w)\}$. However, straightforward calculation reveals that this bracket is ill defined. Therefore, we begin by calculating the bracket $\{T^\varepsilon(z), T^\eta(w)\}$, where $T^\varepsilon(z)$ and $T^\eta(w)$ satisfy (2.6) at different points ε and η,

$$\{T^\varepsilon_{ij}(z), T^\eta_{kl}(w)\} = \sum_{mnps} \int d^2z' d^2w' \frac{\delta T^\varepsilon_{ij}(z)}{\delta A_{mn}(z')} \frac{\delta T^\eta_{kl}(w)}{\delta A_{ps}(w')} \{A_{mn}(z'), A_{ps}(w')\}. \tag{2.7}$$

To calculate the functional derivative $\delta T^\varepsilon_{ij}(z)/\delta A_{mn}(z')$, we consider a variation of (2.5),

$$X(z) = t(z)D - Dt(z) - k\,\bar\partial t(z) + d, \tag{2.8}$$

where $X(z) = T^{-1}(z)\,\delta A(z) T(z)$, $t(z) = T^{-1}(z)\,\delta T(z)$, and $d = \delta D$.

From (2.8), we obtain

$$\frac{\delta D_i}{\delta A_{kl}(z)} = \frac{1}{(\tau - \overline{\tau})} T^{-1\,\varepsilon}_{ik}(z) T^\varepsilon_{li}(z). \tag{2.9}$$

Let us introduce the following function $\Phi(z,s)$ of two complex variables:

$$\Phi(z,s) = \frac{\sigma(z+s)}{\sigma(z)\sigma(s)} e^{-2\zeta(1/2)zs} e^{2\pi i s(z-\bar z)/(\tau-\overline{\tau})}. \tag{2.10}$$

Here $\sigma(z)$ and $\zeta(z)$ are the Weierstrass σ- and ζ-functions with periods equal to 1 and τ. The function $\Phi(z,s)$ is the only double-periodic solution to the following equation:

$$\bar\partial \Phi(z,s) + \frac{2\pi i s}{\tau - \overline{\tau}} \Phi(z,s) = 2\pi i \delta(z). \tag{2.11}$$

It is also convenient to define $\Phi(z,0)$ as follows:

$$\Phi(z,0) = \lim_{\varepsilon \to 0} \left(\Phi(z,\varepsilon) - \frac{1}{\varepsilon}\right) = \zeta(z) - 2\zeta\Big(\frac{1}{2}\Big)z + 2\pi i \frac{z - \bar z}{\tau - \overline{\tau}}.$$

This function solves the equation

$$\bar\partial \Phi(z,0) = 2\pi i\, \delta(z) - \frac{2\pi i}{\tau - \overline{\tau}}.$$

We introduce the notation

$$q_{ij} \equiv q_i - q_j, \quad \text{where } q_i = \frac{\tau - \overline{\tau}}{2\pi i k} D_i.$$

Using these functions, one can write the solution to (2.8) obeying the condition $t(\varepsilon)e = 0$ [**20**],

$$t(z) = \kappa \sum_{i,j} \int d^2w (\Phi(\varepsilon - w, q_{ij}) X_{ij}(w) E_{ii} - \Phi(z - w, q_{ij}) X_{ij}(w) E_{ij}). \tag{2.12}$$

In what follows, we denote $(2\pi i k)^{-1}$ by κ.

Varying (2.12) with respect to $A_{mn}(w)$, one obtains

$$\frac{\delta T^\varepsilon_{ij}(z)}{\delta A_{mn}(w)} = \kappa \Big(\sum_k \Phi(\varepsilon - w, q_{jk}) T^\varepsilon_{ij}(z) T^{-1\,\varepsilon}_{jm}(w) T^\varepsilon_{nk}(w) - \sum_k \Phi(z - w, q_{kj}) T^\varepsilon_{ik}(z) T^{-1\,\varepsilon}_{km}(w) T^\varepsilon_{nj}(w) \Big).$$

To compute the bracket (2.7), one needs the following relation between $T^\varepsilon(z)$ and $T^\eta(z)$:

$$T^\varepsilon(z) = T^\eta(z) H^{\eta\varepsilon}, \tag{2.13}$$

where $H^{\eta\varepsilon}$ is a constant diagonal matrix.

By direct computation, one finds that

$$\frac{1}{\kappa}\{T_1^\varepsilon(z), T_2^\eta(w)\} = T_1^\varepsilon(z)T_2^\eta(w)(H_1^{\varepsilon\eta}H_2^{\eta\varepsilon}r_{12}^{\varepsilon\eta}(z,w) - \alpha f^{\varepsilon\eta}(z,w)), \tag{2.14}$$

where

$$\begin{aligned} r_{12}^{\varepsilon\eta}(z,w) = &\sum_{ij}\Phi(\varepsilon-\eta, q_{ij})E_{ii}\otimes E_{jj} + \sum_{ij}\Phi(z-w, q_{ij})E_{ij}\otimes E_{ji} \\ &- \sum_{ij}\Phi(z-\eta, q_{ij})E_{ij}\otimes E_{jj} + \sum_{ij}\Phi(w-\varepsilon, q_{ij})E_{jj}\otimes E_{ij}, \end{aligned} \tag{2.15}$$

$$f^{\varepsilon\eta}(z,w) = \Phi(\varepsilon-\eta,0) + \Phi(\omega-\varepsilon,0) + \Phi(z-\omega,0) - \Phi(z-\eta,0). \tag{2.16}$$

The bracket (2.14) has an r-matrix form with the r-matrix depending not only on the coordinates q_i, but also on the additional variables H.

In the limit $\eta \to \varepsilon$, a singularity in (2.14) arises. This shows that the variable $T(z)$ is not a good candidate for describing the Poisson structure (2.1). However, one can use the freedom to multiply $T(z)$ by any functional of A. Thus, we introduce a new variable,

$$\mathbf{T}^\varepsilon(z) = T^\varepsilon(z)(\det T^\varepsilon(\varepsilon))^\beta. \tag{2.17}$$

We use $\det T^\varepsilon(\varepsilon)$ in the definition of $\mathbf{T}^\varepsilon(z)$ in order to preserve the group structure for the new variables.

Using the Poisson bracket (2.14), one immediately finds that

$$\begin{aligned} \frac{1}{\kappa}\{\mathbf{T}_1^\varepsilon(z), \mathbf{T}_2^\eta(w)\} = \mathbf{T}_1^\varepsilon(z)\mathbf{T}_2^\eta(w)(&H_1^{\varepsilon\eta}H_2^{\eta\varepsilon}r_{12}^{\varepsilon\eta}(z,w) - \alpha f^{\varepsilon\eta}(z,w) \\ &+ \beta I\otimes \operatorname{tr}_3 H_3^{\varepsilon\eta}H_2^{\eta\varepsilon}r_{32}^{\varepsilon\eta}(\varepsilon,w) \\ &+ \beta \operatorname{tr}_3 H_1^{\varepsilon\eta}H_3^{\eta\varepsilon}r_{13}^{\varepsilon\eta}(z,\eta)\otimes I \\ &+ \beta^2 \operatorname{tr}_{34} H_3^{\varepsilon\eta}H_4^{\eta\varepsilon}r_{34}^{\varepsilon\eta}(\varepsilon,\eta)I\otimes I), \end{aligned} \tag{2.18}$$

since $f(\varepsilon,w) = f(z,\eta) = 0$.

Now, we pass to the limit as $\eta \to \varepsilon$.[1] To do this, one must take into account the following behavior of $H^{\varepsilon\eta}$ when η tends to ε: $H^{\varepsilon\eta} = 1 + (\varepsilon-\eta)h + o(\varepsilon-\eta)$, where h is a constant diagonal matrix that is a functional of A. It turns out that there exists a unique choice for α and β, namely, $\alpha = 1/N$, $\beta = -1/N$, for which the singularities are canceled and there is no contribution from the matrix h. In the limit $\eta \to \varepsilon = 0$,[2] for these values of α and β, we obtain

$$\frac{1}{\kappa}\{\mathbf{T}_1(z), \mathbf{T}_2(w)\} = \mathbf{T}_1(z)\mathbf{T}_2(w)\mathbf{r}_{12}(z,w). \tag{2.19}$$

Here the limiting $\mathbf{r}$-matrix reads

$$\mathbf{r}_{12}(z,w) = r_{12}(z,w) - s(z)\otimes I + I\otimes s(w) - \frac{1}{N}\Phi(z-w,0)I\otimes I, \tag{2.20}$$

[1] Without loss of generality, we assume that ε and η are real.

[2] The ε-dependence can be easily restored by shifting both z and w by ε.

where

$$r(z,w)=\sum_{i\neq j}\Phi(q_{ij})E_{ii}\otimes E_{jj}+\sum_{ij}\Phi(z-w,q_{ij})E_{ij}\otimes E_{ji}$$

$$-\sum_{ij}\Phi(z,q_{ij})E_{ij}\otimes E_{jj}+\sum_{ij}\Phi(w,q_{ij})E_{jj}\otimes E_{ij}, \tag{2.21}$$

$$s(z)=\frac{1}{N}\sum_{ij}(\Phi(q_{ij})E_{ii}-\Phi(z,q_{ij})E_{ij}). \tag{2.22}$$

Here the function $\Phi(q_{ij})$ denotes the regular part of $\Phi(\varepsilon,q_{ij})$ as $\varepsilon\to 0$,

$$\Phi(q_{ij})=\zeta(q_{ij})-2\zeta(1/2)q_{ij},\quad i\neq j,\qquad \Phi(q_{ii})=0.$$

Note that both $\mathbf{r}$ and r are skew-symmetric, $\mathbf{r}_{12}(z,w)=-\mathbf{r}_{21}(w,z)$.

THEOREM 2.2. *The* $\mathbf{r}$*-matrix* (2.20) *is an* N*-parameter solution of the classical Yang–Baxter equation,*

$$[[\mathbf{r},\mathbf{r}]]\equiv[\mathbf{r}_{12}(z_1,z_2),\mathbf{r}_{13}(z_1,z_3)+\mathbf{r}_{23}(z_2,z_3)]+[\mathbf{r}_{13}(z_1,z_3),\mathbf{r}_{23}(z_2,z_3)]=0. \tag{2.23}$$

PROOF. Let us write the following equation, which follows from the Jacobi identity for the bracket (2.18):

$$\begin{aligned}&[\mathbf{r}_{12}^{\varepsilon\eta}(z,w),\mathbf{r}_{13}^{\varepsilon\rho}(z,s)+\mathbf{r}_{23}^{\eta\rho}(w,s)]+[\mathbf{r}_{13}^{\varepsilon\rho}(z,s),\mathbf{r}_{23}^{\eta\rho}(w,s)]\\&+\frac{1}{\kappa}T_3^{\rho\ -1}(s)\{T_3^{\rho}(s),\mathbf{r}_{12}^{\varepsilon\eta}(z,w)\}+\frac{1}{\kappa}T_1^{\varepsilon\ -1}(z)\{T_1^{\varepsilon}(z),\mathbf{r}_{23}^{\eta\rho}(w,s)\}\\&-\frac{1}{\kappa}T_2^{\eta\ -1}(w)\{T_2^{\eta}(w),\mathbf{r}_{13}^{\varepsilon\rho}(z,s)\}=0,\end{aligned} \tag{2.24}$$

where

$$\begin{aligned}\mathbf{r}_{12}^{\varepsilon\eta}(z,w)&=H_1^{\varepsilon\eta}H_2^{\eta\varepsilon}r_{12}^{\varepsilon\eta}(z,w)-\alpha f^{\varepsilon\eta}(z,w)+\beta I\otimes\operatorname{tr}_3 H_3^{\varepsilon\eta}H_2^{\eta\varepsilon}r_{32}^{\varepsilon\eta}(z,w)\\&+\beta\operatorname{tr}_3 H_1^{\varepsilon\eta}H_3^{\eta\varepsilon}r_{13}^{\varepsilon\eta}(z,w)\otimes I+\beta^2\operatorname{tr}_{34}H_3^{\varepsilon\eta}H_4^{\eta\varepsilon}r_{34}^{\varepsilon\eta}(z,w)I\otimes I.\end{aligned} \tag{2.25}$$

To study (2.24), one needs to know the Poisson bracket of $\mathbf{T}^{\varepsilon}(z)$ with $H^{\eta\rho}$. This bracket can be easily derived from Eqs. (2.13), (2.14), and (2.17):

$$\begin{aligned}&\frac{1}{\kappa}\mathbf{T}_1^{\varepsilon\ -1}(z)\{\mathbf{T}_1^{\varepsilon}(z),H_2^{\eta\rho}\}\\&=\sum_{i\neq j}H_i^{\varepsilon\rho}H_j^{\eta\varepsilon}\Phi(\varepsilon-\rho,q_{ij})\Big(E_{ii}-\frac{1}{N}I\Big)\otimes E_{jj}\\&\quad-\sum_{i\neq j}H_i^{\varepsilon\eta}H_j^{\eta\varepsilon}H_j^{\eta\rho}\Phi(\varepsilon-\eta,q_{ij})\Big(E_{ii}-\frac{1}{N}I\Big)\otimes E_{jj}\\&\quad-\sum_{i\neq j}H_i^{\varepsilon\rho}H_j^{\eta\varepsilon}\Phi(z-\rho,q_{ij})E_{ij}\otimes E_{jj}\\&\quad+\sum_{i\neq j}H_i^{\varepsilon\eta}H_j^{\eta\varepsilon}H_j^{\eta\rho}\Phi(z-\eta,q_{ij})E_{ij}\otimes E_{jj}\\&\quad+\sum_{j}\frac{1}{N}H_j^{\eta\rho}(\Phi(\varepsilon-\rho,0)+\Phi(\eta-\varepsilon,0)\\&\qquad+\Phi(z-\eta,0)-\Phi(z-\rho,0))\Big(E_{jj}-\frac{1}{N}I\Big)\otimes E_{jj}.\end{aligned} \tag{2.26}$$

Equation (2.24) holds at arbitrary values of all parameters. Without loss of generality, one can set $\rho = 0$ and $\eta = a\varepsilon$, where a and ε are real. Now let us perform the following change of variables $H_i^{\varepsilon\eta}$:

$$H_i^{\varepsilon\eta} = 1 + h_i^{\varepsilon\eta}$$

and consider the expansion on the left-hand side of (2.24) in powers of ε and h.

Note that the matrix $\mathbf{r}^{\varepsilon\eta}(z,w)$ has the following expansion in powers of h as $\varepsilon - \eta \to 0$:

$$\mathbf{r}^{\varepsilon\eta}(z,w) = \mathbf{r}(z,w) + \mathbf{r}^{(1)}_{\text{reg}}(z,w,\varepsilon)h + \mathbf{r}^{(2)}(z,w,\varepsilon)h^2 + o(\varepsilon - \eta), \tag{2.27}$$

where $\mathbf{r}(z,w)$ is given by (2.20). The matrix $\mathbf{r}^{(1)}_{\text{reg}}(z,w,\varepsilon)$ is regular as $\varepsilon \to 0$, and the ε-dependence of $\mathbf{r}^{(2)}(z,w,\varepsilon)$ is unessential. Substituting (2.27) into the bracket $\mathbf{T}^{-1}\{\mathbf{T},\mathbf{r}\}$, one obtains

$$\mathbf{T}^{-1}\{\mathbf{T},\mathbf{r}\} = \mathbf{r}^{(1)}_{\text{reg}}(\varepsilon)\mathbf{T}^{-1}\{\mathbf{T},h\} + \mathbf{r}^{(2)}(\varepsilon)h\mathbf{T}^{-1}\{\mathbf{T},h\}. \tag{2.28}$$

Clearly, Eq. (2.24) should be satisfied in any order in h and ε. Since we are interested in finding an equation for $\mathbf{r}$, we consider only the terms independent of h and ε in the expansion of Eq. (2.24). The terms of zero order in h and ε occurring in (2.24) come from $\mathbf{r}^{\varepsilon\eta}$ and from the first term in (2.28). However, from the explicit expression for the bracket $\{\mathbf{T},H\}$, one can see that $\{\mathbf{T},h\}|_{h=0} = o(\varepsilon)$. Now, taking into account that $\mathbf{r}^{(1)}_{\text{reg}}(\varepsilon)$ is regular as $\varepsilon \to 0$, we conclude that the last three terms in (2.24) give no contribution. Thus, in the zero order in h and ε, Eq. (2.24) reduces to the CYBE for $\mathbf{r}(z,w)$. □

Let us note that, as one might expect, the condition $\det \mathbf{T}(0) = 1$ is compatible with the bracket (2.19), since $\det \mathbf{T}(z)$ is a central element of algebra (2.19).

Note that the choice $\alpha = 1/N$ corresponds to the case in which only the subalgebra $sl(N)(z,\bar z)$ is centrally extended. In terms of $\mathbf{T}(z)$ ($\beta = -1/N$), the boundary condition reads $\mathbf{T}(0)e = \lambda e$ and $\det \mathbf{T}(0) = 1$. One can also check that the field $A(z)$ defined by (2.5) with $\mathbf{T}(z)$ replacing $T(z)$ obeys the Poisson algebra (2.1).

The next step is to consider a special parameterization for the field $g(z)$. To this end, we introduce $\widetilde{A}(z)$:

$$\widetilde{A} = gAg^{-1} - k\bar\partial g g^{-1} + \frac{k}{N}\operatorname{tr}\bar\partial g g^{-1}. \tag{2.29}$$

One can check that $\widetilde{A}(z)$ Poisson-commutes with $A(w)$ and obeys the following Poisson algebra:

$$\begin{aligned}\{\widetilde{A}_1(z),\widetilde{A}_2(w)\} &= -\frac{1}{2}\,[C,\widetilde{A}_1(z) - \widetilde{A}_2(w)]\,\delta(z-w)\\ &\quad + k\Big(C - \frac{1}{N}I\Big)\frac{\partial}{\partial\bar z}\,\delta(z-w),\\ \{\widetilde{A}_1(z),g_2(w)\} &= Cg_2(w)\,\delta(z-w).\end{aligned} \tag{2.30}$$

Now we factor $\widetilde{A}(z)$ in the same manner as for $A(z)$,

$$\widetilde{A}(z) = \mathbf{U}(z)D\mathbf{U}^{-1}(z) - k\,\bar\partial\mathbf{U}(z)\mathbf{U}^{-1}(z), \tag{2.31}$$

where $\mathbf{U}(z)$ satisfies the boundary condition $\mathbf{U}(0)e = \lambda e$ and $\det \mathbf{U}(0) = 1$. Obviously, $\mathbf{U}(z)$ Poisson-commutes with $\mathbf{T}(w)$ and satisfies the Poisson algebra

$$\frac{1}{\kappa}\{\mathbf{U}_1(z), \mathbf{U}_2(w)\} = -\mathbf{U}_1(z)\mathbf{U}_2(w)\mathbf{r}_{12}(z, w). \tag{2.32}$$

From (2.29) and (2.31) one can find the following representation for the field g:

$$g(z) = (\det g(z))^{1/N}\mathbf{U}(z)\mathbb{P}\mathbf{T}^{-1}(z), \tag{2.33}$$

where $\mathbb{P}$ is a constant diagonal matrix.

Computing the determinants on both sides of Eq. (2.33), we obtain

$$\det \mathbb{P} = \det(\mathbf{T}(z)/\mathbf{U}(z)).$$

Since the left-hand side does not depend on z and $\det \mathbf{T}(0) = \det \mathbf{U}(0) = 1$, we obtain $\det \mathbb{P} = 1$ and $\det \mathbf{T}(z) = \det \mathbf{U}(z)$.

Calculating the Poisson brackets of $\mathbb{P}$ with $\mathbb{P}$ and $Q = \operatorname{diag}(q_1, \dots, q_N)$ as above, we find that

$$\{\mathbb{P}_1, \mathbb{P}_2\} = 0, \tag{2.34}$$

$$\frac{1}{\kappa}\{Q_1, \mathbb{P}_2\} = \mathbb{P}_2\Big(\sum_i E_{ii} \otimes E_{ii} - \frac{1}{N} I \otimes I\Big). \tag{2.35}$$

In fact, this means that

$$\log \mathbb{P}_i = p_i - \frac{1}{N}\sum_i p_i,$$

where the p_i are canonically conjugate to q_i.

For the remaining Poisson brackets of $\mathbb{P}$ with the fields $\mathbf{T}$, $\mathbf{U}$, we have

$$\frac{1}{\kappa}\{\mathbf{T}_1(z), \mathbb{P}_2\} = \mathbf{T}_1(z)\mathbb{P}_2\bar{\mathbf{r}}_{12}(z), \tag{2.36}$$

$$\frac{1}{\kappa}\{\mathbf{U}_1(z), \mathbb{P}_2\} = \mathbf{U}_1(z)\mathbb{P}_2\bar{\mathbf{r}}_{12}(z). \tag{2.37}$$

Here

$$\bar{\mathbf{r}}_{12}(z) = \bar{r}_{12}(z) - s(z) \otimes I - \frac{1}{N} I \otimes \sum_{ij} \Phi(q_{ij})E_{jj}, \tag{2.38}$$

where we have introduced the $\bar{r}$-matrix

$$\bar{r}_{12}(z) = \sum_{ij} \Phi(q_{ij})E_{ii} \otimes E_{jj} - \sum_{ij} \Phi(z, q_{ij})E_{ij} \otimes E_{jj}. \tag{2.39}$$

To complete the description of the classical Poisson structure on the cotangent bundle, we present the Poisson bracket of $\det g$ with other variables,

$$\frac{1}{\kappa}\{Q, \det g(w)\} = \det g(w), \qquad \{\mathbb{P}, \det g(w)\} = 0,$$

$$\frac{1}{\kappa}\{\mathbf{T}(z), \det g(w)\} = -\det g(w)\mathbf{T}(z)(\Phi(z-w, 0) + \Phi(w, 0)),$$

$$\frac{1}{\kappa}\{\mathbf{U}(z), \det g(w)\} = -\det g(w)\mathbf{U}(z)(\Phi(z-w, 0) + \Phi(w, 0)).$$

Recall that the Jacobi identity for the bracket (2.19) reduces to the classical Yang–Baxter equation for the $\mathbf{r}$-matrix. As for the Poisson relations (2.36) and

(2.37), one finds that the Jacobi identity is equivalent to the following equation of the second order in $\bar{\mathbf{r}}$:

$$[\bar{\mathbf{r}}_{12}(z), \bar{\mathbf{r}}_{13}(z)] - \mathbb{P}_3^{-1}\{\bar{\mathbf{r}}_{12}(z), \mathbb{P}_3\} + \mathbb{P}_2^{-1}\{\bar{\mathbf{r}}_{13}(z), \mathbb{P}_2\} = 0 \tag{2.40}$$

and the equation involving $\mathbf{r}$ and $\bar{\mathbf{r}}$,

$$[\mathbf{r}_{12}(z,w), \bar{\mathbf{r}}_{13}(z) + \bar{\mathbf{r}}_{23}(w)] + [\bar{\mathbf{r}}_{13}(z), \bar{\mathbf{r}}_{23}(w)] - \mathbb{P}_3^{-1}\{\bar{\mathbf{r}}_{12}(z,w), \mathbb{P}_3\} = 0. \tag{2.41}$$

One can check by straightforward calculations that the matrices $\mathbf{r}$ and $\bar{\mathbf{r}}$ given by (2.20) and (2.38) do solve these equations.

Finally, we note that the fields $A(z)$ and $g(z)$ defined by (2.5) and (2.33) obey the Poisson relations (2.1)–(2.3).

Let us note that the Poisson relation for the generator $\mathbf{W}(z) = \mathbf{T}^{-1}(z)\mathbf{U}(z)$ turns out to be the Sklyanin bracket,

$$\frac{1}{\kappa}\{\mathbf{W}_1(z), \mathbf{W}_2(w)\} = [\mathbf{r}_{12}(z,w), \mathbf{W}_1(z)\mathbf{W}_2(w)], \tag{2.42}$$

which, therefore, defines the structure of a Poisson–Lie group. This group is an infinite-dimensional analog of the Frobenius group that appeared in [**18**], where the Poisson–Lie group structure was related to the existence of a nondegenerate 2-cocycle on the corresponding Lie algebra. It would be interesting to find a similar interpretation in the infinite-dimensional case.

2.2. Classical L-operator. In this subsection, we define a special function on the cotangent bundle, which we call the *classical L-operator*. The motivation for treating this function as an L-operator is that the Poisson algebra of L is equivalent to the one found in [**9**] for the L-operator of the elliptic RS model.

Denote by L the following function:

$$L(z) = \mathbf{T}^{-1}(z)g(z)\mathbf{T}(z) = (\det g(z))^{1/N}\mathbf{T}^{-1}(z)\mathbf{U}(z)\mathbb{P}. \tag{2.43}$$

Using the formulas of 2.1, one can easily derive the Poisson bracket of L and Q,

$$\frac{1}{\kappa}\{Q_1, L_2(z)\} = L_2(z)\sum_i E_{ii} \otimes E_{ii}, \tag{2.44}$$

and the Poisson algebra of the L-operator,

$$\begin{aligned} \frac{1}{\kappa}\{L_1(z), L_2(w)\} &= \mathbf{r}_{12}(z,w)L_1(z)L_2(w) \\ &\quad + L_1(z)L_2(w)(\bar{\mathbf{r}}_{12}(z) - \bar{\mathbf{r}}_{21}(w) - \mathbf{r}_{12}(z,w)) \\ &\quad + L_1(z)\bar{\mathbf{r}}_{21}(w)L_2(w) - L_2(w)\bar{\mathbf{r}}_{12}(z)L_1(z). \end{aligned} \tag{2.45}$$

Clearly, the generators Q and L form a Poisson subalgebra in the Poisson algebra of the cotangent bundle. An important feature of this subalgebra is that $I_n(z) = \operatorname{tr} L^n(z)$ forms a set of mutually commuting variables.

Just as in the finite-dimensional case [**18**], from (2.44), we can see that the L-operator admits the following factorization: $L(z) = W(z)P$, where Q and $\log P$ are canonically conjugate variables, the W-algebra coincides with (2.42), and the bracket of W and P is

$$\frac{1}{\kappa}\{W_1(z), P_2\} = -P_2[\bar{\mathbf{r}}_{12}(z), W_1(z)]. \tag{2.46}$$

Everything we need to quantize the L-operator algebra (2.45) is now prepared. The problem of quantization has been reduced to finding the quantum $\mathbf{R}$- and $\overline{\mathbf{R}}$-matrices satisfying the quantum analogs of Eqs. (2.23), (2.40), and (2.41),

$$\mathbf{R}_{12}(z_1,z_2)\mathbf{R}_{21}(z_2,z_1)=1, \tag{2.47}$$

$$\mathbf{R}_{12}(z_1,z_2)\mathbf{R}_{13}(z_1,z_3)\mathbf{R}_{23}(z_2,z_3)=\mathbf{R}_{23}(z_2,z_3)\mathbf{R}_{13}(z_1,z_3)\mathbf{R}_{12}(z_1,z_2), \tag{2.48}$$

$$\mathbf{R}_{12}(z_1,z_2)\overline{\mathbf{R}}_{13}(z_1)\overline{\mathbf{R}}_{23}(z_2)=\overline{\mathbf{R}}_{23}(z_2)\overline{\mathbf{R}}_{13}(z_1)P_3^{-1}\mathbf{R}_{12}(z_1,z_2)P_3, \tag{2.49}$$

$$\overline{\mathbf{R}}_{12}(z)P_2^{-1}\overline{\mathbf{R}}_{13}(z)P_2=\overline{\mathbf{R}}_{13}(z)P_3^{-1}\overline{\mathbf{R}}_{12}(z)P_3. \tag{2.50}$$

These matrices are assumed to have standard behaviors near $\hbar=0$,

$$\mathbf{R}=1+\hbar\mathbf{r}+o(\hbar), \qquad \overline{\mathbf{R}}=1+\hbar\bar{\mathbf{r}}+o(\hbar),$$

where $\hbar$ is the quantization parameter.

The problem as formulated seems to be rather complicated, due to the presence of the s-matrix in the classical $\mathbf{r}$- and $\bar{\mathbf{r}}$-matrices. However, the Poisson algebra (2.45) possesses an important property that allows one to avoid the problem at hand. Namely, the matrix $s(z)$ coming both in $\mathbf{r}$ and $\bar{\mathbf{r}}$ drops out of the right-hand side of (2.45). Therefore, the following proposition holds.

PROPOSITION 2.1. *The Poisson algebra of the L-operator* (2.43) *has the form*

$$\begin{aligned}\frac{1}{\kappa}\{L_1(z),L_2(w)\}&=r_{12}(z,w)L_1(z)L_2(w)\\&\quad-L_1(z)L_2(w)(r_{12}(z,w)+\bar r_{21}(w)-\bar r_{12}(z))\\&\quad+L_1(z)\bar r_{21}(w)L_2(w)-L_2(w)\bar r_{12}(z)L_1(z).\end{aligned} \tag{2.51}$$

If we denote the sum

$$r_{12}^F(z,w)=r_{12}(z,w)+\bar r_{21}(w)-\bar r_{12}(z) \tag{2.52}$$

by r_{12}^F, then, using (2.21) and (2.39), we obtain

$$r_{12}^F(z-w)=-\sum_{ij}\Phi(q_{ij})E_{ii}\otimes E_{jj}+\sum_{ij}\Phi(z-w,q_{ij})E_{ij}\otimes E_{ji}. \tag{2.53}$$

In this expression, one can recognize the elliptic solution to the classical Gervais–Neveu–Felder equation [**16, 15**]:

$$\begin{aligned}&[r_{12}^F(z_1-z_2),r_{13}^F(z_1-z_3)+r_{23}^F(z_2-z_3)]\\&\quad+[r_{13}^F(z_1-z_3),r_{23}^F(z_2-z_3)]-P_3^{-1}\{r_{12}^F(z_1-z_2),P_3\}\\&\quad+P_2^{-1}\{r_{13}^F(z_1-z_3),P_2\}-P_1^{-1}\{r_{23}^F(z_2-z_3),P_1\}=0.\end{aligned} \tag{2.54}$$

In fact, r^F emerges as the semiclassical limit of the quantum R-matrix found in [**15**].

The absence of the s-matrix in the resulting L-operator algebra and the appearance of the r^F-matrix show that a closed system of equations involving only r- and $\bar r$-matrices in the classical case, and R- and $\overline{R}$-matrices in the quantum case, may exist. In 2.3 below, we find the desired system of equations and describe the Poisson structure for which these equations are the consistency equations for the Jacobi identity.

Note that the algebra (2.51) coincides with the one obtained in [**20**] using the Hamiltonian reduction procedure. A mere similarity transformation of L turns the algebra (2.51) into the one previously found in [**9**]. In contrast to [**9**], where (2.51) was derived by direct calculations using the particular form of the L-operator for the RS model, our treatment does not depend on the particular form of L.

2.3. Simplest representation of the L-operator algebra. To find the L-operator as a function of the canonical variables, we employ the Hamiltonian reduction procedure. As noted at the beginning of the present section, the Poisson structure (2.1)–(2.3) admits the Hamiltonian action of the current group.

LEMMA 2.3. *The momentum map with values in the dual space to the Lie algebra of G,*

$$M(z) = k\,\bar\partial g(z)g^{-1}(z) + A(z) - g(z)A(z)g^{-1}(z),$$

generates the action of the current group.

We fix the value of $M(z)$ as follows [**20**]:

$$M(z) = -\frac{2\pi ik}{\tau-\bar\tau}\gamma + 2\pi ik\,\delta_\varepsilon(z)\,\frac{1-e^{-2\pi ix}}{2\pi i}\,K. \tag{2.55}$$

Here γ (the coupling constant) and ε are arbitrary complex numbers,

$$\delta_\varepsilon(z) = \frac{1}{\tau-\bar\tau}\,\delta_\varepsilon(x)\,\delta(y),$$

$$\delta_\varepsilon(x) = \frac{1}{\varepsilon}\Big(\theta\Big(x+\frac{\varepsilon}{2}\Big)-\theta\Big(x-\frac{\varepsilon}{2}\Big)\Big) = \frac{1}{4\pi^2 i\varepsilon}\sum_{n=-\infty}^{n=+\infty}\frac{1}{n}(e^{2\pi in\varepsilon/2}-e^{-2\pi in\varepsilon/2})e^{2\pi inx},$$

K is a constant matrix, $K = e\otimes e^t$, and $z = x+\tau y$.

To obtain the finite-dimensional reduced phase space, one must consider the limit as ε goes to zero. To treat this limit, we employ the same strategy as in [**19**]. We multiply both sides of (2.55) by $g(z)$, which gives

$$\begin{aligned} &k\,\bar\partial g(z) + A(z)g(z) - g(z)A(z) + \frac{2\pi ik}{\tau-\bar\tau}\gamma g(z) \\ &\quad = 2\pi ik\,\delta_\varepsilon(z)K\,\frac{1-e^{-2\pi ix}}{2\pi i}\,g(z). \end{aligned} \tag{2.56}$$

The left-hand side of this equation does not contain any explicit dependence on ε. As to the right-hand side, as ε tends to zero, $\delta_\varepsilon(z)$ becomes proportional to $\delta(x)\delta(y)$, and the right-hand side is well defined only if the function $(2\pi i)^{-1}(1-e^{-2\pi ix})g(x,0)$ is well defined at $x=0$. In this case,

$$\lim_{\varepsilon\to0}\delta_\varepsilon(z)\,\frac{1-e^{-2\pi ix}}{2\pi i}\,g(z) = \delta(z)Z, \quad \text{where } Z = \frac{1-e^{-2\pi ix}}{i}\,g(x,0)|_{x=0}.$$

Thus, we can define the constraint surface as the solution of the equation

$$k\,\bar\partial g(z) + [A(z), g(z)] + \frac{2\pi ik}{\tau-\bar\tau}\gamma g(z) = 2\pi ik\,\delta(z)KZ. \tag{2.57}$$

In fact, the procedure described above corresponds to fixing the linearized moment map

$$M(z) = k\,\bar\partial g(z) + [A(z), g(z)]$$

that satisfies the Poisson algebra

$$\{M_1(z), M_2(w)\} = [C, M_1(z)g_1(z)]\delta(z-w).$$

To describe the reduced phase space, we begin with the following differential equation:

$$k\,\bar\partial g(z) + [D, g(z)] + \frac{2\pi ik}{\tau-\bar\tau}\gamma g(z) = 2\pi ik\,\delta(z)Y, \tag{2.58}$$

where D is a constant diagonal matrix and Y is an arbitrary constant matrix. Equation (2.58) can be solved similarly to (2.8); we obtain

$$g(z) = \sum_{ij} \Phi(z, q_{ij} + \gamma) Y_{ij} E_{ij}. \tag{2.59}$$

Now we turn to the moment map equation (2.57). By using a generic gauge transformation, we can diagonalize the field A. Equation (2.57) takes the form of (2.58):

$$k\,\bar\partial g'(z) + [D, g'(z)] + \frac{2\pi i k}{\tau - \bar\tau}\gamma g'(z) = 2\pi i k\,\delta(z) K' Z', \tag{2.60}$$

where

$$A(z) = T(z) D T^{-1}(z) - k\,\bar\partial T(z) T^{-1}(z), \qquad g(z) = T(z) g'(z) T^{-1}(z)$$

for some T, and $Z' = x g'(x,0)|_{x=0}$. We also have

$$K' = T^{-1}(0) K T(0) = T^{-1}(0) e \otimes e^t T(0) = f \otimes v^t, \qquad \langle f, v\rangle = N,$$

i.e., $f = T^{-1}(0)e$ and $e^t T(0) = v^t$. According to (2.59),

$$g'(z) = \sum_{ij} \Phi(z, q_{ij} + \gamma)(K'Z')_{ij} E_{ij}. \tag{2.61}$$

Taking the value of $xg'(x,0)$ at $x = 0$, we arrive at the compatibility condition

$$Z' = K'Z' = f \otimes v^t Z', \qquad \langle f, v\rangle = N.$$

The solution to this equation is $Z' = f \otimes l^t$, where l is an arbitrary vector. Now it is easy to find Z:

$$Z = T(0) Z T^{-1}(0) = T(0) f \otimes l^t T(0)^{-1} = e \otimes l^t T^{-1}(0) = e \otimes c^t.$$

Thus, we obtain

$$k\,\bar\partial g(z) + [A(z), g(z)] + \frac{2\pi i k}{\tau - \bar\tau}\gamma g(z) = 2\pi i k (e \otimes c^t)\,\delta(z). \tag{2.62}$$

To summarize, Eq. (2.57) has a solution for any field A and, for any field g such that $xg(x,0)|_{x=0}$, is of the form $e \otimes c^t$. For a fixed field A and a vector c, this solution is unique. Note that, in general, $\langle c, e\rangle \neq N$.

The form of the right-hand side of (2.60) shows that the isotropy group of this equation is

$$G_{\text{isot}} = \{T(z) \in G(z, \bar z) \mid T(0)e = \lambda e,\ \lambda \in \mathbb{C}\}.$$

This group transforms a solution of (2.60) into another, so that the reduced phase space is defined as

$$\mathcal{P}_{\text{red}} = \frac{\text{all solutions of (2.57)}}{G_{\text{isot}}}.$$

The group G_{isot} is large enough to diagonalize the field A and, hence, we can parametrize the reduced phase space by the section (D, L), where L is a solution of (2.57) with $A = D$. One can easily see that $\mathcal{P}_{\text{red}}$ is finite-dimensional and its dimension is equal to $2N$, i.e., N coordinates of D plus N coordinates of the vector c. By virtue of Eq. (2.61), the corresponding L-operator has the following form:

$$L(z) = \sum_{ij} \Phi(z, q_{ij} + \gamma) c_j E_{ij}. \tag{2.63}$$

It follows from the results obtained in the previous subsection that $c_j = b_j P_j$, where b_j are functions of q only. The operator (2.63) must satisfy the Poisson algebra (2.51), which implies the following nontrivial Poisson relations on c_j:

$$\kappa\{q_i, c_j\} = c_j \delta_{ij}, \tag{2.64}$$

$$\kappa\{c_i, c_j\} = c_i c_j (2\zeta(q_{ij}) - \zeta(q_{ij}+\gamma) - \zeta(q_{ij}-\gamma)). \tag{2.65}$$

To obtain (2.64) and (2.65), we note that

$$L_{ii}(z) = \Phi(z,\gamma) c_i, \tag{2.66}$$

so, in order to calculate the bracket $\{c_i, c_j\}$, it is sufficient to examine only the $\{L_{ii}, L_{jj}\}$ bracket. From (2.51), for $i \neq j$, we obtain

$$\frac{1}{\kappa}\{L_{ii}(z), L_{jj}(z)\} = L_{ji}(z) L_{ij}(w) \Phi(z-w, q_{ij}) - L_{ij}(z) L_{ji}(w) \Phi(z-w, q_{ji}). \tag{2.67}$$

It follows from this equation that

$$\begin{aligned} \frac{1}{\kappa}\{c_i, c_j\} = c_i c_j (&\Phi(z, q_{ji}+\gamma)\Phi(w, q_{ij}+\gamma)\Phi(z-w, q_{ij}) \\ &- \Phi(z, q_{ij}+\gamma)\Phi(w, q_{ji}+\gamma)\Phi(z-w, q_{ji}))\frac{1}{\Phi(z,\gamma)\Phi(w,\gamma)}. \end{aligned} \tag{2.68}$$

Using the elliptic identity (see (A.3) from Appendix A), we obtain (2.65). Equation (2.64) is proven similarly.

To estimate the representation of b_j in terms of q_i, one may look at the rational case. Indeed, in the rational limit, the algebra (3.28) tends to the one obtained in [**18**], where the coefficients b_j were found to be

$$b_j = \prod_{a \neq j} \frac{q_{aj}+\gamma}{q_{aj}}. \tag{2.69}$$

Therefore, it is natural to assume that the elliptic analog of (2.69) is given by

$$b_j = \prod_{a \neq j} \Phi(\gamma, q_{aj}). \tag{2.70}$$

By direct calculation, one can verify that (2.70) holds.

It is worth mentioning that the coefficients b_j are not uniquely determined, since one can perform a canonical transformation of the (q,p)-variables. For instance, the variables

$$\tilde{b}_j = \prod_{a \neq j} \frac{\sigma(q_{aj}+\gamma)}{\sigma(\gamma)\sigma(q_{aj})} \tag{2.71}$$

are related to b_j by the canonical transformation $q_i \to q_i$ and

$$P_i \to \exp\left\{\alpha \sum_a q_{ai}\right\} P_i, \quad \text{where } \alpha = 2\zeta\left(\frac{1}{2}\right)\gamma - 2\pi i \frac{\gamma - \bar{\gamma}}{\tau - \bar{\tau}}.$$

The relation of the L-operator (2.63) to the one found in [**10**] is clarified in Section 3.

2.4. Quadratic Poisson algebra with derivatives. In subsection 2.1 we obtained the matrices $\mathbf{r}$ and $\bar{\mathbf{r}}$ obeying the system of equations (2.23), (2.40), and (2.41). Clearly, these equations are not satisfied when r and $\bar{r}$ are replaced by $\mathbf{r}$ and $\bar{\mathbf{r}}$. However, computing the left-hand side of these equations after this substitution, we arrive at a surprisingly simple result:

$$
\begin{aligned}
&[r_{12}(z_1,z_2), r_{13}(z_1,z_3)+r_{23}(z_2,z_3)] + [r_{13}(z_1,z_3), r_{23}(z_2,z_3)] \\
&\quad = -(\partial_1+\partial_2)r_{12}(z_1,z_2) + (\partial_1+\partial_3)r_{13}(z_1,z_3) - (\partial_2+\partial_3)r_{23}(z_2,z_3),
\end{aligned} \tag{2.72}
$$

$$
\begin{aligned}
&[\bar{r}_{12}(z), \bar{r}_{13}(z)] - P_3^{-1}\{\bar{r}_{12}(z), P_3\} + P_2^{-1}\{\bar{r}_{13}(z), P_2\} \\
&\quad = -\partial(\bar{r}_{12}(z) - \bar{r}_{13}(z)),
\end{aligned} \tag{2.73}
$$

$$
\begin{aligned}
&[r_{12}(z_1,z_2), \bar{r}_{13}(z_1)+\bar{r}_{23}(z_2)] + [\bar{r}_{13}(z_1), \bar{r}_{23}(z_2)] - P_3^{-1}\{r_{12}(z_1,z_2), P_3\} \\
&\quad = -(\partial_1+\partial_2)r_{12}(z_1,z_2) + \partial_1\bar{r}_{13}(z_1) - \partial_2\bar{r}_{23}(z_2).
\end{aligned} \tag{2.74}
$$

Here $\partial \equiv \partial/\partial x$, where $x = \operatorname{Re} z$. Note that Eqs. (2.40) and (2.41) involve $\mathbb{P}$. However, since all of the matrices depend only on the difference $q_{ij} = q_i - q_j$, we simply replace $\mathbb{P}$ by P.

Comparing Eqs. (2.72)–(2.74) for r and $\bar{r}$ with (2.23), (2.40), and (2.41) for $\mathbf{r}$ and $\bar{\mathbf{r}}$, we come to the conclusion that the $s(z)$ matrix appearing in $\mathbf{r}$ and $\bar{\mathbf{r}}$ effectively plays the role of the derivative with respect to the spectral parameter.

It is worth mentioning that Eqs. (2.72)–(2.74) obtained for r and $\bar{r}$ can be rewritten in the same form as Eqs. (2.23), (2.40), and (2.41) if we replace r and $\bar{r}$ by $r_{12} - \partial_1 + \partial_2$ and $\bar{r}_{12} - \partial_1$. In particular, for (2.72), we have

$$
\begin{aligned}
&[r_{12} - \partial_1 + \partial_2, r_{13} - \partial_1 + \partial_3] + [r_{13} - \partial_1 + \partial_3, r_{23} - \partial_2 + \partial_3] \\
&\qquad + [r_{12} - \partial_1 + \partial_2, r_{23} - \partial_2 + \partial_3] = 0.
\end{aligned} \tag{2.75}
$$

Thus, $r_{12} - \partial_1 + \partial_2$ is a matrix first-order differential operator satisfying the standard classical Yang–Baxter equation.

PROPOSITION 2.2. *The Jacobi identity for the Poisson algebra generated by the fields $T(z)$, $U(z)$, Q, and P,*

$$\frac{1}{\kappa}\{T_1(z), T_2(w)\} = T_1(z)T_2(w)r_{12}(z,w) + T_1'T_2 - T_1T_2', \tag{2.76}$$

$$\frac{1}{\kappa}\{U_1(z), U_2(w)\} = -U_1(z)U_2(w)r_{12}(z,w) - U_1'U_2 + U_1U_2', \tag{2.77}$$

$$\frac{1}{\kappa}\{T_1(z), P_2\} = P_2T_1(z)\bar{r}_{12}(z) + P_2T_1'(z), \tag{2.78}$$

$$\frac{1}{\kappa}\{U_1(z), P_2\} = P_2U_1(z)\bar{r}_{12}(z) + P_2U_1'(z), \tag{2.79}$$

$$\{Q_1, P_2\} = P_2\sum_i E_{ii}\otimes E_{ii}, \qquad \{Q_1, T_2\} = \{Q_1, U_2\} = 0, \tag{2.80}$$

where $T' = \partial T$, is equivalent to Eqs. (2.72)–(2.74).

It is also worth mentioning that the Poisson structure (2.76)–(2.80) is not compatible with the boundary condition $T(0)e = \lambda e$.

Let us note that there exists a Poisson subalgebra of the Poisson algebra (2.76)–(2.80), formed by the generators

$$A(z) = T(z)DT^{-1}(z) - k\,\bar{\partial}T(z)T^{-1}(z), \qquad g(z) = U(z)PT^{-1}(z),$$

which coincides with the Poisson algebra of the cotangent bundle with central charge $\alpha = 0$.

Defining the L-operator as $L(z) = T^{-1}(z)g(z)T(z) = T^{-1}(z)U(z)P$, we obtain the previously obtained algebra (2.51). As in 2.3 above, the commutativity of $I_n(z)$ follows from the commutativity of $g(z)$.

The main advantage of the Poisson algebra (2.76)–(2.80) is that it can be easily quantized.

3. Quantization

3.1. Quantum R-matrices. In this section, following the ideology of the quantum inverse scattering method [**24, 25**], we quantize the classical r- and $\bar{r}$-matrices and derive the quantum L-operator algebra.

We start with the quantization of relations (2.72)–(2.74). Let $T(z)$, $U(z)$ be matrix generating functions that are formal Fourier series in the variables x and y,

$$T(z) = \sum_{mn} T_{mn} e^{2\pi i(mx+ny)}, \qquad U(z) = \sum_{mn} U_{mn} e^{2\pi i(mx+ny)}.$$

Here, $z = x + \tau y$.

DEFINITION 3.1. Denote by $\mathcal{A}$ the free associative unital algebra over the field $\mathbf{C}$, generated by matrix elements of the Fourier modes of $T(z)$, $U(z)$ and by the entries of the diagonal matrices P and Q, modulo the relations

$$T_1(z)T_2(w-\hbar) = T_2(w)T_1(z-\hbar)R_{12}(-\hbar, z, w), \tag{3.1}$$

$$U_1(z)U_2(w+\hbar) = U_2(w)U_1(z+\hbar)R_{12}(\hbar, z, w), \tag{3.2}$$

$$T_1(z+\hbar)P_2\overline{R}_{12}(\hbar, z) = P_2T_1(z), \tag{3.3}$$

$$U_1(z+\hbar)P_2\overline{R}_{12}(\hbar, z) = P_2U_1(z), \tag{3.4}$$

$$[Q_1, P_2] = -\hbar P_2 \sum_i E_{ii} \otimes E_{ii}, \tag{3.5}$$

$$[T_1(z), U_2(w)] = [T_1(z), Q_2] = [U_1(z), Q_2] = [P_1, P_2] = [Q_1, Q_2] = 0.$$

In Definition 3.1, $R(\hbar, z, w)$ and $\overline{R}_{12}(\hbar, z)$ are double-periodic matrix functions of spectral parameters. These functions also depend on the coordinates q_i and have the following semiclassical behavior at $\hbar = 0$:

$$R = 1 + \hbar r + o(\hbar), \qquad \overline{R} = 1 + \hbar\bar{r} + o(\hbar). \tag{3.6}$$

The next step is to find the conditions on R and $\overline{R}$ that ensure the consistency of the defining relations for $\mathcal{A}$. In what follows, we often use $R(z, w)$ as a shorthand notation for $R(\hbar, z, w)$.

First, we write the compatibility condition for the algebra (3.1) or (3.2), which reduces to the quantum Yang–Baxter equation with the spectral parameters shifted by $\hbar$,

$$\begin{aligned} &R_{12}(z,w)R_{13}(z-\hbar, s-\hbar)R_{23}(w,s) \\ &\quad = R_{23}(w-\hbar, s-\hbar)R_{13}(z,s)R_{12}(z-\hbar, w-\hbar). \end{aligned} \tag{3.7}$$

As in the classical case, one can introduce the matrix differential operator

$$\mathcal{R}(z,w) = e^{\hbar\,\partial/\partial w} R(z,w) e^{-\hbar\,\partial/\partial z}.$$

In terms of this operator, Eq. (3.7) reads as the standard quantum Yang–Baxter equation

$$\mathcal{R}_{12}(z,w)\mathcal{R}_{13}(z,s)\mathcal{R}_{23}(w,s)=\mathcal{R}_{23}(w,s)\mathcal{R}_{13}(z,s)\mathcal{R}_{12}(z,w). \tag{3.8}$$

Relation (3.1) also requires the "unitarity" condition for R,

$$R_{12}(z,w)R_{21}(w,z)=1. \tag{3.9}$$

Similarly, we find the following compatibility conditions for (3.3):

$$P_3^{-1}\overline{R}_{12}(z)P_3\overline{R}_{13}(z-\hbar)=P_2^{-1}\overline{R}_{13}(z)P_2\overline{R}_{12}(z-\hbar), \tag{3.10}$$

$$P_3^{-1}R_{12}(z,w)P_3\overline{R}_{13}(z-\hbar)\overline{R}_{23}(w)=\overline{R}_{23}(w-\hbar)\overline{R}_{13}(z)R_{12}(z-\hbar,w-\hbar). \tag{3.11}$$

Now, taking into account (3.6), one can easily see that in the semiclassical limit,

$$-\frac{1}{\kappa}\{\cdot,\cdot\}=\lim_{\hbar\to 0}\frac{1}{\hbar}[\cdot,\cdot],$$

relations (3.1)–(3.5) determine the Poisson structure (2.76)–(2.80), while Eqs. (3.7), (3.10), and (3.11) become (2.72), (2.73), and (2.74), respectively, in the order $\hbar^2$. In the first order in $\hbar$, the unitarity condition (3.9) requires r to be skew-symmetric. Hence, the algebra $\mathcal{A}$ with defining relations (3.1)–(3.5), where R and $\overline{R}$ are the solutions of (3.7)–(3.11) obeying (3.6), is a quantization of the Poisson structure (2.76)–(2.80).

Now we are in position to find the matrices R and $\overline{R}$ explicitly.

THEOREM 3.2. *A solution to Eqs.* (3.7), (3.9)–(3.11) *with a semiclassical asymptotic behavior* (3.6) *is given by the following quantum dynamical R-matrices R and $\overline{R}$:*

$$\begin{aligned} f(z,w)R(\hbar,z,w)=&\sum_{ij}\Phi(\hbar,q_{ij})E_{ii}\otimes E_{jj}+\sum_{ij}\Phi(z-w,q_{ij})E_{ij}\otimes E_{ji}\\ &-\sum_{ij}\Phi(z+\hbar,q_{ij})E_{ij}\otimes E_{jj}\\ &+\sum_{ij}\Phi(w,q_{ij})E_{jj}\otimes E_{ij}, \end{aligned} \tag{3.12}$$

$$\begin{aligned} \frac{1}{\sigma(\hbar)}\overline{R}_{12}(\hbar,z)=&\sum_{i\neq j}\Phi(\hbar,q_{ij})E_{ii}\otimes E_{jj}-\sum_{i\neq j}\Phi(z+\hbar,q_{ij})E_{ij}\otimes E_{jj}\\ &-\Phi(z+\hbar,-\hbar)\sum_{i}E_{ii}\otimes E_{ii}, \end{aligned} \tag{3.13}$$

where $f(z,w)=\sqrt{\mathcal{P}(\hbar)-\mathcal{P}(z-w)}$.

PROOF. We start with the R-matrix, for which we assume the following natural ansatz:

$$\begin{aligned} fR(\hbar,z,w)=&\sum_{ij}\Phi(\hbar_1,q_{ij}+\hbar_2)E_{ii}\otimes E_{jj}+\sum_{ij}\Phi(z-w+\hbar_3,q_{ij}+\hbar_4)E_{ij}\otimes E_{ji}\\ &-\sum_{ij}\Phi(z+\hbar_5,q_{ij}+\hbar_6)E_{ij}\otimes E_{jj}\\ &+\sum_{ij}\Phi(w+\hbar_7,q_{ij}+\hbar_8)E_{jj}\otimes E_{ij}. \end{aligned} \tag{3.14}$$

This form is compatible with the structure of the classical r-matrix. Here $\hbar_1, \dots, \hbar_8$ are arbitrary parameters specified by Eqs. (3.7) and (3.9), and f is a scalar function that may depend only on $\hbar_i$ and the spectral parameters. It turns out that the parameters h_i are almost uniquely fixed by the unitarity condition (3.9). Substituting (3.14) into (3.9) and using the elliptic function identities, we obtain

$$\hbar_2 = \hbar_3 = \hbar_4 = \hbar_6 = \hbar_8 = 0, \qquad \hbar_5 = \hbar_1 + \hbar_7, \qquad f^2(z, w) = \mathcal{P}(\hbar_1) - \mathcal{P}(z - w),$$

where $\mathcal{P}(z)$ is the Weierstrass $\mathcal{P}$-function. Now it is a matter of direct calculation to check that Eq. (3.7) holds for $\hbar_1 = \hbar$. The remaining parameter $\hbar_7$ is inessential, since it corresponds to an arbitrary common shift of the spectral parameters z and w. Below, we choose $\hbar_7 = 0$. Therefore, the obtained solution to (3.7) and (3.6) reads as (3.12).

One must be careful in the definition of $R(-\hbar, z, w)$. This matrix is defined by (3.12) with the substitutions $\hbar \to -\hbar$ and $f \to -f$. Therefore, $R(\hbar, z, w)$ and $R(-\hbar, z, w)$ are related as follows:

$$R_{12}(-\hbar, z, w) = R_{21}(\hbar, w - \hbar, z - \hbar). \tag{3.15}$$

To find the $\overline{R}$-matrix, we adopt the following ansatz:

$$\begin{aligned} \frac{1}{\sigma(\hbar)}\overline{R}_{12}(\hbar, z) &= \sum_{ij} \Phi(\hbar_1, q_{ij} + \hbar_2) E_{ii} \otimes E_{jj} \\ &\quad - \sum_{ij} \Phi(z + h_3, q_{ij} + \hbar_4 + \delta_{ij}\hbar_5) E_{ij} \otimes E_{jj}, \end{aligned} \tag{3.16}$$

which has almost the same matrix structure as the classical $\bar{r}$-matrix. Since (3.10) is easier to deal with than (3.11), we first substitute (3.16) into (3.10), thus obtaining $\overline{R}$:

$$\begin{aligned} \frac{1}{\sigma(\hbar)}\overline{R}_{12}(\hbar, z) &= \sum_{i\neq j} \Phi(\hbar, q_{ij}) E_{ii} \otimes E_{jj} - \sum_{i\neq j} \Phi(z + \hbar_3, q_{ij}) E_{ij} \otimes E_{jj} \\ &\quad - \Phi(z + \hbar_3, -\hbar) \sum_i E_{ii} \otimes E_{ii}, \end{aligned} \tag{3.17}$$

where h_3 remains unfixed.

Equation (3.11) involves both the R- and $\overline{R}$-matrices and is independent of (3.7) and (3.10). One can verify by direct calculations that R and $\overline{R}$ given by (3.12) and (3.17) also satisfy (3.11) so long as $\hbar_3 = \hbar$.

One can easily check that in the case of real $\hbar$, the matrices R and $\overline{R}$ have the proper semiclassical behavior (3.6). □

In what follows, we also need the matrix $\overline{R}^{-1}$,

$$\frac{1}{\sigma(\hbar)}\overline{R}_{12}^{-1}(\hbar, z) = -\sum_{ij} \Phi(-\hbar, q_{ij} + \hbar) E_{ii} \otimes E_{jj} + \sum_{ij} \Phi(z, q_{ij} + \hbar) E_{ij} \otimes E_{jj}. \tag{3.18}$$

It would be of interest to mention that just as in the rational case, one can introduce the formal variable $W(z) = T^{-1}(z)U(z)$ without the spectral parameter [**18**], with the permutation relations following from (3.1)–(3.5),

$$R_{12}(z, w) W_1(z) W_2(w + \hbar) = W_2(w) W_1(z + \hbar) R_{12}(z, w), \tag{3.19}$$

$$W_1(z + \hbar) P_2 \overline{R}_{12}(z) = P_2 \overline{R}_{12}(z) W_1(z). \tag{3.20}$$

Similarly to the rational case, it is natural to treat Eq. (3.19) as the defining relation of the quantum elliptic Frobenius group.

3.2. Quantum L-operator algebra. Just as in the classical case, we introduce a new variable:

$$L(z) = T^{-1}(z)U(z)P = W(z)P, \tag{3.21}$$

which we call a *quantum L-operator*. Using the relations of the algebra $\mathcal{A}$, we can formally derive the following algebraic relations that are satisfied by the quantum L-operator:

$$[Q_1, L_2(z)] = -\hbar L_2(z) \sum_i E_{ii} \otimes E_{ii}, \tag{3.22}$$

$$\begin{aligned} &R_{12}(z,w)L_1(z)\overline{R}_{21}(w)L_2(w) \\ &\quad = L_2(w)\overline{R}_{12}(z)L_1(z)\overline{R}_{21}(w-\hbar)R_{12}(z-\hbar, w-\hbar)\overline{R}_{12}^{-1}(z-\hbar). \end{aligned} \tag{3.23}$$

Although L has the form $L(z) = W(z)P$, we cannot reconstruct relations (3.19) and (3.20) for W and P from (3.22) and (3.23). Therefore, in what follows, we do not assume any relations on W and P.

Let us define

$$R_{12}^F(z,w) = \overline{R}_{21}(w)R_{12}(z,w)\overline{R}_{12}^{-1}(z). \tag{3.24}$$

Then, using the explicit form of the R- and $\overline{R}$-matrices and the elliptic function identities, we obtain

$$\begin{aligned} fR_{12}^F(z-w) = &-\sum_{i\neq j} \Phi(-\hbar, q_{ij})E_{ii} \otimes E_{jj} + \sum_{i\neq j} \Phi(z-w, q_{ij})E_{ij} \otimes E_{ji} \\ &+ \Phi(z-w,\hbar)\sum_i E_{ii} \otimes E_{ii}, \end{aligned} \tag{3.25}$$

which is nothing but the R-matrix of Felder [**15**], i.e., an elliptic solution to the quantum Gervais–Neveu–Felder equation [**16, 15**]:

$$\begin{aligned} &P_1^{-1}R_{23}^F(w-s)P_1R_{13}^F(z-s)P_3^{-1}R_{12}^F(z-w)P_3 \\ &\quad = R_{12}^F(z-w)P_2^{-1}R_{13}^F(z-s)P_2R_{23}^F(w-s). \end{aligned} \tag{3.26}$$

Recall that one feature of R^F is the "weight zero" condition

$$[P_1P_2, R_{12}^F(z-w)] = 0. \tag{3.27}$$

Developing R^F in powers of $\hbar$, we have $R^F = 1 + \hbar r^F + o(\hbar)$, where r^F is given by (2.53).

Let us stress that in our consideration, R^F is used to obtain the explicit form of R and $\overline{R}$, and that the Gervais–Neveu–Felder equation does not follow from system (3.7)–(3.11). Formula (3.24) shows that the matrix $\overline{R}$ plays the role of the twist, which transforms the matrix $R(z,w)$ (a particular solution of (3.7)) into a solution of (3.26) that depends only on the difference $z-w$.

Thus, we have proven the following proposition.

PROPOSITION 3.1. *The quantum L-operator* (3.21) *obeys the following algebra*:

$$R_{12}(z,w)L_1(z)\overline{R}_{21}(w)L_2(w) = L_2(w)\overline{R}_{12}(z)L_1(z)R_{12}^F(z-w). \tag{3.28}$$

The quantum L-operator algebra seems to be automatically compatible, since $\mathcal{A}$ is compatible. However, a simple analysis shows that $\mathcal{A}$ and the algebra (3.28) admit different structures of representations. In particular, the simplest representation for L, which we present below, does not realize the algebra (3.19), (3.20). Therefore, we find it necessary to give a direct proof of the compatibility of (3.28). In this way, we come across Eq. (3.26) and discover a new relation involving R^F and $\overline{R}$. To this end, let us multiply both sides of (3.28) by

$$P_2^{-1}\overline{R}_{31}(s+\hbar)P_2\overline{R}_{32}(s)L_3(s)$$

and then use (3.28), transforming the string $L_1 \cdots L_2 \cdots L_3$ into $L_3 \cdots L_2 \cdots L_1$. For the left-hand side, we have

$$\begin{aligned}
&R_{12}(z,w)L_1\overline{R}_{21}(w)L_2P_2^{-1}\overline{R}_{31}(s+\hbar)P_2\overline{R}_{32}(s)L_3\\
&\quad= R_{12}(z,w)L_1\overline{R}_{21}(w)\overline{R}_{31}(s+\hbar)L_2\overline{R}_{32}(s)L_3\\
&\quad= R_{12}(z,w)L_1\overline{R}_{21}(w)\overline{R}_{31}(s+\hbar)R_{32}(s,w)L_3\overline{R}_{23}(w)L_2R^F_{23}(w-s)\\
&\quad= R_{12}(z,w)R_{32}(s+\hbar,w+\hbar)L_1\overline{R}_{31}(s)L_3P_3^{-1}\\
&\qquad\times\overline{R}_{21}(w+\hbar)P_3\overline{R}_{23}(w)L_2R^F_{23}(w-s)\\
&\quad= R_{12}(z,w)R_{32}(s+\hbar,w+\hbar)R_{31}(s,z)L_3\overline{R}_{13}(z)L_1\\
&\qquad\times R^F_{13}(z-s)P_3^{-1}\overline{R}_{21}(w+\hbar)P_3\overline{R}_{23}(w)L_2R^F_{23}(w-s).
\end{aligned}\tag{3.29}$$

At this point, we interrupt the calculations to mention that the next step implies that we can somehow push R^F through $P_3^{-1}\overline{R}_{21}(w+\hbar)P_3\overline{R}_{23}(w)$. This can be done by virtue of the following new relation involving R^F and $\overline{R}$:

$$R^F_{12}(z)P_2^{-1}\overline{R}_{31}(w)P_2\overline{R}_{32}(w-\hbar) = P_1^{-1}\overline{R}_{32}(w)P_1\overline{R}_{31}(w-\hbar)R^F_{12}(z),\tag{3.30}$$

which can be checked directly by using the explicit forms (3.25) and (3.13) of R^F and $\overline{R}$, respectively.

Now we pursue the calculation (3.29), with relation (3.30) at hand:

$$\begin{aligned}
&R_{12}(z,w)R_{32}(s+\hbar,w+\hbar)R_{31}(s,z)L_3\overline{R}_{13}(z)\overline{R}_{23}(w+\hbar)\\
&\qquad\times L_1\overline{R}_{21}(w)L_2P_2^{-1}R^F_{13}(z-s)P_2R^F_{23}(w-s).
\end{aligned}$$

As for the right-hand side, the same technique yields

$$\begin{aligned}
&L_2\overline{R}_{12}(z)L_1R^F_{12}(z-w)P_2^{-1}\overline{R}_{31}(s+\hbar)P_2\overline{R}_{32}(s)L_3\\
&\quad= L_2\overline{R}_{12}(z)\overline{R}_{32}(s+\hbar)L_1\overline{R}_{31}(s)L_3P_3^{-1}R^F_{12}(z-w)P_3\\
&\quad= L_2\overline{R}_{12}(z)\overline{R}_{32}(s+\hbar)R_{31}(s,z)L_3\overline{R}_{13}(z)L_1R^F_{13}(z-s)P_3^{-1}R^F_{12}(z-w)P_3\\
&\quad= R_{31}(s+\hbar,z+\hbar)L_2\overline{R}_{32}(s)L_3P_3^{-1}\overline{R}_{12}(z+\hbar)P_3\\
&\qquad\times\overline{R}_{13}(z)L_1R^F_{13}(z-s)P_3^{-1}R^F_{12}(z-w)P_3\\
&\quad= R_{31}(s+\hbar,z+\hbar)R_{32}(s,w)L_3\overline{R}_{23}(w)\overline{R}_{13}(z+\hbar)L_2\overline{R}_{12}(z)L_1\\
&\qquad\times P_1^{-1}R^F_{23}(w-s)P_1R^F_{13}(z-s)P_3^{-1}R^F_{12}(z-w)P_3\\
&\quad= R_{31}(s+\hbar,z+\hbar)R_{32}(s,w)R_{12}(z+\hbar,w+\hbar)L_3\overline{R}_{13}(z)\overline{R}_{23}(w+\hbar)L_1\\
&\qquad\times\overline{R}_{21}(w)L_2R^F_{21}(w-z)P_1^{-1}R^F_{23}(w-s)P_1R^F_{13}(z-s)P_3^{-1}R^F_{12}(z-w)P_3.
\end{aligned}\tag{3.31}$$

Therefore, comparing the resulting expressions, we conclude that the compatibility condition for the L-operator algebra (3.28) reduces to four equations: (3.7), (3.11), (3.25), and (3.30).

The existence of the Poisson commuting functions $I_n(z)$ in the classical case implies that the commuting family must exist in the quantum case as well. This should be an intrinsic property of the algebra (3.28) itself, without referring to the explicit form of its representations. Let us demonstrate the commutativity of the simplest quantities, $\operatorname{tr} L(z)$ and $\operatorname{tr} L^{-1}(z)$, postponing the discussion of the general case to the next section. To this end, we need one more relation involving the matrices R^F, R, and $\overline{R}$.

By analogy with the rational case, it is useful to introduce the variable $g(z) = U(z)PT^{-1}(z)$. Calculation of the commutator $[g_1(z), g_2(w)]$ with the help of the defining relations of $\mathcal{A}$ results in

$$\begin{aligned}[g_1(z), g_2(w)] = U_2(w)U_1(z+\hbar)(&R_{12}(\hbar, z, w)P_1\overline{R}_{21}(w)P_2\overline{R}_{12}(z-\hbar)R_{12}(-\hbar, z, w) \\ &- P_2\overline{R}_{12}(z)P_1\overline{R}_{21}(w-\hbar))T_2^{-1}(w-\hbar)T_1^{-1}(z).\end{aligned}$$

When the spectral parameter is absent, the algebra $\mathcal{A}$ allows one to establish a connection with the quantum cotangent bundle (see [**18**] for details). Then, in particular, the quantity $[g_1, g_2]$ is equal to zero. In the case at hand, we cannot construct a subalgebra of $\mathcal{A}$ that is isomorphic to the quantum cotangent bundle. However, one should note that in the elliptic case, the commutativity of $g(z)$ and $g(w)$ follows from the identity

$$R_{12}(\hbar, z, w)P_1\overline{R}_{21}(w)P_2\overline{R}_{12}(z-\hbar)R_{12}(-\hbar, z, w) = P_2\overline{R}_{12}(z)P_1\overline{R}_{21}(w-\hbar).$$

Using the definition of R^F, Eq. (3.15), and the "weight zero" condition (3.27), the last formula can be written in the following elegant form:

$$R_{12}(z, w) = P_2\overline{R}_{12}(z)P_2^{-1}R_{12}^F(z-w)P_1\overline{R}_{21}^{-1}(w)P_1^{-1}. \tag{3.32}$$

Identity (3.32) plays the primary role in proving the commutativity of the family $\operatorname{tr} L(z)$. To prove this commutativity, let us multiply both sides of

$$L_2(w)\overline{R}_{12}(z)L_1(z) = R_{12}(z, w)L_1(z)\overline{R}_{21}(w)L_2(w)R_{21}^F(w-z)$$

by $P_2\overline{R}_{12}^{-1}(z)P_2^{-1}$ and take the trace in the first and second matrix spaces. We have

$$\begin{aligned}&\operatorname{tr}_{12} P_2\overline{R}_{12}^{-1}(z)P_2^{-1}L_2(w)\overline{R}_{12}(z)L_1(z) \\ &\quad = \operatorname{tr}_{12} P_2\overline{R}_{12}^{-1}(z)P_2^{-1}R_{12}(z, w)L_1(z)\overline{R}_{21}(w)L_2(w)R_{21}^F(w-z).\end{aligned} \tag{3.33}$$

It is useful to write $\overline{R}_{12}^{-1}$ in factored form,

$$\overline{R}_{12}^{-1}(z) = \sum_{ij} \overline{R}_{ij}^{-1} \otimes E_{jj}, \tag{3.34}$$

where

$$\overline{R}_{ij}^{-1} = -\sigma(\hbar)\Phi(-\hbar, q_{ij}+\hbar)E_{ii} + \sigma(\hbar)\Phi(z, q_{ij}+\hbar)E_{ij}.$$

Then the left-hand side of (3.33) becomes

$$\sum_{ij} \operatorname{tr}_{12}(P_j\overline{R}_{ij}^{-1}P_j^{-1} \otimes E_{jj})L_2\overline{R}_{12}(z)L_1.$$

Taking into account the fact that $\overline{R}_{12}$ is diagonal in the second matrix space and using the cyclic property of the trace, we obtain

$$\sum_{ij} \operatorname{tr}_{12}(P_j\overline{R}_{ij}^{-1}P_j^{-1} \otimes I)(I \otimes LE_{jj})\overline{R}_{12}(z)L_1.$$

Since $L = WP$, where all the entries of W commute with q_i, we arrive at

$$\sum_{ij} \mathrm{tr}_{12}(P_j \overline{R}_{ij}^{-1} P_j^{-1} \otimes I)(I \otimes WP_j E_{jj})\overline{R}_{12}(z)L_1$$

$$= \sum_{ij} \mathrm{tr}_{12}\, L_2(\overline{R}_{ij}^{-1} \otimes E_{jj})\overline{R}_{12}(z)L_1 = \mathrm{tr}\, L(w)\, \mathrm{tr}\, L(z).$$

As for the right-hand side of (3.33), we use identity (3.32) to rewrite it as

$$\mathrm{tr}_{12}\, R_{12}^F(z-w)P_1\overline{R}_{21}^{-1}(w)P_1^{-1}L_1(z)\overline{R}_{21}(w)L_2(w)R_{21}^F(w-z).$$

Having in mind that $\overline{R}_{21}$ is diagonal in the first matrix space and taking into account property (3.27), one can easily see that under the trace sign, the matrix R^F can be pushed to the right, where it cancels with R_{21}^F. Therefore, we obtain

$$\mathrm{tr}_{12}\, P_1\overline{R}_{21}^{-1}(w)P_1^{-1}L_1(z)\overline{R}_{21}(w)L_2(w). \tag{3.35}$$

Now, applying the technique we used above for the left-hand side of (3.33) to this expression, we conclude that Eq. (3.35) is equal to $\mathrm{tr}\, L(z)\, \mathrm{tr}\, L(w)$. Thus, we have proved that $\mathrm{tr}\, L(z)$ commutes with $\mathrm{tr}\, L(w)$. Similarly, one can prove that $\mathrm{tr}\, L^{-1}(z)$ commutes with $\mathrm{tr}\, L^{-1}(w)$ and with $\mathrm{tr}\, L(w)$.

Next, we give an example of the simplest representation of the algebra (3.22) and (3.28) associated with the elliptic RS model.

PROPOSITION 3.2. *The L-operator*

$$L(z) = \sum_{ij} \Phi(z, q_{ij}+\gamma)b_j P_j E_{ij}, \quad \text{where } b_j = \prod_{a\neq j}\Phi(\gamma, q_{aj}) \tag{3.36}$$

and the parameter γ is the coupling constant of the elliptic RS model, satisfies the algebra (3.28).

PROOF. Straightforward calculations. □

Let us note that the L-operator of the same form (3.36) appeared in Section 2 as the result of the Hamiltonian reduction procedure applied to $T^\star\widehat{GL}(N)(z,\bar{z})$. Thus, we see that the quantization procedure respects the form of the L-operator if the proper ordering of the canonical variables is chosen.

We call this L the *quantum L-operator of the elliptic RS model.* Indeed, taking the Hamiltonian to be $H = \mathrm{tr}\, L(z)$, one can see that the quantum canonical transformation of the form

$$P_i^R = \prod_{a\neq i}\left(\frac{\sigma(q_{ai}+\gamma)}{\sigma(q_{ai})}\right)^{1/2} P_i \prod_{a\neq i}\left(\frac{\sigma(q_{ai})}{\sigma(q_{ai}-\gamma)}\right)^{1/2}, \tag{3.37}$$

where P_i^R is the momentum in the Ruijsenaars Hamiltonian, turns H into the first integral S_1 from the Ruijsenaars commuting family [**10**]. Moreover, after the canonical transformation (3.37), the L-operator (3.36) coincides, in essence, with the classical L-operator of the RS model.

In Section 4, we demonstrate that the generating function for the commuting family, in terms of L, can be written as

$$I(z,\mu) = :\det(L(z)-\mu): = \sum_k (-\mu)^{N-k} I_k(z), \tag{3.38}$$

where the normal ordering $:\ :$ means that all momentum operators are pushed to the right.

4. Connection with the fundamental relation $RLL = LLR$

In this section, we establish the connection of the quantum L-operator algebra (3.28) to the fundamental relation of the type $RLL = LLR$.

In [**21**], the operators from the Ruijsenaars commuting family were obtained by using a special representation $\widehat{L}$ of the algebra

$$R^B_{12}(z-w)\widehat{L}_1(z)\widehat{L}_2(w) = \widehat{L}_2(w)\widehat{L}_1(z)R^B_{12}(z-w), \tag{4.1}$$

where $R^B(z)$ is the Belavin R-matrix, which is an elliptic solution to the quantum Yang–Baxter equation [**26**]. The explicit form of R^B we use here can be found in [**27**]. For the reader's convenience, we briefly describe the construction of $\widehat{L}$ [**28, 29**].

Denote by $\mathbf{h}^*$ the weight space for $sl_N(\mathbb{C})$ that can be realized in $\mathbb{C}^N$, with the basis ϵ_i, $\langle\epsilon_i, \epsilon_j\rangle = \delta_{ij}$, as the orthogonal complement to $\sum_{i=1}^N \epsilon_i$. Let $\bar{\epsilon}_k$ be the orthogonal projection of ϵ_k:

$$\bar{\epsilon}_k = \epsilon_k - \frac{1}{N}\sum_{i=1}^N \epsilon_i.$$

For each $q \in \mathbf{h}^*$, one can introduce the intertwining vectors [**32, 33**]

$$\phi(z)_q^{q+\hbar\bar{\epsilon}_k}{}_j = \theta_j\Big(\frac{z}{N} - \langle q, \bar{\epsilon}_k\rangle\Big)\frac{1}{i\eta(\tau)}, \tag{4.2}$$

where

$$\theta_j(z) = \sum_{n\in N/2-j+N\mathbb{Z}} \exp 2\pi i\bigg[n\Big(z+\frac{1}{2}\Big) + \frac{n^2}{2N}\tau\bigg]$$

and $\eta(\tau) = p^{1/24}\prod_{m=1}^\infty(1-p^m)$ is the Dedekind eta function with $p = \exp 2\pi i\tau$.

Following [**28**], we denote by $\bar{\phi}(z)_q^{q+\hbar\bar{\epsilon}_k\, j}$ the entries of the matrix inverse to $\phi(z)_q^{q+\hbar\bar{\epsilon}_k}{}_j$. Then the orthogonality relations read as follows:

$$\sum_{j=1}^n \bar{\phi}(z)_q^{q+\hbar\bar{\epsilon}_k\, j}\phi(z)_q^{q+\hbar\bar{\epsilon}_{k'}}{}_j = \delta_{kk'}, \qquad \sum_{k=1}^n \phi(z)_q^{q+\hbar\bar{\epsilon}_k}{}_j\bar{\phi}(z)_q^{q+\hbar\bar{\epsilon}_k\, j'} = \delta_{jj'}. \tag{4.3}$$

Below, the following formula [**21**] is intensively used:

(4.4)
$$\sum_{m=1}^n \bar{\phi}(z)_{q'}^{q'+\hbar\bar{\epsilon}_j\, m}\phi(z)_q^{q+\hbar\bar{\epsilon}_i}{}_m = \frac{\theta(z+\langle q', \bar{\epsilon}_j\rangle - \langle q, \bar{\epsilon}_i\rangle)}{\theta(z)}\prod_{j'\neq j}\frac{\theta(\langle q', \bar{\epsilon}_{j'}\rangle - \langle q, \bar{\epsilon}_i\rangle)}{\theta(\langle q', \bar{\epsilon}_{j'}\rangle - \langle q', \bar{\epsilon}_j\rangle)}.$$

Here $\theta(z)$ is the Jacobi θ-function,

$$\theta(z) = -\sum_n e^{2\pi i(z+1/2)(n+1/2)+i\pi\tau(n+1/2)^2} = \theta'(0)e^{-\zeta(1/2)z^2}\sigma(z).$$

It was shown in [**29, 30**] that the $\widehat{L}$-operator

$$\widehat{L}_{ij}(z) = \sum_{k=1}^N \phi(z+\gamma N)_q^{q+\hbar\bar{\epsilon}_k}{}_i\bar{\phi}(z)_q^{q+\hbar\bar{\epsilon}_k\, j}e^{\hbar\bar{\epsilon}_k\,\partial/\partial q_k}, \tag{4.5}$$

acting on the space of functions on $\mathbf{h}^*$, satisfies relation (4.1). This $\widehat{L}$ is an $(N\times N)$-generalization of the (2×2)-dimensional Sklyanin L-operator [**31**].

The intertwining vectors $\phi(z)_q^{q+\hbar\bar{\epsilon}_k}{}_j$ appearing in the definition of $\widehat{L}$ relate the matrix R^B with the Boltzmann weights for the $A_{n-1}^{(1)}$ face model. Recall [**33**] that the nonzero Boltzmann weights depending on the spectral parameter z are explicitly given by

$$(4.6)\qquad \begin{aligned} &\check{W}\begin{bmatrix} & q+\hbar\bar{\epsilon}_i & \\ q & z & q+2\hbar\bar{\epsilon}_i \\ & q+\hbar\bar{\epsilon}_i & \end{bmatrix} = \frac{\theta(z+\hbar)}{\theta(\hbar)}, \\ &\check{W}\begin{bmatrix} & q+\hbar\bar{\epsilon}_i & \\ q & z & q+\hbar(\bar{\epsilon}_i+\bar{\epsilon}_j) \\ & q+\hbar\bar{\epsilon}_i & \end{bmatrix} = \frac{\theta(-z+q_{ij})}{\theta(q_{ij})} \qquad (i\neq j), \\ &\check{W}\begin{bmatrix} & q+\hbar\bar{\epsilon}_i & \\ q & z & q+\hbar(\bar{\epsilon}_i+\bar{\epsilon}_j) \\ & q+\hbar\bar{\epsilon}_j & \end{bmatrix} = \frac{\theta(z)}{\theta(\hbar)}\,\frac{\theta(\hbar+q_{ij})}{\theta(q_{ij})} \qquad (i\neq j), \end{aligned}$$

where $q_{ij} = \langle q, \bar{\epsilon}_i - \bar{\epsilon}_j\rangle$.

The relation between R^B and the face weights is

$$(4.7)\qquad \begin{aligned} &\sum_{i'j'} R^B(z-w)_{ij}^{i'j'}\,\phi(z)_q^{q+\hbar\bar{\epsilon}_k}{}_{i'}\,\phi(w)_{q+\hbar\bar{\epsilon}_k}^{q+\hbar(\bar{\epsilon}_k+\bar{\epsilon}_m)}{}_{j'} \\ &= \sum_s \phi(w)_q^{q+\hbar\bar{\epsilon}_s}{}_j\,\phi(z)_{q+\hbar\bar{\epsilon}_s}^{q+\hbar(\bar{\epsilon}_k+\bar{\epsilon}_m)}{}_i\,\check{W}\begin{bmatrix} & q+\hbar\bar{\epsilon}_k & \\ q & z-w & q+\hbar(\bar{\epsilon}_k+\bar{\epsilon}_m) \\ & q+\hbar\bar{\epsilon}_s & \end{bmatrix}. \end{aligned}$$

In what follows, we use the concise notation

$$W_s^k[k+m] = \check{W}\begin{bmatrix} & q+\hbar\bar{\epsilon}_k & \\ q & z-w & q+\hbar(\bar{\epsilon}_k+\bar{\epsilon}_m) \\ & q+\hbar\bar{\epsilon}_s & \end{bmatrix}.$$

Then, the dual relation to (4.7) reads

$$(4.8)\qquad \begin{aligned} &\sum_{i'j'} \bar{\phi}(w)_q^{q+\hbar\bar{\epsilon}_k\,j'}\,\bar{\phi}(z)_{q+\hbar\bar{\epsilon}_k}^{q+\hbar(\bar{\epsilon}_k+\bar{\epsilon}_m)\,i'}\,R^B(z-w)_{i'j'}^{ij} \\ &= \sum_s W_k^s[k+m]\,\bar{\phi}(z)_q^{q+\hbar\bar{\epsilon}_s\,i}\,\bar{\phi}(w)_{q+\hbar\bar{\epsilon}_s}^{q+\hbar(\bar{\epsilon}_k+\bar{\epsilon}_m)\,j}. \end{aligned}$$

The other L-operator, $\widetilde{L}$, was introduced in [**21**]. It is related to $\widehat{L}$ in the following way:

$$(4.9)\qquad \begin{aligned} \widehat{L}_{ij}(z) \to \widetilde{L}_{ij}(z) &= \sum_{i'j'} \bar{\phi}(z)_q^{q+\hbar\bar{\epsilon}_i\,i'}\,\phi(z)_q^{q+\hbar\bar{\epsilon}_j}{}_{j'}\,\widehat{L}_{i'j'}(z) \\ &= \frac{\theta(z+\gamma+q_{ij})}{\theta(z)}\prod_{n\neq i}\frac{\theta(\gamma+q_{nj})}{\theta(q_{ni})}\,e^{\hbar\bar{\epsilon}_j\,\partial/\partial q_j}. \end{aligned}$$

In what follows, we also need to remove the nonholomorphic dependence on the spectral and quantization parameters from the quantum L-operator algebra (3.28). This can be achieved by considering the following transformation of the L-operator:

$$(4.10)\qquad L(z) \to e^{\alpha(z)Q}e^{-\beta Q}L(z)e^{\beta Q}e^{-\alpha(z)Q},$$

where $\alpha(z)$ is an arbitrary function of the spectral parameter and β is a complex number. Since the transformed L-operator also has the structure WP, the relation

$$L_2(w)e^{\alpha(z)Q_1} = e^{\alpha(z)Q_1} L_2(w) e^{\hbar\alpha(z)r_0} \tag{4.11}$$

holds. Here $r_0 \equiv \sum_i E_{ii} \otimes E_{ii}$.

Recalling that the L-operator (3.36) satisfies the quantum L-operator algebra (3.28) and using (4.11), one can easily determine the algebra satisfied by the transformed L,

$$\check{R}_{12}(z,w)L_1(z)\check{\overline{R}}_{21}(w)L_2(w) = L_2(w)\check{\overline{R}}_{12}(z)L_1(z)\check{R}^F_{12}(z-w),$$

where the matrices $\check{R}$, $\check{\overline{R}}$, and $\check{R}^F$ are

$$\check{R}_{12}(z,w) = e^{-\alpha(z)Q_1-\alpha(w)Q_2+\beta Q_2} R_{12}(z,w) e^{\alpha(z)Q_1+\alpha(w)Q_2-\beta Q_1}, \tag{4.12}$$

$$\check{\overline{R}}_{12}(z) = e^{\hbar\alpha(z)r_0-\alpha(z)Q_1+\beta Q_2}\overline{R}_{12}(z)e^{\alpha(z)Q_1-\beta Q_1}, \tag{4.13}$$

$$\begin{aligned}\check{R}^F_{12}(z,w) &= e^{-\alpha(z)Q_1-\alpha(w)Q_2+(\hbar/2)(\alpha(w)-\alpha(z))r_0+\beta Q_1} R^F_{12}(z,w) \\ &\quad\times e^{\alpha(z)Q_1+\alpha(w)Q_2+(\hbar/2)(\alpha(w)-\alpha(z))r_0-\beta Q_2}.\end{aligned} \tag{4.14}$$

Since the transformation in question preserves the form of the quantum L-operator algebra, the transformed matrices $\check{R}$, $\check{\overline{R}}$, and $\check{R}^F$ also satisfy all of the compatibility conditions. In particular, the transformation (4.14) defines another solution of (3.26). For $\beta = 0$, this was observed in [**14**].

For the particular choice

$$\beta = 2\pi i\,\frac{h-\bar{h}}{\tau-\bar{\tau}}, \qquad \alpha(z) = 2\pi i\,\frac{z-\bar{z}}{\tau-\bar{\tau}} + \beta,$$

we find that the matrices $\check{R}$, $\check{\overline{R}}$, and $\check{R}^F$ are given by the same formulas (3.12), (3.13), and (3.25), where the function

$$\Phi(z,s) = \frac{\theta(z+s)}{\theta(z)\theta(s)}\,\theta'(0)e^{2\pi i\,(z-\bar{z})/(\tau-\bar{\tau})}$$

is replaced by the meromorphic function $\theta(z+s)/(\theta(z)\theta(s))$. The transformation with such a choice of α and β transforms (up to an inessential multiplier) the L-operator (3.36) into

$$L_{ij}(z) = \frac{\theta(z+q_{ij}+\gamma)}{\theta(z)\theta(q_{ij}+\gamma)} \prod_{n\neq j} \frac{\theta(q_{nj}+\gamma)}{\theta(q_{nj})}\,\mathbb{P}_j, \tag{4.15}$$

which is a quasi-periodic meromorphic matrix function of the spectral parameter.

$$L(z+1) = L(z), \qquad L(z+\tau) = e^{-2\pi i(\gamma+\hbar)}e^{-2\pi i Q}L(z)e^{2\pi i Q}.$$

We assume that the L-operator is of the form $W\mathbb{P}$, where $\mathbb{P}_i = e^{\hbar\bar{\epsilon}_i\partial/\partial q_i}$. When substituting $\mathbb{P}$ for P, all of the consistency conditions are preserved, because the R-matrices depend only on the difference $q_i - q_j$. Thus, Eq. (3.28), with the R-matrices defined in terms of $\Phi(z,s) = \theta(z+s)/(\theta(z)\theta(s))$, pertains to the meromorphic version of the quantum L-operator algebra, while (4.15) provides us with its particular meromorphic representation. In what follows, we use only this meromorphic version.

Comparing (4.15) to (4.9), we find that L and $\widetilde{L}$ are related in the following way:

$$\widetilde{L}_{ij}(z) = \frac{\prod_{n\neq j}\theta(q_{nj})}{\prod_{n\neq i}\theta(q_{ni})} L_{ij}(z). \tag{4.16}$$

Since the combined transformation (4.9), (4.16) from $\widehat{L}$ to L depends only on q, we arrive at the following theorem.

THEOREM 4.1. *Any representation L of the quantum L-operator algebra* (3.28) *is gauge equivalent to some representation $\widehat{L}$ of* (4.1), *with the gauge equivalence defined as*

$$\widehat{L}_{ij}(z) = \sum_{i'j'} \phi(z)_q^{q+\hbar\bar{\epsilon}_{i'}}{}_i \bar{\phi}(z)_q^{q+\hbar\bar{\epsilon}_{j'}}{}^j \frac{\prod_{n\neq j'}\theta(q_{nj'})}{\prod_{n\neq i'}\theta(q_{ni'})} L_{i'j'}(z). \tag{4.17}$$

PROOF. Assume $\widehat{L}$ to be an abstract L-operator satisfying the algebra (4.1), and introduce $\widetilde{L}$ by Eq. (4.9). Assume that $\widetilde{L}$ has the structure $W\mathbb{P}$, where the entries of the diagonal matrix $\mathbb{P}$ are $\mathbb{P}_i = e^{\hbar\bar{\epsilon}_i\partial/\partial q_i}$ and the entries of W commute with q_i. Then, substituting $\widehat{L}$ expressed via $\widetilde{L}$ in (4.1) and performing a straightforward calculation using (4.8), we find the algebra satisfied by $\widetilde{L}$,

$$\begin{aligned}
&\sum_{\substack{i'j'\\abcd}} W_k^a[a+c]\delta_{a+c,i+k}\bar{\phi}(z)_q^{q+\hbar\bar{\epsilon}_a}{}^{i'}\phi(z)_q^{q+\hbar\bar{\epsilon}_b}{}_{i'}\bar{\phi}(w)_{q+\hbar\bar{\epsilon}_a}^{q+\hbar(\bar{\epsilon}_a+\bar{\epsilon}_c)}{}^{j'}\\
&\qquad\qquad\times\phi(w)_{q+\hbar\bar{\epsilon}_j}^{q+\hbar(\bar{\epsilon}_j+\bar{\epsilon}_d)}{}_{j'}\widetilde{L}_{bj}(z)\widetilde{L}_{dl}(w)\\
&\quad=\sum_{\substack{i'j'\\abcd}} W_a^j[j+l]\delta_{j+l,a+c}\bar{\phi}(w)_q^{q+\hbar\bar{\epsilon}_k}{}^{i'}\phi(w)_q^{q+\hbar\bar{\epsilon}_d}{}_{i'}\\
&\qquad\qquad\times\bar{\phi}(z)_{q+\hbar\bar{\epsilon}_k}^{q+\hbar(\bar{\epsilon}_k+\bar{\epsilon}_i)}{}^{j'}\bar{\phi}(z)_{q+\hbar\bar{\epsilon}_c}^{q+\hbar(\bar{\epsilon}_c+\bar{\epsilon}_b)}{}_{j'}\widetilde{L}_{dc}(w)\widetilde{L}_{ba}(z).
\end{aligned}$$

Performing the summation in i' and j' with the help of (4.4), we obtain

$$\begin{aligned}
&\sum_{abc} W_k^b[b+a]\delta_{a+b,i+k}\frac{\theta(w+q_{ac}+\hbar\delta_{ab}-\hbar\delta_{jc})}{\theta(w)}\\
&\qquad\qquad\times\prod_{n\neq a}\frac{\theta(q_{nc}+\hbar\delta_{nb}-\hbar\delta_{jc})}{\theta(q_{na}+\hbar\delta_{nb}-\hbar\delta_{ab})}\widetilde{L}_{bj}(z)\widetilde{L}_{cl}(w)\\
&\quad=\sum_{abc} W_a^j[j+l]\delta_{j+l,a+c}\frac{\theta(z+q_{ib}+\hbar\delta_{ik}-\hbar\delta_{bc})}{\theta(z)}\\
&\qquad\qquad\times\prod_{n\neq i}\frac{\theta(q_{nb}+\hbar\delta_{nk}-\hbar\delta_{bc})}{\theta(q_{ni}+\hbar\delta_{nk}-\hbar\delta_{ik})}\widetilde{L}_{kc}(w)\widetilde{L}_{ba}(z).
\end{aligned}$$

Let us introduce an operator L by inverting (4.16). Substituting this L into the last formula, taking into account the nonzero components of the face weights, and multiplying both sides by the function

$$\frac{\prod_{n\neq k}\theta(q_{nk})}{\prod_{n\neq j}\theta(q_{nj})}\frac{\prod_{n\neq i}\theta(q_{ni}+\hbar\delta_{nk}-\hbar\delta_{ik})}{\prod_{n\neq l}\theta(q_{nl}+\hbar\delta_{nj}-\hbar\delta_{jl})},$$

we finally arrive at the following algebra satisfied by L:

(4.18)

$$\begin{aligned}
&\sum_s W_k^k[2k]\delta_i^k \frac{\theta(w+q_{ks}+\hbar-\hbar\delta_{js})}{\theta(w)} \frac{\prod_{n\neq k}\theta(q_{ns}-\hbar\delta_{js})}{\prod_{n\neq s}\theta(q_{ns}+\hbar\delta_{nj}-\hbar\delta_{js})} L_{kj}(z)L_{sl}(w) \\
&\quad + \sum_s W_k^k[k+i] \frac{\theta(w+q_{is}-\hbar\delta_{js})}{\theta(w)} \frac{\prod_{n\neq i}\theta(q_{ns}+\hbar\delta_{nk}-\hbar\delta_{js})}{\prod_{n\neq s}\theta(q_{ns}+\hbar\delta_{nj}-\hbar\delta_{js})} L_{kj}(z)L_{sl}(w) \\
&\quad + \sum_s W_k^i[i+k] \frac{\theta(q_{ki}+\hbar)}{\theta(q_{ki}-\hbar)} \frac{\theta(w+q_{ks}-\hbar\delta_{js})}{\theta(w)} \\
&\qquad\qquad \times \frac{\prod_{n\neq k}\theta(q_{ns}+\hbar\delta_{ni}-\hbar\delta_{js})}{\prod_{n\neq s}\theta(q_{ns}+\hbar\delta_{nj}-\hbar\delta_{js})} L_{ij}(z)L_{sl}(w) \\
&= \sum_s W_j^j[j+l]\left(\frac{\theta(q_{lj}+\hbar)}{\theta(q_{lj}-\hbar)}+2\delta_{lj}\right) \frac{\theta(z+q_{is}+\hbar\delta_{ik}-\hbar\delta_{ls})}{\theta(z)} \\
&\qquad\qquad \times \frac{\prod_{n\neq i}\theta(q_{ns}+\hbar\delta_{nk}-\hbar\delta_{ls})}{\prod_{n\neq s}\theta(q_{ns}+\hbar\delta_{nl}-\hbar\delta_{ls})} L_{kl}(w)L_{sj}(z) \\
&\quad + \sum_s W_l^j[j+l] \frac{\theta(z+q_{is}+\hbar\delta_{ik}-\hbar\delta_{js})}{\theta(z)} \\
&\qquad\qquad \times \frac{\prod_{n\neq i}\theta(q_{ns}+\hbar\delta_{nk}-\hbar\delta_{js})}{\prod_{n\neq s}\theta(q_{ns}+\hbar\delta_{nj}-\hbar\delta_{js})} L_{kj}(w)L_{sl}(z).
\end{aligned}$$

Here $i\neq k$ in the second and the third lines. The ratio of the products of the theta-functions in each term in (4.18) allows one to sum over s, e.g., when $i\neq k$, we have

$$\begin{aligned}
&\frac{\prod_{n\neq i}\theta(q_{ns}+\hbar\delta_{nk}-\hbar\delta_{js})}{\prod_{n\neq s}\theta(q_{ns}+\hbar\delta_{nj}-\hbar\delta_{js})} \\
&\quad = \delta_{is}\left(\delta_{ij}\left.\frac{\theta(q_{ki})}{\theta(q_{ki}-\hbar)}\right|_{j\neq k} + \delta_{kj}(1-\delta_{ij}) + \left.\frac{\theta(q_{ki}+\hbar)\theta(q_{ji})}{\theta(q_{ki})\theta(q_{ji}+\hbar)}\right|_{\substack{j\neq i\\ j\neq k}}\right) \\
&\quad + \delta_{ks}(1-\delta_{kj})\left(\delta_{ij}\frac{\theta(\hbar)}{\theta(q_{ik}+\hbar)} + \left.\frac{\theta(q_{jk})\theta(\hbar)}{\theta(q_{jk}+\hbar)\theta(q_{ik})}\right|_{j\neq i}\right) \\
&\quad + \delta_{js}\left.\frac{\theta(q_{kj})\theta(\hbar)}{\theta(q_{kj}-\hbar)\theta(q_{ji}+\hbar)}\right|_{j\neq i}.
\end{aligned}$$

To compare (4.18) with (3.28), we rewrite (3.28) with the help of (3.32) in the following form:

$$\begin{aligned}
&R^F_{12}(z-w)P_1\overline{R}_{21}^{-1}(w)P_1^{-1}L_1(z)\overline{R}_{21}(w)L_2(w) \\
&\quad = P_2\overline{R}_{12}^{-1}(z)P_2^{-1}L_2(w)\overline{R}_{12}(z)L_1(z)R^F_{12}(z-w).
\end{aligned} \tag{4.19}$$

In component form, the algebra (4.19) is presented in Appendix B. Comparing the components of (4.18) to those of (4.19), we can establish that they coincide up to the overall multiplicative factor $\theta(z-w)\theta(\hbar)^2$. Thus, we have shown that the transformation (4.17) maps any representation of the algebra (3.28) to a representation of (4.1). □

The established connection means that the algebra (3.28) possesses a family of N commuting integrals, as does the algebra (4.1), and that the formula of determinant type (3.38) for the commuting family proved in [**21**] is also valid for the L-operator (3.36).

5. Conclusion

In this paper, we described the dynamical R-matrix structure of the quantum elliptic RS model. The quantum L-operator algebra possesses a family of commuting operators. It turns out that this algebra has a surprisingly simple structure and can be analyzed explicitly in component form. Furthermore, one can hope that the problem of finding new representations of the algebra will be simpler than the corresponding problem for the algebra of the type $RLL = LLR$.

There are several interesting problems to be discussed.

First, recall that in the classical case, we obtained two different Poisson algebras that led to the same classical L-operator algebra. Only one of them was quantized. It is desirable to quantize the second one and to show that the corresponding quantum L-operator algebra is isomorphic to the algebra obtained in this paper.

The elliptic RS model we have dealt with corresponds to the A_{N-1} root system. It seems possible to extend our approach to other root systems and to derive the corresponding L-operator algebras. To this end, one should find a proper parameterization of the corresponding cotangent bundle.

Generalizing our approach to the cotangent bundle over a centrally extended group of smooth mappings from a higher-genus Riemann surface into a Lie group, one may expect to obtain new integrable systems.

It is known that the CM systems admit spin generalizations [**34**]–[**36**]. Recently, the spin generalization was found for the elliptic RS model [**37**]. However, the Hamiltonian formulation for the model has been established only for the simplest rational case [**38**]. One may hope that the spin models can arise as higher representations of the L-operator algebra in our approach.

Probably the most interesting and complicated problem is to separate the variables for the quantum elliptic RS model. Up to now, only the classical n-particle case has been solved explicitly [**39**]. One could expect that the L-operator algebra obtained in this paper may shed some light on the problem.

The authors are grateful to M. A. Olshanetsky and N. A. Slavnov for valuable discussions.

Appendix A

Here we present the basic elliptic function identities, formulated as a set of functional relations on $\Phi(z, w)$ [**8**]:

$$\Phi(z,x)\Phi(w,y) = \Phi(z,x-y)\Phi(z+w,y) + \Phi(z+w,x)\Phi(w,y-x), \tag{A.1}$$

$$\begin{aligned}\Phi(z,x)\Phi(z,y) = \Phi(z,x+y)(\Phi(z,0) + \Phi(x,0)&\\ + \Phi(y,0) - \Phi(z+x+y,0)),&\end{aligned} \tag{A.2}$$

$$\begin{aligned}&\Phi(z-w,a-b)\Phi(z,x+b)\Phi(w,y+a)\\ &\qquad - \Phi(z-w,x-y)\Phi(z,y+a)\Phi(w,x+b)\\ &= \Phi(z,x+a)\Phi(w,y+b)(\Phi(a-b,0) + \Phi(x+b,0)\\ &\qquad - \Phi(x-y,0) - \Phi(a+y,0)).\end{aligned} \tag{A.3}$$

Equation (A.2) is the limiting case of (A.1) where $w \to z$, and (A.3) is a consequence of (A.1) and (A.2). Note that the exponential term

$$\exp\left\{ -2\zeta\left(\frac{1}{2}\right)zs + 2\pi i s\,\frac{z-\bar{z}}{\tau-\bar{\tau}}\right\}$$

in $\Phi(z,s)$, as well as the linear term

$$-2\zeta\left(\frac{1}{2}\right)z + 2\pi i\,\frac{z-\bar{z}}{\tau-\bar{\tau}}$$

in $\Phi(z,0)$, is irrelevant since it drops out from (A.1)–(A.2).

To establish the unitarity relation for R, one also needs an identity involving the Weierstrass $\mathcal{P}$-function:

$$\Phi(z,s)\Phi(z,-s) = \mathcal{P}(z) - \mathcal{P}(s).$$

To prove Eq. (2.72), the following relation between the derivatives of Φ is useful:

$$\frac{\partial\Phi(z,q_{ij})}{\partial z} = \frac{\partial\Phi(z,q_{ij})}{\partial q_{ij}} - (\Phi(z,0)-\Phi(q_{ij}))\Phi(z,q_{ij}).$$

Appendix B

In this appendix, we present the quantum L-operator algebra

$$\begin{aligned} R^F_{12}(z-w)P_1\overline{R}^{-1}_{21}(w)P_1^{-1}L_1(z)\overline{R}_{21}(w)L_2(w) \\ = P_2\overline{R}^{-1}_{12}(z)P_2^{-1}L_2(w)\overline{R}_{12}(z)L_1(z)R^F_{12}(z-w) \end{aligned} \tag{B.1}$$

in component form. The left-hand side of (B.1) has the form

$$\begin{aligned} &\sum_{i\neq k;\,j,l} \Phi(\hbar,q_{ki})\Phi(\hbar,q_{ik})\Phi(\hbar,q_{kj}-\hbar)L_{ij}(z)L_{kl}(w)E_{ij}\otimes E_{kl} \\ &\quad - \sum_{i\neq k;\,j,l} \Phi(\hbar,q_{ki})\Phi(\hbar,q_{ij}-\hbar)\Phi(w,q_{kj}-\hbar)L_{ij}(z)L_{jl}(w)E_{ij}\otimes E_{kl} \\ &\quad + \sum_{i\neq k;\,j,l} \Phi(\hbar,q_{ki})\Phi(w,q_{ki})\Phi(\hbar,q_{ij}-\hbar)L_{ij}(z)L_{il}(w)E_{ij}\otimes E_{kl} \\ &\quad + \sum_{i\neq k;\,j,l} \Phi(z-w,q_{ik})\Phi(\hbar,q_{ki})\Phi(\hbar,q_{ij}-\hbar)L_{kj}(z)L_{il}(w)E_{ij}\otimes E_{kl} \\ &\quad - \sum_{i\neq k;\,j,l} \Phi(z-w,q_{ik})\Phi(\hbar,q_{kj}-\hbar)\Phi(w,q_{ij}-\hbar)L_{kj}(z)L_{jl}(w)E_{ij}\otimes E_{kl} \\ &\quad + \sum_{i\neq k;\,j,l} \Phi(z-w,q_{ik})\Phi(w,q_{ik})\Phi(\hbar,q_{kj}-\hbar)L_{kj}(z)L_{kl}(w)E_{ij}\otimes E_{kl} \\ &\quad + \sum_{j,k,l} \Phi(z-w,\hbar)\Phi(w,\hbar)\Phi(\hbar,q_{kj}-\hbar)L_{kj}(z)L_{kl}(w)E_{kj}\otimes E_{kl} \\ &\quad - \sum_{j,k,l} \Phi(z-w,\hbar)\Phi(w,\hbar)\Phi(w+\hbar,q_{kj}-\hbar)L_{kj}(z)L_{jl}(w)E_{kj}\otimes E_{kl}. \end{aligned}$$

The right-hand side of (B.1) reads

$$\sum_{i\neq k;j\neq l} \Phi(\hbar, q_{ki})\Phi(\hbar, q_{il}-\hbar)\Phi(\hbar, q_{lj})L_{kl}(w)L_{ij}(z)E_{ij}\otimes E_{kl}$$
$$+\sum_{i\neq k;\, j,l} \Phi(\hbar, q_{ki})\Phi(\hbar, q_{ij}-\hbar)\Phi(z-w, q_{lj}+\hbar\delta_{lj})L_{kj}(w)L_{il}(z)E_{ij}\otimes E_{kl}$$
$$-\sum_{i\neq k;\, j\neq l} \Phi(\hbar, q_{kl}-\hbar)\Phi(z, q_{il}-\hbar)\Phi(\hbar, q_{lj})L_{kl}(w)L_{lj}(z)E_{ij}\otimes E_{kl}$$
$$-\sum_{i\neq k;\, j,l} \Phi(\hbar, q_{kj}-\hbar)\Phi(z, q_{ij}-\hbar)\Phi(z-w, q_{lj}+\hbar\delta_{lj})L_{kj}(w)L_{jl}(z)E_{ij}\otimes E_{kl}$$
$$+\sum_{i\neq k;\, j\neq l} \Phi(z, q_{ik})\Phi(\hbar, q_{kl}-\hbar)\Phi(\hbar, q_{lj})L_{kl}(w)L_{kj}(z)E_{ij}\otimes E_{kl}$$
$$+\sum_{i\neq k;\, j,l} \Phi(z, q_{ik})\Phi(\hbar, q_{kj}-\hbar)\Phi(z-w, q_{lj}+\hbar\delta_{lj})L_{kj}(w)L_{kl}(z)E_{ij}\otimes E_{kl}$$
$$+\sum_{j\neq l;\, i} \Phi(z, \hbar)\Phi(\hbar, q_{il}-\hbar)\Phi(\hbar, q_{lj})L_{il}(w)L_{ij}(z)E_{ij}\otimes E_{il}$$
$$+\sum_{i,j,l} \Phi(z, \hbar)\Phi(\hbar, q_{ij}-\hbar)\Phi(z-w, q_{lj}+\hbar\delta_{lj})L_{ij}(w)L_{il}(z)E_{ij}\otimes E_{il}$$
$$-\sum_{j\neq l;\, i} \Phi(z, \hbar)\Phi(z+\hbar, q_{il}-\hbar)\Phi(\hbar, q_{lj})L_{il}(w)L_{lj}(z)E_{ij}\otimes E_{il}$$
$$-\sum_{i,j,l} \Phi(z, \hbar)\Phi(z+\hbar, q_{ij}-\hbar)\Phi(z-w, q_{lj}+\hbar\delta_{jl})L_{ij}(w)L_{jl}(z)E_{ij}\otimes E_{il}.$$

References

[1] O. Babelon and C. M. Viallet, *Hamiltonian structures and Lax equations*, Phys. Lett. B **237** (1990), 411–416.

[2] J. Avan and M. Talon, *Classical R-matrix structure for the Calogero model*, Phys. Lett. B **303** (1993), 33–37.

[3] J. Avan, O. Babelon, and M. Talon, *Construction of the classical R-matrices for the Toda and Calogero models* Algebra i Analiz **6** (1994), no. 2, 67–89; English transl. in St.-Petersburg Math. J. **6** (1995).

[4] E. K. Sklyanin, *Dynamic r-matrices for the elliptic Calogero-Moser models*, Algebra i Analiz **6** (1994), no. 2, 227–237; English transl. in St.-Petersburg Math. J. **6** (1995).

[5] H. W. Braden and T. Suzuki, *Takashi R-matrices for elliptic Calogero-Moser models*, Lett. Math. Phys. **30** (1994), 147–158.

[6] J. Avan and G. Rollet, *The classical r-matrix for the relativistic Ruijsenaars–Schneider system*, Phys. Lett. A **212** (1996), 50–54.

[7] Yu. B. Suris, *Why is the Ruijsenaars-Schneider hierarchy governed by the same R-operator as the Calogero-Moser one?*, `hep-th/9602160`, Phys. Lett. A **225** (1997), 253–262.

[8] F. W. Nijhoff, V. B. Kuznetsov, E. K. Sklyanin, and O. Ragnisco, *Dynamical r-matrix for the elliptic Ruijsenaars–Schneider model*, `solv-int/9603006`, J. Phys. A **29** (1996), L333–L340.

[9] Yu. B. Suris, *Elliptic Ruijsenaars-Schneider and Calogero-Moser hierarchies are governed by the same r-matrix*, `solv-int/9603011`.

[10] S. N. Ruijsenaars, *Complete integrability of relativistic Calogero–Moser systems and elliptic function identities*, Comm. Math. Phys. **110** (1987), 191–213.

[11] M. A. Olshanetsky and A. M. Perelomov, *Classical integrable finite-dimensional systems related to Lie algebras*, Phys. Rep. **71** (1981), 313–400.

[12] A. Gorsky and N. Nekrasov, *Hamiltonian systems of Calogero type and two dimensional Yang–Mills theory*, Nuclear Phys. B **414** (1994), 213–238; *Relativistic Calogero–Moser model as gauged WZW theory*, Nuclear Phys. B **436** (1995), 582–608; A. Gorsky, *Integrable many*

body systems in the field theories, Teoret. Mat. Fiz. **103** (1995), no. 3, 437–460; English transl. Theoret. and Math. Phys. **103** (1995), 681–700.

[13] G. E. Arutyunov and P. B. Medvedev, *Generating equation for r-matrices related to dynamic systems of Calogero type*, `hep-th/9511070`; Phys. Lett. A **223** (1996), 66–74.

[14] J. Avan, O. Babelon, and E. Billey, *The Gervais–Neveu–Felder equation and the quantum Calogero–Moser systems*, Comm. Math. Phys. **178** (1996), 281–299; `hep-th/9505091`.

[15] G. Felder, *Conformal field theory and integrable systems associated with elliptic curves*, Proc. International Congress Math. Vol. 1, 2 (Zurich, 1994), Birkhäuser, Basel, 1995, pp. 1247–1255.

[16] J. L. Gervais and A. Neveu, *Novel triangle relation and absence of tachyons in Liouville string field theory*, Nuclear Phys. B **238** (1984), 125–141.

[17] O. Babelon, D. Bernard, and E. Billey, *A quasi-Hopf algebra interpretation of quantum 3-j and 6-j symbols and difference equations*, Phys. Lett. B **375** (1996), 89–97; `q-alg/9511019`.

[18] G. E. Arutyunov and S. A. Frolov, *Quantum dynamic R-matrices and quantum Frobenius group*, `q-alg/9610009`, Comm. Math. Phys. **191** (1998), 15–29.

[19] G. E. Arutyunov, S. A. Frolov, and P. B. Medvedev, *Elliptic Ruijsenaars–Schneider model via the Poisson reduction of the affine Heisenberg double*, `hep-th/9607170`, J. Phys. A **30** (1997), 5051–5063.

[20] ———, *Elliptic Ruijsenaars–Schneider model from the cotangent bundle over the two-dimensional current group*, `hep-th/9608013`, J. Math. Phys. **38** (1997), 5682–5689.

[21] K. Hasegawa, *Ruijsenaars' commuting difference operators as commuting transfer matrices*, `q-alg/9512029`, Comm. Math. Phys. **187** (1997), 289–325.

[22] P. I. Etingof and I. B. Frenkel, *Central extensions of current groups in two dimensions*, Comm. Math. Phys. **165** (1994), 429–444.

[23] F. Falceto and K. Gawedski, *Chern-Simons states at genus one*, Comm. Math. Phys. **159** (1994), 549–579.

[24] L. D. Faddeev, *Integrable models in $(1 + 1)$-dimensional quantum field theory*, Recent Advances in Field Theory and Statistical Mechanics (J. B. Zuber and R. Stora, eds.) (Les Houches Summer School Proc., Session XXXIX, 1982), Elsevier, 1984, pp. 561–608.

[25] P. P. Kulish and E. K. Sklyanin, *Quantum spectral transform method. Recent developments*, Integrable Quantum Field Theories (J. Hietarinta and C. Montonen, eds.), Lecture Notes Phys., Vol. 151, Springer-Verlag, Berlin, 1982, pp. 61–119.

[26] A. A. Belavin, *Dynamical symmetry of integrable quantum systems*, Nuclear Phys. B **180** **[FS2]** (1981), 189–200.

[27] M. P. Richey and C. A. Tracy, *Z_n Baxter model: symmetries and the Belavin parametrization*, J. Statist. Phys. **42** (1986), 311–348.

[28] Y. Quano and A. Fujii, *Yang–Baxter equation for $A^{(1)}_{n-1}$ broken $\mathbb{Z}_N$ models*, Modern Phys. Lett. A **8** (1993), 1585–1597.

[29] K. Hasegawa, *On the crossing symmetry of the elliptic solution of the Yang–Baxter equation and a new L-operator for Belavin's solution*, J. Phys. A **26** (1993), 3211–3228.

[30] ———, *L-operator for Belavin's R-matrix acting on the space of theta functions*, J. Math. Phys. **35** (1994), 6158–6171.

[31] E. K. Sklyanin, *On some algebraic structures related to the Yang–Baxter equation. Representations of the quantum algebra*, Funktsional. Anal. i Prilozhen. **17** (1983), no. 4, 34–48; English transl., Functional Anal. Appl. **17** (1983), no. 4, 273–284.

[32] R. J. Baxter, *Eight-vertex model in lattice statistics and one-dimensional anisotropic Heisenberg chain. I. Some fundamental eigenvectors*, Ann. Phys. **76** (1973) 1–24; 25–47; 48–71.

[33] M. Jimbo, T. Miwa, and M. Okado, *Solvable lattice models whose states are dominant integral weights of $A^{(1)}_{n-1}$*, Lett. Math. Phys. **14** (1987) 123–131; *Local state probabilities of solvable lattice models: an $A^{(1)}_{n-1}$ family*, Nuclear Phys. B **300** **[FS22]** (1988), 74–108.

[34] I. M. Krichever, O. Babelon, E. Billey, and M. Talon, *Spin generalization of Calogero–Moser system and the matrix KP equation*, Topics in Topology and Mathematical Physics, Amer. Math. Soc. Transl. Ser. 2, Vol. 170, Amer. Math. Soc., Providence, RI, 1995, pp. 83–119, `hep-th/9411160`.

[35] N. Nekrasov, *Holomorphic bundles and many-body systems*, `hep-th/9503157`, Comm. Math. Phys. **180** (1996), no. 3, 587–603.

[36] B. Enriquez and V. Rubtsov, *Hitchin systems, higher Gauden operators and r-matrices*, Math. Res. Lett. **3** (1996), no. 3, 343–358.

[37] I. M. Krichever and A. V. Zabrodin, *Spin generalizations of the Ruijsenaars–Schneider model, non-Abelian 2D Toda chain and representations of Sklyanin algebra*, Uspekhi Mat. Nauk **50** (1995), no. 6, 3–56; English transl. in Russian Math. Surveys **50** (1995), no. 6, 1101–1150, hep-th/9505039.

[38] G. E. Arutyunov and S. A. Frolov, *On the Hamiltonian structure of the spin Ruijsenaars–Schneider model*, hep-th/9703119, J. Phys. A **31** (1998), 4203–4216.

[39] V. B. Kuznetsov, F. W. Nijhoff, and E. K. Sklyanin, *Separation of variables for the Ruijsenaars system*, solv-int/9701004, Comm. Math. Phys. **189** (1997), 855–877.

Steklov Mathematical Institute, 8, Gubkina str., GSP-1, 117966, Moscow, Russia
E-mail address: arut@genesis.mi.ras.ru

E-mail address: chekhov@genesis.mi.ras.ru

E-mail address: frolov@genesis.mi.ras.ru

Amer. Math. Soc. Transl.
(2) Vol. **191**, 1999

Some Examples of Quantum Groups in Higher Genus

B. Enriquez and V. Rubtsov

ABSTRACT. This is a survey of our construction of current algebras associated with complex curves and rational differentials. We also study in detail two classes of examples. The first is the case of a rational curve with differentials $z^n dz$; these algebras are "building blocks" for the quantum current algebras introduced in our earlier work. The second is the case of a curve X of genus greater than 1 endowed with a regular differential having only double zeros.

In our papers [**10, 11**], we introduced a family of quasi-Hopf algebras associated with complex curves and rational differentials. These algebras are the quantizations of quasi-Lie bialgebra structures that were defined by V. Drinfeld in [**6**].

Our purpose here is to first present (Section 1) a survey of the constructions of [**10, 11**]. After that, we present some examples.

Let us first recall the construction of the quasi-Lie bialgebras of [**6**]. We fix a semisimple Lie algebra $\bar{\mathfrak{g}}$, a rational curve X, and a nonzero rational form ω on it. We denote by S the set of points of X consisting of all poles and zeros of ω, and by $\mathcal{K}_S$ the direct sum of local fields of X at the points of S. We define R as the subring of $\mathcal{K}_S$ formed by the Laurent expansions at the points of S of the regular functions on $X - S$. Then $\mathcal{K}_S$ is endowed with a scalar product defined by ω, and R is then a maximal isotropic subspace of $\mathcal{K}_S$. We fix a maximal isotropic supplementary space Λ to R in $\mathcal{K}_S$. The Lie algebra $\bar{\mathfrak{g}} \otimes \mathcal{K}_S$, being endowed with the product pairing, possesses a direct sum decomposition

$$\bar{\mathfrak{g}} \otimes \mathcal{K}_S = (\bar{\mathfrak{g}} \otimes R) \oplus (\bar{\mathfrak{g}} \otimes \Lambda)$$

into isotropic subspaces, whose first summand is a Lie subalgebra. This defines quasi-Lie bialgebra structures on $\bar{\mathfrak{g}} \otimes \mathcal{K}_S$ and $\bar{\mathfrak{g}} \otimes R$: we have maps $\delta_{\mathcal{K}_S}$ and δ_R from $\bar{\mathfrak{g}} \otimes \mathcal{K}_S$ and $\bar{\mathfrak{g}} \otimes R$ to their second exterior power, and an element ϕ of $\wedge^3(\bar{\mathfrak{g}} \otimes R)$, satisfying certain compatibility conditions (see [**6**]). The problem of their quantization means we must construct quasi-Hopf algebras $U_\hbar(\bar{\mathfrak{g}} \otimes \mathcal{K}_S)$ and $U_\hbar(\bar{\mathfrak{g}} \otimes R)$ deforming their enveloping algebras, coproducts on these algebras deforming extensions of $\delta_{\mathcal{K}_S}$ and of δ_R, and elements of the tensor cubes of these algebras, satisfying the quasi-Hopf algebra axioms.

1991 *Mathematics Subject Classification.* Primary 17B37, 81R50.

V. R. acknowledges support of grants RFBR-95-01-01101, INTAS-96-196 and No. 96-15-96455 for support of scientific schools.

In [**10**], we solved this problem for double extensions of these quasi-Lie bialgebra structures (these extensions are defined by the usual cocycle on current algebras, and by a derivation), in the special case $\bar{\mathfrak{g}} = \mathfrak{sl}_2$. For this, we defined semi-infinite twists of these structures, in the spirit of Drinfeld's *new realizations*, and quantized them (see [**5, 10**]). These quantizations, $(U_\hbar\mathfrak{g}, \Delta)$ and $(U_\hbar\mathfrak{g}, \bar{\Delta})$, are related by some twist operation. In [**11**], we constructed a subalgebra $U_\hbar\mathfrak{g}_R$ of $U_\hbar\mathfrak{g}$ deforming $U\mathfrak{g}_R$. Using the coideal properties of this algebra with respect to Δ and $\bar{\Delta}$, we reduced the problem of finding a quasi-Hopf structure on $U_\hbar\mathfrak{g}$, preserving $U_\hbar\mathfrak{g}_R$, to a certain decomposition problem on the twist F; this problem was solved using some results on Hopf duality. We close the section by a result (Proposition 1.12 and Corollary 1.1) characterizing the zeros and poles of the structure function $q(z,w)$ defining $U_\hbar\mathfrak{g}$.

In Section 2, we come back on the quantization problem for the non-extended quasi-Lie bialgebra structures. We remark that there, due to the absence of derivations, many more relations are possible for the algebra $U_\hbar\mathfrak{g}$. We describe these relations, and show how the construction of quasi-Hopf algebras described above can be generalized in that situation.

Let us comment here on the applications of the constructions of Section 1 in genera 0 and 1. In genus zero, and with $\omega = dz$, this construction agrees with the quantum currents presentation of double Yangians; in [**8**], we derived another expression of Khoroshkin–Tolstoy twists ([**17**]) relating Drinfeld's coproduct for the double Yangians with the usual (L-operator) coproduct.

In the elliptic case, with ω regular, these algebras are related to Felder's elliptic quantum groups ([**7**]). In both situations, the problem was to suitably refine the decomposition of F, using additional conditions provided by algebras "opposite" to the regular one.

Our goal in the next sections is to present some other examples of the construction of Sections 1, 2.

In Section 3, we treat the case of a rational curve with differential $z^n dz$. The corresponding algebra is denoted $U_{z^n dz}\mathfrak{g}$. We compute its structure coefficients $q(z,w)$ in Proposition 3.1. We also study some properties of $U_{z^n dz}\mathfrak{g}$: making use of the $\mathbb{Z}_{n+1}$-symmetry of the situation, we give a presentation of the vertex relations of $U_{z^n dz}\mathfrak{g}$ not involving the $(n+1)$st roots as in the expression of $q(z,w)$, which allows us to give a meaning to this algebra of vertex relations for complex values of the deformation parameter, without any completion procedure. We then show that, as for the Yangian algebra, the quantum algebras of this family are isomorphic for all nonzero values of the quantum parameter.

In Theorem 3.1, we show that the algebras $U_{z^n dz}\mathfrak{g}$ are "building blocks" for the quantum current algebras $U_\hbar\mathfrak{g}$ constructed in Section 1, i.e., each algebra $U_\hbar\mathfrak{g}$ is isomorphic to the tensor product of algebras $U_{z^n dz}\mathfrak{g}$ with their centers identified.

In Section 4, we turn to the case of a curve X of genus greater than 1, when the differential ω is regular and has only double zeros. The existence of such a form is a well-known fact from Riemann surface theory (see e.g. [**20, 12**]), and is equivalent to the existence of odd theta-characteristics.

In this situation, we first construct isotropic supplementaries in local fields, using functions on X that are multivalued along the b-cycles (4.1.1). We then compute the Green kernel of this decomposition (4.1.2). Applying results of Section 2, we then present relations for quantizations of the centerless versions of these algebras (4.2). We remark that after a finite twist, the above decomposition has a part consisting of regular functions on X minus some points. From this we derive the construction of a regular subalgebra in the quantum current algebra (4.2.3). We show (Remark 4.16) how these results may be extended to doubly extended quasi-Lie bialgebra structures.

The zero-level relations might have special interest when the shift parameter $\hbar h$ (which belongs to the Jacobian of X) is nonformal and torsion, like what happens for elliptic quantum groups with torsion parameter ([**16**]).

In both algebras, we also construct deformations of the enveloping algebra of $\bar{\mathfrak{g}} \otimes (\mathcal{K}_\delta \oplus \mathcal{O}_{S'})$, where δ is the set of zeros of ω, S' are other marked points on X, and $\mathcal{K}_\delta$ and $\mathcal{O}_{S'}$ are the sums of the local fields and rings at these points. This might be useful for constructing induced modules.

Now let us discuss possible continuations of the present work. We did not examine degenerations of our constructions; of special interest should be rational or elliptic curves with points identified. In particular, in the latter case, and with $\omega = dz$, the two types of algebras studied here (which are defined either by applying q^∂ to variables in the structure relations, or by some shifts in theta-functions) should coincide.

We also hope that Corollary 1.1 will help treat quantum conformal blocks in higher genus as it was done by B. Feigin and A. Stoyanovsky in [**15**] in the case of the affine Kac–Moody algebra $\widehat{\mathfrak{sl}}_2$, and in [**3**] in the quantum affine case.

Finally, we also would like to mention the papers [**14, 13**], which should be closely connected with some problems left open in [**11**]: identification of the twists of [**8**] with those of Khoroshkin–Tolstoy in [**17**]; and construction of quantum Serre relations in higher genus.

We would like to thank J. Ding, B. Feigin, G. Felder, K. Gawedzki, K. Hasegawa, M. Jimbo, Y. Kosmann-Schwarzbach, H. Konno, and J. Shiraishi for discussions related to the subjectmatter of this paper. B. E. would like to thank T. Miwa for the invitation to RIMS, where a part of this work was carried out in a very stimulating atmosphere. V. R. would also like to thank M. Audin and M. Rosso for their invitation to Université Louis Pasteur Strasbourg–I.

1. Review of quantum current algebras and quasi-Hopf algebras in higher genus

1.1. Manin pairs, triples and classical twists. Let X be a smooth, connected, compact complex curve, and ω be a nonzero meromorphic differential on X. Let S be a finite set of points of X containing the set S_0 of its zeros and poles. For each $s \in S$, let $\mathcal{K}_s$ be the local field at s; we set

$$\mathcal{K} = \bigoplus_{s \in S} \mathcal{K}_s.$$

Let R be the ring of meromorphic functions on X regular outside S; the ring R can be viewed as a subring of $\mathcal{K}$. The ring R is endowed with the discrete topology and $\mathcal{K}$ with its usual (formal series) topology. On $\mathcal{K}$ let us define the bilinear form

$$\langle f, g \rangle_{\mathcal{K}} = \sum_{s \in S} \operatorname{res}_s (fg\omega),$$

and the derivation $\partial f = df/\omega$. We shall use the notation $\mathfrak{x}(A) = \mathfrak{x} \otimes A$, for any ring A over $\mathbb{C}$ and complex Lie algebra $\mathfrak{x}$.

Let $\bar{\mathfrak{g}} = \mathfrak{sl}_2(\mathbb{C})$. On $\bar{\mathfrak{g}}(\mathcal{K})$ define the bilinear form $\langle\ ,\ \rangle_{\bar{\mathfrak{g}}(\mathcal{K})}$ by setting

$$\langle x \otimes \epsilon, y \otimes \eta \rangle_{\bar{\mathfrak{g}}(\mathcal{K})} = \langle x, y \rangle_{\bar{\mathfrak{g}}} \langle \epsilon, \eta \rangle_{\mathcal{K}}, \qquad x, y \in \bar{\mathfrak{g}},\ \epsilon, \eta \in \mathcal{K},$$

where $\langle\ ,\ \rangle_{\bar{\mathfrak{g}}}$ is the Killing form of $\bar{\mathfrak{g}}$; define the derivation $\partial_{\bar{\mathfrak{g}}(\mathcal{K})}$ by setting

$$\partial_{\bar{\mathfrak{g}}(\mathcal{K})}(x \otimes \epsilon) = x \otimes \partial \epsilon, \qquad x \in \bar{\mathfrak{g}},\ \epsilon \in \mathcal{K},$$

and the cocycle c given by

$$c(\xi, \eta) = \langle \xi, \partial_{\bar{\mathfrak{g}}(\mathcal{K})} \eta \rangle_{\bar{\mathfrak{g}}(\mathcal{K})}.$$

Let $\hat{\mathfrak{g}}$ be the central extension of $\bar{\mathfrak{g}}(\mathcal{K})$ by this cocycle. We then have

$$\hat{\mathfrak{g}} = \bar{\mathfrak{g}}(\mathcal{K}) \oplus \mathbb{C}K,$$

with a bracket for which K is central, and $[\xi, \eta] = ([\bar{\xi}, \bar{\eta}], c(\bar{\xi}, \bar{\eta})K)$, for any $\xi, \eta \in \hat{\mathfrak{g}}$ with first components $\bar{\xi}$, $\bar{\eta}$.

Let us denote by $\partial_{\hat{\mathfrak{g}}}$ the derivation of $\hat{\mathfrak{g}}$ defined by $\partial_{\hat{\mathfrak{g}}}(\xi, 0) = (\partial_{\bar{\mathfrak{g}}(\mathcal{K})}\xi, 0)$ and $\partial_{\hat{\mathfrak{g}}}(K) = 0$.

Let $\mathfrak{g}$ be the skew product of $\hat{\mathfrak{g}}$ with $\partial_{\hat{\mathfrak{g}}}$. We have $\mathfrak{g} = \hat{\mathfrak{g}} \oplus \mathbb{C}D$, with bracket such that $\hat{\mathfrak{g}} \to \mathfrak{g}$, $\xi \mapsto (\xi, 0)$ is a Lie algebra morphism, and $[D, (\xi, 0)] = (\partial_{\hat{\mathfrak{g}}}(\xi), 0)$ for $\xi \in \hat{\mathfrak{g}}$.

View $\bar{\mathfrak{g}}(\mathcal{K})$ as a subspace of $\mathfrak{g} = \hat{\mathfrak{g}} \oplus \mathbb{C}D = \bar{\mathfrak{g}}(\mathcal{K}) \oplus \mathbb{C}K \oplus \mathbb{C}D$ by $\xi \mapsto (\xi, 0, 0)$. On $\mathfrak{g}$ define the pairing $\langle\ ,\ \rangle_{\mathfrak{g}}$ by

$$\langle K, D \rangle_{\mathfrak{g}} = 1, \qquad \langle K, \bar{\mathfrak{g}}(\mathcal{K}) \rangle_{\mathfrak{g}} = \langle D, \bar{\mathfrak{g}}(\mathcal{K}) \rangle_{\mathfrak{g}} = 0,$$
$$\langle \xi, \eta \rangle_{\mathfrak{g}} = \langle \xi, \eta \rangle_{\bar{\mathfrak{g}}(\mathcal{K})}, \qquad \xi, \eta \in \bar{\mathfrak{g}}(\mathcal{K}).$$

Endow $\bar{\mathfrak{g}}(\mathcal{K})$ with the bracket $\langle\ ,\ \rangle_{\bar{\mathfrak{g}}(\mathcal{K})}$. The subspace $\bar{\mathfrak{g}}(R) \subset \bar{\mathfrak{g}}(\mathcal{K})$ is a maximal isotropic subalgebra of $\bar{\mathfrak{g}}(\mathcal{K})$. Drinfeld's Manin pair is $(\bar{\mathfrak{g}}(\mathcal{K}), \bar{\mathfrak{g}}(R))$ (see [**6**]). In [**10**], we introduced the following extension of this pair. Let $\mathfrak{g}_R = \bar{\mathfrak{g}}(R) \oplus \mathbb{C}D$; $\mathfrak{g}_R \subset \mathfrak{g}$ is a maximal isotropic subalgebra of $\mathfrak{g}$. The extended Drinfeld's Manin pair of [**10**] is then $(\mathfrak{g}, \mathfrak{g}_R)$.

In [**10**], we also introduced the following Manin triple. Let Λ be a Lagrangian complement to R in $\mathcal{K}$, commensurable with $\bigoplus_{s \in S} \mathcal{O}_s$ (where $\mathcal{O}_s$ is the completed local ring at s). Let $\mathfrak{n}_+ = \mathbb{C}e$, $\mathfrak{n}_- = \mathbb{C}f$, $\mathfrak{h} = \mathbb{C}h$. Let

$$\mathfrak{g}_+ = \mathfrak{h}(R) \oplus \mathfrak{n}_+(\mathcal{K}) \oplus \mathbb{C}D, \qquad \mathfrak{g}_- = (\mathfrak{h} \otimes \Lambda) \oplus \mathfrak{n}_-(\mathcal{K}) \oplus \mathbb{C}K;$$

put $\mathfrak{g} = \mathfrak{g}_+ \oplus \mathfrak{g}_-$. Then both $\mathfrak{g}_+$ and $\mathfrak{g}_-$ are maximal isotropic subalgebras of $\mathfrak{g}$. The Manin triple is then $(\mathfrak{g}, \mathfrak{g}_+, \mathfrak{g}_-)$.

We shall also consider the following Manin triple, which we may consider as being obtained from the previous one by the action of the nontrivial element of the Weyl group of $\bar{\mathfrak{g}}$. Let

$$\bar{\mathfrak{g}}_+ = \mathfrak{h}(R) \oplus \mathfrak{n}_-(\mathcal{K}) \oplus \mathbb{C}D, \qquad \bar{\mathfrak{g}}_- = (\mathfrak{h} \otimes \Lambda) \oplus \mathfrak{n}_+(\mathcal{K}) \oplus \mathbb{C}K;$$

then $(\mathfrak{g}, \bar{\mathfrak{g}}_+, \bar{\mathfrak{g}}_-)$ again forms a Manin triple.

According to [**4**], to each of the Manin triples $(\mathfrak{g}, \mathfrak{g}_+, \mathfrak{g}_-)$ and $(\mathfrak{g}, \bar{\mathfrak{g}}_+, \bar{\mathfrak{g}}_-)$ is associated a Lie bialgebra structure on $\mathfrak{g}$; denote by $\delta, \bar{\delta} \colon \mathfrak{g} \to \mathfrak{g} \hat{\otimes} \mathfrak{g}$ the corresponding cocycle maps.

Let $\mathfrak{g}_\Lambda = (\bar{\mathfrak{g}} \otimes \Lambda) \oplus \mathbb{C}K \subset \mathfrak{g}$; the algebra $\mathfrak{g}_\Lambda$ is a Lagrangian complement of $\mathfrak{g}_R$ in $\mathfrak{g}$. It induces a Lie quasi-bialgebra structure on $\mathfrak{g}_R$, and from [**1**] it follows also that there is a Lie quasi-bialgebra structure on $\mathfrak{g}$, associated to the Manin pair $(\mathfrak{g}, \mathfrak{g}_R)$ and to $\mathfrak{g}_\Lambda$; we denote by $\delta_R \colon \mathfrak{g} \to \mathfrak{g} \hat{\otimes} \mathfrak{g}$ the corresponding cocycle map.

These Lie (quasi-)bialgebra structures on $\mathfrak{g}$ are related by the following classical twist operations.

Let $(e^i)_{i\in\mathbb{N}}$, $(e_i)_{i\in\mathbb{N}}$ be dual bases of R and Λ; we choose them is such a way that e_i tends to 0 when i tends to ∞. Let ϵ^i, ϵ_i, $i\in\mathbb{Z}$, be dual bases of $\mathcal{K}$, defined by $\epsilon_i = e_i$, $\epsilon^i = e^i$, $i \geqslant 0$, $\epsilon_i = e^{-i-1}$, $\epsilon^i = e_{-i-1}$, $i<0$.

LEMMA 1.1 (see [**11**]). *Let* $f = \sum_{i\in\mathbb{Z}} e[\epsilon^i]\otimes f[\epsilon_i]$; $f = f_1 + f_2$, *with*

$$f_1 = \sum_{i\in\mathbb{N}} e[e_i]\otimes f[e^i] \quad and \quad f_2 = \sum_{i\in\mathbb{N}} e[e^i]\otimes f[e_i].$$

For $\xi\in\mathfrak{g}$, *we have*

$$\delta_R(\xi) = \delta(\xi) + [f_1, \xi\otimes 1 + 1\otimes\xi], \qquad \bar\delta(\xi) = \delta_R(\xi) + [f_2, \xi\otimes 1 + 1\otimes \xi].$$

1.2. Results on Green kernels.

NOTATION. For $a = a(z,w)$ a function of two variables z and w, we denote by $a^{(21)}$ the function $a^{(21)}(z,w) = a(w,z)$.

Let us fix dual bases $(e^i)_{i\in\mathbb{N}}, (e_i)_{i\in\mathbb{N}}$ of R and Λ. Let $G\in R\widehat{\otimes}\Lambda$ be the series

$$G = \sum_i e^i\otimes e_i;$$

it is called the *Green kernel* of (X,ω,S,Λ). Note that $R\widehat{\otimes} k$ is an algebra continuing G. Let $\gamma = (\partial\otimes 1)G - G^2$; then

LEMMA 1.2 (see [**10**]). γ *belongs to* $R\otimes R$.

Let $\hbar$ be a formal variable and let $T\colon k[[\hbar]]\to k[[\hbar]]$ be the operator given by $T = \sinh(\hbar\partial)/\hbar\partial$. We shall use the notation $q = e^\hbar$.

Let $(\gamma_i)_{i\geqslant 0}$ be a set of free variables, and $\phi,\psi\in\hbar\mathbb{C}[\gamma_i][[\hbar]]$ be the solutions of the equation

$$\frac{\partial\psi}{\partial\hbar} = D\psi - 1 - \gamma_0\psi^2, \qquad \frac{\partial\phi}{\partial\hbar} = D\phi - \gamma_0\psi,$$

where $D = \sum_{i\geqslant 0}\gamma_{i+1}\partial/\partial\gamma_i$; then

PROPOSITION 1.1 (see [**10**], Proposition 3).

$$\sum_{i\in\mathbb{N}} \frac{q^\partial - 1}{\partial} e^i\otimes e_i = \phi(\hbar, (\partial^i\otimes 1)\gamma) - \ln(1 + G\psi(\hbar, (\partial^i\otimes 1)\gamma)).$$

This proposition implies the next one.

PROPOSITION 1.2 (see [**10**]). *For certain elements* $\phi\in(R\otimes R)[[\hbar]]$, $\psi_+,\psi_-\in\hbar(R\otimes R)[[\hbar]]$, *we have the following identities in* $(R\widehat{\otimes}\mathcal{K})[[\hbar]]$:

$$\sum_i Te^i\otimes e_i = \phi + \frac{1}{2\hbar}\ln\frac{1+G\psi_-}{1+G\psi_+}, \qquad \sum_i e^i\otimes Te_i = -\phi^{(21)} + \frac{1}{2\hbar}\ln\frac{1-G\psi_+^{(21)}}{1-G\psi_-^{(21)}}.$$

LEMMA 1.3 (see [**10**]). *The expression*

$$\sum_i Te^i\otimes e_i - e^i\otimes Te_i$$

belongs to $S^2(R)[[\hbar]]$. *We shall denote by* τ *any element of* $(R\otimes R)[[\hbar]]$ *such that*

$$\tau + \tau^{(21)} = \sum_i Te^i\otimes e_i - e^i\otimes Te_i. \tag{1.1}$$

Note that $\sum_i Te^i \otimes e_i$ is well defined in $(R \widehat{\otimes} \mathcal{K})[[\hbar]]$, because e_i tends to zero as i tends to infinity. Since ∂ is a continuous map from $\mathcal{K}$ to itself, the same is true for the sequence $\partial^k e_i$. So $\sum_i e^i \otimes Te_i$ is well defined in the same space; $\sum_{i\in\mathbb{Z}} T\epsilon^i \otimes \epsilon_i - \epsilon^i \otimes T\epsilon_i$ is well defined in $(\mathcal{K} \overline{\otimes} \mathcal{K})[[\hbar]]$ for the same reasons.

Define

$$q(z,w) = q^{2(\tau-\phi)} \frac{1+\psi_+ G}{1+\psi_- G}(z,w). \tag{1.2}$$

1.3. Hopf algebras $(U_\hbar\mathfrak{g}, \Delta)$, $(U_\hbar\mathfrak{g}, \bar{\Delta})$ and the twist connecting them. In [**10**], we introduced a Hopf algebra $(U_\hbar\mathfrak{g}, \Delta)$ quantizing $(\mathfrak{g}, \delta)$.

This algebra is the quotient of $T(\mathfrak{g})\hat{}[[\hbar]]$ by the following relations. Let e, f, h be the Chevalley basis of $\mathfrak{sl}_2(\mathbb{C})$. Denote in $T(\mathfrak{g})\hat{}[[\hbar]]$, the element $x \otimes \epsilon \in \bar{\mathfrak{g}}(\mathcal{K}) \subset \mathfrak{g}$ of $\mathfrak{g}$ by $x[\epsilon]$, and for $r \in R$ let $h^+[r] = h[r]$, $h^-[\lambda] = h[\lambda]$. Introduce the generating series

$$e(z) = \sum_{i\in\mathbb{Z}} e[\epsilon_i]\epsilon^i(z), \qquad f(z) = \sum_{i\in\mathbb{Z}} f[\epsilon_i]\epsilon^i(z),$$
$$h^+(z) = \sum_{i\in\mathbb{N}} h^+[e^i]e_i(z), \qquad h^-(z) = \sum_{i\in\mathbb{N}} h^-[e_i]e^i(z).$$

The Cartan fields are arranged in the series

$$K^+(z) = e^{((T+U)h^+)(z)}, \qquad K^-(z) = e^{-\hbar h^-(z)};$$

here U is the linear operator from Λ to $R[[\hbar]]$ defined by $U(\lambda) = \langle \tau, 1 \otimes \lambda \rangle$. The relations for $U_\hbar\mathfrak{g}$ are the coefficients of

$$[K^+(z), K^+(w)] = 0, \qquad (K^+(z), K^-(w)) = \frac{q(z,w)}{q(z, q^{-K\partial}(w))}, \tag{1.3}$$

$$(K^-(z), K^-(w)) = \frac{q(q^{-K\partial}(z), q^{-K\partial}(w))}{q(z,w)}, \tag{1.4}$$

$$(K^+(z), e(w)) = q(z,w), \qquad (K^-(z), e(w)) = q(w, q^{-K\partial}(z)), \tag{1.5}$$

$$(K^+(z), f(w)) = q(w,z), \qquad (K^-(z), f(w)) = q(z,w), \tag{1.6}$$

$$\begin{aligned}(z_s - w_s)(1+\psi_+G)(z,w)e(z)e(w) \\ = (z_s - w_s)e^{2(\tau-\phi)(z,w)}(1+\psi_-G)(z,w)e(w)e(z),\end{aligned} \tag{1.7}$$

$$\begin{aligned}(z_s - w_s)e^{2(\tau-\phi)(z,w)}(1+\psi_-G)(z,w)f(z)f(w) \\ = (z_s - w_s)(1+\psi_+G)(z,w)f(w)f(z),\end{aligned} \tag{1.8}$$

$$[e(z), f(w)] = \delta(z,w)K^+(z) - \delta(z, q^{-K\partial}(w))K^-(w)^{-1}, \tag{1.9}$$

$$[D, x^\pm(z)] = -(\partial x^\pm)(z) + \hbar(Ah^+)(z)x^\pm(z), \tag{1.10}$$

$$[D, K^\pm(z)] = -(\partial K^\pm)(z) + \hbar(B^\pm h^+)(z)K^\pm(z). \tag{1.11}$$

K is central. We used the standard notation (a,b) for the group commutator $aba^{-1}b^{-1}$; we also set $\delta(z,w) = G(z,w) + G^{(21)}(z,w)$. Here A and $B^\pm$ are operators from Λ to $R[[\hbar]]$; A is defined by $A(\lambda) = T((\partial\lambda)_R) + \partial(U\lambda) - U((\partial\lambda)_\Lambda)$; formulas for $B^\pm$ can be extracted from [**10**].

The formulas

$$\Delta(K) = K \otimes 1 + 1 \otimes K, \tag{1.12}$$

$$\begin{aligned} \Delta(h^+[r]) &= h^+[r] \otimes 1 + 1 \otimes h^+[r], \\ \Delta(h^-(z)) &= h^-(z) \otimes 1 + 1 \otimes (q^{-K_1\partial} h^-)(z), \end{aligned} \tag{1.13}$$

$$\Delta(e(z)) = e(z) \otimes K^+(z) + 1 \otimes e(z), \tag{1.14}$$

$$\Delta(f(z)) = f(z) \otimes 1 + K^-(z)^{-1} \otimes (q^{-K_1\partial} f)(z), \tag{1.15}$$

$$\Delta(D) = D \otimes 1 + 1 \otimes D + \sum_{i \in \mathbb{N}} \frac{\hbar}{4} h^+[e^i] \otimes h^+[Ae_i], \tag{1.16}$$

$r \in R$, for the coproduct, and

$$\varepsilon(h^+[r]) = \varepsilon(h^-[\lambda]) = \varepsilon(x[\epsilon]) = \varepsilon(D) = \varepsilon(K) = 0, \tag{1.17}$$

$x = e, f$, $r \in R$, $\lambda \in \Lambda$, $\epsilon \in \mathcal{K}$, for the counit, define a topological (with respect to the completion introduced above) Hopf algebra structure on $U_\hbar \mathfrak{g}$.

The coalgebra structure of $U_\hbar \bar{\mathfrak{g}}$ is defined by the coproduct

$$\bar{\Delta}(K) = K \otimes 1 + 1 \otimes K, \tag{1.18}$$

$$\begin{aligned} \bar{\Delta}(h^+[r]) &= h^+[r] \otimes 1 + 1 \otimes h^+[r], \\ \bar{\Delta}(\bar{h}^-(z)) &= h^-(z) \otimes 1 + 1 \otimes (\bar{q}^{-K_1\partial} h^-)(z), \end{aligned} \tag{1.19}$$

$$\bar{\Delta}(e(z)) = (q^{\bar{K}_2\partial}(e \otimes K^-))(z) + 1 \otimes e(z), \tag{1.20}$$

$$\bar{\Delta}(f(z)) = f(z) \otimes 1 + K^+(z) \otimes f(z), \tag{1.21}$$

$$\bar{\Delta}(\bar{D}) = \bar{D} \otimes 1 + 1 \otimes \bar{D} + \sum_{i \in \mathbb{N}} \frac{\hbar}{4} h^+[e^i] \otimes h^+[Ae_i], \tag{1.22}$$

$r \in R$, and the counit

$$\bar{\varepsilon}(h^+[r]) = \bar{\varepsilon}(h^-[\lambda]) = \bar{\varepsilon}(x[\epsilon]) = \bar{\varepsilon}(D) = \bar{\varepsilon}(K) = 0, \tag{1.23}$$

$x = e, f$, $r \in R$, $\lambda \in \Lambda$, $\epsilon \in \mathcal{K}$.

THEOREM 1.4. *The pairs* $(U_\hbar \mathfrak{g}, \Delta)$ *and* $(U_\hbar \mathfrak{g}, \bar{\Delta})$ *defined by the above relations are Hopf algebras quantizing the Lie bialgebras* $(\mathfrak{g}, \delta)$ *and* $(\mathfrak{g}, \bar{\delta})$.

This result means in particular that $U_\hbar \mathfrak{g}$ is a flat deformation of the enveloping algebra $U\mathfrak{g}$. This follows from the following Poincaré–Birkhoff–Witt-type (PBW) result:

PROPOSITION 1.3 (see [**11**], Proposition 4.1). *Let A be an algebra with generators $x_n, x \in \mathbb{Z}$, generating the series $x(z) = \sum_{n \in \mathbb{Z}} x_n z^{-n}$, and relations defined by the modes of*

$$\Big(z - w + \sum_{i \geqslant 1} \hbar^i a_i(z,w)\Big) x(z) x(w) = \Big(z - w + \sum_{i \geqslant 1} \hbar^i b_i(z,w)\Big) x(w) x(z),$$

for a_i and b_i series of $\mathbb{C}[[z,w]][z^{-1}, w^{-1}]$. Set

$$a(z,w) = z - w + \sum_{i \geqslant 1} \hbar^i a_i(z,w), \qquad b(z,w) = z - w + \sum_{i \geqslant 1} \hbar^i b_i(z,w).$$

Then if the series a_i and b_i satisfy $a(z,w)a(w,z) = b(z,w)b(w,z)$, it follows that A is a flat deformation of the symmetric algebra in the variables x_n, $n \in \mathbb{Z}$.

Theorem 1.4 follows from a double construction. More precisely, define $U_\hbar\mathfrak{h}_R$, respectively $U_\hbar\mathfrak{h}_\Lambda$, as the subalgebras of $U_\hbar\mathfrak{g}$ generated by D and $h^+[r]$, $r \in R$, respectively K and $h[\lambda]$, $\lambda \in \Lambda$; and define $U_\hbar\mathfrak{n}_+$, respectively $U_\hbar\mathfrak{n}_-$, as the subalgebras of $U_\hbar\mathfrak{g}$ generated by $e[\epsilon]$, $\epsilon \in \mathcal{K}$, respectively $f[\epsilon]$, $\epsilon \in \mathcal{K}$. Set

$$U_\hbar\mathfrak{g}_+ = U_\hbar\mathfrak{h}_R U_\hbar\mathfrak{n}_+, \qquad U_\hbar\mathfrak{g}_- = U_\hbar\mathfrak{h}_\Lambda U_\hbar\mathfrak{n}_-,$$
$$U_\hbar\bar{\mathfrak{g}}_+ = U_\hbar\mathfrak{h}_R U_\hbar\mathfrak{n}_-, \qquad U_\hbar\bar{\mathfrak{g}}_- = U_\hbar\mathfrak{h}_\Lambda U_\hbar\mathfrak{n}_+.$$

Then we have

PROPOSITION 1.4 (see [**10**]). *$U_\hbar\mathfrak{g}_\pm$ and $U_\hbar\bar{\mathfrak{g}}_\pm$ are subalgebras of $U_\hbar\mathfrak{g}$, while $(U_\hbar\mathfrak{g}_\pm, \Delta)$ and $(U_\hbar\bar{\mathfrak{g}}_\pm, \bar\Delta)$ are Hopf subalgebras of $(U_\hbar\mathfrak{g}, \Delta)$ and $(U_\hbar\mathfrak{g}, \bar\Delta)$. Moreover, $(U_\hbar\mathfrak{g}_+, \Delta)$ and $(U_\hbar\mathfrak{g}_-, \Delta')$ are Hopf dual, while $(U_\hbar\bar{\mathfrak{g}}_+, \bar\Delta)$ and $(U_\hbar\bar{\mathfrak{g}}_-, \bar\Delta')$ and $(U_\hbar\mathfrak{g}, \Delta)$ and $(U_\hbar\mathfrak{g}, \bar\Delta)$ are the corresponding Drinfeld doubles.*

LEMMA 1.5. *The restriction of the Hopf pairing between $(U_\hbar\mathfrak{g}_+, \Delta)$ and $(U_\hbar\mathfrak{g}_-, \Delta')$ to $U_\hbar\mathfrak{n}_+ \times U_\hbar\mathfrak{n}_-$ agrees up to permutation of factors with the restriction to $U_\hbar\mathfrak{n}_- \times U_\hbar\mathfrak{n}_+$ of the Hopf pairing between $(U_\hbar\bar{\mathfrak{g}}_+, \bar\Delta)$ and $(U_\hbar\bar{\mathfrak{g}}_-, \bar\Delta')$.*

Let us define the completion $U_\hbar\mathfrak{g} \,\overline{\otimes}\, U_\hbar\mathfrak{g}$ as follows. Let $I_N \subset U_\hbar\mathfrak{g}$ be the left ideal generated by $x[\epsilon]$, $\epsilon \in \prod_{s\in S} z_s^N \mathbb{C}[[z_s]]$. Define $U_\hbar\mathfrak{g} \,\overline{\otimes}\, U_\hbar\mathfrak{g}$ as the inverse limit of the $U_\hbar\mathfrak{g}^{\otimes 2}/I_N \otimes U_\hbar\mathfrak{g} + U_\hbar\mathfrak{g} \otimes I_N$ (where the tensor products are $\hbar$-adically completed). Then $U_\hbar\mathfrak{g} \,\overline{\otimes}\, U_\hbar\mathfrak{g}$ is clearly a completion of $U_\hbar\mathfrak{g}^{\hat\otimes 2}$.

PROPOSITION 1.5. *Let (α^i), (α_i) be bases of $U_\hbar\mathfrak{n}_+$ and $U_\hbar\mathfrak{n}_-$, dual for the pairing of Lemma 1.5. Set*

$$F = \sum_i \alpha^i \otimes \alpha_i; \tag{1.24}$$

F belongs to $U_\hbar\mathfrak{g} \,\overline{\otimes}\, U_\hbar\mathfrak{g}$, and is a twist transforming $U_\hbar\mathfrak{g}$ into $U_\hbar\bar{\mathfrak{g}}$. More precisely,

$$\mathrm{Ad}(F)(\Delta(x)) = \bar\Delta(x), \tag{1.25}$$

for each $x \in U_\hbar\mathfrak{g}$.

1.4. Universal R-matrices and Hopf algebra pairings. Let $U_\hbar\mathfrak{g}_\pm$ be the subalgebras of $U_\hbar\mathfrak{g}$ generated by $\mathfrak{g}_\pm$. These are Hopf subalgebras of $U_\hbar\bar{\mathfrak{g}}$, dual to each other if $U_\hbar\mathfrak{g}_-$ is endowed with the coproduct opposite to Δ. Then $U_\hbar\mathfrak{g}_+$ and $U_\hbar\mathfrak{g}_-$ are dual to each other, and $U_\hbar\mathfrak{g}$ is the corresponding double algebra.

The pairing $\langle\ ,\ \rangle_{U_\hbar\mathfrak{g}}$ between $U_\hbar\mathfrak{g}_+$ and $U_\hbar\mathfrak{g}_-$ is defined by

$$\langle h^+[r], h^-[\lambda]\rangle_{U_\hbar\mathfrak{g}} = \frac{2}{\hbar}\langle r, \lambda\rangle_k, \qquad \langle e[\epsilon], f[\eta]\rangle_{U_\hbar\mathfrak{g}} = \frac{1}{\hbar}\langle \epsilon, \eta\rangle_k,$$

for $\epsilon, \eta \in k$, $r \in R$, $\lambda \in \Lambda$,

$$\langle D, K\rangle_{U_\hbar\mathfrak{g}} = 1, \qquad \langle D, \bar{\mathfrak{g}}(\mathcal{K})\rangle_{U_\hbar\mathfrak{g}} = \langle K, \bar{\mathfrak{g}}(\mathcal{K})\rangle_{U_\hbar\mathfrak{g}} = 0,$$

and will be a Hopf algebra pairing, $U_\hbar\mathfrak{g}_-$ being endowed with the coproduct opposite to the one given by its embedding in $U_\hbar\mathfrak{g}$.

Let $U_\hbar\bar{\mathfrak{g}}_\pm$ be the subalgebras of $U_\hbar\bar{\mathfrak{g}}$ generated by $\bar{\mathfrak{g}}_\pm$. These are Hopf subalgebras of $U_\hbar\bar{\mathfrak{g}}$, dual to each other if $U_\hbar\bar{\mathfrak{g}}_+$ is given the coproduct opposite to $\bar\Delta$. The pairing $\langle\ ,\ \rangle_{U_\hbar\bar{\mathfrak{g}}}$ between $U_\hbar\bar{\mathfrak{g}}_-$ and $U_\hbar\bar{\mathfrak{g}}_+$ is defined by the formulas

$$\langle h^-[\lambda], h^+[r]\rangle_{U_\hbar\bar{\mathfrak{g}}} = \frac{2}{\hbar}\langle r, \lambda\rangle_k, \qquad \langle e[\epsilon], f[\eta]\rangle_{U_\hbar\bar{\mathfrak{g}}} = \frac{1}{\hbar}\langle \epsilon, \eta\rangle_k,$$

for $\epsilon, \eta \in k$, $r \in R$, $\lambda \in \Lambda$,

$$\langle D, K\rangle_{U_\hbar\bar{\mathfrak{g}}} = 1, \qquad \langle D, \bar{\mathfrak{g}}(\mathcal{K})\rangle_{U_\hbar\bar{\mathfrak{g}}} = \langle K, \bar{\mathfrak{g}}(\mathcal{K})\rangle_{U_\hbar\bar{\mathfrak{g}}} = 0.$$

PROPOSITION 1.6 (see [**11**], Proposition 6.1). *The Hopf algebras $U_\hbar\mathfrak{g}$ and $U_\hbar\bar{\mathfrak{g}}$ are quasi-triangular, and their universal R-matrices are respectively*

$$(1.26) \quad \mathcal{R} = q^{D\otimes K} q^{\frac{1}{2}\sum_{i\in\mathbb{N}} h^+[e^i]\otimes h^-[e_i]} F, \qquad \bar{\mathcal{R}} = F^{21} q^{D\otimes K} q^{\frac{1}{2}\sum_{i\in\mathbb{N}} h^+[e^i]\otimes h^-[e_i]}.$$

In fact, $\mathcal{R}$ and $\bar{\mathcal{R}}$ represent the identity for the pairings $\langle\ ,\ \rangle_{U_\hbar\mathfrak{g}}$ and $\langle\ ,\ \rangle_{U_\hbar\bar{\mathfrak{g}}}$. Note that

$$\bar{\mathcal{R}} = F^{21}\mathcal{R}F^{-1} = e^{\hbar\sum_{i\in\mathbb{Z}} f[\epsilon^i]\otimes e[\epsilon_i]} e^{\frac{\hbar}{2}D\otimes K} e^{\frac{\hbar}{2}\sum_{i\in\mathbb{N}} h^+[e^i]\otimes h^-[e_i]}.$$

1.5. Regular subalgebra $U_\hbar\mathfrak{g}_R$. Let $U_\hbar\mathfrak{g}_R$ be the subalgebra of $U_\hbar\mathfrak{g}$ generated by $x[r]$, $x = e, f, h$, $r \in R$. Then:

PROPOSITION 1.7 (see [**11**], Section 5.2). *The inclusion of $U_\hbar\mathfrak{g}_R$ in $U_\hbar\mathfrak{g}b$ is a flat deformation of that of $U\mathfrak{g}_R$ in $U\mathfrak{g}$.*

Moreover, $U_\hbar\mathfrak{g}_R$ has the following coideal properties with respect to Δ and $\bar{\Delta}$:

$$(1.27) \qquad \Delta(U_\hbar\mathfrak{g}_R) \subset U_\hbar\mathfrak{g} \otimes U_\hbar\mathfrak{g}_R, \qquad \bar{\Delta}(U_\hbar\mathfrak{g}_R) \subset U_\hbar\mathfrak{g}_R \otimes U_\hbar\mathfrak{g}$$

(see [**11**], Proposition 5.4).

1.6. Decomposition of F. Now we would like to decompose F defined in (1.24) as a product

$$(1.28) \qquad F_2F_1 \quad \text{with } F_1 \in U_\hbar\mathfrak{g} \,\widehat{\otimes}\, U_\hbar\mathfrak{g}_R,\ F_2 \in U_\hbar\mathfrak{g}_R \,\widehat{\otimes}\, U_\hbar\mathfrak{g}.$$

The interest of this decomposition lies in the following proposition:

PROPOSITION 1.8. *For any decomposition* (1.28), *the map* $\mathrm{Ad}(F_1) \circ \Delta$ *defines an algebra morphism from $U_\hbar\mathfrak{g}_R$ to $U_\hbar\mathfrak{g}_R \,\widehat{\otimes}\, U_\hbar\mathfrak{g}_R$ (where the tensor product is completed over $\mathbb{C}[[\hbar]]$).*

This follows at once from the coideal properties (1.27).

Let us now try to decompose F according to (1.28). Let (m_i), respectively (m_i') be a basis of $U_\hbar\mathfrak{g}$ as a left, respectively right, $U_\hbar\mathfrak{g}_R$-module. Assume that $m_0 = m_0' = 1$. Due to the form of F_1 and F_2, we have the decompositions

$$F_2 = \sum_i (1 \otimes m_j')F_2^{(j)}, \quad F_1 = \sum_i F_1^{(i)}(m_i \otimes 1), \qquad F_1^{(i)}, F_2^{(j)} \in U_\hbar\mathfrak{g}_R^{\widehat{\otimes}2}.$$

It follows that

$$(1.29) \qquad F = \sum_i F_2F_1^{(i)}(m_i \otimes 1) = \sum_j (1 \otimes m_j')F_2^{(j)}F_1.$$

Now let Π, respectively Π', be the left, respectively right, $U_\hbar\mathfrak{g}_R$-module morphisms from $U_\hbar\mathfrak{g}$ to $U_\hbar\mathfrak{g}_R$ such that $\Pi(m_i) = 0$ for $i \neq 0$, $\Pi(1) = 1$, and $\Pi'(m_i') = 0$ for $i \neq 0$, $\Pi'(1) = 1$.

It follows from (1.29) that

$$(1.30) \qquad F_2F_1^{(0)} = (\Pi \otimes 1)F, \qquad F_2^{(0)}F_1 = (1 \otimes \Pi')F.$$

Equation (1.30) determines the possible values of F_1 and F_2, up to right, respectively left, multiplication by elements of $U_\hbar\mathfrak{g}_R^{\widehat{\otimes}2}$.

PROPOSITION 1.9 (see [**11**], Proposition 7.2). *Let*

$$F_{\Pi,\Pi'} = [(\Pi \otimes 1)F]^{-1} F [(1 \otimes \Pi')F]^{-1};$$

then

$$F_{\Pi,\Pi'} \in U_\hbar \mathfrak{g}_R^{\widehat{\otimes} 2}. \tag{1.31}$$

This is the key point of the construction of [**11**]; the proof uses Hopf duality arguments. We prove that $F^{-1}[(\Pi \otimes 1)F]$ belongs to $U_\hbar\mathfrak{g} \otimes U_\hbar\mathfrak{g}_R$, and that $[(1 \otimes \Pi')F]F^{-1}$ belongs to $U_\hbar\mathfrak{g}_R \otimes U_\hbar\mathfrak{g}$. To do that, the idea is to compute the annihilators of the nilpotent parts of $U_\hbar\mathfrak{g}_R$ for the pairings $\langle\ ,\ \rangle_{U_\hbar\mathfrak{g}}$ and $\langle\ ,\ \rangle_{U_\hbar\bar{\mathfrak{g}}}$; these annihilators are left and right ideals. Then we pair the second factor of $F^{-1}[(\Pi \otimes 1)F]$, and the first factor of $[(1 \otimes \Pi')F]F^{-1}$ with this annihilator, and use the Hopf algebra pairing rules, as well as the algebraic properties of Π and Π', to show that these pairings give zero.

From the above proposition the solution of the decomposition problem (1.28) follows:

PROPOSITION 1.10. *Any decomposition of F according to* (1.28) *is of the form*

$$F_2 = [(\Pi \otimes 1)F]b, \qquad F_1 = a[(1 \otimes \Pi')F],$$

with $a, b \in U_\hbar\mathfrak{g}_R^{\widehat{\otimes} 2}$ satisfying $ab = F_{\Pi,\Pi'}$.

1.7. Quasi-Hopf structures on $U_\hbar\mathfrak{g}$ and $U_\hbar\mathfrak{g}_R$. Let us choose a solution (F_1, F_2) of (1.28). Consider the algebra morphism $\Delta_R\colon U_\hbar\mathfrak{g} \to U_\hbar\mathfrak{g}^{\widehat{\otimes} 2}$, defined by

$$\Delta_R = \mathrm{Ad}(F_1) \circ \Delta = \mathrm{Ad}(F_2^{-1}) \circ \bar{\Delta}; \tag{1.32}$$

define

$$\Phi = F_1^{23}(1 \otimes \Delta)(F_1)[F_1^{12}(\Delta \otimes 1)(F_1)]^{-1}. \tag{1.33}$$

PROPOSITION 1.11. *Φ belongs to $U_\hbar\mathfrak{g}_R^{\widehat{\otimes} 3}$.*

Let

$$u_R = m(1 \otimes S)(F_1), \tag{1.34}$$

where m is the multiplication in $U_\hbar\mathfrak{g}$, and S is the antipode of $(U_\hbar\mathfrak{g}, \Delta)$.

THEOREM 1.6. *The algebra $U_\hbar\mathfrak{g}$, endowed with the coproduct Δ_R, associator Φ, counit ε, antipode $S_R = \mathrm{Ad}(u_R) \circ S$, respectively defined in* (1.32), (1.33), (1.17), (1.34), *and the R-matrix*

$$\mathcal{R}_R = [a^{21}(\Pi' \otimes 1)(F^{21})] q^{D \otimes K} q^{\frac{1}{2}\sum_{i \in \mathbb{N}} h^+[e^i] \otimes h^-[e_i]} [(\Pi \otimes 1)(F) F_{\Pi,\Pi'} a^{-1}],$$

is a quasi-triangular quasi-Hopf algebra. It contains $U_\hbar\mathfrak{g}_R$ as a sub-quasi-Hopf algebra.

1.8. Properties of $q(z,w)$. In this section, we determine the location of zeros and poles of the function $q(z,w)$ that are near the diagonal.

First recall that for an element $f \in \mathcal{K}$, the product $(f \otimes 1 - 1 \otimes f)G$ belongs to $\bigoplus_{s,t \in S} \mathbb{C}((z_s, w_t))$. Here z_s is a local coordinate at $s \in S$, and $\mathbb{C}((z_s, w_t)) = \mathbb{C}[[z_s, w_t]][z_s^{-1}, w_t^{-1}]$.

PROPOSITION 1.12. *Let z be the element of $\mathcal{K}$ defined as $(z_s)_{s \in S}$. Then for some $i \in 1 + \hbar \bigoplus_{s,t \in S} \mathbb{C}((z_s, w_t))[[\hbar]]$, we have*

$$z - q^{-\partial} w = i \cdot [z - w + (z - w) G \psi(\hbar, (\partial^i \otimes 1)\gamma)].$$

PROOF. Let $\alpha = (z-w)G$; then α belongs to $\bigoplus_{s,t\in S}\mathbb{C}((z_s,w_t))$. Let us first show that if we replace z by $q^{-\partial}w$, the expression $z-w+\alpha(z,w)\psi(\hbar,(\partial_z^i\gamma(z,w))$ vanishes.

The result of this substitution is a formal series $u(w,\hbar)\in\mathcal{K}[[\hbar]]$. It satisfies the equation

$$\frac{\partial u}{\partial\hbar}(\hbar,w) = \Big[-\partial z + (-\partial_z\alpha(z,w))\psi(\hbar,z,w) - \alpha(z,w)\partial_z\psi(\hbar,z,w) + \alpha(z,w)\,\frac{\partial\psi}{\partial\hbar}(\hbar,z,w)\Big]\Big|_{z=q^{-\partial}w}. \tag{1.35}$$

Since

$$\partial_z\alpha(z,w) = (z-w)\gamma(z,w) + [(\partial z)+\alpha(z,w)]G,$$

we have $m(\alpha)+\partial z = 0$ (m being the multiplication map of $\mathcal{K}$); it follows that the equation $(z-w)\xi = \partial z + \alpha(z,w)$ has a solution in $\bigoplus_{s,t\in S}\mathbb{C}((z_s,w_t))$, which we denote by

$$\frac{\partial z+\alpha(z,w)}{z-w}.$$

The left-hand side of (1.35) is now equal to

$$(-(\partial z+\alpha(z,w))(1+G\psi)(z,w) - (z-w+(\alpha\psi)(\hbar,z,w))(\gamma\psi)(\hbar,z,w))|_{z=q^{-\partial}w};$$

that is,

$$-\Big((\gamma\psi)(\hbar,z,w) + \frac{\partial z+\alpha(z,w)}{z-w}\Big)\Big|_{z=q^{-\partial}w}u(\hbar,w).$$

It follows that the series $u(\hbar,w)$ satisfies the equation

$$\frac{\partial u}{\partial\hbar}(\hbar,w) = v(\hbar,w)u(\hbar,w), \tag{1.36}$$

where v is equal to

$$-\Big((\gamma\psi)(\hbar,z,w) + \frac{\partial z+\alpha(z,w)}{z-w}\Big)\Big|_{z=q^{-\partial}w},$$

and so belongs to $\mathcal{K}[[\hbar]]$.

Since we have $u(0,w)=0$, Eq. (1.36) implies that u is identically zero.

Let us now recall Lemma 4.2 of [**11**]:

LEMMA 1.7. *Let $z-w+E$ belong to $z-w+\hbar\mathbb{C}((z,w))[[\hbar]]$. Then there exist unique $e\in\hbar\mathbb{C}((w))[[\hbar]]$ and $\kappa_E\in 1+\hbar\mathbb{C}((z,w))[[\hbar]]$ such that*

$$z-w+E = \kappa_E(z-w+e). \tag{1.37}$$

Consider the case where $z-w+E = z-w+\alpha(z,w)\psi(\hbar,z,w)$; the fact that $u=0$ implies that e must satisfy $z-w+e(w) = z-q^{-\partial}w$. Lemma 1.7 then proves the proposition. □

Proposition 1.12 now implies

COROLLARY 1.1. *The function $q(z,w)$ defined by (1.2) vanishes for $z=q^{-\partial}w$, and its inverse vanishes for $z=q^{\partial}w$.*

2. Quantum currents and quasi-Hopf algebras at level zero

In this section, we shall show how one may define a large family of algebras quantizing the "non-doubly extended" (i.e., original) quasi-Lie bialgebra structures. These algebras are defined by relations similar to those of Theorem 1.4, using functions $q(z,w)$ not necessarily having the zeros and poles structure described by Corollary 1.1.

Let again $\mathcal{K} = R \oplus \Lambda$ be some decomposition of $\mathcal{K}$ into isotropic subspaces, and $G = \sum_{i \geqslant 0} e^i \otimes e_i$, where e^i, e_i are dual bases of R and Λ. We shall also set $\epsilon_i = e_i$, $\epsilon_{-i-1} = e^i$ for $i \geqslant 0$.

Let $a(z,w)$ and $b(z,w)$ belong to $1 + \hbar(R \otimes R)[[\hbar]]$ and $\hbar(R \otimes R)[[\hbar]]$, and define the algebra $U_{a,b}\mathfrak{g}$ as the algebra generated by $e[\epsilon_i]$, $f[\epsilon_i]$, $h[\epsilon_i]$, with generating series

$$e(z) = \sum_{i \in \mathbb{Z}} e[\epsilon_i]\epsilon_{-i-1}(z), \qquad f(z) = \sum_{i \in \mathbb{Z}} f[\epsilon_i]\epsilon_{-i-1}(z),$$
$$h^+(z) = \sum_{i \geqslant 0} h[e^i]e_i(z), \qquad h^-(z) = \sum_{i \geqslant 0} h[e_i]e^i(z),$$

and the relations

$$[h[\epsilon_i], h[\epsilon_j]] = 0, \tag{2.1}$$

for any i, j,

$$K^+(z)e(w)K^+(z)^{-1} = \frac{a + bG^{(21)}}{a^{(21)} - b^{(21)}G^{(21)}}(z,w)e(w), \tag{2.2}$$

$$K^-(z)e(w)K^-(z)^{-1} = \frac{a^{(21)} + b^{(21)}G}{a - bG}(z,w)e(w), \tag{2.3}$$

$$K^+(z)f(w)K^+(z)^{-1} = \frac{a^{(21)} - b^{(21)}G^{(21)}}{a + bG^{(21)}}(z,w)f(w), \tag{2.4}$$

$$K^-(z)f(w)K^-(z)^{-1} = \frac{a - bG}{a^{(21)} + b^{(21)}G}(z,w)f(w), \tag{2.5}$$

$$\begin{aligned}&(\alpha(z) - \alpha(w))(a(w,z) + b(w,z)G(z,w))e(z)e(w) \\ &\quad = (\alpha(z) - \alpha(w))(a(z,w) - b(z,w)G(z,w))e(w)e(z),\end{aligned} \tag{2.6}$$

$$\begin{aligned}&(\alpha(z) - \alpha(w))(a(z,w) - b(z,w)G(z,w))f(z)f(w) \\ &\quad = (\alpha(z) - \alpha(w))(a(w,z) + b(w,z)G(z,w))f(w)f(z),\end{aligned} \tag{2.7}$$

for any element α of $\mathcal{K}$,

$$[e(z), f(w)] = \delta(z,w)(K^+(z) - K^-(z)^{-1}), \tag{2.8}$$

where $\delta(z,w) = G(z,w) + G(w,z)$,

$$K^+(z) = \exp\Big(\sum_i h[e^i](1+V)e_i(z)\Big), \qquad K^-(z) = \exp\Big(\sum_i h[e_i]e^i(z)\Big),$$

and V is the linear operator from Λ to R defined as follows: let B be the linear map from Λ to $\mathcal{K}$ defined by

$$B(\lambda) = \Big\langle \log \frac{a + bG^{(21)}}{a^{(21)} - b^{(21)}G^{(21)}}, \mathrm{id} \otimes \lambda \Big\rangle,$$

and let $B = B_R + B_\Lambda$, where B_R and B_Λ are the compositions of B with the projections on R and Λ. Then $B_\Lambda = \hbar\,\mathrm{id}_\Lambda + o(\hbar)$, $B_R = O(\hbar)$, so that B_Λ is invertible, and we set $V = B_R \circ B_\Lambda^{-1}$. In other words, if we set

$$\log \frac{a + bG^{(21)}}{a^{(21)} - b^{(21)}G^{(21)}} = \sum_i A_i \otimes e^i,$$

we shall have $V((A_j)_\Lambda) = (A_j)_R$. Therefore the relations (2.2), (2.4) have correct functional properties with respect to z and can be written as

$$[h[e^i], e(z)] = B_\Lambda(e^i)(z)e(z), \qquad [h[e^i], f(z)] = -B_\Lambda(e^i)(z)f(z),$$

for $r \in R$.

The algebra $U_{a,b}\mathfrak{g}$ has coproducts $\Delta_{a,b}$ and $\bar{\Delta}_{a,b}$ defined by

$$\Delta_{a,b}(h[\epsilon_i]) = h[\epsilon_i] \otimes 1 + 1 \otimes h[\epsilon_i], \tag{2.9}$$

$$\begin{aligned} \Delta_{a,b}(e(z)) &= e(z) \otimes K^+(z) + 1 \otimes e(z), \\ \Delta_{a,b}(f(z)) &= f(z) \otimes 1 + K^-(z)^{-1} \otimes f(z), \end{aligned} \tag{2.10}$$

and

$$\bar{\Delta}_{a,b}(h[\epsilon_i]) = h[\epsilon_i] \otimes 1 + 1 \otimes h[\epsilon_i], \tag{2.11}$$

$$\begin{aligned} \bar{\Delta}_{a,b}(e(z)) &= e(z) \otimes 1 + K^-(z)^{-1} \otimes e(z), \\ \bar{\Delta}_{a,b}(f(z)) &= f(z) \otimes K^+(z) + 1 \otimes f(z). \end{aligned} \tag{2.12}$$

The algebra $U_{a,b}\mathfrak{g}$ shares all the properties of $U_\hbar\mathfrak{g}$ that were used in [**11**] for constructing a quasi-Hopf structure on a certain subalgebra of this algebra:

THEOREM 2.1. *$(U_{a,b}\mathfrak{g}, \Delta_{a,b})$ is a Hopf algebra; it is a flat deformation of the enveloping algebra of $\mathfrak{g}$. It is the double of its subalgebras $U_{a,b}\mathfrak{g}_\pm$, generated by the modes of $K^+(z)$, $e(z)$, respectively $K^-(z)$, $f(z)$. The Hopf pairing between these algebras is defined by*

$$\langle h[e^i], h[e_{-j-1}]\rangle = \langle e^i, B_\Lambda e_i\rangle_{\mathcal{K}}, \quad i,j \geqslant 0, \qquad \langle e[\epsilon_i], f[\epsilon_j]\rangle = \delta_{i,-j-1}, \quad i,j \in \mathbb{Z},$$

and the universal R-matrix of $(U_{a,b}\mathfrak{g}, \Delta_{a,b})$ is then equal to

$$\mathcal{R} = \exp\left(\frac{1}{2}\sum_{i \geqslant 0} h[e^i] \otimes h[B_\Lambda e_i]\right) F_{a,b};$$

here $F_{a,b} = \sum_i \alpha^i \otimes \alpha_i$, where (α^i), (α_i) are dual bases of the subalgebras $U_{a,b}\mathfrak{n}_+$ and $U_{a,b}\mathfrak{n}_-$ of $U_{a,b}\mathfrak{g}_+$ and $U_{a,b}\mathfrak{g}_-$, respectively generated by $e[\epsilon]$, $\epsilon \in \mathcal{K}$, and $f[\epsilon]$, $\epsilon \in \mathcal{K}$.

The same statements hold with $\Delta_{a,b}$ replaced by $\bar{\Delta}_{a,b}$. The subalgebras are then generated by the modes of $K^-(z)$, $e(z)$, respectively $K^+(z)$, $f(z)$. The Hopf pairing is defined by the same formulas as above, and the universal R-matrix is equal to

$$\bar{\mathcal{R}} = F_{a,b}^{21} \exp\left(\frac{1}{2}\sum_{i \geqslant 0} h[e^i] \otimes h[B_\Lambda e_i]\right).$$

Moreover, $\bar{\Delta}_{a,b}$ is obtained from $\Delta_{a,b}$ by the twist $\bar{\Delta}_{a,b} = F_{a,b}\Delta_{a,b}F_{a,b}^{-1}$.

The subalgebra $U_{a,b}\mathfrak{g}_R$ spanned by the $x[e^i]$, $x = e, f, h$, $i \geqslant 0$, is a flat deformation of the enveloping algebra of $\mathfrak{g}_R = \bar{\mathfrak{g}} \otimes R$. It satisfies

$$\Delta_{a,b}(U_{a,b}\mathfrak{g}_R) \subset U_{a,b}\mathfrak{g} \otimes U_{a,b}\mathfrak{g}_R, \qquad \bar{\Delta}_{a,b}(U_{a,b}\mathfrak{g}_R) \subset U_{a,b}\mathfrak{g}_R \otimes U_{a,b}\mathfrak{g}.$$

The procedure of [11] then can be followed to obtain a decomposition of F, which will serve to define a quasi-Hopf algebra structure on $U_{a,b}\mathfrak{g}_R$.

REMARK 2.2. *Dependence of* $U_{a,b}\mathfrak{g}$ *on* (a,b). Clearly, $U_{a,b}\mathfrak{g}$ depends on (a,b) only through

$$q_{a,b}(z,w) = \left(\frac{a + bG^{(21)}}{a^{(21)} - b^{(21)}G^{(21)}}\right)(z,w);$$

in particular, we have $q_{a,b} = q_{sa,sb}$, if $s \in (R \otimes R)[[\hbar]]$ is invertible and symmetric in (z,w), and also $q_{a,b} = q_{a-cG,b+c}$, if c is a function of $R \otimes R$ vanishing on the diagonal of X.

On the other hand, if we have $q_{a,b} = \lambda q_{a',b'}$ for some invertible element λ in $(R \otimes R)[[\hbar]]$, there is an isomorphism between the algebras $U_{a,b}\mathfrak{g}$ and $U_{a',b'}\mathfrak{g}$, defined by multiplying the fields $e(z)$ and $f(z)$ by suitable combinations of $K^+(z)$ and $K^-(z)$.

REMARK 2.3. The condition that $U_{a,b}\mathfrak{g}$ can be embedded in some algebra "with derivation and central extension" quantizing its double extension seems to impose severe constraints on a and b. Indeed, a natural way to achieve this is to use the relations of [11], Section 8. These relations imply in particular that we have

$$K^+(z)K^-(w)K^+(z)^{-1}K^-(w)^{-1} = \frac{q_{a,b}(z,w)}{q_{a,b}(z,q^{K\partial}w)},$$
$$[D, K^+(z)] = -\partial_z K^+(z) + (Ah^+)(z)K^+(z),$$
$$[D, K^-(z)] = -\partial_z K^-(z) + (Bh^+)(z)K^-(z),$$

where A and B are finite rank operators from Λ to R, and K is the central generator; this implies in particular that the set of zeros and poles of $q(z,w)$ is stable under the diagonal action of ∂ on $X \times X$. In the case studied in [11], these sets are $\{(x, q^{\partial}x), x \in X\}$ and the diagonal of $X \times X$.

3. Examples on a rational curve

3.1. Manin pairs. In this section, we shall consider the following situation. An integer $N \geqslant 2$ is fixed, and $\mathcal{K} = \mathbb{C}((z))$, $\omega = z^{N-1}dz$.

3.1.1. If N is odd, write $N = 2n+1$, with n a nonnegative integer. Let us set $R = z^{-n-1}\mathbb{C}[z^{-1}]$, $\Lambda = z^{-n}\mathbb{C}[[z]]$. Then R is a maximal isotropic subring of $\mathcal{K}$ for the pairing induced by ω, and Λ is a maximal isotropic supplementary supspace.

Dual bases of R and Λ are $e^i = z^{-n-i-1}$, $e_i = z^{i-n}$ for $i \geqslant 0$. We then have

$$G = \sum_{i \geqslant 0} z^{-n-i-1} \otimes z^{i-n} = \frac{(zw)^{-n}}{z-w},$$

expanded for w near 0.

We then construct a Manin pair as follows: we define $\mathfrak{g}$ as the Lie algebra $(\bar{\mathfrak{g}} \otimes \mathcal{K}) \oplus \mathbb{C}K \oplus \mathbb{C}D$, where the central and cocentral extensions are defined as in 1.1 above and endowed with the usual scalar product. The Lie subalgebra $\mathfrak{g}_R = (\bar{\mathfrak{g}} \otimes R) \oplus \mathbb{C}K$ is then maximal isotropic for this scalar product. This defines a Manin pair. Quasi-Lie bialgebra structures are then defined on $\mathfrak{g}$ and on $\mathfrak{g}_R$ by the choice of the isotropic complement $\mathfrak{g}_\Lambda = (\bar{\mathfrak{g}} \otimes \Lambda) \oplus \mathbb{C}D$ of $\mathfrak{g}_R$.

3.1.2. Suppose N is even. Write $N = 2(n+1)$, with n a nonnegative integer. Let $\tilde{\mathfrak{g}}$ be the semidirect product $\tilde{\mathfrak{g}} = \mathfrak{g} \oplus \mathbb{C}\tilde{h}$, where $\mathfrak{g} \subset \tilde{\mathfrak{g}}$ is a Lie algebra embedding, and the action of $\tilde{h}$ on $\mathfrak{g}$ is such that $\tilde{h} - h \otimes z^{-n}$ is central. Extend the scalar product $\langle\ ,\ \rangle_{\mathfrak{g}}$ on $\mathfrak{g}$ to a scalar product $\langle\ ,\ \rangle_{\tilde{\mathfrak{g}}}$ on $\tilde{\mathfrak{g}}$ by the rules $\langle \tilde{h}, \mathfrak{g}\rangle_{\tilde{\mathfrak{g}}} = 0$, $\langle \tilde{h}, \tilde{h}\rangle_{\tilde{\mathfrak{g}}} + \langle h \otimes z^{-n}, h \otimes z^{-n}\rangle_{\tilde{\mathfrak{g}}} = 0$.

A quasi-Lie bialgebra structure on $\tilde{\mathfrak{g}}$ is then defined as follows: let $\mathfrak{g}_R^{(0)}$ and $\mathfrak{g}_\Lambda^{(0)}$ be the subspaces of $\mathfrak{g}$ equal to $(\bar{\mathfrak{g}} \otimes z^{-n-1}\mathbb{C}[z^{-1}]) \oplus \mathbb{C}K$ and $(\bar{\mathfrak{g}} \otimes z^{1-n}\mathbb{C}[[z]]) \oplus \mathbb{C}D$, and define $\tilde{\mathfrak{g}}_R$ and $\tilde{\mathfrak{g}}_\Lambda$ as the direct sums of $\mathbb{C}(\tilde{h} - h \otimes z^{-n})$ and $\mathbb{C}(\tilde{h} + h \otimes z^{-n})$ with the images of $\mathfrak{g}_R^{(0)}$ and $\mathfrak{g}_\Lambda^{(0)}$ in $\tilde{\mathfrak{g}}$. Then $\tilde{\mathfrak{g}}_R$ is a maximal isotropic Lie subalgebra of $\tilde{\mathfrak{g}}$, and $\tilde{\mathfrak{g}}_\Lambda$ is an isotropic complement. This defines Lie quasi-bialgebra structures on $\tilde{\mathfrak{g}}$ and $\tilde{\mathfrak{g}}_R$.

By the natural projection of $\tilde{\mathfrak{g}}$ to $\mathfrak{g}$, these structures define quasi-Lie bialgebra structures on $\mathfrak{g}$ and on $\mathfrak{g}_R = (\bar{\mathfrak{g}} \otimes z^{-n}\mathbb{C}[z^{-1}]) \oplus \mathbb{C}K$; the structure on $\mathfrak{g}$ is not a double one.

NOTATION. Here and later, we shall use the notation $z_\lambda = (z^N + \lambda N\hbar)^{1/N}$. We have $z_\lambda = q^{\lambda\partial}(z)$, where ∂ is the derivation defined by ω.

3.2. The functions $q(z,w)$**.** Define the series $\phi(z,w)$ of $\mathbb{C}[[z^{-1}]]((w))[[\hbar]]$ as the expansion of $\log((z_1 - w)/(z-w))$. Set

$$\phi(z,w) = \sum_{p,q\in\mathbb{Z}} a_{pq} z^p w^q, \qquad \phi_{w^{>-n}}(z,w) = \sum_{q>-n, p\in\mathbb{Z}} a_{pq} z^p w^q.$$

It is easy to check that $\phi_{w^{>-n}}(z,w)$ belongs to

$$z^{-n}w^{-1}\mathbb{C}[[z^{-1}, w^{-1}, \hbar]].$$

PROPOSITION 3.1. 1) *Let the notation be as in* 3.1.1. *Set* $N = 2n+1$. *There exists some linear operator* $U\colon \Lambda \to \hbar R[[\hbar]]$ *such that*

$$\left(1 \otimes \left(\frac{q^{\partial} - q^{-\partial}}{\partial} + U\right)\right) G = \log\frac{z_1 - w}{z - w_1} + \phi_{z^{>-n}}(z,w) - \phi_{w^{>-n}}(w,z). \tag{3.1}$$

2) *Let the notation be as in* 3.1.2. *Let* $N = 2n+2$. *For a certain linear operator* $U\colon z^{-n}\mathbb{C}[[z]] \to z^{-n}\mathbb{C}[z^{-1}][[\hbar]]$, (3.1) *holds.*

PROOF. Let us prove 1). First we shall show that

$$D_\hbar = G - (q^{-\partial} \otimes q^{-\partial})G \tag{3.2}$$

belongs to $(R \otimes R)[[\hbar]]$.

We have

$$(\partial \otimes 1 + 1 \otimes \partial)G \in R \otimes R. \tag{3.3}$$

This follows from Lemma 1.2. To show (3.3), we may also compute explicitly

$$(\partial \otimes 1)G = -G^2 + \gamma \quad \text{with } \gamma = -z^{-2n}w^{-n}\frac{1}{z-w}\left[\frac{z^{-n} - w^{-n}}{z-w} + nz^{-n-1}\right],$$

which belongs to $R \otimes R$ because the term in brackets vanishes for $z = w$, and

$$(1 \otimes \partial)G = G^2 - \gamma^{(21)},$$

so that $(\partial \otimes 1 + 1 \otimes \partial)G = \gamma - \gamma^{(21)}$ belongs to $R \otimes R$.

Now R is stable under ∂, so that

$$D_\hbar = [(q^{-\partial} \otimes q^{-\partial}) - 1]G = \frac{q^{-(\partial\otimes 1+1\otimes\partial)} - 1}{\partial \otimes 1 + 1 \otimes \partial}(\partial \otimes 1 + 1 \otimes \partial)G$$

also belongs to $R \otimes R$.

Therefore $(q^{\partial} \otimes 1)(D_\hbar)$ also belongs to $(R \otimes R)[[\hbar]]$. It follows that for some linear operator $V_+ \colon \Lambda \to R[[\hbar]]$, we have

$$(q^{\partial} \otimes 1)(D_\hbar) = (1 \otimes V_+)(G).$$

Therefore

$$(1 \otimes (q^{-\partial} + V_+))G = (q^{\partial} \otimes 1)\left(\frac{z^{-n}w^{-n}}{z-w}\right) = \frac{z_1^{1-N}}{z_1 - w} - \left(\frac{z_1^{1-N}}{z_1 - w}\right)_{w>-n};$$

let U_+ be the unique linear operator from Λ to $\hbar R[[\hbar]]$ such that $\partial_\hbar U_+ = V_+$. Integrating in $\hbar$, we obtain

$$(3.4) \qquad \left(1 \otimes \left(\frac{1 - q^{-\partial}}{\partial} + U_+\right)\right)G = \log\frac{z_1 - w}{z - w} - \phi_{w>-n}(z, w).$$

In the same way we may construct a linear operator U_- from Λ to $\hbar R[[\hbar]]$ such that

$$(3.5) \qquad \left(1 \otimes \left(\frac{q^{\partial} - 1}{\partial} + U_-\right)\right)G = \log\frac{z - w}{z - w_1} + \phi_{z>-n}(w, z).$$

To obtain the statement of the proposition, we then set $U = U_+ + U_-$.

The proof of 2) is similar. □

Let us choose U as in Proposition 3.1. We then find that

$$(3.6) \qquad q(z, w) = \exp(\phi_{z>-n}(z, w) - \phi_{w>-n}(w, z))\frac{(z^N + \hbar N)^{1/N} - w}{z - (w^N + \hbar N)^{1/N}}.$$

3.3. The algebra $U_{\hbar, z^{N-1}dz}\mathfrak{g}$. We denote by $U_{\hbar, z^{N-1}dz}\mathfrak{g}$ the Hopf algebra resulting from the construction of Theorem 1.4. It contains a regular subalgebra generated by the x_i, $i \leqslant -n$.

In what follows, we shall set

$$U_{\hbar, z^{-n-2}dz}\mathfrak{g} = U_{\hbar, z^{n}dz}\mathfrak{g}$$

for $n \geqslant 0$, and $U_{\hbar, z^{-1}dz}\mathfrak{g}$ equal to the quantum affine algebra attached to $\mathfrak{g}$.

The interest in the algebras $U_{\hbar, z^{N-1}dz}\mathfrak{g}$ lies in the following fact.

THEOREM 3.1. *Let $U_\hbar\mathfrak{g}$ be the algebra of Theorem 1.4, attached to the data (X, ω, S) and let $U_\hbar\mathfrak{g}'$ be its subalgebra with the same generators except D. For each point s of S, let n_s be the order of the zero or pole of ω at s. Then $U_\hbar\mathfrak{g}'$ is isomorphic to the quotient $\bigotimes_{s\in S} U_{\hbar, z^{n_s}dz}\mathfrak{g}'/(K^{(s)} - K^{(t)})$, where $K^{(s)}$ is the central generator of the sth factor.*

PROOF. The argument is similar to that of the Introduction in [**9**]: at each point s fix a coordinate z_s such that ω is locally expressed by $z_s^{n_s}dz_s$. Then for each s, we have a specialization morphism ev_s from $U_\hbar\mathfrak{g}'$ to $U_{\hbar, z^{n_s}dz}\mathfrak{g}'$, sending each $x[\epsilon_t]$ to $\delta_{st}x[\epsilon_t]$ if $\epsilon_t \in \mathcal{K}_t$, and K to K. Fix a coproduct Δ_R for $U_\hbar\mathfrak{g}'$ as in 1.7 above. We have then an algebra morphism $\Delta_R^{(\mathrm{card}\, S)}$ from $U_\hbar\mathfrak{g}'$ to $(U_\hbar\mathfrak{g}')^{\otimes \mathrm{card}\, S}$, defined by

$$\Delta_R^{(\mathrm{card}\, S)} = (\Delta_R \otimes \mathrm{id}^{\otimes \mathrm{card}\, S - 1}) \circ \cdots \circ \Delta_R.$$

Choose an order of the points of S and compose $\Delta_R^{(\mathrm{card}\, S)}$ with $\bigotimes_{s\in S} \mathrm{ev}_s$. The resulting map is an algebra morphism from $U_\hbar\mathfrak{g}'$ to $\bigotimes_{s\in S} U_{\hbar,z^{n_s}dz}\mathfrak{g}'$. The fact that it gives an isomorphism after composition with projection to the quotient by the ideal generated by the $K^{(s)} - K^{(t)}$ follows from the inspection of its classical limit. □

The algebras $U_{\hbar,z^{N-1}dz}\mathfrak{g}$ also have the property that for any nonzero complex λ, $U_{\lambda\hbar,z^{N-1}dz}\mathfrak{g}$ is isomorphic to $U_{\hbar,z^{N-1}dz}\mathfrak{g}$; this follows from the fact that (writing the formal parameter in indices) $q_{\alpha^N\hbar}(\alpha z, \alpha w) = q_\hbar(z, w)$. This generalizes the property that Yangians are isomorphic for all nonzero values of the deformation parameter.

REMARK 3.2. In the framework of the preceding sections, one should consider the curve $X = \mathbb{C}P^1$ with differential ω and marked points 0 and ∞. The resulting algebra is then nothing but the tensor square of $U_{\hbar,z^{N-1}dz}\mathfrak{g}$.

REMARK 3.3. If we complete $U_{\hbar,z^{N-1}dz}\mathfrak{g}$ with respect to the ideals generated by the elements $x[z^{-i}]$, $i \geqslant n$, $x = e, f, h$, the relations defining it make sense for complex values of $\hbar$.

3.4. Another presentation of the vertex relations of $U_{\hbar,z^{N-1}dz}\mathfrak{g}$. It is easy to see that after we multiply the generating series $e(z)$ by a suitable Cartan fields, it will satisfy the equation

$$((z^N + N\hbar)^{1/N} - w)\tilde{e}(z)\tilde{e}(w) = (z - (w^N + N\hbar)^{1/N})\tilde{e}(w)\tilde{e}(z).$$

We shall now show how this relation can be written avoiding the use of Nth roots.

Let us denote by $\mathbb{Z}_N$ the group $\mathbb{Z}/N\mathbb{Z}$ and by μ_N the group of Nth roots of unity in $\mathbb{C}$. Let us decompose the field $\tilde{e}(z)$ as

$$\tilde{e}(z) = \sum_{\alpha\in\mathbb{Z}_N} e^{(\alpha)}(z) \quad \text{with } e^{(\alpha)}(\zeta z) = \zeta^\alpha e^{(\alpha)}(z), \tag{3.7}$$

for $\zeta \in \mu_N$. We also set $e^{(\alpha)}(z) = z^\alpha E^{(\alpha)}(z^n)$, where we denote by $\bar{\alpha}$ the representative in $[0, N-1]$ of the element α of $\mathbb{Z}_N$, and we abuse notation by writing $z^\alpha = z^{\bar{\alpha}}$.

For $a, b \in \mathbb{Z}_N$, define $r(a, b)$ as the number (equal to 0 or 1) such that $\bar{a} + \bar{b} = \overline{a+b} + r(a, b)N$; this is the carryover for the addition of $\bar{a}$ and $\bar{b}$ modulo N.

PROPOSITION 3.2. *Relation* (3.7) *is equivalent to the system of relations*

(3.8)

$$(Z - W + N\hbar) \sum_{\alpha\in\mathbb{Z}_N} Z^{r(N-p-\alpha,\alpha)}(W + N\hbar)^{r(p+\alpha-1,q-\alpha)} E^{(\alpha)}(Z)E^{(q-\alpha)}(W)$$
$$= (Z - W - N\hbar) \sum_{\alpha\in\mathbb{Z}_N} (Z + N\hbar)^{r(N-p-\alpha,\alpha)} W^{r(p+\alpha-1,q-\alpha)} E^{(q-\alpha)}(W)E^{(\alpha)}(Z).$$

PROOF. Write (3.7) as

$$\frac{\tilde{e}(z)\tilde{e}(w)}{z - w_1} = \frac{\tilde{e}(w)\tilde{e}(z)}{z_1 - w}.$$

This implies that for $p \in \mathbb{Z}_N$ we have

$$\sum_{\zeta\in\mu_N} \frac{\zeta^p\tilde{e}(\zeta z)\tilde{e}(w)}{\zeta z - w_1} = \sum_{\zeta\in\mu_N} \frac{\zeta^p\tilde{e}(w)\tilde{e}(\zeta z)}{\zeta z_1 - w}.$$

Since

$$\sum_{\zeta\in\mu_N}\frac{\zeta^p}{\zeta z-w}=\frac{nw^{p-1}z^{N-p}}{z^N-w^N},$$

it follows that

$$\sum_{\zeta\in\mu_N}\frac{\zeta^p\tilde{e}(\zeta z)}{\zeta z-w}=\frac{N}{z^N-w^N}\Big(\sum_{\alpha\in\mathbb{Z}_N}z^{N-p-\alpha}w^{p+\alpha-1}e^{(\alpha)}(z)\Big).$$

Therefore

$$\frac{N}{z^N-w_1^N}\Big(\sum_{\alpha\in\mathbb{Z}_N}z^{N-p-\alpha}w_1^{p+\alpha-1}e^{(\alpha)}(z)\Big)\sum_{\beta\in\mathbb{Z}_N}e^{(\beta)}(w)$$
$$=\frac{N}{z_1^N-w^N}\Big(\sum_{\beta\in\mathbb{Z}_N}e^{(\beta)}(w)\Big)\Big(\sum_{\alpha\in\mathbb{Z}_N}z_1^{N-p-\alpha}w^{p+\alpha-1}e^{(\alpha)}(z)\Big).$$

Separating isotypic components for the action of $\mathbb{Z}_N$ in the variable w, for each $q\in\mathbb{Z}_N$ we obtain

$$\frac{N}{z^N-w^N-N\hbar}\sum_{\alpha\in\mathbb{Z}_N}z^{N-p-\alpha}w_1^{p+\alpha-1}e^{(\alpha)}(z)e^{(q-\alpha)}(w)$$
$$=\frac{N}{z^N-w^N+N\hbar}\sum_{\alpha\in\mathbb{Z}_N}z_1^{N-p-\alpha}w^{p+\alpha-1}e^{(q-\alpha)}(w)e^{(\alpha)}(z),$$

so that in terms of the fields $X^{(\alpha)}$ we can write

$$\frac{N}{z^N-w^N-N\hbar}\sum_{\alpha\in\mathbb{Z}_N}z^{N-p-\alpha}z^{\alpha}w_1^{p+\alpha-1}w_1^{q-\alpha}E^{(\alpha)}(z^N)E^{(q-\alpha)}(w^N)$$
$$=\frac{N}{z^N-w^N+N\hbar}\sum_{\alpha\in\mathbb{Z}_N}z_1^{N-p-\alpha}z_1^{\alpha}w^{p+\alpha-1}w^{q-\alpha}E^{(q-\alpha)}(w^N)E^{(\alpha)}(z^N),$$

or, setting $Z=z^N$ and $W=w^N$,

$$\frac{N}{Z-W-N\hbar}\sum_{\alpha\in\mathbb{Z}_N}Z^{r(N-p-\alpha,\alpha)}(W+N\hbar)^{r(p+\alpha-1,q-\alpha)}E^{(\alpha)}(Z)E^{(q-\alpha)}(W)$$
$$=\frac{N}{Z-W+N\hbar}\sum_{\alpha\in\mathbb{Z}_N}(Z+N\hbar)^{r(n-p-\alpha,\alpha)}W^{r(p+\alpha-1,q-\alpha)}E^{(q-\alpha)}(W)E^{(\alpha)}(Z),$$

which is (3.8).

The above arguments can easily be reversed to prove the proposition. □

REMARK 3.4. We may construct an algebra $A_{\hbar,z^{N-1}dz}$ with the generators $E_i^{(\alpha)}$, $i\in\mathbb{Z}$, arranged in the series

$$E^{(\alpha)}(Z)=\sum_{i\in\mathbb{Z}}E_i^{(\alpha)}Z^{-i},$$

subject to the above relations (3.8). As we have seen, for $\hbar$ a formal parameter, this algebra is isomorphic with the part of $U_{\hbar,z^{N-1}dz}$ generated by the field $\tilde{e}(z)$. This is also true in the case when $\hbar$ is complex, after we complete $U_{\hbar,z^{N-1}dz}$ as in Remark 3.3. However, since $\hbar$ appears polynomially in the defining relations of $A_{\hbar,z^{N-1}dz}$, they make sense without completing the algebra, even when $\hbar$ is complex.

REMARK 3.5. The algebra $A_{\hbar,z^{N-1}dz}$ has an obvious morphism to the upper nilpotent subalgebra of the double Yangian $DY(\mathfrak{sl}_2)$, defined by $E^{(\alpha)}(Z) \mapsto \delta_{\alpha 0}e(Z)$.

REMARK 3.6. Proposition 3.2 can easily be extended to the case of mixed vertex relations

$$(z_\lambda - w)x(z)y(w) = (z - w_\lambda)y(w)x(z).$$

4. Genus > 1 examples associated to odd theta-characteristics

Let X be a smooth curve of genus greater than 1. Let ω be a regular form on X all of whose zeros are double. The existence of such a form follows from that of a nonsingular odd theta-characteristic, i.e., from the existence of an effective divisor doubly equivalent to the canonical divisor (e.g. see [**20**], Lemma 1, p. 3.208, or [**12**]). Let $\delta = \sum_{i=1}^{g-1} \delta_i$ be this effective divisor; we then have $\mathrm{div}(\omega) = 2\delta$. Let $\mathcal{L}_\delta$ be the line bundle associated with δ. Then we have $\mathcal{L}_\delta^{\otimes 2} = K$ and $h^0(\mathcal{L}_\delta) = 1$.

Let us also recall the properties of the vector of Riemann constants ([**12, 20**]). Let $\mathrm{Jac}^n(X)$ be the degree n component of the Jacobian of X. View the basic theta-function θ as a quasi-periodic function on a cover of $\mathrm{Jac}^0(X)$. We denote in the same way points of X and their image in $\mathrm{Jac}^1(X)$. Then for some vector Δ of $\mathrm{Jac}^{g-1}(X)$, we have

$$\theta\Big(-\Delta + \sum_{i=1}^{g-1} y_i \Big) = 0, \tag{4.1}$$

for any collection of $g-1$ points y_i of X. Moreover, the zero set of θ is equal to $\{-\Delta + \sum_{i=1}^{g-1} y_i,\ y_i \in X\}$.

4.1. Quasi-Lie bialgebras. Here we shall consider some Manin pairs in which the Lie subalgebra will be formed by currents regular at some points, as it was done in [**7**] in genus 1.

More precisely, let S' be a set of points of X not containing any δ_i, and let us define

$$\mathcal{K} = \bigoplus_{s \in S'} \mathcal{K}_s, \qquad \mathcal{O} = \bigoplus_{s \in S'} \mathcal{O}_s,$$

and the pairing $\langle\ ,\ \rangle_{\mathcal{K}}$ on $\mathcal{K}$ by

$$\langle f, g \rangle_{\mathcal{K}} = \sum_{s \in S'} \mathrm{res}_s(fg\omega).$$

$\mathcal{O}$ is clearly an isotropic subalgebra of $\mathcal{K}$.

4.1.1. *Isotropic subspaces.* We define an isotropic subspace L of $\mathcal{K}$ as follows. Fix a system $(a_i, b_i)_{i=1,\dots,g}$ of a- and b-cycles on X. Denote by $\widetilde{X}$ the universal cover of X, and by γ_{a_i} and γ_{b_i} the deck transformations associated to a_i and b_i. Let $X^{(a)}$ be the quotient of $\widetilde{X}$ by the equivalences $z \sim \gamma_{a_i} z$. Let us fix lifts a_i of the a-cycles in $X^{(a)}$, which we also denote by a_i, and let X_0 be the fundamental domain in $X^{(a)}$ bounded by the a_i and the $\gamma_{b_i}(a_i)$. We identify local fields at points of S' with the local fields at their lifts in X_0, and denote by $\widetilde{S}'$ the lift of S' to the fundamental domain.

Define L as the set of expansions at the points of $\widetilde{S}'$ of the functions f such that

$$f(\gamma_{a_i} z) = f(z), \qquad f(\gamma_{b_i} z) = f(z) + c_i(f),$$

f is regular except at the points of $\widetilde{S}'$, has (at most) simple poles at the lifts of the δ_i, and satisfies

$$\int_{a_i} f\omega = -\frac{c_i(f)}{2}\int_{a_i}\omega.$$

We can generalize this construction of isotropic subspaces of $\mathcal{K}$ as follows. Let V be a vector subspace of $\mathbb{C}^g$. Define L_V as the set of expansions at the points of $\widetilde{S}'$ of the functions f defined on $X^{(a)}$ such that $f(\gamma_{a_i}z) = f(z)$, $(f(\gamma_{b_i}z) - f(z))_{i=1,\dots,g}$ belongs to V, and the periods condition

$$\sum_i \alpha_{ij}\Big(\int_{a_i} f\omega + \int_{\gamma_{b_i}(a_i)} f\omega\Big) = 0, \qquad j = 1,\dots,s, \tag{4.2}$$

where $(\alpha_{ij})_{i=1,\dots,g}$, $j = 1,\dots,s$, are the coordinates of a basis of V.

We then have:

LEMMA 4.1. *For each subspace V of $\mathbb{C}^g$, the space L_V is isotropic for $\langle\ ,\ \rangle_{\mathcal{K}}$.*

PROOF. Let f, g belong to L_V. By the residues theorem, $\langle f, g\rangle_{\mathcal{K}}$ is equal to

$$-\sum_{i=1}^{g-1}\operatorname{res}_{\delta_i}(fg\omega) - \sum_{i=1}^{g}\Big(\int_{a_i} fg\omega - \int_{\gamma_{b_i}(a_i)} fg\omega\Big); \tag{4.3}$$

by the simple poles conditions on f and g, the first sum in (4.3) vanishes. On the other hand, we can set

$$f(\gamma_{b_i}z) - f(z) = \sum_{j=1}^{s}\lambda_{ij}(f)\alpha_{ij}, \qquad g(\gamma_{b_i}z) - g(z) = \sum_{j=1}^{s}\lambda_{ij}(g)\alpha_{ij}; \tag{4.4}$$

we then have

$$\sum_{i=1}^{g}\alpha_{ij}\Big(2\int_{a_i} f\omega + \sum_{j=1}^{s}\lambda_{ij}(f)\alpha_{ij}\int_{\gamma_{b_i}(a_i)}\omega\Big) = 0, \tag{4.5}$$

$$\sum_{i=1}^{g}\alpha_{ij}\Big(2\int_{a_i} g\omega + \sum_{j=1}^{s}\lambda_{ij}(g)\alpha_{ij}\int_{\gamma_{b_i}(a_i)}\omega\Big) = 0, \tag{4.6}$$

$j = 1,\dots,s$. It follows from (4.4) that the second sum of (4.3) is equal to

$$\sum_{i=1}^{g}\int_{a_i}\Big(f + \sum_{j=1}^{s}\lambda_{ij}(f)\alpha_{ij}\Big)\Big(g + \sum_{j=1}^{s}\lambda_{ij}(g)\alpha_{ij}\Big)\omega - \int_{a_i} fg\omega; \tag{4.7}$$

multiplying (4.5) by $\lambda_{ij}(g)$ and (4.6) by $\lambda_{ij}(f)$ and summing both sets of equations, we find that (4.7) vanishes. Therefore (4.3) vanishes. □

The spaces L_V differ from L only by finite-dimensional pieces (that is, their projection to L parallel to $\mathcal{O}$ has finite kernel and cokernel). For $V = \mathbb{C}^g$, we have $L_V = L$.

REMARK 4.2. *Lagrangian supplementaries associated with bundles.* Fix a family $(g_i)_{i=1,\dots,g}$ of elements of G. We may consider the subspace $L_{(g_i)}$ of $\mathfrak{g}\otimes\mathcal{K}$ formed by the expansions at the points of S' of the functions f from $\widetilde{X}$ to $\mathfrak{g}$, such that f has simple poles at δ,

$$f(\gamma_{a_i}z) = f(z), \qquad f(\gamma_{b_i}z) = \operatorname{Ad}(g_i)(f(z)) + x_i(f),$$

and

$$\int_{a_i} f\omega = -\frac{1}{2}\operatorname{Ad}(g_i^{-1})x_i(f)\int_{a_i}\omega.$$

The sum of the residues of f at the points of δ is then $\sum_i(1-\operatorname{Ad}(g_i^{-1}))(x_i(f))$ (which need not be zero). The space $L_{(g_i)}$ is an isotropic subspace of $\mathfrak{g}\otimes\mathcal{K}$.

In the case of sums of line bundles, we obtain the analogs of the spaces L_λ of [**7**].

It would be interesting to understand if the r-matrix associated with these supplementaries satisfies some variant of the dynamical Yang–Baxter equation.

4.1.2. *Green kernels.* In this section, we shall consider the case when V is one-dimensional; set $V = \mathbb{C}h$, $h \in \mathbb{C}^g$.

Let us set, for z, w in $X^{(a)}$,

$$G_h(z,w) = \frac{\partial_h\theta(z-w+\delta-\Delta)}{\theta(z-w+\delta-\Delta)}.$$

This function has the following properties:

PROPOSITION 4.1. *The function $G_h(z,w)$ is antisymmetric in z and w. It has poles when the projections on X of z and w are equal, or one of them is equal to some δ_i. For any z, w, we have*

$$G_h(\gamma_{b_i}z,w) = G_h(z,w) - h_i. \tag{4.8}$$

Near the diagonal $z = w$, the function $G_h(z,w)$ has the expansions

$$G_h(z,w) = \frac{C(h)}{\int_z^w \omega} + O(1), \tag{4.9}$$

for some constant $C(h)$, which is nonzero if and only if h does not belong to the linear span of the V_{δ_i}; for any point x of X, we denote by V_x some tangent vector at x of the embedding of X in its Jacobian.

PROOF. The function defined on $\mathbb{C}^g$ by $\mathbf{z}\mapsto\theta(-\Delta+\delta+\mathbf{z})$ is odd, so that $\mathbf{z}\mapsto\partial_h\theta/\theta(-\Delta+\delta+\mathbf{z})$ is also odd; it follows that $G_h(z,w)$ is antisymmetric.

Recall that

$$\theta(\mathbf{z}+A_i) = \text{const}\cdot\theta(\mathbf{z}), \qquad \theta(\mathbf{z}+B_i) = \text{const}\cdot e^{-z_i}\theta(\mathbf{z}),$$

where $\mathbf{z} = (z_i)_{1\leqslant i\leqslant g}$, A_i are the basis vectors of $\mathbb{C}^g$ and B_i is the vector

$$\left(\int_{b_i}\omega_1,\dots,\int_{b_i}\omega_g\right),$$

the forms ω_i being holomorphic differentials such that $\int_{a_i}\omega_j = \delta_{ij}$. Taking the logarithmic derivative, we find that

$$\frac{\partial_h\theta}{\theta}(\mathbf{z}+A_i) = \frac{\partial_h\theta}{\theta}(\mathbf{z}), \qquad \frac{\partial_h\theta}{\theta}(\mathbf{z}+B_i) = \frac{\partial_h\theta}{\theta}(\mathbf{z}) - h_i$$

if h has the components $(h_1,\dots,h_g)$. Formula (4.8) follows from these identities.

Finally, to prove (4.9), we need the following result:

LEMMA 4.3. *The expression $\alpha(z)\,dz = d_z\theta(z-w+\delta-\Delta)|_{z=w}$ is a 1-form on $X^{(a)}$, defined as the restriction on the diagonal of $(X^{(a)})^2$ of $d_z\theta(z-w+\delta-\Delta)$ (which is a 1-form in z and a function in w).*

This 1-form is proportional to the lift to $X^{(a)}$ of ω: we have $\alpha(z)\,dz = \kappa\omega$, with $\kappa\neq 0$.

PROOF. Let us study the transformation properties of $\alpha(z)\,dz$ when z is taken to $\gamma_{b_i}(z)$. We have

$$\theta(\gamma_{b_i}(z) - \gamma_{b_i}(w) + \delta - \Delta) = e^{\int_z^w \omega_i}\theta(z - w + \delta - \Delta),$$

and therefore

$$d_z\theta(\gamma_{b_i}(z) - \gamma_{b_i}(w) + \delta - \Delta) = e^{\int_z^w \omega_i} d_z\theta(z - w + \delta - \Delta) + d_z(e^{\int_z^w \omega_i})\theta(z - w + \delta - \Delta);$$

since $\theta(z - w + \delta - \Delta)$ vanishes for $z = w$ and $e^{\int_z^w \omega_i}$ is equal to 1 for $z = w$, we obtain

$$d_z\theta(\gamma_{b_i}(z) - \gamma_{b_i}(w) + \delta - \Delta)|_{z=w} = d_z\theta(z - w + \delta - \Delta)|_{z=w},$$

so that $\alpha(z)\,dz$ is invariant under all deck transformations γ_{b_i}, and is therefore the lift of some 1-form $\widetilde{\alpha}(z)\,dz$ defined on X.

Now let us determine this 1-form. Obviously, $\widetilde{\alpha}(z)\,dz$ is regular on X. On the other hand, we have the following expansion of $\theta(z - w + \delta - \Delta)$ for z and w in the vicinity of some δ_i (see [**20, 12**]):

$$\theta(z - w + \delta - \Delta) = z_i w_i (z_i - w_i) a(z_i, w_i),$$

where z_i, w_i are the coordinates of z and w at δ_i and $a(z, w)$ is regular and nonzero at $(0, 0)$. We then have

$$d_z\theta(z - w + \delta - \Delta) = z_i w_i a(z_i, w_i)\,dz_i + (z_i - w_i)(z_i w_i a_{z_i}(z_i, w_i) + w_i a(z_i, w_i)),$$

so that

$$d_z\theta(z - w + \delta - \Delta)|_{z=w} = z_i^2 a(z_i, z_i)\,dz_i,$$

and $\alpha(z)\,dz$ has a double pole at δ_i. It follows that $\widetilde{\alpha}(z)\,dz$ also has a double pole at δ_i, and is therefore proportional to ω. □

Since the function $(z, w) \mapsto \theta(z - w + \delta - \Delta)$ vanishes on the diagonal $z = w$, it follows from this lemma that $\theta(z - w + \delta - \Delta)$ is equivalent to $\kappa \int_z^w \omega$ near $z = w$. When $\partial_h\theta(\delta - \Delta)$ is not equal to zero, $G_h(z, w)$ is then equivalent to $\partial_h\theta(\delta - \Delta)/(\kappa \int_z^w \omega)$, whence we obtain (4.9) with $C(h) = \kappa^{-1}\partial_h\theta(\delta - \Delta)$.

Before we study the vanishing of $C(h)$, we establish the following lemma:

LEMMA 4.4. *The vectors V_{δ_i}, $i = 1, \dots, g - 1$, are independent in $\mathbb{C}^g$; for x a generic point of X, the vector V_x does not belong to $\bigoplus_{i=1}^{g-1} \mathbb{C}V_{\delta_i}$.*

PROOF. Consider the map $\sigma \colon X^{(g)} \to \mathrm{Jac}^g(X)$, defined by $\sigma((y_i)) = \sum_i y_i$. By [**12**], p. 6, and [**19, 18**], this map has rank g at the point (y_i) if and only if $h^1(\sum_i y_i) = 0$. This is the case for the point $\sum_i \delta_i + x$ if x is a generic point of X. Indeed, $h^1(\sum_i \delta_i + x)$ is then equal, by Serre duality, to $h^0(\sum_i \delta_i - x)$, which is zero for x generic (because $h^0(\sum_i \delta_i)$ is equal to 1). The tangent space to the image of σ at this point is the span of V_x and of the vectors V_{δ_i}. It follows that these vectors are independent provided x is generic. □

Let us now study the vanishing of $C(h)$. Clearly $C(h)$ vanishes for h equal to some V_{δ_i}, because we have $\partial_{V_{\delta_i}}\theta(\delta - \Delta) = (d/dt)\theta(\sum_{j \neq i} \delta_j + \delta_i(t) - \Delta)|_{t=0}$, where $t \mapsto \delta_i(t)$ is some coordinate map from a neighborhood of 0 to that of δ_i; on the other hand, $\theta(\sum_{j \neq i} \delta_j + \delta_i(t) - \Delta)$ vanishes identically, because of (4.1).

On the other hand, $C(h)$ does not vanish for $h = V_Q$ if Q is some point of X distinct from the δ_i. Indeed, we have $\partial_{V_Q}\theta(\delta - \Delta) = (d/dt)\theta(Q(t) - Q + \delta - \Delta)|_{t=0}$, where $t \mapsto Q(t)$ is a coordinate map from a neighborhood of 0 to that of Q; from

Lemma 4.3 it now follows that $\partial_{V_Q}\theta(\delta - \Delta)$ is equal to ω_Q/dt and is therefore not zero.

In view of the first part of Lemma 4.4, it follows that the linear form $C(h)$ vanishes if and only if h belongs to $\bigoplus_{i=1}^{g-1}\mathbb{C}V_{\delta_i}$, whence the last part of the proposition. □

REMARK 4.5. It follows from the proof above that the second statement of Lemma 4.4 can be made more precise: if Q is a point of X distinct from the δ_i, $i = 1, \dots, g-1$, then V_Q does not belong to $\bigoplus_{i=1}^{g-1}\mathbb{C}V_{\delta_i}$.

This statement can be translated as follows: for any point x of X, let z_x be some local coordinate at x. The ring of adeles $\mathbb{A}$ of X is the restricted product of the formal series fields $\mathbb{C}((z_x))$. The function field $\mathbb{C}(X)$ of X is embedded in $\mathbb{A}$ by taking Laurent expansions of any function at each point of X. We denote by $\mathcal{O}_{\mathbb{A}}$ the subring of $\mathbb{A}$ of integral adeles from the restricted product of the formal series rings $\mathbb{C}[[z_x]]$. The first cohomology ring $H^1(X, \mathcal{O}_X)$ is defined by $H^1(X, \mathcal{O}_X) = \mathbb{A}/\mathbb{C}(X) + \mathcal{O}_{\mathbb{A}}$; it is a vector space of dimension g. For x in X, let us denote by z_x^{-1} the element of $\mathbb{A}$ with x-component z_x^{-1} and other components zero.

Then the classes of the elements $z_{\delta_i}^{-1}$ and z_x^{-1} form a basis of $H^1(X, \mathcal{O}_X)$.

Now our aim is to prove first that for h not in the span of the vectors V_{δ_i}, the spaces $L_{\mathbb{C}h}$ and $\mathcal{O}$ are supplementary, and then that G_h is the corresponding Green function.

Let us expand $G_h(z, w)$ for w near S'. We then obtain a certain element G_h of $\mathcal{O} \otimes (L_{\mathbb{C}h} + \mathbb{C}1)$. This element satisfies

$$G_h + G_h^{(21)} = \kappa C(h) \sum_i \epsilon^i \otimes \epsilon_i,$$

for ϵ^i, ϵ_i dual bases of $\mathcal{K}$ associated with ω; indeed, we have

$$G_h + G_h^{(21)} = C(h)\left[\left(\int_z^w \omega\right)^{-1} + \left(\int_w^z \omega\right)^{-1}\right] = C(h)\kappa\alpha(z)^{-1}\delta(z - w);$$

on the other hand, recall that $\omega = \kappa^{-1}\alpha(z)\,dz$, so that $\sum_i \epsilon^i \otimes \epsilon_i = \alpha(z)^{-1}\delta(z - w)$.

This implies that $L_{\mathbb{C}h} + \mathcal{O} = \mathcal{K}$. Now, since both $L_{\mathbb{C}h}$ and $\mathcal{O}$ are isotropic and since the scalar product on $\mathcal{K}$ is nondegenerate, their intersection is reduced to zero; therefore we have shown that $L_{\mathbb{C}h}$ and $\mathcal{O}$ are supplementary.

Let us denote by $\bar{G}_h$ the Green function associated with this decomposition. The function G_h is an element of $\mathcal{O} \otimes L_{\mathbb{C}h}$, and it satisfies

$$\bar{G}_h + \bar{G}_h^{(21)} = \sum_i \epsilon^i \otimes \epsilon_i.$$

Then the difference between $G_h - \kappa C(h)\bar{G}_h$ is antisymmetric, and it belongs to $(L_{\mathbb{C}h} + \mathbb{C}1) \otimes \mathcal{O}$. Since the intersection of $L_{\mathbb{C}h} + \mathbb{C}1$ and $\mathcal{O}$ is reduced to the constants, this difference is equal to zero.

Therefore we have the following theorem.

THEOREM 4.6. *For $h \notin \bigoplus_i \mathbb{C}V_{\delta_i}$, the spaces $L_{\mathbb{C}h}$ and $\mathcal{O}$ are supplementary. The Green function associated with the Lagrangian decomposition $\mathcal{K} = \mathcal{O} \oplus L_{\mathbb{C}h}$ is*

$$\bar{G}_h(z, w) = (\kappa C(h))^{-1}\frac{\partial_h\theta}{\theta}(z - w + \delta - \Delta).$$

REMARK 4.7. For the case in which $V = 0$, the space L_V consists of the rational functions on X regular outside $S' \cup \delta$ and having simple poles at δ_i. We then have $L_0 \cap \mathcal{O} = \mathbb{C}1$. Indeed, this intersection consists of the rational functions on X with at most simple poles at δ. This space is exactly $H^0(X, \mathcal{L}_\delta)$, which is 1-dimensional, and therefore consists of the constants.

4.2. Quantum algebras at level zero.

4.2.1. *Quasi-Hopf algebra* $U_{\hbar,h}\mathfrak{g}_{S'}$. For any s of S', let z_s be a local coordinate at s and ∂_{z_s} the derivation d/dz_s. In what follows, we shall denote by (z_s) the point of X with coordinate z_s.

For s in S' and $i \geqslant 0$, let us set

$$l_i^{(s)}(w) = \frac{1}{i!}\partial_{z_s}^i \frac{\partial_h \theta}{\theta}(z - w + \delta - \Delta)_{z=s}.$$

Then Theorem 4.6 implies that $(l_i^{(s)})_{i\geqslant 0, s\in S'}$ is a basis of $L_{\mathbb{C}h}$ dual to the basis $(z_s^i)_{i\geqslant 0,\, s\in S'}$ of $\mathcal{O}'_S$.

PROPOSITION 4.2. *Let* $U_{\hbar,h}\mathfrak{g}_{S'}$ *be the algebra with generators* $x[z_s^i], x[l_i^{(s)}]$, $s \in S'$, $i \geqslant 0$, $x = e, f, h$, *generating series*

$$x^{(s)}(z_s) = \sum_{i\geqslant 0} x[l_i^{(s)}] z_s^i + \sum_{i\geqslant 0,\, t\in S'} x[z_t^i] l_i^{(t)}((z_s)), \qquad x = e, f,$$

$$h^+(z) = \sum_{i\geqslant 0} h[z_s^i] l_i^{(s)}(z), \qquad h^{-(s)}(z_s) = \sum_{i\geqslant 0} h[l_i^{(s)}] z_s^i,$$

and relations

$$[h[\alpha], h[\beta]] = 0, \tag{4.10}$$

for any α, β,

$$K^+(z) e^{(s)}(w_s) K^+(z)^{-1} = \frac{\theta(z - (w_s) - \hbar h + \delta - \Delta)}{\theta(z - (w_s) + \hbar h + \delta - \Delta)} e^{(s)}(w_s), \tag{4.11}$$

$$K^{-(s)}(z_s) e^{(t)}(w_t) K^{-(s)}(z_s)^{-1} = \frac{\theta((z_s) - (w_t) + \hbar h + \delta - \Delta)}{\theta((z_s) - (w_t) - \hbar h + \delta - \Delta)} e^{(t)}(w_t), \tag{4.12}$$

$$K^+(z) f^{(s)}(w_s) K^+(z)^{-1} = \frac{\theta(z - (w_s) + \hbar h + \delta - \Delta)}{\theta(z - (w_s) - \hbar h + \delta - \Delta)} f^{(s)}(w_s), \tag{4.13}$$

$$K^{-(s)}(z_s) f^{(t)}(w_t) K^{-(s)}(z_s)^{-1} = \frac{\theta((z_s) - (w_t) - \hbar h + \delta - \Delta)}{\theta((z_s) - (w_t) + \hbar h + \delta - \Delta)} f^{(t)}(w_t), \tag{4.14}$$

$$\begin{aligned} &\theta((z_s) - (w_t) + \hbar h + \delta - \Delta) e^{(s)}(z_s) e^{(t)}(w_t) \\ &\quad = \theta((z_s) - (w_t) - \hbar h + \delta - \Delta) e^{(t)}(w_t) e^{(s)}(z_s), \end{aligned} \tag{4.15}$$

$$\begin{aligned} &\theta((z_s) - (w_t) - \hbar h + \delta - \Delta) f^{(s)}(z_s) f^{(t)}(w_t) \\ &\quad = \theta((z_s) - (w_t) + \hbar h + \delta - \Delta) f^{(t)}(w_t) f^{(s)}(z_s), \end{aligned} \tag{4.16}$$

and

$$[e^{(s)}(z_s), f^{(t)}(w_t)] = \delta_{st}\delta(z_s - w_t)(K^+((z_s)) - K^{-(s)}(z_s)^{-1}), \tag{4.17}$$

with $K^+(z), K^{-(s)}(z_s)$ *defined as in Section 2, and the coproduct* $\Delta_{S',\delta}$ *defined by* (2.9), (2.10), *is a quantization of the double Lie bialgebra structure on* $\bar{\mathfrak{g}} \otimes \mathcal{K}_{S'}$ *defined by the decomposition*

$$\bar{\mathfrak{g}} \otimes \mathcal{K}_{S'} = (\bar{\mathfrak{h}} \otimes \mathcal{O}_S \oplus \bar{\mathfrak{n}}_+ \otimes \mathcal{K}_{S'}) \oplus (\bar{\mathfrak{h}} \otimes L_{\mathbb{C}h} \oplus \bar{\mathfrak{n}}_- \otimes \mathcal{K}_{S'}).$$

Theorem 2.1 *can be applied to it to define a quantization of the double quasi-Lie bialgebra structure on* $\bar{\mathfrak{g}} \otimes \mathcal{K}_{S'}$ *defined by the decomposition*

$$\bar{\mathfrak{g}} \otimes \mathcal{K}_{S'} = (\bar{\mathfrak{g}} \otimes \mathcal{O}_{S'}) \oplus (\bar{\mathfrak{g}} \otimes L_{\mathbb{C}h}).$$

PROOF. This follows from Theorem 4.6, the expansion

$$\frac{\theta(z-w+\hbar h+\delta-\Delta)}{\theta(z-w+\delta-\Delta)} = a(z,w) + \hbar\frac{\partial_h\theta}{\theta}(z-w+\delta-\Delta)b(z,w),$$

where $a(z,w)$ and $b(z,w)$ belong to $\mathcal{O}\widehat{\otimes}\mathcal{O}[[\hbar]]$, and Theorem 2.1. □

4.2.2. *The algebra* $U_{\hbar,h}\mathfrak{g}_{S',\delta}$. Let $L_{\mathbb{C}h}^{S',\delta}$ be the space of functions f defined on $\widetilde{X}$, regular outside the lifts of S' and δ (here and later, we shall also denote by δ the support $\{\delta_i\}$ of δ), such that the differences $f(\gamma_{b_i}z) - f(z)$ are constant and form a vector proportional to $h = (h_i)_{1\leqslant i\leqslant g}$, and such that

$$\sum_{i=1}^{g} h_i\Big(\int_{a_i} f\omega + \int_{\gamma_{b_i}(a_i)} f\omega\Big) = 0.$$

On the other hand, let $\widetilde{\mathcal{O}}_{S',\delta}$ be the direct sum $\mathcal{O}_{S'} \oplus (\bigoplus_{i=1}^{g-1} z_{\delta_i}^{-1}\mathcal{O}_{\delta_i})$.

PROPOSITION 4.3. *Endow* $\mathcal{K}_{S',\delta}$ *with the scalar product defined by*

$$\langle\phi,\psi\rangle_{\mathcal{K}_{S',\delta}} = \sum_{\alpha\in S'\cup\delta} \operatorname*{res}_{\alpha}(\phi\psi\omega).$$

The spaces $L_{\mathbb{C}h}^{S',\delta}$ *and* $\widetilde{\mathcal{O}}_{S',\delta}$ *are isotropic supplementary subspaces in* $\mathcal{K}_{S',\delta}$. *The Green function associated to this decomposition is given by the collection of expansions in* w, *near each point of* $S'\cup\delta$, *of the function*

$$\widetilde{G}_h(z,w) = \frac{\partial_h\theta}{\theta}(z-w+\delta-\Delta).$$

PROOF. The argument showing that $L_{\mathbb{C}h}^{S',\delta}$ is isotropic is similar to the argument used for $L_{\mathbb{C}h}$. On the other hand, since ω is regular on S' and has double poles at the δ_i, the space $\widetilde{\mathcal{O}}_{S',\delta}$ is also isotropic. It also follows from Remark 4.5 that the direct sum of these spaces is $\widetilde{\mathcal{O}}_{S',\delta}$. This proves the first part of the proposition.

To prove its second part, let us expand $\widetilde{G}_h(z,w)$ on w near each point of $S'\cup\delta$. Since for fixed z, the function $w\mapsto\theta(z-w+\delta-\Delta)$ either vanishes to first order (for w near δ_i) or is nonzero (for w near S'), this expansion will be the series $\sum_{\iota a} f_\lambda\otimes o_\lambda$, with o_λ in $\widetilde{\mathcal{O}}_{S',\delta}$. On the other hand, as a function of z, $\widetilde{G}_h(z,w)$ is regular for z outside δ and w, and has the functional properties (4.8), so that the functions f_λ belong to $L_{\mathbb{C}h}^{S',\delta}\oplus\mathbb{C}1$.

Now let us compare the resulting expansion of $\widetilde{G}_h(z,w)$ with the Green function $G_{S',\delta}$ of the decomposition $L_{\mathbb{C}h}^{S',\delta}\oplus\widetilde{\mathcal{O}}_{S',\delta}$. As is Proposition 4.3, we can check that the sums $\widetilde{G}_h+\widetilde{G}_h^{(21)}$ and $G_{S',\delta}+G_{S',\delta}^{(21)}$ coincide with the same delta-functions. We conclude that the difference $\widetilde{G}_h - G_{S',\delta}$ is antisymmetric. Since it belongs to the tensor square of the intersection of $\widetilde{\mathcal{O}}_{S',\delta}$ and $L_{\mathbb{C}h}^{S',\delta}\oplus\mathbb{C}1$, which is $\mathbb{C}1$, this difference is zero. □

Now let us define $(U_{\hbar,h}\mathfrak{g}_{S',\delta},\Delta)$ as the algebra defined by the generators and relations of Proposition 4.2, with S', $L_{\mathbb{C}h}$, $\mathcal{O}_{S'}$ replaced by $S'\cup\delta$, $L_{\mathbb{C}h}^{S',\delta}$, $\widetilde{\mathcal{O}}_{S',\delta}$.

LEMMA 4.8. *The algebra $U_{\hbar,h}\mathfrak{g}_{S',\delta}$ is a flat deformation of the enveloping algebra of $\mathfrak{g}_{S',\delta} = \bar{\mathfrak{g}} \otimes \mathcal{K}_{S',\delta}$.*

PROOF. We first prove:

LEMMA 4.9. *The subalgebra of $U_{\hbar,h}\mathfrak{g}_{S',\delta}$ generated by $e[\phi]$, $\phi \in \mathcal{K}_{S',\delta}$, is a flat deformation of the corresponding classical subalgebra.*

PROOF. Let us first consider the subalgebras A_α generated by the elements $e[\phi]$, $\phi \in \mathcal{K}_\alpha$, $\alpha \in S' \cup \delta$. The function $(z,w) \mapsto \theta(z - w + \delta - \Delta)$ has the following behavior: for z, w near some δ_i, we have

$$\theta(z - w + \delta - \Delta) = z_i w_i (z_i - w_i)\phi_i(z_i, w_i),$$

with $\phi_i(z_i, w_i)$ invertible in $\mathbb{C}[[z_i, w_i]][[\hbar]]$; for z, w near some $s \in S'$, we have

$$\theta(z - w + \delta - \Delta) = (z_s - w_s)\phi_s(z_s, w_s),$$

with $\phi_s(z_s, w_s)$ invertible in $\mathbb{C}[[z_s, w_s]][[\hbar]]$. After we divide by $z_i w_i \phi(z_i, w_i)$ in the first case, and by $\phi(z_s, w_s)$ in the second case, the relations between the fields $e^{(\alpha)}(z_\alpha)$ have the form of the vertex relations of Proposition 1.3. It follows that each algebra A_α is a flat deformation of the corresponding classical subalgebra.

Now let us study the relations between the various $e[\phi_\alpha]$, $\phi_\alpha \in \mathcal{K}_\alpha$. Recall that the function $(z,w) \mapsto \theta(z - w + \delta - \Delta)$ has simple zeros for w equal to z, or for z or w equal to one of the δ_i. It follows that, after we divide the relation between $e^{(\alpha)}(z_\alpha)$ and $e^{(\beta)}(z_\beta)$, by z_α if α belongs to δ, and by z_β if β belongs to δ, we obtain a relation of the form

$$e^{(\alpha)}(z_\alpha)e^{(\beta)}(w_\beta) = f_{\alpha\beta}(z_\alpha, w_\beta)e^{(\beta)}(w_\beta)e^{(\alpha)}(z_\alpha),$$

with $f_{\alpha\beta}(z_\alpha, w_\beta)$ invertible in $\mathbb{C}[[z_\alpha, w_\beta]][[\hbar]]$. It follows that monomials in the $e[\phi_\alpha]$, $\phi_\alpha \in \mathcal{K}_\alpha$, can be expressed as sums of monomials with the α's occurring in prescribed order. □

Therefore the subalgebra of $U_{\hbar,h}\mathfrak{g}_{S',\delta}$ generated by $e[\alpha]$, $\alpha \in \mathcal{K}_{S',\delta}$, is a flat deformation of the corresponding classical subalgebra. One may then obtain a similar result for the subalgebra generated by the fields $f[\alpha]$. The result then follows from the triangular decomposition of $U_{\hbar,h}\mathfrak{g}_{S',\delta}$. □

The Hopf algebra $(U_{\hbar,h}\mathfrak{g}_{S',\delta}, \Delta)$ is then the quantization of the double structure on $\bar{\mathfrak{g}} \otimes \mathcal{K}_{S',\delta}$ given by the decomposition

$$\bar{\mathfrak{g}} \otimes \mathcal{K}_{S',\delta} = (\bar{\mathfrak{h}} \otimes L_{\mathbb{C}h}^{S',\delta} \oplus \bar{\mathfrak{n}}_+ \otimes \mathcal{K}_{S',\delta}) \oplus (\bar{\mathfrak{h}} \otimes \widetilde{\mathcal{O}}_{S',\delta} \oplus \bar{\mathfrak{n}}_- \otimes \mathcal{K}_{S',\delta}).$$

Since $\widetilde{\mathcal{O}}_{S',\delta}$ is not a subring of $\mathcal{K}_{S',\delta}$, we cannot expect to find a corresponding subalgebra of $U_{\hbar,h}\mathfrak{g}_{S',\delta}$. However, we have:

PROPOSITION 4.4. *The subalgebra of $U_{\hbar,h}\mathfrak{g}_{S',\delta}$ generated by $x[\phi_i]$, $\phi_i \in \mathcal{K}_{\delta_i}$, and $x[o_s]$, $o_s \in \mathcal{O}_s$, $x = e, f, h$, is a flat deformation of the enveloping algebra of $\bar{\mathfrak{g}} \otimes (\mathcal{K}_\delta \oplus \mathcal{O}_{S'})$.*

PROOF. Let us first prove that a similar statement is true for the subalgebra N_+ of $U_{\hbar,h}\mathfrak{g}_{S',\delta}$ generated by $e[\phi_i]$, $\phi_i \in \mathcal{K}_{\delta_i}$, and the $e[o_s]$, $o_s \in \mathcal{O}_s$. The commutation relations between $e[\phi_\alpha]$, $\phi_\alpha \in \mathcal{K}_\alpha$, and $e[\phi_\beta]$, $\phi_\beta \in \mathcal{K}_\beta$, for $\alpha \neq \beta$, are of the form

$$e[z_\alpha^n]e[z_\beta^m] = \sum_{n',m' \in \mathbb{Z}} a(\alpha,\beta)_{nm}^{n'm'} e[z_\beta^{m'}]e[z_\alpha^{n'}],$$

and the summation is over $n' \geqslant 0$ (respectively $m' \geqslant 0$) if α is in δ and $n \geqslant 0$ (respectively β is in δ and $m \geqslant 0$). It follows that a basis of $U_{\hbar,h}\mathfrak{g}_{S',\delta}$ is given by the products of bases for its subalgebras generated by $e[z_s^n]$, $n \in \mathbb{Z}$, $s \in S'$, and $e[z_i^n]$, $n \geqslant 0$. These subalgebras are flat deformations of their classical analogs: this is because $e[z_s^n]$, $n \in \mathbb{Z}$, are subject to vertex relations, and because we have Yangian-type commutation relations between $e[o_s]$, $o_s \in \mathcal{O}_s$.

Let us now prove that the algebra N_+ is stable under the adjoint action of $h[\phi_i]$, $\phi_i \in \mathcal{K}_{\delta_i}$, and of $h[o_s]$, $o_s \in \mathcal{O}_s$. From the identity (4.11) it follows that

$$[(1+V)h^+(z), e^{(\alpha)}(z_\alpha)] = \log \frac{\theta(z-(z_\alpha)+\hbar h+\delta-\Delta)}{\theta(z-(z_\alpha)-\hbar h+\delta-\Delta)} e^{(\alpha)}(z_\alpha), \tag{4.18}$$

for α in S' or δ. We can then check the following two facts:

1) The adjoint action of any $h[\phi]$, $\phi \in \mathcal{O}_s$, preserves the linear space spanned by the $e[\phi_i]$, $\phi_i \in \mathcal{K}_{\delta_i}$, and the $e[z_t^n]$, $n \geqslant 0$, t in S'. This is because (4.18) yields, for α is S' or δ,

$$[h^+[z_s^n], e^{(\alpha)}(z_\alpha)] = \operatorname*{res}_{z_s=0}\left(\log \frac{\theta((z_s)-(z_\alpha)+\hbar h+\delta-\Delta)}{\theta((z_s)-(z_\alpha)-\hbar h+\delta-\Delta)} z_s^n\, dz_s\right) e^{(\alpha)}(z_\alpha),$$

and the function on the right is regular for $z_\alpha = 0$ if α is in S'.

2) The adjoint action of any $h[\phi]$, $\phi \in \mathcal{K}_{\delta_i}$, preserves the linear space spanned by $e[\phi_j]$, $\phi_j \in \mathcal{K}_{\delta_j}$, and $e[o_s]$, $o_s \in \mathcal{O}_s$. The first statement is clear; to show the second one, we first prove as in item 1) above that the adjoint action of any $h[\phi]$, $\phi \in z_i^{-1}\mathcal{O}_i$, preserves the linear space formed by $e[o_s]$, $o_s \in \mathcal{O}_s$; after that, (4.12) implies

$$[h^{-(i)}(z_i), e^{(s)}(z_s)] = \log \frac{\theta((z_i)-(z_s)+\hbar h+\delta-\Delta)}{\theta((z_i)-(z_s)-\hbar h+\delta-\Delta)} e^{(s)}(z_s).$$

Now let $(r_i^{(k)}, r_s^{(k)})_{k\geqslant 0}$ be the dual basis in $L_{\mathbb{C}h}^{S',\delta}$ to $(z_i^{k-1}, z_s^k)_{k\geqslant 0}$. Each r_i^k is then multivalued on X, it has poles of order $-k-2$ at δ_i, of order 1 at most at each s, and it is regular at the other points. We have

$$[h[r_i^{(k)}], e^{(s)}(z_s)] = \operatorname*{res}_{z_i=0} z_i^2\, dz_i\, r_i^{(k)}(z_i) \log \frac{\theta((z_i)-(z_s)+\hbar h+\delta-\Delta)}{\theta((z_i)-(z_s)-\hbar h+\delta-\Delta)} e^{(s)}(z_s),$$

and since the function in the right-hand side is regular for $z_s = 0$, the adjoint action of $h[r_i^{(k)}]$ preserves $\{e[o_s], o_s \in \mathcal{O}_s\}$. Since any element $\mathcal{K}_i$ can be obtained as a linear combination of $h[r_i^{(k)}]$ and $h[\lambda]$, $\lambda \in (\bigoplus_i z_{\delta_i}^{-1}\mathcal{O}_{\delta_i}) \oplus (\bigoplus_{s\in S'} \mathcal{O}_s)$, the same result is true for any $h[\lambda], \lambda \in \mathcal{K}_{\delta_i}$.

After that, we prove that the subalgebra N_+ of $U_{\hbar,h}\mathfrak{g}_{S',\delta}$ generated by $f[\phi]$, ϕ in $\mathcal{O}_s$ and in $\mathcal{K}_{\delta_i}$, is a flat deformation of its classical analog, using the same reasoning as for N_+. Finally, any commutator $[e[\phi], f[\psi]]$, ϕ, ψ in the sums of $\mathcal{O}_s$ and $\mathcal{K}_{\delta_i}$, is expressed as the product of $h[\phi]$, ϕ in $\mathcal{O}_s$ or $\mathcal{K}_{\delta_i}$. □

REMARK 4.10. Following the proof of Theorem 3.1, one can show that the algebra $U_{\hbar,h}\mathfrak{g}_{S',\delta}$ is isomorphic to a quotient (with central elements identified) of the tensor product $DY(\mathfrak{sl}_2)_0^{\prime\otimes \operatorname{card} S'} \otimes U_{\hbar,h}\mathfrak{g}_{S'}$, where $U_{\hbar,h}\mathfrak{g}_{S'}$ is the algebra corresponding to an empty S', and $DY(\mathfrak{sl}_2)_0'$ is the double Yangian algebra (with no derivation nor central element). The subalgebra of Proposition 4.4 can then be identified with $Y(\mathfrak{sl}_2)^{\otimes \operatorname{card} S'} \otimes U_{\hbar,h}\mathfrak{g}_{S'}$, where $Y(\mathfrak{sl}_2)$ is the Yangian subalgebra of $DY(\mathfrak{sl}_2)_0'$ generated by the nonnegative modes generators.

4.2.3. *Regular subalgebra in $U_{\hbar,h}\mathfrak{g}_{S',\delta}$.* Let us define, for ϵ in $\mathcal{K}_{S',\delta}$, the element $\bar{h}[\epsilon]$ as the Cartan element of $U_{\hbar,h}\mathfrak{g}_{S',\delta}$ such that

$$[\bar{h}[\epsilon], e^\alpha(z_\alpha)] = 2\epsilon^{(\alpha)}(z_\alpha)e^\alpha(z_\alpha),$$

for any α in $S' \cup \delta$. If ϕ is a regular function on $\widetilde{X} - \widetilde{S}'$, let us set

$$\bar{h}[\phi] = \sum_{\alpha \in S' \cup \delta} \bar{h}[\phi^{(\alpha)}],$$

where $\phi^{(\alpha)}$ is the image in $\mathcal{K}_{S',\delta}$ of the element of $\mathcal{K}_\alpha$ given by expansion of ϕ near $\widetilde{\alpha}$.

Let P be a point of S'. We set

$$x(z) = \exp\left(\frac{1}{4}\log\frac{\theta(z-P+\hbar h+\delta-\Delta)}{\theta(z-P-\hbar h+\Delta+\delta)}\bar{h}[1] - \frac{1}{4}\bar{h}\left[\log\frac{\theta(\cdot-P+\hbar h+\delta-\Delta)}{\theta(\cdot-P-\hbar h+\delta-\Delta)}\right]\right),$$

$$\bar{e}^{(\alpha)}(z_\alpha) = e^{(\alpha)}(z_\alpha)x(z_\alpha), \qquad \bar{f}^{(\alpha)}(z_\alpha) = f^{(\alpha)}(z_\alpha)x(z_\alpha), \tag{4.19}$$

$$\bar{K}^+(z) = K^+(z)x(z)^2, \qquad \bar{K}^{-(\alpha)}(z_\alpha) = K^{-(\alpha)}(z_\alpha)x((z_\alpha))^2. \tag{4.20}$$

LEMMA 4.11. *Set*

$$q_0(z,w) = \frac{\theta(z-w+\hbar h+\delta-\Delta)}{\theta(z-w-\hbar h+\delta-\Delta)}, \qquad q_\pm(z,w) = \theta(z-w\pm\hbar h+\delta-\Delta),$$

$$q(z,w) = \frac{q_0(z,w)}{q_0(z,P)q_0(P,w)}, \qquad \widetilde{q}_\pm(z,w) = \frac{q_\pm(z,w)}{q_\pm(z,P)q_\pm(P,w)}.$$

Then $\bar{e}^{(\alpha)}(z_\alpha)$, $\bar{f}^{(\alpha)}(z_\alpha)$, $\bar{K}^+(z)$, and $\bar{K}^{-(\alpha)}(z_\alpha)$ satisfy the relations

$$\bar{K}^+(z)\bar{e}^{(\alpha)}(w_\alpha)\bar{K}^+(z)^{-1} = q(z,(w_\alpha))\bar{e}^{(\alpha)}(w_\alpha),$$
$$\bar{K}^+(z)\bar{f}^{(\alpha)}(w_\alpha)\bar{K}^+(z)^{-1} = q(z,(w_\alpha))^{-1}\bar{f}^{(\alpha)}(w_\alpha),$$
$$\bar{K}^{-(\alpha)}(z_\alpha)\bar{e}^{(\beta)}(w_\beta)\bar{K}^{-(\beta)}(z_\beta)^{-1} = q((z_\alpha),(w_\beta))^{-1}\bar{e}^{(\beta)}(w_\beta),$$
$$\bar{K}^{-(\alpha)}(z_\alpha)\bar{f}^{(\beta)}(w_\beta)\bar{K}^{-(\beta)}(z_\beta)^{-1} = q((z_\alpha),(w_\beta))\bar{f}^{(\beta)}(w_\beta),$$
$$[\bar{e}^{(\alpha)}(z_\alpha), \bar{f}^{(\beta)}(w_\beta)] = \delta_{\alpha\beta}\delta(z_\alpha - w_\beta)(\bar{K}^+((z_\alpha)) - \bar{K}^{-(\alpha)}(z_\alpha)),$$
$$\widetilde{q}_+((z_\alpha),(w_\beta))\bar{e}^{(\alpha)}(z_\alpha)\bar{e}^{(\beta)}(w_\beta) = \widetilde{q}_-((z_\alpha),(w_\beta))\bar{e}^{(\beta)}(w_\beta)\bar{e}^{(\alpha)}(z_\alpha),$$
$$\widetilde{q}_-((z_\alpha),(w_\beta))\bar{f}^{(\alpha)}(z_\alpha)\bar{f}^{(\beta)}(w_\beta) = \widetilde{q}_+((z_\alpha),(w_\beta))\bar{f}^{(\beta)}(w_\beta)\bar{f}^{(\alpha)}(z_\alpha).$$

PROOF. This follows from the identities

$$x(z)e^{(\alpha)}(w_\alpha)x(z)^{-1} = \left(\frac{\theta(z-P+\hbar h+\delta-\Delta)}{\theta(z-P-\hbar h+\delta-\Delta)} : \frac{\theta((w_\alpha)-P+\hbar h+\delta-\Delta)}{\theta((w_\alpha)-P-\hbar h+\delta-\Delta)}\right)^{1/2} e^{(\alpha)}(w_\alpha),$$

$$x(z)f^{(\alpha)}(w_\alpha)x(z)^{-1} = \left(\frac{\theta(z-P-\hbar h+\delta-\Delta)}{\theta(z-P+\hbar h+\delta-\Delta)} : \frac{\theta((w_\alpha)-P-\hbar h+\delta-\Delta)}{\theta((w_\alpha)-P+\hbar h+\delta-\Delta)}\right)^{1/2} f^{(\alpha)}(w_\alpha).$$

□

Let $R_{S',\delta}$ be the algebra of functions on X regular outside $S' \cup \delta$. Then the intersection of $R_{S',\delta}$ and $L_{\mathbb{C}h}^{S',\delta}$ has codimension 1 in each of these spaces. A supplementary subspace to this intersection in $L_{\mathbb{C}h}^{S',\delta}$ is spanned by

$$e_0(z) = \frac{\partial_h \theta}{\theta}(z - P + \delta - \Delta) + c,$$

where c is a constant.

Let us define Σ as the direct sum of $\mathbb{C}e_0$ and the orthogonal complement to e_0 in $\widetilde{\mathcal{O}}_{S',\delta}$. Then:

LEMMA 4.12. *The spaces $R_{S',\delta}$ and Σ are isotropic supplementaries in $\mathcal{K}_{S',\delta}$. The corresponding Green kernel is proportional to the collection of expansions in w near the points of $S' \cup \delta$ of the function*

$$G_R(z,w) = \frac{\partial_h \theta}{\theta}(z - w + \delta - \Delta) - \frac{\partial_h \theta}{\theta}(z - P + \delta - \Delta) + \frac{\partial_h \theta}{\theta}(w - P + \delta - \Delta).$$

PROOF. Let us set $e_{-1} = 1$; then the pairing $\langle e_0, e_{-1} \rangle_{\mathcal{K}_{S',\delta}}$ is nonzero (otherwise $R_{S',\delta} + L_{\mathbb{C}h}^{S',\delta}$ would be isotropic). Let us complete e_0 and e_{-1} to dual bases $(e_i)_{i \geqslant 0}$ and $(e_{-i-1})_{i \geqslant 0}$ of $R_{S',\delta}$ and $L_{\mathbb{C}h}^{S',\delta}$. Then bases for $R_{S',\delta}$ and Σ are $(e_{-1}, e_i)_{i>0}$ and $(e_0, e_{-i-1})_{i>0}$, so that these spaces are supplementary to each other. The difference between the Green function for this decomposition and $\mathcal{K}_{S',\delta} = L_{\mathbb{C}h}^{S',\delta} \oplus \widetilde{\mathcal{O}}_{S',\delta}$ is just the difference $e_0 \otimes e_{-1} - e_{-1} \otimes e_0$. □

We can therefore construct another double quasi-Lie bialgebra structure on $\bar{\mathfrak{g}} \otimes \mathcal{K}_{S',\delta}$ based on this decomposition:

$$\bar{\mathfrak{g}} \otimes \mathcal{K}_{S',\delta} = (\bar{\mathfrak{g}} \otimes R_{S',\delta}) \oplus (\bar{\mathfrak{g}} \otimes \widetilde{\mathcal{O}}_{S',\delta})$$

and its usual infinite twist

$$\bar{\mathfrak{g}} \otimes \mathcal{K}_{S',\delta} = (\bar{\mathfrak{h}} \otimes R_{S',\delta} \oplus \bar{\mathfrak{n}}_+ \otimes \mathcal{K}_{S',\delta}) \oplus (\bar{\mathfrak{h}} \otimes \widetilde{\mathcal{O}}_{S',\delta} \oplus \bar{\mathfrak{n}}_- \otimes \mathcal{K}_{S',\delta}). \tag{4.21}$$

Set

$$q_{0+}(z,w) = \frac{\theta(z - w - \hbar h + \delta - \Delta)}{\theta(z - w + \delta - \Delta)}.$$

We then have

$$q_{0+}(\gamma_{a_i} z, w) = q_{0+}(z, \gamma_{a_i} w) = q_{0+}(z,w), \tag{4.22}$$

$$q_{0+}(\gamma_{b_i} z, w) = e^{-h_i} q_{0+}(z,w), \qquad q_{0+}(z, \gamma_{b_i} w) = e^{h_i} q_{0+}(z,w), \tag{4.23}$$

so that the function

$$(z,w) \mapsto \frac{q_{0+}(z,w)}{q_{0+}(z,P) q_{0+}(P,w)}$$

is single-valued on the complement of the diagonal of $X - (P \cup \delta)$.

LEMMA 4.13. *Set*

$$q_{0+}(z,w) = \frac{\theta(z - w - \hbar h + \delta - \Delta)}{\theta(z - w + \delta - \Delta)}.$$

Then for some a, b in $R_{S',\delta}^{\otimes 2}[[\hbar]]$, we have

$$\frac{q_{0+}(z,w)}{q_{0+}(z,P) q_{0+}(P,w)} = a(z,w) + b(z,w) G_R(z,w).$$

PROOF. For z close to w, we have the expansions $G_P(z,w) = C(h)/\int_z^w \omega + O(1)$, and

$$\frac{q_{0+}(z,w)}{q_{0+}(z,P)q_{0+}(P,w)} = \frac{1}{q_{0+}(z,P)q_{0+}(P,z)} \frac{\theta(\hbar h)}{\kappa \int_z^w \omega} + O(1),$$

by the remark following Lemma 4.3.

Then let us set

$$b(z,w) = \frac{\theta(\hbar h)/(C(h)\kappa)}{q_{0+}(z,P)q_{0+}(P,z)}.$$

By (4.22) and (4.23), this function is single-valued on X; since its only poles are at P and δ, this is a series in $\hbar$ with coefficients in $R_{S',\delta}^{\otimes 2}$. Set

$$a(z,w) = \frac{q_{0+}(z,w)}{q_{0+}(z,P)q_{0+}(P,w)} - b(z,w)G_R(z,w);$$

this is again a single-valued function on the complement of the diagonal of the square $(X-(\delta \cup P))^2$, which is also regular on the diagonal. $\square$

According to Theorem 2.1, we can then define a quantization $(U_{\hbar,h}\tilde{\mathfrak{g}}_{S',\delta}, \widetilde{\Delta}_{S',\delta})$ of the quasi-Lie bialgebra structure on $\bar{\mathfrak{g}}\otimes\mathcal{K}_{S',\delta}$ defined by the decomposition (4.21), using the functions a and b of Lemma 4.13. Denote by $\tilde{e}^{(\alpha)}(z_\alpha)$, $\tilde{f}^{(\alpha)}(z_\alpha)$, $\widetilde{K}^+(z)$, and $\widetilde{K}^{-(\alpha)}(z_\alpha)$ the generating fields of this algebra, similar to the fields $e(z)$, $f(z)$, $K^-(z)$, and $K^+(z)$ of Section 2 (note the inversion of indices of the fields K).

PROPOSITION 4.5. *The map i defined by* (4.19) *and* (4.20), *taking the fields $\bar{e}^{(\alpha)}(z_\alpha)$, $\bar{f}^{(\alpha)}(z_\alpha)$, $\bar{K}^+(z)$, and $\bar{K}^{-(\alpha)}(z_\alpha)$ to the fields $\tilde{e}^{(\alpha)}(z_\alpha)$, $\tilde{f}^{(\alpha)}(z_\alpha)$, $\widetilde{K}^+(z)$, and $\widetilde{K}^{-(\alpha)}(z_\alpha)$ respectively, defines an algebra isomorphism $U_{\hbar,h}\mathfrak{g}_{S',\delta} \to U_{\hbar,h}\mathfrak{g}_{S',\delta}$. The coproducts $\Delta_{S',\delta}$ and $\widetilde{\Delta}_{S',\delta}$ are then connected by the twist transformation*

$$\widetilde{\Delta}_{S',\delta}(i(x)) = F_0(i\otimes i)(\Delta_{S',\delta}(x))F_0^{-1},$$

for any x in $U_{\hbar,h}\mathfrak{g}_{S',\delta}$, where

$$F_0 = \exp\left[\frac{1}{8}\left(\bar{h}\left[\log\frac{\theta(\cdot - P + \hbar h + \delta - \Delta)}{\theta(\cdot - P - \hbar h + \delta - \Delta)}\right]\otimes \bar{h}[1] \right.\right.$$
$$\left.\left. - \bar{h}[1]\otimes\bar{h}\left[\log\frac{\theta(\cdot - P + \hbar h + \delta - \Delta)}{\theta(\cdot - P - \hbar h + \delta - \Delta)}\right]\right)\right].$$

PROOF. Let us first check that the relations defining i are consistent. For ϵ in $\mathcal{K}_{S',\delta}$, let $\tilde{h}[\epsilon]$ be the Cartan element of $U_{\hbar,h}\tilde{\mathfrak{g}}_{S',\delta}$ such that

$$[\tilde{h}[\epsilon], \tilde{e}^\alpha(z_\alpha)] = \epsilon^{(\alpha)}(z_\alpha)\tilde{e}^\alpha(z_\alpha).$$

For some functions $\lambda(z,w_\beta)$, $\lambda^{(\alpha)}(z_\alpha,w_\beta)$, we have

$$[\log\bar{K}^+(z), e^{(\beta)}(w_\beta)] = \lambda(z,w_\beta)e^{(\beta)}(w_\beta),$$
$$[\log\bar{K}^{-(\alpha)}(z_\alpha), e^{(\beta)}(w_\beta)] = \lambda^{(\alpha)}(z_\alpha,w_\beta)e^{(\beta)}(w_\beta),$$

as well as

$$[\log\widetilde{K}^+(z), \tilde{e}^{(\beta)}(w_\beta)] = \lambda(z,w_\beta)\tilde{e}^{(\beta)}(w_\beta),$$
$$[\log\widetilde{K}^{-(\alpha)}(z_\alpha), \tilde{e}^{(\beta)}(w_\beta)] = \lambda^{(\alpha)}(z_\alpha,w_\beta)\tilde{e}^{(\beta)}(w_\beta).$$

Expanding

$$\lambda(z, w_\beta) = \sum_i a_i(z) b_i(w_\beta), \qquad \lambda^{(\alpha)}(z_\alpha, w_\beta) = \sum_i a_i^{(\alpha)}(z_\alpha) b_i^{(\alpha)}(w_\beta),$$

we obtain

$$\log \bar{K}^+(z) = \sum_i \bar{h}[b_i] a_i(z), \qquad \log \bar{K}^{-(\alpha)}(z_\alpha) = \sum_i \bar{h}[b_i^{(\alpha)}] a_i^{(\alpha)}(z_\alpha),$$

$$\log \widetilde{K}^+(z) = \sum_i \tilde{h}[b_i] a_i(z), \qquad \log \widetilde{K}^{-(\alpha)}(z_\alpha) = \sum_i \tilde{h}[b_i^{(\alpha)}] a_i^{(\alpha)}(z_\alpha).$$

It follows that the generating formulas for $i(\bar{K}^+(z))$ and $i(\bar{K}^{-(\alpha)}(z_\alpha))$ are consistent and yield $i(\bar{h}[\epsilon]) = \tilde{h}[\epsilon]$, for any ϵ in $\mathcal{K}_{S',\delta}$.

The relations of Lemma 4.11 then imply that i is an algebra morphism. The conjugation identity is checked directly. □

For $x = \bar{e}, \bar{f}, \bar{h}$, and $\phi \in \mathcal{K}_{S',\delta}$, set $x[\phi] = \sum_{\alpha \in S' \cup \delta} \mathrm{res}_\alpha(x(z)\phi(z)\omega)$. Then:

COROLLARY 4.1. *The subalgebra of $U_{\hbar,h}\mathfrak{g}_{S',\delta}$ generated by $\bar{e}[r]$, $\bar{f}[r]$, and $\bar{K}^+[r]$, for r in $R_{S',\delta}$, is a flat deformation of the enveloping algebra of $\bar{\mathfrak{g}} \otimes R_{S',\delta}$.*

PROOF. After we apply i, this follows from the PBW result of Theorem 2.1 on regular subalgebras of the algebras $U_{a,b}\mathfrak{g}$. □

REMARK 4.14. The field $K^+(z)$ satisfies the functional equations

$$K^+(\gamma_{a_i} z) = K^+(z), \qquad K^+(\gamma_{b_i} z) = K^+(z)\bar{h}[1]^{-\hbar h_i},$$

analogous to relation (44) of [**7**].

REMARK 4.15. It is easy to specialize Proposition 4.5 to obtain an isomorphism between the centerless versions of the elliptic algebras of [**10**] and [**7**].

REMARK 4.16. By analogy with [**7**], one may construct a "centrally extended" version $U_{\hbar,h}\hat{\mathfrak{g}}_{S',\delta}$, of the algebra $U_{\hbar,h}\mathfrak{g}_{S',\delta}$, with additional central generator K and relations (4.10), (4.17), and (4.13) replaced by

$$[K^+(z), K^+(z')] = [K^{-(\alpha)}(z), K^{-(\beta)}(z')] = 0,$$

$$K^+(z)K^{-(\alpha)}(z_\alpha)K^+(z)^{-1}K^{-(\alpha)}(z_\alpha)^{-1}$$
$$= \frac{\theta(z - (z_\alpha) + \hbar h + \delta - \Delta)}{\theta(z - (z_\alpha) - \hbar h + \delta - \Delta)} \frac{\theta(z - (z_\alpha) + \hbar h(K-1) + \delta - \Delta)}{\theta(z - (z_\alpha) + \hbar h(K+1) + \delta - \Delta)},$$

$$[e^{(\alpha)}(z_\alpha), f^{(\beta)}(w_\beta)] = \delta_{\alpha\beta}(\delta(z, (w_\beta))K^+((z_\alpha)) - \delta(z, (w_\beta); K)K^{-(\beta)}((w_\beta))^{-1}),$$

$$K^{-(\alpha)}(z_\alpha)e^{(\beta)}(w_\beta)K^{-(\alpha)}(z_\alpha)^{-1} = \frac{\theta((z_\alpha) - w_\beta - \hbar(K+1)h + \delta - \Delta)}{\theta(z_\alpha - w_\beta - \hbar(K-1)h)} e^{(\beta)}(w_\beta),$$

where

$$\delta(z, w) = \frac{\partial_h \theta}{\theta}(z - w + \delta - \Delta) + \frac{\partial_h \theta}{\theta}(w - z + \delta - \Delta),$$

$$\delta(z, w; K) = \frac{\partial_h \theta}{\theta}(z - w + \hbar K h + \delta - \Delta) + \frac{\partial_h \theta}{\theta}(w - z + \hbar K h + \delta - \Delta).$$

The construction in Proposition 4.5 and Corollary 4.1 of a deformation of the enveloping algebra of $\bar{\mathfrak{g}} \otimes R_{S',\delta}$ in $U_{\hbar,h}\mathfrak{g}_{S',\delta}$ may be extended to $U_{\hbar,h}\hat{\mathfrak{g}}_{S',\delta}$.

However, it seems difficult to construct a coproduct on $U_{\hbar,h}\hat{\mathfrak{g}}_{S',\delta}$, because the maps $z \mapsto z_K$, where z_K is the solution "close to z" of $\theta(z_K - z + \hbar K h + \delta - \Delta) = 0$, do not satisfy $(z_{K_1})_{K_2} = z_{K_1+K_2}$.

REMARK 4.17. *Quantization of double extensions.* A Hopf algebra $U_\hbar\tilde{\mathfrak{g}}_{S',\delta}$ quantizing the doubly extended quasi-Lie bialgebra

$$[(\bar{\mathfrak{h}} \otimes L^{\mathbb{C}h}_{S',\delta}) \oplus (\bar{\mathfrak{n}}_+ \otimes \mathcal{K}_{S',\delta}) \oplus \mathbb{C}D] \oplus [(\bar{\mathfrak{h}} \otimes \widetilde{\mathcal{O}}_{S',\delta}) \oplus (\bar{\mathfrak{n}}_- \otimes \mathcal{K}_{S',\delta}) \oplus \mathbb{C}K]$$

may be obtained by replacing, in the defining relations of the above remark, $q(z,w)$ by

$$q_\partial(z,w) = \frac{\theta(q^\partial z - w + \delta - \Delta)}{\theta(z - q^\partial w + \delta - \Delta)}$$

and the shifts by $\hbar K h$ by actions of $q^{K\partial}$ on the variables. One may then extend the construction of the subalgebra deforming the enveloping algebra of the algebra $\bar{\mathfrak{g}} \otimes (\mathcal{O}_{S'} \oplus \mathcal{K}_\delta) \oplus \mathbb{C}K$ to this situation. One may also construct and twist this Hopf structure to obtain a quantization of

$$[(\bar{\mathfrak{h}} \otimes R^{\mathbb{C}h}_{S',\delta}) \oplus (\bar{\mathfrak{n}}_+ \otimes \mathcal{K}_{S',\delta}) \oplus \mathbb{C}D] \oplus [(\bar{\mathfrak{h}} \otimes \Sigma) \oplus (\bar{\mathfrak{n}}_- \otimes \mathcal{K}_{S',\delta}) \oplus \mathbb{C}K],$$

which will be isomorphic to the one defined in Section 1. This is because the structure coefficient

$$\widetilde{q}_\delta(z,w) = \frac{q_\partial(z,w)}{q_\partial(z,P)q_\partial(P,w)}$$

has the following properties: it vanishes for $z = q^\partial w$, is single-valued for z, w on X, and it has its poles only for $z = w$ or z, w in $\delta \cup \{P\}$.

One may then construct a deformation of the enveloping algebra of $R_{S',\delta}$ in the usual way in $U_\hbar\tilde{\mathfrak{g}}_{S',\delta}$. Theorem 3.1 then implies that $U_\hbar\mathfrak{g}'_{S',\delta}$ is isomorphic to the quotient of $U_{\hbar,z^2dz}\mathfrak{g}'^{\otimes(g-1)} \otimes DY(\mathfrak{sl}_2)'^{\otimes \mathrm{card}\, S'}$ by the identification of the central generators.

REMARK 4.18. If one replaces δ by an arbitrary effective divisor $\sum_i x_i$ in the definition of $U_{\hbar,h}\mathfrak{g}_{S',\delta}$, the resulting algebra is no longer a flat deformation of $\widehat{\mathfrak{sl}}_2$. For example, if one replaces the vertex relations by

$$\theta\Bigg(\sum_{i=1}^{g-1} x_i + (z_\alpha) - (w_\beta) + \hbar h - \Delta\Bigg) e^{(\alpha)}(z_\alpha) e^{(\beta)}(w_\beta)$$
$$= -\theta\Bigg(\sum_{i=1}^{g-1} x_i + (w_\beta) - (z_\alpha) + \hbar h - \Delta\Bigg) e^{(\beta)}(w_\beta) e^{(\alpha)}(z_\alpha),$$

one obtains, for $\alpha = \beta = x_i$,

$$(w_i(z_i - w_i) + \hbar \cdots) e^{(i)}(z_i) e^{(i)}(w_i) = (z_i(z_i - w_i) + \hbar \cdots) e^{(i)}(w_i) e^{(i)}(z_i),$$

which are the relations for a flat deformation of the affinization of the Lie superalgebra $\mathfrak{osp}(2|1)$ (we owe this remark to B. Feigin).

References

[1] M. Bangoura and Y. Kosmann-Schwarzbach, *The double of a Jacobian Lie quasi-bialgebra*, Lett. Math. Phys. **28** (1993), 13–29.

[2] A. A. Beilinson, B. L. Feigin, and B. Mazur, *Algebraic field theory*, Preprint.

[3] J. Ding and B. Feigin, *Quantum current operators* (III): *commutative quantum current operators, semi-infinite construction and functional models*, q-alg/9612009.

[4] V. G. Drinfeld, *Quantum groups*, Proc. Int. Congress Math. (Berkeley, 1986), Vol. 1, Amer. Math. Soc., Providence, RI, 1988, pp. 798–820.

[5] ———, *A new realization of Yangians and quantized affine algebras*, Dokl. Akad. Nauk SSSR, **296** (1987), no. 1, 13–17; English transl., Soviet Math. Dokl. **36** (1988), 212–216.

[6] ———, *Quasi-Hopf algebras*, Algebra Analiz **1** (1989), no. 6, 114–148; English transl., Leningrad Math. J. **1** (1990), no. 6, 1419–1457.

[7] B. Enriquez and G. Felder, *Elliptic quantum groups $E_{\tau,\eta}(\mathfrak{sl}_2)$ and quasi-Hopf algebras*, Comm. Math. Phys. **195** (1998), 651–689.

[8] ———, *A construction of Hopf algebra cocycles for the double Yangian $DY(\mathfrak{sl}_2)$*, J. Phys. A **31** (1998), 2401–2413.

[9] ———, *Coinvariants for Yangian doubles and quantum Knizhnik–Zamolodchikov equations*, q-alg/9706012; Internat. Math. Res. Notices (to appear).

[10] B. Enriquez and V. Rubtsov, *Quantum groups in higher genus and Drinfeld's new realizations method ($\mathfrak{sl}_2$ case)*, Ann. Sci. École Norm. Sup. (4), **30** (1997), 821–846.

[11] ———, *Quasi-Hopf algebras associated with $\mathfrak{sl}_2$ and complex curves*, Israel J. Math. (to appear).

[12] J. Fay, *Theta functions on Riemann surfaces*, Lecture Notes in Math. Vol. 352, Springer-Verlag, Berlin–Heidelberg, 1973.

[13] B. Feigin, M. Jimbo, T. Miwa, A. Odesski, and Ya. Pugai, *Algebra of screening operators for the deformed W_n algebra*, q-alg/9702029 Comm. Math. Phys. **191** (1998), 501–541.

[14] B. Feigin and A. Odesski, *A family of elliptic algebras*, Internat. Math. Res. Notices **11** (1997), 531–539.

[15] B. Feigin and A. Stoyanovsky, *A realization of the modular functor in the space of differentials and the geometric approximation of the moduli space of G-bundles*, Funktsional. Anal. i Prilozhen. **28** (1994), no. 4, 42–65; English. transl., Functional Anal. Appl. **28** (1994), no. 4, 257–275.

[16] G. Felder and A. Varchenko, *On representations of the elliptic quantum group $E_{\tau,\eta}(\mathfrak{sl}_2)$*, q-alg/9601003; Comm. Math. Phys. **181** (1996), 741–761.

[17] S. M. Khoroshkin and V. N. Tolstoy, *On Drinfeld's realization of quantum affine algebras*, J. Geom. Phys. **11** (1993), 445–452.

[18] J. Lewittes, *Riemann surfaces and theta functions*, Acta Math. **111** (1964), 37–61.

[19] A. Mayer, *On the Jacobi inversion theorem*, Princeton Univ. Thesis, 1961.

[20] D. Mumford, *Tata lectures on Theta.* II, Progress in Math., Vol. 43, Birkhäuser, Boston, 1984.

[21] J.-P. Serre, *Groupes algébriques et corps de classes*, Hermann, Paris, 1959.

Centre de Mathématiques, Ecole Polytechnique, URA 169 du CNRS, 91128 Palaiseau, France; FIM, ETH-Zentrum, HG G46, CH-8092 Zürich, Switzerland

Dépt. de Mathématiques, Univ. d'Angers, 2, Bd. Lavoisier, 49045 Angers, France; ITEP, 25 Bol. Cheremushkinskaia, 117259 Moscou, Russia

Amer. Math. Soc. Transl.
(2) Vol. **191**, 1999

Poisson Structure on Moduli of Flat Connections on Riemann Surfaces and the r-Matrix

V. V. Fock and A. A. Rosly

ABSTRACT. We consider the space of graph connections (lattice gauge fields) which can be endowed with a Poisson structure in terms of a *ciliated fat graph.* (A ciliated fat graph is a graph with a fixed linear order of ends of edges at each vertex.) However our aim is to study the Poisson structure on the moduli space of locally flat vector bundles on a Riemann surface with holes (i.e., with boundary). It is shown that this moduli space can be obtained as the quotient of the space of graph connections by the Poisson action of a lattice gauge group endowed with a Poisson–Lie structure. The present paper contains as a part an updated version of a 1992 preprint which we decided still deserves publishing. We have removed some obsolete unessential remarks and added some newer remarks.

1. Introduction

The moduli space of flat G-bundles on a Riemann surface is the classical phase space for Chern–Simons gauge theory and, thus, it is in a sense the classical limit of WZNW conformal field theory. This means that by quantizing it, one can get a space of quantum states, which turns out to be isomorphic to the space of conformal blocks of the corresponding WZNW theory. This statement has been checked by several authors (see [**8, 20**]) with the help of different quantization methods. On the other hand, the moduli spaces of flat bundles, as well as closely related moduli spaces of holomorphic bundles (see [**17**]), have attracted much attention from the purely mathematical point of view (see [**6, 15**]).

In Section 3 we discuss in detail the canonical Poisson structure on the moduli space of flat bundles on Riemann surfaces with holes (i.e., with boundary). In Section 4 we construct a Poisson structure on the space of graph connections in such a way that the action of the graph gauge group is Poissonian with respect to an appropriate nontrivial Poisson–Lie structure. The considerations of this section are inspired mainly by constructions from [**19, 2, 1**], where a discrete analog of the current algebra was suggested and investigated. Then we prove that the quotient of the space of graph connections by the gauge group coincides with the moduli space of flat connections on a Riemann surface determined by the graph.

1991 *Mathematics Subject Classification.* Primary 58D27, 58F05.

This work was partially supported by RFBR grant no. 95-01-01101 and by grant no. 96-15-96455 for the support of scientific schools.

One of the main aims of our preprint [**11**] (see also [**12**]) was to give a description of the moduli space of flat connections in a form ready for quantization. We are not going to discuss this problem here because, since then, considerable progress has been made in this direction; see [**3, 9, 13**]. The interested reader can find details and discussions in those papers. Note only that the result of quantization is a noncommutative algebra which has WZNW model conformal blocks as its representation space and is functorial with respect to the embeddings of surfaces.

For the readers interested in a more understandable and detailed presentation, we would like to recommend the excellent survey article by M. Audin [**7**].[1]

2. Ciliated fat graphs and Poisson manifolds

The moduli space of flat connections on a compact Riemann surface is by definition a subquotient of the topologically trivial space of all connections. This description is useful also since a nontrivial Poisson manifold (which is the moduli space, or an orbifold, to be more precise) is represented as the result of the reduction of a trivial symplectic manifold (see Section 3 for details). Unlike the former, the latter has plenty of convenient parameterizations. The only disadvantage of this description is that the space of connections is infinite-dimensional. In this paper (Section 4), we consider an alternative description of the moduli space in which the role of the space of all connections on a Riemann surface is played by a finite-dimensional manifold. The idea is quite familiar both from lattice gauge theory and from Čech cohomology. Namely, consider a triangulation of a compact Riemann surface S (with boundary, in general). Then we get a graph formed by the vertices and the edges of this triangulation. By a graph connection (or lattice gauge field) we mean an assignment of an element of the gauge group G to each (oriented) edge. The group of lattice gauge transformations $\mathcal{G}^l$ acting on the space of graph connections in a natural way is simply the product of several copies of G, one copy for each vertex of the graph. A flat graph connection satisfies the condition that the monodromies around all the faces of the triangulation are equal to $1 \in G$. (The monodromy is the product of group elements corresponding to the consecutive edges of a face, whatever shape of faces is used. One has to take into account only the orientations of the edges in an obvious way.) Now, it is a standard assertion that the moduli space of (smooth) flat connections on S is isomorphic to the space of flat graph connections modulo graph gauge transformations. (This is in fact nothing but the statement in Čech cohomology that this space is represented by $H^1(S, G)$.) Dealing with a surface with holes amounts to saying that some faces of the triangulation are left empty and one does not have to require anything about the corresponding monodromies. It is important to note that if a graph l is obtained from a triangulation of a surface S, it can be endowed with an additional structure which (together with the graph itself) contains all the information about the topology of the surface. We suppose that S is oriented. The orientation of S induces a cyclic order of the ends of the edges incident to each vertex. A graph l with a given cyclic order at each vertex is called a *fat graph.* If S has at least one hole, the most economical way is to consider a fat graph with all the faces empty, which is always possible. Conversely, given a fat graph l, the corresponding surface can be restored by replacing edges of l by strips glued together at vertices

[1]It is worth looking at [**7**] not only because of the very transparent presentation in it, but also because of the very nice pictures there.

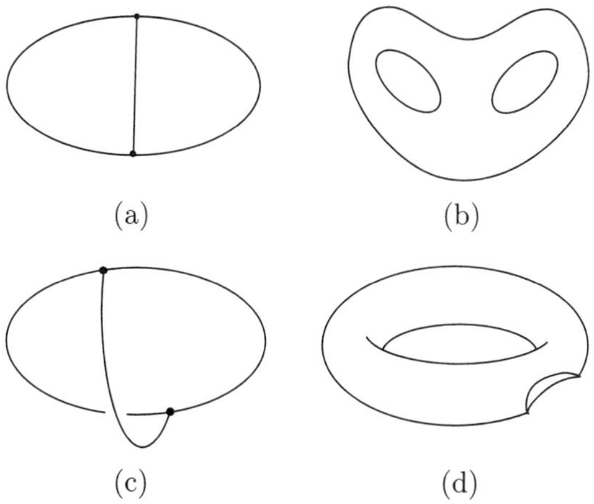

FIGURE 1. Examples of fat graphs and surfaces corresponding to them. The cyclic orders at vertices are understood to be counterclockwise. The graph (a) gives a disk with two holes (b). The graph (c) gives a torus with one hole (d).

respecting the cyclic order (cf. Figure 1). Summarizing, in order to describe the moduli space $\mathcal{M}$ of flat connections on a surface S with holes, we choose a fat graph corresponding to S (this choice is not unique) and consider the quotient of the space of graph connections $\mathcal{A}^l$ by the action of graph gauge transformations, $\mathcal{M} = \mathcal{A}^l/\mathcal{G}^l$.

Having described the moduli space as a manifold, we are now interested in describing its Poisson structure. Let us forget for a moment that we can define a Poisson structure on $\mathcal{M}$ by reduction of the space of all (smooth) connections on S, and try instead to define a Poisson structure on $\mathcal{A}^l$ in such a way that it can be pulled down on $\mathcal{M}$.[2] We would like to have a Poisson structure on $\mathcal{A}^l$ such that the projection $\mathcal{A}^l \to \mathcal{M}$ will be a Poisson map. This can be achieved if $\mathcal{G}^l$ will act on $\mathcal{A}^l$ in a Poisson way (see [**19**] for the definition of Poisson group actions on Poisson manifolds). To do this, we must first define a Poisson–Lie structure on $\mathcal{G}^l$ itself. The group of graph gauge transformations $\mathcal{G}^l$ is the direct product of several copies of G, with one copy for each vertex of l. Let us define the Poisson structure on $\mathcal{G}^l$ as the direct product of Poisson structures on each copy of G in $\mathcal{G}^l$. The latter can be defined independently at each vertex. (To define a Poisson structure on G, one must choose a classical r-matrix.) Now we look for a Poisson structure on $\mathcal{A}^l$. The requirement that the action of $\mathcal{G}^l$ is Poisson is almost sufficient to determine the Poisson structure on $\mathcal{A}^l$. The ambiguity amounts in fact to choosing a linear order of ends of edges at each vertex. Therefore, instead of fat graphs, we must deal with graphs with linear order. Let us call such graphs *ciliated fat graphs*. A ciliated fat graph can be regarded as a fat graph with an additional structure (the fat graph underlying a given ciliated fat one is restored uniquely). This additional structure (linear order at each vertex) can be represented by picturing the underlying fat graph on a sheet of paper in such a way that the

[2]As will be proved in Section 4, in this way we obtain the same Poisson structure as the one defined by the reduction procedure from smooth connections.

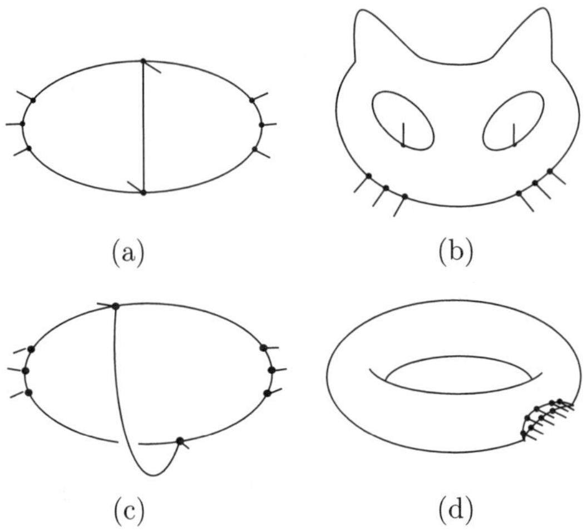

FIGURE 2. Examples of ciliated fat graphs and of the corresponding ciliated surfaces. Cilia are indicated by small strokes at the vertices. The graph (a) gives a disk with two holes (b). The graph (c) gives a torus with one hole (d).

cyclic order is everywhere, say, counterclockwise and by placing at each vertex a small cilium separating the minimal and the maximal end incident to that vertex. As we mentioned, a fat graph defines a surface, more precisely an oriented surface with holes (Figure 1); a ciliated fat graph, similarly, defines a ciliated surface, that is an oriented surface with holes and with some points marked on the boundary (Figure 2). Thus for every ciliated fat graph we have an associated Poisson manifold, namely the space of graph connections endowed with an r-matrix Poisson structure. Of course it may happen that two different ciliated graphs give isomorphic Poisson manifolds of graph connections. In particular, one can show that the isomorphism class of the arising Poisson manifold depends only on the diffeomorphism class of the corresponding ciliated surface.

It may be worth mentioning some distinguished examples of graphs and corresponding Poisson manifolds. The Poisson manifold corresponding to the graph consisting only of two vertices and one edge (Figure 3a) coincides with the Poisson–Lie group G, provided that the r-matrices chosen at the vertices are related by the operation of permutation of tensor factors ($r = r_{12} \mapsto r_{21}$). With the same condition on r-matrices, the graph consisting of two vertices and two edges connecting them (Figure 3b) yields the manifold $G \times G$ endowed with a Poisson–Lie structure coinciding with that of the double $D \simeq G \times G$. If we take the same r-matrices at two vertices, we get D_+ as our Poisson manifold (see [**19**] for the definitions of doubles). Finally, the graph consisting of one vertex and one edge (Figure 3c) corresponds to the Poisson manifold G^*, the dual Poisson–Lie group.

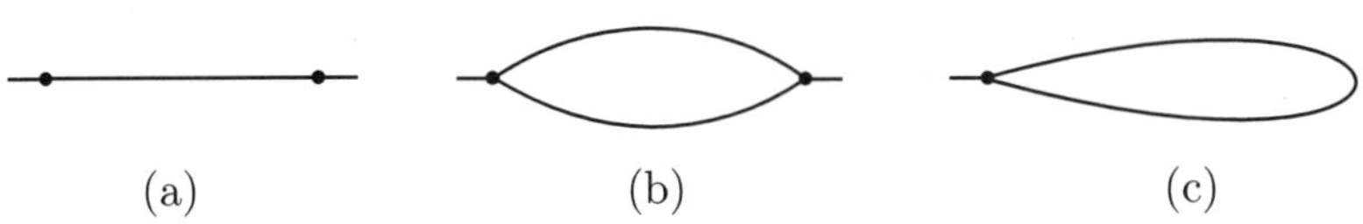

FIGURE 3. The graphs corresponding to (a) the Poisson–Lie group G, (b) its double $D \simeq G \times G$, (c) its dual Poisson–Lie group G^*.

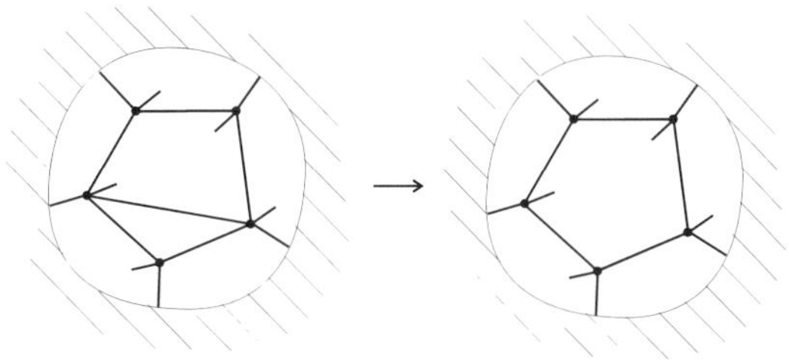

FIGURE 4. Operation of erasing an edge. The shaded region represents the remainder of the graph.

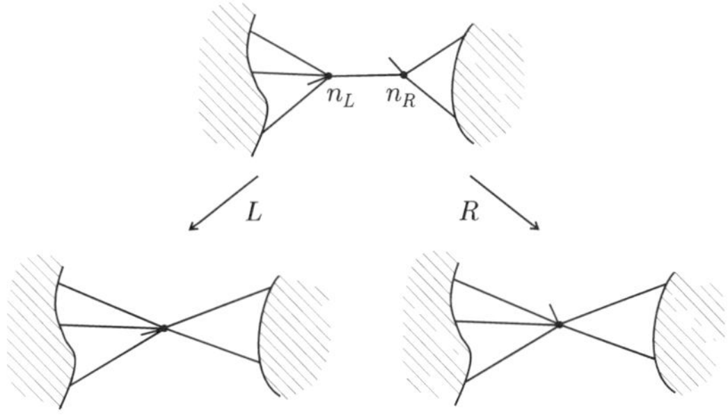

FIGURE 5. Operations of contractions of an edge. L and R are the two different ways of contraction. L corresponds to factoring by gauge transformations at the vertex n_R. R corresponds to factoring by gauge transformations at the vertex n_L.

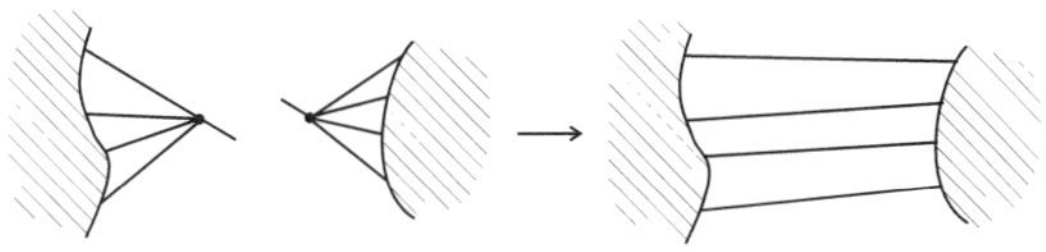

FIGURE 6. Operation of gluing graphs.

The following operations with graphs are important to discuss: (i) erasing an edge (Figure 4), (ii) contracting an edge (Figure 5), (iii) gluing graph(s) (Figure 6), and (iv) adding a loop (see Section 4). The linear orders at the vertices affected by such an operation descend from those of the original graph in a more or less obvious way (cf. Figures 4–6). We must only mention that there are in fact two ways to contract an edge, which differ in what happens to the cilia. The operation of gluing deserves some explanation. Given two vertices on a graph with the same number N of ends of edges incident to them, we can form a new graph by erasing both vertices and gluing together the liberated edges. (The kth end liberated at one vertex is to be glued to the $(N-k)$th end at the other vertex.) Note that with the help of this operation one can glue together two different graphs, obtaining a single new one.

For the operations on graphs just described, there exist natural maps between the corresponding spaces of graph connections. These maps are in fact projections in directions shown by the arrows in Figures 4–6. A pleasant feature is that these

(a)

(b)

(c)

(d)

(e)

(f)

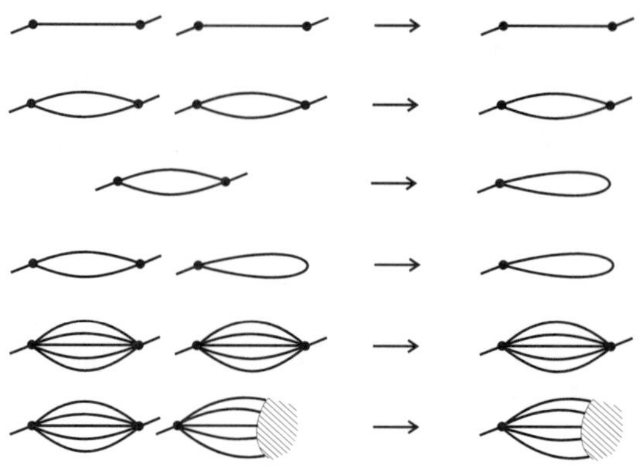

FIGURE 7. Some particular cases of gluing graphs which correspond to natural operations in Poisson–Lie groups: (a) multiplication in G, (b) multiplication in D, (c) projection $D \to G^*$, (d) action of D on G^*, (e) multiplication in the 5-uble, (f) action of the 5-uble on a space of graph connections.

maps turn out to be Poisson maps. More precisely, in the case of gluing, one must require that the r-matrices at the two vertices to be glued be related by a permutation of the tensor factors. Consider, for instance, the map corresponding to gluing together two simplest graphs (Figure 7a), each of which represents the Poisson–Lie group G (an edge with two vertices). The result of gluing is again a graph of the same shape, while the corresponding map of graph connections, $G \times G \to G$, is simply the group product, which is known to be a Poisson map. Similarly, gluing together the graphs representing D gives the Poisson map $D \times D \to D$ (Figure 7b) corresponding to the group multiplication. Contracting one of two edges of the graph D (Figure 7c), one obtains the Poisson map $D \to G^*$. As a Poisson manifold, the dual group G^* can be identified with the coset D/G_Δ, where G_Δ is the diagonal subgroup in $D \simeq G \times G$ (cf. [**19**]). The isomorphism of G^* and the coset D/G_Δ shows that there is a Poisson action of D on G^*, i.e., a Poisson map $D \times G^* \to G^*$, which again can be described by gluing graphs (as shown in Figure 7d). Looking at the above pictures suggests the following generalization of the notion of double. Namely, we can define a Poisson–Lie group, called in general a *polyuble*,[3] by the ciliated fat graph consisting of two vertices and several edges connecting them (as in the case of the double, the r-matrices at the two vertices must be related by the operation of permutation of tensor factors, while the order of ends should be opposite; Figure 7e). An immediate observation is that on the space of graph connections $\mathcal{A}^l$ for an arbitrary ciliated fat graph l there is a Poisson action of the polyuble $P_n(G)$ adjusted to each vertex, where n is the number of legs at that vertex (see Figure 7f). Thus the space $\mathcal{A}^l$ is a homogeneous space for the group P^l, which is the direct product (in the sense of Poisson groups) of the polyubles $P_n(G)$. Note also that the group $\mathcal{G}^l$ of graph gauge transformations that gives us the moduli space $\mathcal{M} = \mathcal{A}^l/\mathcal{G}^l$ is a Poisson subgroup in P^l. (Any individual polyuble, disregarding for the moment the Poisson structure,

[3]We dedicate the Poisson–Lie groups of this type to I. V. Polyubin.

is the product $G \times \cdots \times G$ and contains the diagonal subgroup, which turns out to be a Poisson subgroup.)

Finally, it is worth mentioning that some particular cases of Poisson manifolds defined by graphs have been considered in the literature. Namely, the Poisson manifold of graph connections on a graph corresponding to the boundary of a polygon was suggested in [**19**] as a discrete approximation of the current algebra coadjoint space. (See also [**2, 1**], where this discrete approximation was used to investigate the WZNW conformal model.)

3. Poisson structure of moduli spaces

In this section we shall describe a Poisson structure on the space of flat connections modulo gauge transformations on Riemann surfaces with holes by means of a reduction of the space of all smooth connections on them.

Let S be an oriented compact Riemann surface with holes. Let $\mathcal{A}$ be the space of smooth G-connections on it, where G is a reductive complex Lie group whose Lie algebra $\mathfrak{g}$ is endowed with a chosen nondegenerate invariant quadratic form, which we denote by tr. The space $\mathcal{A}$ is in a natural way a symplectic manifold with the symplectic structure

$$\Omega = \int_S \operatorname{tr}(\delta A \wedge \delta A), \tag{3.1}$$

where $A \in \mathcal{A}$ is a $\mathfrak{g}$-valued 1-form on S, δ is the external differential on $\mathcal{A}$, and $\wedge$ is a shorthand way to denote the wedge product both on $\mathcal{A}$ and on S. This symplectic structure is well known to be invariant with respect to the gauge transformations

$$A \mapsto g^{-1} A g + g^{-1} dg, \tag{3.2}$$

where g is a G-valued function on S.

Now let us try to define the momentum mapping for this action. One can easily check that the infinitesimal gauge transformation ϵ is generated by the Hamiltonian function

$$H_\epsilon = \int_S \operatorname{tr}(\epsilon(dA + A \wedge A)) + \int_{\partial S} \operatorname{tr}(\epsilon A). \tag{3.3}$$

The Hamiltonian generating the given transformation is defined only up to an additive constant, and therefore the Poisson brackets between them, in general, reproduce the commutation relations between the elements of the gauge algebra only up to a cocycle:

$$\{H_{\epsilon_1}, H_{\epsilon_2}\} = H_{[\epsilon_1, \epsilon_2]} + c(\epsilon_1, \epsilon_2). \tag{3.4}$$

In our case

$$c(\epsilon_1, \epsilon_2) = \int_{\partial S} \operatorname{tr}(\epsilon_1 d\epsilon_2). \tag{3.5}$$

One can prove that this cocycle is nontrivial, and therefore we can define the momentum mapping not for the algebra of gauge transformations itself, but only for its central extension by the 2-cocycle (3.5).

Let $\mathfrak{g}^S$ denote the algebra of gauge transformations centrally extended by (3.5) and let $\mathcal{G}^S$ be the corresponding group. The space $\mathfrak{g}^S$ is the space of pairs (ϵ, z), where ϵ is a $\mathfrak{g}$-valued function and z is a complex number. Let us consider the space $(\mathfrak{g}^S)^*$ consisting of triples (R, B, x), where R is a $\mathfrak{g}$-valued 2-form on S, B is a $\mathfrak{g}$-valued 1-form on the boundary of S and x is a complex number. There is a nondegenerate pairing $\langle\ ,\ \rangle$ between $\mathfrak{g}^S$ and $(\mathfrak{g}^S)^*$,

$$\langle (R, B, x), (\epsilon, z)\rangle = \int_S \operatorname{tr}(\epsilon R) + \int_{\partial S} \operatorname{tr}(\epsilon B) + zx. \tag{3.6}$$

The momentum map for the action of $\mathfrak{g}^S$ can be defined now as the mapping $\mathcal{A} \to (\mathfrak{g}^S)^*$ given by the curvature and by the restriction of the connection form to the boundary:

$$A \mapsto (dA + A \wedge A, A|_{\partial S}, 1). \tag{3.7}$$

Now consider the Hamiltonian reduction of $\mathcal{A}$ with respect to $\mathcal{G}_0^S$, the group of gauge transformations that are equal to the identity on the boundary and yield the space of flat connections on S modulo gauge transformations from $\mathcal{G}_0^S$,

$$\mathcal{M}_0 = \{A \in \mathcal{A} \mid dA + A \wedge A = 0\}/\mathcal{G}_0^S. \tag{3.8}$$

The space $\mathcal{M}_0$ can also be regarded as the space of values of flat connections restricted to the boundary. It is well known that the space of G-connections on the circle can be identified with the coadjoint space of the affine Kac–Moody algebra with the standard Kirillov–Kostant Poisson structure. The following proposition shows that these two Poisson structures are related:

PROPOSITION 3.1. *The mapping from the space $\mathcal{M}_0$ to the Kac–Moody coadjoint representation space sending flat connections on the Riemann surface S to their restrictions to a component of the boundary is a Poisson mapping.*

PROOF. This mapping is essentially the momentum mapping for the action of gauge transformations. □

Now let us consider the quotient of the space $\mathcal{M}_0$ by the whole group $\mathcal{G}^S$ (the group $\mathcal{G}^S$ acts on $\mathcal{M}_0$ because the subgroup $\mathcal{G}_0^S$ of gauge transformations equal to the identity on the boundary is normal in $\mathcal{G}^S$). The quotient space,

$$\mathcal{M} = \{A \in \mathcal{A} \mid dA + A \wedge A = 0\}/\mathcal{G}_1^S, \tag{3.9}$$

is a finite-dimensional Poisson manifold. Its symplectic leaves are in one-to-one correspondence with the coadjoint orbits of the centrally extended group of gauge transformations, which in turn are parameterized by the conjugacy classes of monodromies around the holes. Thus we have

PROPOSITION 3.2. *The space $\mathcal{M}$ of flat G-connections modulo gauge transformations on a Riemann surface with holes inherits a Poisson structure from the space of all (smooth) G-connections. The symplectic leaves of this structure are parameterized by the conjugacy classes of monodromies around holes.*

4. Graph connections

In this section we shall construct a Poisson structure on the space $\mathcal{A}^l$ of graph connections in such a way that the lattice gauge group $\mathcal{G}^l$ endowed with a nontrivial r-matrix Lie–Poisson structure acts on $\mathcal{A}^l$ in a Poisson way.

Let l be a ciliated fat graph, which is homotopy equivalent to a Riemann surface S with holes. Denote by $E(l)$ the set of ends of edges of l and by $N(l)$ the set of its vertices. Each element of $N(l)$ corresponds to the subset of $E(l)$ of ends of edges incident to a given vertex. In what follows we shall identify elements of $N(l)$ with the corresponding subsets. The mapping which sends each end α of each edge to the opposite end $\alpha^\vee$ of the same edge is an involution of the set $E(l)$. The ciliated fatness of l defines an ordering inside each $n \in N(l)$. One can easily see that such data (a set divided into ordered subsets and an involution on it without fixed points) unambiguously define a ciliated fat graph. Let $[\alpha]$ be the vertex containing α and $[\alpha, \alpha^\vee]$ be the edge linking α and $\alpha^\vee$.

By a *graph connection* on a graph l we mean an assignment of an element $\mathbf{A}_\alpha$ of the group G to each $\alpha \in E(l)$ such that[4]

$$\mathbf{A}_{\alpha^\vee} = \mathbf{A}_\alpha^{-1}. \tag{4.1}$$

The lattice gauge group $\mathcal{G}^l$ is a product of finite dimensional groups G—one copy for each vertex of the graph. The group $\mathcal{G}^l$ acts on $\mathcal{A}^l$ in a natural way:

$$\mathbf{A}_\alpha \mapsto \mathbf{g}_{\alpha^\vee}^{-1} \mathbf{A}_\alpha \mathbf{g}_\alpha. \tag{4.2}$$

The space of graph connections can be regarded as the quotient space of the space of flat connections on a surface S. Indeed, let us take the surface S corresponding to the graph l and embed the graph into it so that S is contractible to the image of l. Then for a (smooth) connection A on S, we can construct a graph connection on l by assigning to $\alpha \in E(l)$ the parallel transport operator along the edge linking $\alpha^\vee$ and α. This graph connection does not change if we transform the connection A by a gauge transformation equal to the identity at the vertices. It is clear that every graph connection can be continuously extended to the surface, and therefore the space of graph connections $\mathcal{A}^l$ can be represented as the quotient

$$\mathcal{A}^l \cong \{A \in \mathcal{A} \mid dA + A \wedge A = 0\}/\mathcal{G}_1^S, \tag{4.3}$$

where $\mathcal{G}_1^S$ is the group of gauge transformations equal to the identity at the vertices. Of course, this representation is defined only up to the action of the graph gauge group, and therefore the isomorphism between the spaces $\mathcal{M}$ and $\mathcal{A}^l/\mathcal{G}^l$ is canonical.

This isomorphism shows that although the space $\mathcal{A}^l$ has so far no *a priori* Poisson structure, the space $\mathcal{A}^l/\mathcal{G}^l$ has one. Our aim is to introduce a Poisson structure on $\mathcal{A}^l$ compatible with that on $\mathcal{A}^l/\mathcal{G}^l$ and with the graph gauge group action.

For each vertex n of the graph let us fix a classical r-matrix $r(n) \in \mathfrak{g} \otimes \mathfrak{g}$, that is to say, a solution of the classical Yang–Baxter equation:

$$[r_{12}(n), r_{13}(n)] + [r_{12}(n), r_{23}(n)] + [r_{13}(n), r_{23}(n)] = 0 \tag{4.4}$$

such that

$$\frac{1}{2}(r_{12}(n) + r_{21}(n)) = t, \tag{4.5}$$

where $t \in \mathfrak{g} \otimes \mathfrak{g}$ is the quadratic Casimir element:

$$t = \sum e_i \otimes e_i, \tag{4.6}$$

[4]Perhaps it would be more natural to assign a group element to each edge, as we did in Section 2 above, rather than to each end of each edge. However, then we would have to choose some orientations of the edges. For this we would need a definition of the Poisson manifold $\mathcal{A}^l$ which would depend on an *oriented* ciliated fat graph. In such a case, it would be possible to prove that two Poisson manifolds corresponding to two graphs differing only by their orientations are isomorphic. We prefer to get rid of this complication, at the price of a slightly more complicated notation.

where $\{e_i\}$ is an orthonormal basis in $\mathfrak{g}$.[5]

Let us define a bivector field B on $\mathcal{A}^l$ as

$$B = \sum_{n\in N(l)} \left(\sum_{\alpha,\beta\in n;\,\alpha<\beta} r^{ij}(n)\, X_i^\alpha \wedge X_j^\beta + \frac{1}{2}\sum_{\alpha\in n} r^{ij}(n)\, X_i^\alpha \wedge X_j^\alpha \right), \tag{4.7}$$

where $X_i^\alpha = L_i^\alpha - R_i^{\alpha^\vee}$, L_i^α and R_i^α are, respectively, the left- and right-invariant vector fields corresponding to the element $e_i \in \mathfrak{g}$ on the group assigned to $\alpha \in E(l)$, while $r^{ij}(n)$ is the r-matrix at the vertex n written in the basis $\{e_i\}$. Note that the vector fields X_i^α are chosen to be consistent with (4.1).

PROPOSITION 4.1. a) *The bivector B defines a Poisson structure on $\mathcal{A}^l$.*

b) *The group $\mathcal{G}^l$ endowed with the direct product Poisson–Lie structure acts on $\mathcal{A}^l$ in a Poisson way.*

The proof can be obtained by a straightforward verification.

However, sometimes it is more convenient to use other ways of presenting the Poisson bivector (4.7). If one separates explicitly the symmetric part t and skew-symmetric part $r_a = (r_{12} - r_{21})/2$ of the r-matrix, so that $r = r_a + t$, one obtains

$$B = \sum_n \left(r_a^{ij}(n) X_i^\Delta(n) \otimes X_j^\Delta(n) + \sum_{\alpha,\beta\in n} (n,\alpha,\beta) \sum_i X_i^\alpha \otimes X_i^\beta \right), \tag{4.8}$$

where $X_i^\Delta(n) = \sum_{\alpha\in n} X_i^\alpha$ and

$$(n,\alpha,\beta) = \begin{cases} 1, & \alpha > \beta, \\ 0, & \alpha = \beta, \\ -1, & \alpha < \beta, \end{cases} \qquad \text{for } \alpha,\beta \in n. \tag{4.9}$$

Since the vectors $X_i^\Delta(n)$ are tangent to $\mathcal{G}^l$-orbits, one sees that the Poisson bracket induced by Eq. (4.7) on the quotient $\mathcal{M} = \mathcal{A}^l/\mathcal{G}^l$ does not change if the skew-symmetric part r_a of the r-matrix is changed.

Another way of defining the Poisson structure is by giving explicit expressions for the Poisson brackets between matrix elements of $\mathbf{A}_\alpha$ in some representation of the group G; we consider these matrix elements as functions on $\mathcal{A}^l$. (We shall denote matrices representing $\mathbf{A}_\alpha$ and r by the same symbols, $\mathbf{A}_\alpha$ and r, respectively.) We have:

$$\{\mathbf{A}_\alpha \otimes \mathbf{A}_\alpha\} = r_a(1)\,(\mathbf{A}_\alpha \otimes \mathbf{A}_\alpha) + (\mathbf{A}_\alpha \otimes \mathbf{A}_\alpha)\, r_a(2) \tag{4.10}$$

for the case $[\alpha] \neq [\alpha^\vee]$ (here $r(1) = r([\alpha])$, $r(2) = r([\alpha^\vee])$);

$$\begin{aligned} \{\mathbf{A}_\alpha, \mathbf{A}_\alpha\} = {} & r_a\,(\mathbf{A}_\alpha \otimes \mathbf{A}_\alpha) + (\mathbf{A}_\alpha \otimes \mathbf{A}_\alpha)\, r_a \\ & + (1 \otimes \mathbf{A}_\alpha)\, r_{21}\, (\mathbf{A}_\alpha \otimes 1) - (\mathbf{A}_\alpha \otimes 1)\, r\, (1 \otimes \mathbf{A}_\alpha) \end{aligned} \tag{4.11}$$

for the case $[\alpha] = [\alpha^\vee]$, $\alpha < \alpha^\vee$;

$$\{\mathbf{A}_\alpha \otimes \mathbf{A}_\beta\} = r\,(\mathbf{A}_\alpha \otimes \mathbf{A}_\beta) \tag{4.12}$$

for the case $[\alpha] = [\beta] \neq [\alpha^\vee] \neq [\beta^\vee] \neq [\alpha]$, $\alpha < \beta$;

$$\{\mathbf{A}_\alpha, \mathbf{A}_\beta\} = r(1)\,(\mathbf{A}_\alpha \otimes \mathbf{A}_\beta) + (\mathbf{A}_\alpha \otimes \mathbf{A}_\beta)\, r(2) \tag{4.13}$$

[5] Note that although the r-matrix, $r(n)$, is allowed to differ for different vertices, its symmetric part, t, is required to be the same everywhere.

for the case $[\alpha] = [\beta] \neq [\alpha^\vee] = [\beta^\vee]$, $\alpha < \beta$, $\alpha^\vee < \beta^\vee$, $r(1) = r([\alpha])$, $r(2) = r([\alpha^\vee])$; and

$$(4.14)\quad \{\mathbf{A}_\alpha, \mathbf{A}_\beta\} = r\,(\mathbf{A}_\alpha \otimes \mathbf{A}_\beta) + (\mathbf{A}_\alpha \otimes \mathbf{A}_\beta)\,r \\ + (1 \otimes \mathbf{A}_\beta)\,r_{21}\,(\mathbf{A}_\alpha \otimes 1) - (\mathbf{A}_\alpha \otimes 1)\,r\,(1 \otimes \mathbf{A}_\beta)$$

for the case $[\alpha] = [\beta] = [\alpha^\vee] = [\beta^\vee]$ and $\alpha < \beta < \alpha^\vee < \beta^\vee$. Unfortunately, the complete list of all possible configurations of one or two edges and cilia is rather long (there are fourteen of them), and we stop here. The reader can easily observe how to write out expressions for other configurations in a similar way.

As described in Section 2, there exist such operations on ciliated fat graphs as erasing an edge, contracting an edge to a vertex, gluing two vertices of the same valence, and adding a loop. One can also change a graph to another one corresponding to the same ciliated surface. All these transformations induce mappings between the corresponding spaces of graph connections. Let us now describe them explicitly.

Erasing an edge (Figure 4). This operation is the most obvious one. The mapping between graph connections is given simply by forgetting the group element assigned to the edge to be erased.

Contracting an edge (Figure 5). This operation can be applied to an edge with distinct ends (i.e., $[\alpha] \neq [\alpha^\vee]$). Let α be an end of such an edge. (In Figure 5, it is the right one for the projection R and the left one for L.) Take $\mathbf{A}_\alpha$ equal to the identity by applying a gauge transformation (that is, the action of one copy of G) at the vertex $[\alpha^\vee]$. Erase the cilium at the vertex $[\alpha^\vee]$. Then contract the edge $[\alpha, \alpha^\vee]$, leaving the group elements on the other edges unchanged (as they were after the above gauge transformation).

Note that, as shown in Figure 5, this operation depends not only on the edge, but also on the choice of a specific extremity of the edge. To emphasize this, we say that we contract the edge *towards* a vertex, in our case $[\alpha]$.

Gluing two vertices (Figure 6). This operation can be applied to two vertices n and n' having the same valence, i.e., $|n| = |n'|$, and such that their r_a-matrices are opposite, i.e., $r_a(n) = -r_a(n')$. Disconnect the ends of the edges at the vertices and connect them in the order prescribed by the gluing (Figure 6), inserting an arbitrarily ciliated 2-valent vertex at each connection. Until now we left the group elements on the edges unchanged. Now take each inserted 2-vertex and contract towards it one of the two incident edges.

Adding a loop. One can add a loop (an edge $[\alpha, \alpha^\vee]$ with $[\alpha = \alpha^\vee]$ to a vertex between two consecutive ends of edges. Assign the unit group element to the new loop.

Ciliated graphs and ciliated surfaces. As we mentioned several times above, a graph embedded into an oriented surface inherits a fatness (cyclic order of ends of edges at vertices). Assume now that we have a graph embedded into a surface in such a way that the vertices are mapped into the boundary. This graph inherits a ciliated fatness, since there is a canonical linear order of the ends of edges meeting at a boundary point. On the other hand, given a ciliated fat graph embedded into the corresponding surface (that is, assuming that the surface is retractable to the image of the graph), there exists a unique way (up to isotopy) to move its vertices to the boundary so as to reproduce the given ciliation. We just move each vertex to the boundary component which the cilium looks into. If we now erase the edges of the graph and leave the cilia to stick out off the boundary components, we get a *ciliated surface* (e.g., Figures 2b, 2d).

Suppose now that we have two ciliated fat graphs l and l' corresponding to the same ciliated surface. (This means, in particular, that their vertices are identified.) We are going to construct an isomorphism $\mathcal{A}^l \xrightarrow{\sim} \mathcal{A}^{l'}$ between the spaces of graph connections on them. Let $\alpha \in E(l')$ be an end of an edge of l'. Take the edge $[\alpha, \alpha^\vee]$ of l and retract it to the graph l'. We obtain a path on the graph l joining the vertices $[\alpha]$ and $[\alpha^\vee]$ and isotopic to the edge $[\alpha, \alpha^\vee]$. Assign to α the monodromy of the graph connection $\mathcal{A}^l$ along this path. Carrying out this procedure for all $\alpha \in E(l')$, we obtain the desired isomorphism.

Now let us summarize some properties of the spaces of graph connections equipped with the Poisson bracket (4.7).

PROPOSITION 4.2. 1) *The mappings between graph connections corresponding to erasing an edge, contracting an edge towards a vertex, and gluing two vertices are Poisson projections onto the image.*

2) *The isomorphism of the spaces of graph connections for two graphs corresponding to isomorphic ciliated surfaces is an isomorphism of Poisson manifolds.*

3) *Adding a loop is a Poisson embedding.*

The proof of the proposition is a straightforward and not very complicated explicit calculation that we omit here. Let us mention only the following statement, useful for the proof as well as by itself.

Let f be a face of a ciliated fat graph l such that there are no cilia looking into f. Let $\mathcal{A}^l(h, f)$ be the set of graph connections with the monodromy around the face f of l conjugate to $h \in G$. Then $\mathcal{A}^l(h, f)$ is a Poisson submanifold in $\mathcal{A}^l$.

Let us now proceed to the relation between the space of graph connections and the space of ordinary connections.

PROPOSITION 4.3. *The quotient of the space of graph connections by the graph gauge group is isomorphic as a Poisson manifold to the quotient of the space of flat connections on the corresponding Riemann surface by the gauge group, i.e.,*

$$\mathcal{A}^l/\mathcal{G}^l \cong \mathcal{M}. \tag{4.15}$$

REMARK 4.1. Let us note that this statement shows that all the ambiguities in the construction of the space $\mathcal{A}^l$, such as choices of ordering and of r-matrices, do not influence the Poisson structure of its quotient by the gauge group. The latter depends only on the cyclic order and on the symmetric part t of the r-matrices (cf. (4.5)). This cannot be otherwise, because these are just the data defining the Poisson manifold $\mathcal{M}$ by (3.1), provided the surface S there is defined by the ciliated fat graph l here and the invariant scalar product tr there is defined by the Casimir element t here. However, it is impossible to introduce a Poisson structure on $\mathcal{A}^l$ compatible with that on the gauge quotient without fixing nontrivial r-matrices. Note also that topologically these moduli spaces are always isomorphic to products of several copies of the group G modulo the overall G-conjugation, although they are not isomorphic to each other as Poisson manifolds. For example, a sphere with three holes and a torus with one hole give topologically the same spaces, $(G \times G)/\operatorname{Ad} G$, while the Poisson structure for, e.g., $G = SL(2)$ is trivial in the first case and nontrivial in the second.

REMARK 4.2. The description of the moduli space $\mathcal{M}$ of flat connections in the language of graphs has the advantage that it allows us to describe the space of functions on $\mathcal{M}$ rather explicitly by using representation theory. In particular, one can construct a linear basis in the space of regular functions on $\mathcal{M}$ in the following way.

Assign an irreducible representation π_α of G in the space V_α to each $\alpha \in E(l)$ in such a way that $\pi_{\alpha^\vee} = \pi_\alpha^*$, and assign an intertwiner $C_n \in \mathrm{Inv}(\bigotimes_{\alpha\in n} V_\alpha^*)$ to each vertex n. We can consider matrices from $\mathrm{End}\, V_\alpha$ as belonging to $V = \bigotimes_{\alpha\in E(l)} V_\alpha$ and the intertwiners C_n as belonging to its dual, V^*.

For each such data $(l, C_\bullet, \pi_\bullet)$ we can define a function $\psi(l, C_\bullet, \pi_\bullet)$ on $\mathcal{A}^l$ by

$$\psi(l, C_\bullet, \pi_\bullet)(\{\mathbf{A}_\alpha\}) = \left\langle \bigotimes_n C_n, \bigotimes_{\alpha\in E_1(l)} \pi_\alpha(\mathbf{A}_\alpha)\right\rangle, \tag{4.16}$$

where $E_1(l) \subset E(l)$ is a set of ends of edges containing exactly one end of each edge. The ambiguity in the choice of this set is unessential, because $\pi_{\alpha^\vee}(\mathbf{A}_{\alpha^\vee}) = \pi_\alpha(\mathbf{A}_\alpha)$ as an element of $V_\alpha \otimes V_\alpha^*$.

One can easily verify that all such functions are $\mathcal{G}^l$ invariant and that they indeed form a complete set of functions on $\mathcal{M}$. The latter is an obvious consequence of the Peter–Weyl theorem.

PROOF OF PROPOSITION 4.3. The Poisson bracket of two functions Ψ and Φ on the space $\mathcal{A}$ of smooth connections on S can be written as

$$\{\Psi, \Phi\}_S = \int_S \mathrm{tr}\left(\frac{\delta\Psi}{\delta A} \wedge \frac{\delta\Phi}{\delta A}\right). \tag{4.17}$$

To prove the proposition, we must compute the Poisson bivector on $\mathcal{M}$ induced by (4.17) by the reduction procedure described in Section 3 and compare the result with the bivector induced by (4.7).

In order to be able to work with the bracket (4.17) and build a bridge between the smooth and the combinatorial approaches to the Poisson brackets on flat connections, let us first compute the Poisson bracket using the formula (4.17) in one particular case. Let I_1 and I_2 be two oriented intervals embedded into S and intersecting transversally. Using (4.17), let us compute the Poisson bracket between two arbitrary functions Ψ and Φ of the corresponding monodromies, considered as functions on $\mathcal{A}$.

The result of the computation is a function on the space $\mathcal{A}$. However, for our further purposes, we need to compute only the restriction of the result to the connections whose restrictions to the intervals vanish everywhere except for two subintervals containing the ends of I_1 and I_2 and none of their intersection points. In this case the expression for the bracket is especially simple; it depends only on the monodromies along the segments and can be straightforwardly computed from Eq. (4.17):

$$\{\Psi, \Phi\} = t^{ij}(R_i\Psi)(R_j\Phi) \sum_{k\in I_1\cap I_2} \varepsilon(k). \tag{4.18}$$

Here k runs over the intersection points, $\varepsilon(k)$ is 1 or -1 if the first segment crosses the second one from the left or from the right, respectively, $\{R_i\}$ is a basis of the left-invariant vector fields on G and t^{ij} is the matrix of the quadratic Casimir $t \in \mathfrak{g} \otimes \mathfrak{g}$ (e.g., of the one given by (4.5)).

Note that this formula does not give Poisson brackets between functions of monodromy along a single segment. Moreover, formula (4.17) cannot be used to compute such brackets.

Let us recall the definition of the Hamiltonian reduction in the language of Poisson brackets. Let M be a symplectic manifold with symplectic action of the group G, let μ be a momentum map (corresponding to the action of G or of a subgroup of G, $\mathsf{M}_0 = \mu^{-1}(0)$, let $\mathsf{N} = \mathsf{M}_0/\mathsf{G}$ be the reduced space, and let $\pi\colon \mathsf{M}_0 \to \mathsf{N}$ be the canonical projection. The Poisson bracket of two functions Ψ and Φ on N is defined as follows. Let Ψ^* and Φ^* be any two functions on M such that their restrictions to M_0 coincide with $\pi^*\Psi$ and $\pi^*\Phi$ respectively. Then $\{\Psi, \Phi\}(\mathsf{x}) := \{\Psi^*, \Phi^*\}(\mathsf{y})$, for any $\mathsf{x} \in \mathsf{N}$ and any $\mathsf{y} \in \pi^{-1}(\mathsf{x})$. (Note that this procedure includes at least three arbitrary choices: the choice of the functions Ψ^* and Φ^* for given Ψ and Φ and the choice of y for a given x. We are going to make these choices in a way that maximally simplifies the calculations.)

This definition can be applied to our situation. We take the space $\mathcal{A}$ of all connections for M, the space $\mathcal{A}^l$ of flat connections for M_0, and the moduli space $\mathcal{M}$ for N. Our task is to compute the Poisson bracket on $\mathcal{M}$ or, equivalently, between $\mathcal{G}^l$-invariant functions on $\mathcal{A}^l$, and then to compare the result with the one given by (4.7).

Let Ψ and Φ be two arbitrary functions and let $\mathbf{l}$ and $\mathbf{l}'$ be two embeddings of the graph l into the surface such that the images of vertices are disjoint and the images of edges are transversal. Using the mappings $\mathcal{A} \to \mathcal{A}^l$ given by the monodromies along the edges, we can lift Ψ and Φ using $\mathbf{l}$ and $\mathbf{l}'$ respectively to $\mathcal{G}^S$-invariant functions Ψ^* and Φ^* on $\mathcal{A}$.

Now to compute the bracket between Ψ^* and Φ^* we need only to apply the lemma to all intersecting edges of $\mathbf{l}$ and $\mathbf{l}'$. To simplify the computations, we can choose a convenient pair of graph embeddings as well as a convenient flat connection within the given $\mathcal{G}^l$-orbit. (In fact, we need two embeddings $\mathbf{l}$ and $\mathbf{l}'$ since formula (4.17) is not applicable for computing brackets between functions given by one embedding.)

Fix a ciliation on l and embed l in the surface so that all vertices map to the boundary and all cilia look outside the surface. Thus we get our first embedding $\mathbf{l}$. To get the second embedding $\mathbf{l}'$, deform the embedding $\mathbf{l}$ in order to make the edges of $\mathbf{l}$ and $\mathbf{l}'$ transversal and formula (4.18) applicable. Fix a point at the middle of each edge. Then move each vertex along the boundary component a little to the left (if viewed from the outside) together with incident edges keeping the middle points stable and making the number of intersection points between deformed and initial edges as low as possible. Such a deformation is illustrated in Figure 8.

We have one intersection point for any two ends of edges $\alpha \in E(\mathbf{l})$ and $\beta \in E(\mathbf{l}')$ belonging to the same vertex. Let us say that these intersection points are *associated* to this vertex. There is also one intersection point at the middle of each edge, which we associate in an arbitrary way to one of the vertices of the corresponding edge.

Now let us choose a convenient connection within a given $\mathcal{G}^l$-orbit. One can fix a disjoint collection of patches around each vertex of $\mathbf{l}$ in such a way that each patch contains the corresponding vertex of $\mathbf{l}'$ as well as all the segments of edges between these vertices and the intersection points associated to them. Since the patches are disjoint and topologically trivial, one can take the connection on them to be zero.

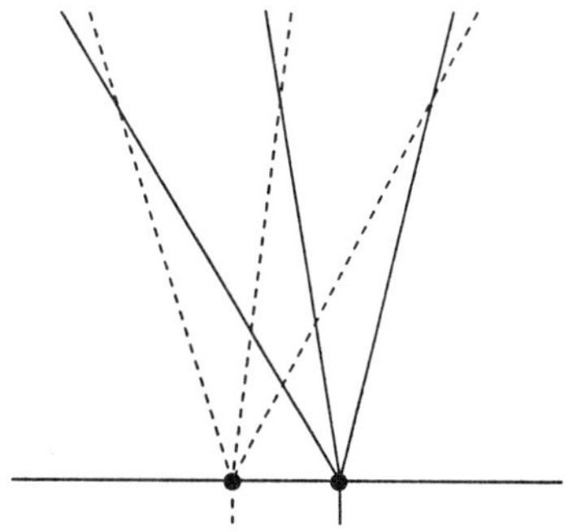

FIGURE 8. A special way of deforming a graph drawn on a surface yielding two transversal graphs; the deformed graph is shown by the dotted line.

Note that since we have chosen the connection to be trivial around the vertices, we can apply formula (4.18). Note also that the intersection points at the middle of the edges give trivial contributions. Finally we get an expression for a bivector giving the Poisson bracket of Ψ and Φ as the following sum over all other intersection points:

$$B = \sum_n \Big(\sum_{i;\, \alpha,\beta \in n;\, \alpha<\beta} X_i^\alpha \wedge X_i^\beta \Big), \tag{4.19}$$

which coincides with (4.8) up to terms vanishing on $\mathcal{G}^l$-invariant functions. □

In this section we described a Poisson structure on $\mathcal{A}^l$, which gave us a description of the Poisson structure on $\mathcal{M}$ as well. As stated in Proposition 3.2 above, it is also possible to characterize the symplectic leaves in $\mathcal{M}$. However, it might be useful to have an explicit description of the symplectic structure on those leaves. For such a description, we refer to the paper by A. Alekseev and A. Malkin [**5**]; see also their work [**4**], where a useful description of the symplectic structure on the symplectic leaves in Poisson–Lie groups is given.

Appendix: Ruijsenaars equations

In this Appendix we describe the geometric meaning of the trigonometric Ruijsenaars Hamiltonian integrable system [**18**], see (A.24) below. This system is a generalization of several integrable systems, such as the rational and trigonometric Calogero systems, rational Ruijsenaars system, and finite Toda chains. All those systems can be obtained from the trigonometric Ruijsenaars system by suitable limiting procedures. Another aspect which makes this system very interesting is its duality property, which means that coordinates and Hamiltonians enter this system symmetrically, i.e., there exists an involution of the phase space interchanging them. This property fails to be present in all the above listed limiting cases, except for the rational Calogero system, where this duality is well known even in the quasiclassical case. The quantum version of the trigonometric Ruijsenaars system is the system of MacDonald difference operators [**16**], and the duality between coordinates and Hamiltonians appears there in the disguise of MacDonald's conjecture, recently proved by Cherednik [**10**] by methods quite different from those described in the present paper. However, we shall not discuss the quantum aspects of this problem here.

We show here that one can interpret the phase space of the trigonometric Ruijsenaars system as a symplectic leaf of the lowest dimension in the moduli space $\mathcal{M}$ of flat $G = SL(k)$ connections on the torus with one hole.[6] The commuting Hamiltonians described in [**18**] are certain conjugation-invariant functions of one of the monodromies, the monodromy around one of the cycles of the torus, while the coordinates are the eigenvalues of the other. This picture shows that the duality is nothing but the action of the element of the mapping class group of the torus interchanging these two cycles.

As a by-product, we can introduce a Poisson bracket, as well as a set of commuting Hamiltonians, on the auxiliary space $G \times G$. The flows generated by the Hamiltonians are particularly simple, and the corresponding Hamiltonian equations can be easily integrated. The projection of the Poisson structure and the Hamiltonians on the quotient $G \times G/\operatorname{Ad} G$ exists and gives exactly the Ruijsenaars Hamiltonian system on restriction to a certain symplectic leaf. This procedure gives a way to solve the Ruijsenaars equation explicitly. Algorithmically, it is of course just the same as the one proposed by S. N. M. Ruijsenaars and H. Schneider [**18**]; we just give a natural geometric meaning to it.

Our aim now is to prove the above statement. For this purpose we must do the following:

1. Compute the canonical Poisson bracket on $\mathcal{M} = G \times G/\operatorname{Ad} G$ (using the technique developed in the main part of the paper).
2. Choose coordinates on $\mathcal{M}$ canonically conjugated (with respect to the Poisson bracket) to the eigenvalues of one of the monodromy operators.
3. Compute a certain function of the other monodromy conjugacy class and verify that this exactly gives the trigonometric Ruijsenaars Hamiltonian.

To describe the symplectic structure on $\mathcal{M}$, choose the ciliated fat graph l consisting of two edges and one vertex with the ciliated fat graph structure as shown in Figure 9 corresponding to the torus with one hole (Figure 10).

The space of graph connections $\mathcal{A}^l$ for such a graph is just the product of two copies of the group G,

$$\mathcal{A}^l = G \times G = \{(A, B)\}, \tag{A.1}$$

where A and B are assigned to the edges of the graph as indicated in Figure 9.

The Poisson brackets on $\mathcal{A}^l$ are given by the following relations implied by the definition (see (4.7)):

$$\{\mathbf{A}, \mathbf{A}\} = r_a(\mathbf{A} \otimes \mathbf{A}) + (\mathbf{A} \otimes \mathbf{A})r_a + (1 \otimes \mathbf{A})r_{21}(\mathbf{A} \otimes 1) - (\mathbf{A} \otimes 1)r(1 \otimes \mathbf{A}), \tag{A.2}$$

$$\{\mathbf{B}, \mathbf{B}\} = r_a(\mathbf{B} \otimes \mathbf{B}) + (\mathbf{B} \otimes \mathbf{B})r_a + (1 \otimes \mathbf{B})r_{21}(\mathbf{B} \otimes 1) - (\mathbf{B} \otimes 1)r(1 \otimes \mathbf{B}), \tag{A.3}$$

$$\{\mathbf{A}, \mathbf{B}\} = r(\mathbf{A} \otimes \mathbf{B}) + (\mathbf{A} \otimes \mathbf{B})r + (1 \otimes \mathbf{B})r_{21}(\mathbf{A} \otimes 1) - (\mathbf{A} \otimes 1)r(1 \otimes \mathbf{B}), \tag{A.4}$$

where $r_a = (r - r_{21})/2$; $\mathbf{A}$ and $\mathbf{B}$ are the matrix functions on $G \times G$ corresponding to A and B, respectively, in the standard k-dimensional representation.

Introduce the standard notation G^* for the group G equipped with the Poisson structure given by (A.2) and corresponding to the graph consisting of just one

[6] The relationship between the Ruijsenaars system and moduli of flat connections on the torus was found by Gorsky and Nekrasov in [**14**].

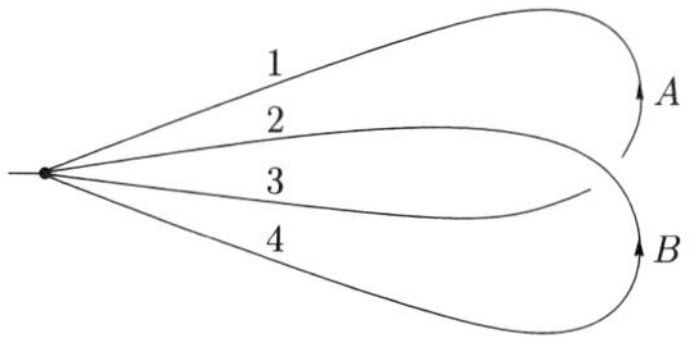

FIGURE 9. The ciliated graph corresponding to a holed torus.

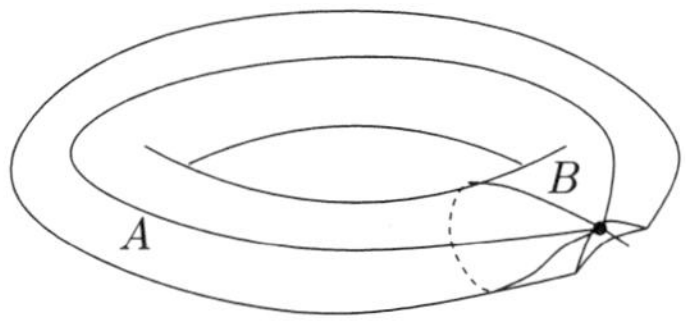

FIGURE 10. The same graph drawn on the holed torus.

loop. Relation (A.2), which coincides, of course, with (4.11), is sometimes called the *reflection equation.*

The projections p_1 and p_2 of $\mathcal{A}^l = G \times G$ onto the first and the second factor, respectively, are obviously Poisson maps $p_{1,2}\colon \mathcal{A}^l \to G^*$.

Now let us restrict ourselves to the case of the standard r-matrix,

$$r = \sum_{\alpha>0} E_\alpha \otimes E_{-\alpha} + \frac{1}{2}\sum_i H_i \otimes H_i. \tag{A.5}$$

In this case one can easily derive from (A.2)–(A.4) the following commutation relations:

$$\{\operatorname{tr}\mathbf{A}^n, \mathbf{A}\} = 0, \qquad \{\operatorname{tr}\mathbf{B}^n, \mathbf{B}\} = 0, \tag{A.6}$$

$$\{\operatorname{tr}\mathbf{A}^n, \mathbf{B}\} = n(\mathbf{A}^n)_0, \tag{A.7}$$

$$\{\operatorname{tr}\mathbf{B}^n, \mathbf{A}\} = n\mathbf{A}(\mathbf{B}^n)_0, \tag{A.8}$$

where $(\mathbf{X})_0$ denotes the traceless part of the matrix $\mathbf{X}$. Therefore, the functions $\operatorname{tr}\mathbf{B}^n$ for $n = 1, \ldots, k-1$, considered as Hamiltonians, generate commuting flows on $\mathcal{A}^l$:

$$\mathbf{B}(t_1, \ldots, t_{k-1}) = \mathbf{B}(0, \ldots, 0), \tag{A.9}$$

$$\mathbf{A}(t_1, \ldots, t_{k-1}) = \mathbf{A}(0, \ldots, 0)\, e^{(t_1\mathbf{B} + \cdots + t_{k-1}\mathbf{B}^{k-1})_0}. \tag{A.10}$$

As was shown in the main part of the paper, the lattice gauge group $\mathcal{G}^l$ acts on $\mathcal{A}^l$ in a Poisson way, and the quotient Poisson manifold coincides with the moduli space $\mathcal{M}$ of smooth flat connections on the Riemann surface corresponding to the fat graph l. In our case, the group $\mathcal{G}^l$ is just G itself (since the graph has only one vertex) and acts on $\mathbf{A}$ and $\mathbf{B}$ by a simultaneous conjugation:

$$g\colon (\mathbf{A}, \mathbf{B}) \mapsto (g\mathbf{A}g^{-1}, g\mathbf{A}g^{-1}). \tag{A.11}$$

The functions $\operatorname{tr}\mathbf{A}^n$ and $\operatorname{tr}\mathbf{B}^n$ are invariant under this action, and therefore they descend to the moduli space $\mathcal{M}$ and generate commuting flows there as well, the trajectories on $\mathcal{M}$ being just the projections of those given by (A.9) and (A.10).

However, the moduli space $\mathcal{M}$ is in our case a Poisson manifold with a degenerate Poisson bracket. The symplectic leaves in $\mathcal{M}$ correspond to connections having a fixed conjugacy class of the monodromy around the hole. In our case, the latter is just the matrix $\mathbf{ABA}^{-1}\mathbf{B}^{-1}$.

Let us recall now that we are actually dealing with the case $G = SL(k)$. Different symplectic leaves in $\mathcal{M}$ have different dimensions, and the lowest dimension among them, in this case, is $2(k-1)$. Those leaves correspond to the monodromy around the hole conjugate to the matrix $x\mathbf{1}+\mathbf{P}$, where $\mathbf{1}$ is the unit matrix, $\operatorname{rk}\mathbf{P} \leqslant 1$, and $x \neq 0$ is the number which parameterizes the set of symplectic leaves of the lowest dimension. (Indeed, the only conjugation invariant of an operator of rank not greater than one is its trace. The latter is $\operatorname{tr}\mathbf{P} = x^{1-k} - x$, since $\det(x\mathbf{1}+\mathbf{P}) = 1$.)

On such leaves, the family of functions $\operatorname{tr}\mathbf{A}^n$, $n = 1, \dots, k-1$, forms a full set of commuting variables. Let us introduce local coordinates on these symplectic leaves in the following way. Let $\lambda_1, \dots, \lambda_k$ be the eigenvalues of the matrix $\mathbf{A}$ and $q_1, \dots, q_k$ be the diagonal matrix elements of $\mathbf{B}$ in a basis in which $\mathbf{A}$ is diagonal. Imposing the condition $\operatorname{rk}\mathbf{P} \leqslant 1$ and conjugating $\mathbf{B}$ by a diagonal matrix, one can bring $\mathbf{B}$ to the form

$$\mathbf{B}^i_j = \frac{\sqrt{q_i q_j}(1-x)}{\lambda_i/\lambda_j - x}. \tag{A.12}$$

The functions λ_i and q_j are locally well-defined functions on the symplectic leaves, and the Poisson brackets between them are

$$\{\lambda_i, \lambda_j\} = 0, \tag{A.13}$$

$$\{q_i, q_j\} = q_i q_j \frac{(\lambda_i + \lambda_j)}{(\lambda_i/\lambda_j - x)(\lambda_j/\lambda_i - x)(\lambda_i - \lambda_j)}, \tag{A.14}$$

$$\{\lambda_i, q_j\} = \lambda_i q_j \delta_{i,j}. \tag{A.15}$$

PROOF OF THE FORMULAS (A.13)–(A.15). To simplify calculations, we assume for a moment that we are working with the group $GL(k)$ rather than $SL(k)$. Having computed the Poisson structure on the quotient space by the lattice gauge group, we can then restrict it to the subspace corresponding to $SL(k)$-connections, since the latter space is a Poisson submanifold in the whole quotient space. The bivector defining the Poisson structure on $\mathcal{A}^l$ for the group $GL(k)$ can be rewritten in the form

$$B = \frac{1}{2} \sum_{i,j,u,v \in \{1\dots 4\}} E^{i(u)}_j \otimes E^{j(v)}_i (\epsilon(u,v) + \epsilon(i,j)), \tag{A.16}$$

where $\epsilon(i,j)$ is $-1, 0$, or 1 if i is less than, equal to, or greater than j, respectively, and $E^{i(u)}_j$ are the standard $gl(k)$ generators acting on the uth end of an edge. (In our case $E^{i(1)}_j$ acts on $\mathbf{A}$ from the left, $E^{i(2)}_j$ acts on $\mathbf{B}$ from the left, $E^{i(3)}_j$ acts on $\mathbf{A}$ from the right and $E^{i(4)}_j$ acts on $\mathbf{B}$ from the right.)

It is not practical, however, to compute the Poisson brackets between λ_i and q_j by using this bivector as it is, because it does not agree with the diagonal form of the matrix $\mathbf{A}$. In order to avoid this difficulty, since we are interested only in computing Poisson brackets of gauge invariant functions, we may change the bivector (A.16) in such a way that it still defines the same Poisson bracket on the space of gauge invariant functions. In other words, there are different ways to write out Poisson brackets on the coset space $G \times G/\operatorname{Ad} G$ in terms of a bivector on the space $G \times G$.

On the other hand, since we know that the projection $\pi\colon G\times G\to G\times G/\operatorname{Ad}G$ is a Poisson map, i.e., the bracket of gauge invariant functions is gauge invariant, it suffices to compute the value of the bracket of two such functions on a submanifold $F\subset G\times G$ that intersects each gauge orbit (i.e., each $\operatorname{Ad}G$-orbit) at least once. By doing this, we can simplify computations by changing the bivector (A.16) by terms vanishing on F. As a prescription, one can formulate the following rule of allowed modifications of the bivector: one can add to any vector $E_j^{i(\alpha)}$ a vector which is tangent to gauge orbits or vanishes on F. The vectors tangent to the gauge orbits are just the generators of the group of gauge transformations, in our case $\sum_{u=1}^4 E_j^{i(u)}$. The vectors vanishing on F (which is in our case the space of connections with $\mathbf{A}$ diagonal) are, for example, $\lambda_j E_j^{i(1)}+\lambda_i E_j^{i(3)}$. Using these rules, one can perform the following replacements:

$$E_j^{i(1)}\rightsquigarrow\frac{\lambda_i}{\lambda_j-\lambda_i}\,(E_j^{i(2)}+E_j^{i(4)}),\tag{A.17}$$

$$E_j^{i(3)}\rightsquigarrow\frac{\lambda_j}{\lambda_i-\lambda_j}\,(E_j^{i(2)}+E_j^{i(4)}).\tag{A.18}$$

By this trick the bivector B can be transformed to the form

$$B'=\sum_{i>j}E_j^{i(2)}\wedge E_i^{j(4)}\frac{\lambda_i+\lambda_j}{2(\lambda_i-\lambda_j)}+\frac{1}{2}\sum_i E_i^{i(2)}\wedge E_i^{i(1)}+E_i^{i(3)}\wedge E_i^{i(4)}.\tag{A.19}$$

Applying this bivector (which now leaves $\mathbf{A}$ diagonal), we get the desired Poisson brackets. □

The form of the brackets given by (A.13)–(A.15) is such that in order to define the variables canonically conjugated to λ_i we can just multiply q_i by factors not depending on q_i. For example, one can take the variables

$$s_i=q_i x^{(n-1)/2}\Big(\prod_{k,\,k\neq i}\frac{(\lambda_k-\lambda_i)(\lambda_i-\lambda_k)}{(\lambda_k-x\lambda_i)(\lambda_i-x\lambda_k)}\Big)^{1/2}.\tag{A.20}$$

One can check by an explicit computation that these new variables s_i have Poisson brackets

$$\{s_i,s_j\}=0,\tag{A.21}$$

$$\{\lambda_i,s_j\}=\lambda_i s_j\delta_{i,j}.\tag{A.22}$$

Using the formula

$$\det\mathbf{B}=x^{n(n-1)/2}\prod_i q_i\prod_{i\neq j}\frac{(\lambda_i-\lambda_j)}{(x\lambda_i-\lambda_j)},\tag{A.23}$$

easily proved by induction, one can express the function $H=\operatorname{tr}(\mathbf{B}+\mathbf{B}^{-1})$ in terms of λ_i and s_i as

$$H=\sum_i(s_i+s_i^{-1})\,x^{(n-1)/2}\Big(\prod_{k,k\neq i}\frac{(\lambda_k-\lambda_i)(\lambda_i-\lambda_k)}{(\lambda_k-x\lambda_i)(\lambda_i-x\lambda_k)}\Big)^{1/2},\tag{A.24}$$

but this function turns out to be exactly the Ruijsenaars Hamiltonian.

Note that the Poisson structure on $\mathcal{A}^l=G\times G$ is nice from various points of view. In particular, it is nondegenerate close to the identity. The action of the group on this space by conjugation is Poisson and has a well-defined momentum map (in the sense of Poisson–Lie groups) $\mu\colon G \times G \to G^*$ such that $\mu\colon (\mathbf{A}, \mathbf{B}) \mapsto \mathbf{A}\mathbf{B}\mathbf{A}^{-1}\mathbf{B}^{-1}$.

References

[1] A. Alekseev, L. Faddeev, and M. Semenov-Tian-Shansky, *Hidden Quantum Group inside Kac–Moody Algebra*, Comm. Math. Phys. **149** (1992), 335–345.

[2] A. Alekseev, L. Faddeev, M. Semenov-Tian-Shansky, and A. Volkov, *The Unraveling of the Quantum Group Structure in WZNW Theory*, Preprint CERN-TH-5981/91, 1991.

[3] A. Yu. Alekseev, H. Grosse, and V. Schomerus, *Combinatorial quantization of the Hamiltonian Chern–Simons theory*, I, II, Comm. Math. Phys. **172** (1995), 317–358; **174** (1996), 561–604.

[4] A. Yu. Alekseev and A. Z. Malkin, *Symplectic structures associated to Lie–Poisson groups*, Comm. Math. Phys. **162** (1994), 147–173.

[5] ———, *Symplectic structure of the moduli space of flat connections on a Riemann surface*, Comm. Math. Phys. **169** (1995), 99–120.

[6] M. Atiyah and R. Bott, *The Yang–Mills equations over a Riemann surface*, Philos. Trans. Roy. Soc. London Ser. A **308** (1983), 523–615.

[7] M. Audin, *Lectures on integrable systems and gauge theory*, Gauge Theory and Symplectic Geometry, Montreal, PQ, 1995, pp. 1–48; NATO Adv. Sci. Inst. Ser. C Math. Phys. Sci., 488, Kluwer Acad. Publ., Dordrecht, 1997, Preprint IRMA-1995/20, 1995.

[8] S. Axelrod, S. Della Pietra, and E. Witten, *Geometric quantization of Chern–Simons gauge theory*, J. Differential Geom. **33** (1991), 787–902.

[9] E. Buffenoir and Ph. Roche, *Two-dimensional lattice gauge theory based on a quantum group*, Comm. Math. Phys. **170** (1995), 669–698.

[10] I. Cherednik, *Double affine Hecke algebra and MacDonald's conjecture*, Ann. of Math. **141** (1995), 191–216.

[11] V. V. Fock and A. A. Rosly, *Poisson structure on moduli of flat connections on Riemann surfaces and r-matrix*, Preprint ITEP-72-92, 1992.

[12] ———, *Flat connections and polyubles*, Teoret. Mat. Fiz. **95** (1993), 228–238; English transl., Theoret. and Math. Phys. **95** (1993), 3195–3206.

[13] S. A. Frolov, *The center of the graph and moduli algebras at roots of* 1, Preprint LMU-TPW 96-11, 1996, `q-alg/9602036`.

[14] A. Gorsky and N. Nekrasov, *Relativistic Calogero–Moser model as gauged WZW theory*, Nuclear Phys. B **436** (1995), 582–608.

[15] N. Hitchin, *Stable bundles and integrable systems*, Duke Math. J. **54** (1987), 97–114.

[16] I. G. MacDonald, *Some conjectures for root systems*, SIAM J. Math. Anal. **13** (1982), 988–1007.

[17] M. S. Narasimhan and C. S. Seshadri, *Stable and unitary vector bundles on Riemann surfaces*, Ann. of Math. **82** (1965), 540–567.

[18] S. N. M. Ruijsenaars and H. Schneider, *A new class of integrable systems and its relation to solitons*, Ann. Physics **170** (1986), 370–405.

[19] M. A. Semenov-Tian-Shansky, *Dressing transformations and Poisson group actions*, Publ. Res. Inst. Math. Sci. **21** (1985), 1237–1260.

[20] E. Witten, *On quantum gauge theory in two dimensions*, Comm. Math. Phys. **141** (1991), 153–209.

Institute of Theoretical and Experimental Physics, B. Cheremushkinskaya 25, 117259, Moscow, Russia

Amer. Math. Soc. Transl.
(2) Vol. **191**, 1999

KZB Equations As a Flat Connection with Spectral Parameter

D. A. Ivanov and A. S. Losev

ABSTRACT. We define the notion of generalized conformal blocks (GCB) associated to a pair consisting of a centrally extended loop algebra G_k and a Riemann surface, and describe their relation to standard conformal blocks (CB): GCB solve Lie-algebraic cocycle equations while CB solve Lie-group cocycle equations. We argue that Knizhnik–Zamolodchikov–Bernard (KZB) equations may be regarded as a connection in the bundle of GCB's over the Teichmüller space, and explain why this connection must have properties similar to the Gauss–Manin connection for the polarized variation of Hodge structure, i.e., it must be a flat connection with spectral parameter respecting certain bilinear form. We also prove these expected properties of KZB equations by a direct calculation.

1. Introduction

Consider the following quadruple: a holomorphic vector bundle E over the base M, a bilinear nondegenerate pairing η in the fibers of the bundle, a connection ∇ in the bundle and a 1-form A taking values in the endomorphisms of E.

For such data and for any $\kappa \in C^*$, we can define a modified connection

$$\nabla(\kappa) = \nabla + \frac{1}{\kappa} A. \tag{1.1}$$

DEFINITION. We shall call such a quadruple a *Griffiths–Saito quadruple* (GS quadruple) if

1) the connection $\nabla(\kappa)$ is flat for any $\kappa \in C^*$;
2) the connection ∇ preserves a bilinear pairing η;
3) the 1-form A takes values in operators that are selfadjoint with respect to the bilinear pairing η.

EXAMPLE 1.1. The standard example [**Gr**] of a GS quadruple is provided by the vector bundle E of the cohomology of a compact Kähler manifold over the

---

1991 *Mathematics Subject Classification.* Primary 81T40.

The work of the first author was supported by RFBR grant 96-02-18046 and by grant 96-15-96455 for support of scientific schools.

The work of the second author was supported by RFBR grant 96-02-18046, by grant 96-15-96455 for support of scientific schools, PYI grant PHY9058501, and DOE grant DE-FG02-92ER40704.

space M of deformations of the complex structures. The connection $\nabla(\kappa)$ is the Gauss–Manin connection, ∇ being a component of this connection preserving the (p,q) type of the cohomology, and the bilinear pairing η is the wedge pairing with the insertion of the Weil operator $(\sqrt{(-1)})^{(p-q)}$ between the forms.

EXAMPLE 1.2. The second example of GS quadruple appears in K. Saito's theory of primitive forms associated to an isolated singularity [**Sa1, Sa2, Los2**].

EXAMPLE 1.3. Frobenius manifolds [**KM, Dub**] (appearing, for instance, in the theory of Gromov–Witten invariants for genus zero) form a GS quadruple, E being the tangent bundle to the Frobenius manifold.

EXAMPLE 1.4. The Knizhnik–Zamolodchikov differential operators [**KZ**]

$$\frac{\partial}{\partial z_i} + \frac{1}{\kappa} \sum_{a=1}^{\dim G} \sum_{j \neq i}^{n} \frac{t_i^a t_j^a}{(z_i - z_j)} \tag{1.2}$$

can be regarded as a connection in the bundle E, the fiber being the space of invariants of the product of n finite-dimensional selfadjoint representations of the Lie algebra $\mathcal{G}$ (attached to marked points z_i on CP_1), over the space $M_{0,n}$ of complex structures on CP_1 with n marked points. The bilinear pairing is the pairing on the space of invariants induced from the canonical bilinear pairing on the selfadjoint representations.

In this paper (based on [**Los1, Iv**]) we generalize Example 1.4 to the Knizhnik–Zamolodchikov–Bernard equations [**B1, B2**]. Namely, we start with a complex Lie group G. The base space M is the Teichmüller space of Riemann surfaces of genus N, the fiber is the space of holomorphic Ad_G-invariant functions on G^N, the connection ∇ is the trivial connection, and the bilinear pairing is given by the integral of the product of two functions over the Nth power of the maximal compact subgroup G_c of G with the Haar measure. Components of the form A are the Bernard operators (second-order differential operators introduced in [**B2**]).

In Section 2 we present a sketch of the construction, explaining why KZB equations together with the pairing introduced above form a GS quadruple.

The flatness of $\nabla(\kappa)$ is demonstrated by constructing its horizontal sections—products of generalized conformal blocks and the square root of the chiral determinant (the partition function in the $\dim G$ copies of the bc-system).

We shall argue that generalized conformal blocks are related to ordinary conformal blocks in the same way as Verma modules are related to integrable representations. Namely, we shall start with a definition of generalized conformal blocks (on a sphere with $2N$ discs cut out) as vectors in the product of $2N$ highest-weight representations of $2N$ copies of the loop algebra. Generalized conformal blocks are defined as vectors annihilated by a parabolic subalgebra in the sum of $2N$ copies of the loop algebra. In the standard definition of conformal blocks, representations are taken to be integrable; this implies that the central charge k of the loop algebra is a positive integer. In constructing generalized conformal blocks we use Verma modules (associated to finite-dimensional representations of G). Generalized conformal blocks on the closed surface are obtained by sewing generalized conformal blocks defined above. Sewing with constant twists associates to each block an Ad_G-invariant function on G^N. We argue that this map is one-to-one. Then we construct a connection on the bundle of generalized conformal blocks using the

Sugawara construction for the energy-momentum tensor. We shall denote covariantly constant sections of this connection by $GCB_\beta(k)$. The "chiral determinant" Z_{bc} as a holomorphic top form on G^N is constructed in a similar spirit, using $\dim G$ copies of a free fermionic bc-system. Thus we arrive at equation (1.1) with κ equal to $k+h^*$, where h^* is the dual Coxeter number of the Lie algebra.

To show that A is selfadjoint, we study the "total" system containing the representation of the loop algebra of level k, the representation of level $-k-2h^*$, and $\dim G$ copies of a free fermionic system. The "total" system contains an operator Q with $Q^2=0$. Moreover, the total energy-momentum tensor of the "total" system is Q-exact. Then it is easy to show that the derivative along the space of complex structures of the product $GCB_\alpha(\kappa)GCB_\beta(-\kappa)Z_{bc}$ is an exact form on G^N. This implies that after a modification of the definition of conformal blocks,

$$GCB^{\text{mod}}_\alpha(\kappa)=GCB_\alpha(\kappa)(Z_{bc})^{1/2}, \tag{1.3}$$

the connection on modified conformal blocks acquires the form $\nabla+\frac{1}{\kappa}A$ for selfadjoint A.

In Section 3 we present an explicit form of the Bernard equations and directly check that they form a GS quadruple on the space of Ad_G-invariant functions. Section 3 is self-contained and formally independent of Section 2.

CONVENTION. Summation over repeated indices is assumed in this text.

2. Generalized conformal blocks and KZB equations on Ad_G-invariant functions on G^N from the representation theory of affine algebras

2.1. Two definitions of conformal blocks. There are two equivalent definitions of conformal blocks: as sections of a line bundle on the moduli space of flat connections, and as "generalized traces" in the representation theory of loop groups,[1] in the spirit of [**MS**] (see also [**V**]). Below we shall show how, in the second definition, one can replace the representation theory of loop groups by the representation theory of loop algebras in order to obtain generalized conformal blocks that make sense for all complex k. We shall see that this framework is appropriate for explaining why the KZB equations form a GS quadruple (1.1).

2.1.1. *The functional-integral motivations and the first definition.* Consider the generating function for current correlators in WZNW theory of level k on a genus N Riemann surface Σ:

$$Z_k(\bar{A})=\int_{h\in \mathrm{Map}(\Sigma,G_c)}\mathcal{D}h\exp C_k(\bar{A},h). \tag{2.1}$$

Here the functional integral is taken over maps h from the Riemann surface to the maximal compact subgroup G_c of a complex group G, while $C_k(\bar{A},g)$ for $g\in \mathrm{Map}(\Sigma,G)$ is given by

$$C_k(\bar{A},g)=kS_{WZNW}(g)+\int_\Sigma \mathrm{Tr}\,(\bar{A}j(g)), \tag{2.2}$$

where $S_{WZNW}(g)$ is the action in WZNW theory [**W, KZ**], the current $j(g)$ is a $(1,0)$-form on the Riemann surface with values in the Lie algebra $\mathcal{G}$ of the group G,

$$j=kg^{-1}\partial g, \tag{2.3}$$

[1] For a survey, see [**K, PS**].

and the generating parameter $\bar{A}$ is a $(0,1)$-form on Σ taking values in the Lie algebra $\mathcal{G}$, while Tr is the Killing pairing in the Lie algebra $\mathcal{G}$. The (formal) measure $\mathcal{D}h$ on $\mathrm{Map}(\Sigma, G_c)$ is assumed left and right invariant.

The main property of $C(g, \bar{A})$ is that it is a group cocycle, see [**Fad, FS, Mick**]. Indeed, take the gauge group $\mathrm{Map}(\Sigma, G)$ and the action of this group on the generating parameters (denoted by $\bar{A}$) considered as the $(0,1)$ part of a connection form in the trivial bundle:

$$\bar{A}^g = g^{-1}\bar{A}g + g^{-1}\bar{\partial}g. \tag{2.4}$$

This action makes the space of smooth functionals of the generating parameter $\bar{A}$ a gauge group module, and $C_k(g, \bar{A})$ is a cocycle of the gauge group with coefficients in this module:

$$C_k(g_1 g_2, \bar{A}) = C_k(g_1, \bar{A}^{g_2}) + C_k(g_2, \bar{A}). \tag{2.5}$$

Let us pick some element g of the gauge group $\mathrm{Map}(\Sigma, G)$. Since C_k is a holomorphic function on the group $\mathrm{Map}(\Sigma, G)$, the shift in $\mathrm{Map}(\Sigma, G)$ in the functional integration from $\mathrm{Map}(\Sigma, G_c)$ to $g\,\mathrm{Map}(\Sigma, G_c)$ does not change the functional integral $Z(\bar{A})$.

Using the cocycle property (2.5) and the invariance of the measure, we find that

$$Z(\bar{A}^g) = \exp(C_k(\bar{A}, g))Z(\bar{A}). \tag{2.6}$$

This leads us to the following definition.

FIRST DEFINITION OF THE CONFORMAL BLOCK. Holomorphic functions $CB_k(\bar{A})$ (of a functional variable $\bar{A}$) are called *conformal blocks* of level k if they satisfy the equation

$$CB_k(\bar{A}^g) = \exp(C_k(\bar{A}, g))CB_k(\bar{A}). \tag{2.7}$$

REMARK. The consistency of equation (2.7) requires that $C_k(\bar{A}, g)$ be a group cocycle.

Thus, the generating function of current correlators in the WZNW functional integral is a conformal block of level k.

Note that $\bar{A}$ induces the structure of a holomorphic G-bundle on the Riemann surface, and an isomorphism between holomorphic bundles is given by the gauge transformations (2.4). Thus, conformal blocks of level k can be regarded as holomorphic sections of the line bundle L^k over the moduli space MH of holomorphic G-bundles, with the line bundle L determined by the cocycle C_1.

The defining equation (2.7) for conformal blocks has an infinitesimal version (obtained by expanding $g(x) = 1 + \epsilon_a(x)t_a + O(\epsilon^2)$):

$$c^a(z; k)CB_k(\bar{A}) = 0, \tag{2.8}$$

where $c^a(z; k)$ is a first order differential operator (Lie-algebraic anomaly consistency condition that goes back to [**WZ**]) on the space of $\bar{A}$'s taking values in $\mathrm{Map}(\Sigma, \mathcal{G})$:

$$c^a(z; k) = \left(\delta^{ac}\frac{\partial}{\partial \bar{z}} + f^{abc}\bar{A}^b(z)\right)\frac{\delta}{\delta \bar{A}^c(z)} - k\partial \bar{A}^a(z), \tag{2.9}$$

and f^{abc} are the structure constants of the Lie algebra $\mathcal{G}$.

REMARKS. 1) For the existence of a group cocycle, k must be integer, while the infinitesimal (Lie-algebraic) condition may be imposed for any k.

2) One can construct a projectively flat connection (see [**ADW, H1, H2**]) in the vector bundle of conformal blocks over the space of complex structures.

2.1.2. *Functional-integral motivation of the second definition of conformal blocks.* The second definition is motivated by attempts to compute the functional integral using holomorphic symmetries generated by holomorphic functions. Compact surfaces have only constants as holomorphic functions, so we start with surfaces with boundary. The functional integral $|I\rangle$ on such a surface is a function of the boundary conditions, so it can be considered as a vector in the space H_{open} of functions of boundary conditions, $|I\rangle \in H_{\mathrm{open}}$. Holomorphic symmetries (see [**BPZ, Va, MS**]) may be regarded as operators $\jmath(\mathrm{Hol})$ acting on H_{open}, such that

$$\jmath(\mathrm{Hol})\,|I\rangle = 0. \tag{2.10}$$

In the WZNW theory, the space H_{open} is a representation of the affine algebra and the operators $\jmath(\mathrm{Hol})$ are constructed from operators representing elements of the affine algebra.

As one can argue from the functional integral, if in the operator $\jmath(\mathrm{Hol})$ we replace a holomorphic function by a meromorphic one, then we get an operator $\jmath(\mathrm{Mer})$ which does not annihilate $|I\rangle$, but creates insertions of the current (2.3) at the positions of the poles of the meromorphic function.

A modification of this procedure will allow us to define multicurrent correlators as a result of applying certain operators, $\jmath(\mathrm{Multimer})$, to $|I\rangle$. By introducing the connection $\bar{A}$ as a generating parameter for multicurrent correlators, we obtain a function $|I(\bar{A})\rangle$ with values in H_{open} (in fact, a formal series in $\bar{A}$), which can be schematically written as

$$|I(\bar{A})\rangle = \jmath(\mathrm{Multimer}, \bar{A})\,|I\rangle. \tag{2.11}$$

Suppose that an open surface is obtained by cutting a closed surface of genus N along the A-cycles. Then the functional integral on the closed surface can be obtained from that on the open surface by "sewing" back, i.e., by integrating over common boundary conditions. Such an operation is represented by the pairing of $|I(\bar{A})\rangle$ with the sewing covector $S \in H^*_{\mathrm{open}}$; thus we schematically get the functional integral $Z(\bar{A})$ in the form

$$Z(\bar{A}) = \langle S|I(\bar{A})\rangle = \langle S|\,\jmath(\mathrm{Multimer}, \bar{A})\,|I\rangle. \tag{2.12}$$

2.1.3. *Plan of the definition of generalized conformal blocks.* The last form, equation (2.12), of the functional integral suggests that $Z(\bar{A})$ can be determined entirely in terms of the representation theory of affine algebras. Namely, we shall do the following:

1) define a space $H_{\mathrm{open}} = H_1 \otimes \cdots \otimes H_{2N}$ corresponding to an open surface with $2N$-component boundary and $2N$ copies of the affine algebra acting in this space;
2) define equations (2.10); the space of solutions to these equations in H_{open} will be called the *space of generalized conformal blocks* for the open surface, and the basis in this space will be denoted by $|\beta\rangle$;
3) define the operators $\jmath(\mathrm{Multimer}, \bar{A})$;
4) define sewing covectors $\langle S_i|$ as elements of $H_i^* \otimes H_{N+i}^*$, and then the total sewing covector $\langle S|$ as $\langle S| = \prod_{i=1}^N \langle S_i|$;

5) define generalized conformal blocks for a closed surface as in (2.12), i.e., as

$$GCB_\beta(\bar{A}) = \langle S|\, j(\text{Multimer}, \bar{A})\,|\beta\rangle. \tag{2.13}$$

REMARK. One can show that generalized conformal blocks $GCB_\beta(\bar{A})$ satisfy the Lie-algebraic anomaly equations (2.8), but they do not satisfy the group cocycle equations (2.7)—that is why we call them *generalized.*

In order to satisfy the group cocycle equations (2.7), k must be restricted to a positive integer, and the representations H_i entering H_{open}, to integrable ones, see [**GW, FGK, B1**]. In this way one will get the second definition of conformal blocks in terms of representation theory. For recent studies in this direction, see [**G, Fe**].

2.2. Definition of generalized conformal blocks on an open surface from the theory of representations of affine algebras.

2.2.1. *Loop algebra, currents and Verma modules.* For a Lie algebra $\mathcal{G}$ and a loop Γ (i.e., a copy of the circle S^1), we define the centrally extended loop algebra $\mathcal{G}_k$ over complex numbers as follows. The algebra $\mathcal{G}_k$ consists of the center k and elements corresponding to smooth maps from Γ to $\mathcal{G}$.

We denote the element corresponding to a map f^a by $\int_\Gamma j^a f^a$, so that

$$\left[\int_\Gamma j^a f_1^a, \int_\Gamma j^b f_2^b\right] = \int_\Gamma f_{abc} j^c f_1^a f_2^b + k \int_\Gamma f_1^a df_2^a. \tag{2.14}$$

This a priori strange-looking notation is well suited for the definition of the current $j^b(z)$ as the limit of $\int_\Gamma j^a \delta^{ab}_{z,\text{reg}}$ when the function $\delta^{ab}_{z,\text{reg}}(w)$ tends to $\delta_{ab}\delta^{(1)}(z-w)$. Due to the properties of δ-functions, the "current" element must be considered as a 1-form at the point z on Γ.

Let us define Verma modules, the representations of the loop algebra that we shall use in this section [**K, PS**]. Consider the loop algebra on an oriented loop Γ. Take a disc such that Γ is its boundary, a point P of this disc and a finite-dimensional representation (V, τ) of the algebra $\mathcal{G}$ ($\tau_a = \tau(t_a)$ are operators in V corresponding to generators t_a of $\mathcal{G}$). Consider the subalgebra in the loop algebra generated by functions that can be holomorphically continued inside the disc, and represent it in V as follows:

$$\tau\left(\int_\Gamma j^a f^a\right) = f^a(P)\tau_a. \tag{2.15}$$

The representations H of $\mathcal{G}_k$ that we use are the representations induced from the above subalgebra represented by (2.15), with the center element represented by a complex number k [**K, PS**]. We call the representation V the "ground state representation" of the Verma module H.

STATEMENT 2.1. *The representations of the loop algebra constructed above depend (up to isomorphism) only on the orientation of* Γ *and on the finite-dimensional representation* (V, τ), *and are independent of the choice of the disc and the choice of the point* P *on it. We call such representations* Verma modules associated to V.

2.2.2. *Generalized conformal blocks on open surfaces.* Let H_i, $i = 1, \dots, 2N$, denote the set of Verma modules associated to finite-dimensional representations V_i, such that $V_{N+i} = V_i^*$.

Let Σ^0 be a Riemann surface of genus zero with $2N$ boundary components $\Gamma_1, \dots, \Gamma_{2N}$; we associate H_i to the boundary Γ_i.

DEFINITION. We define a generalized conformal block (GCB) as an element $|\beta\rangle$ of $H_{\text{open}} = H_1 \otimes \cdots \otimes H_{2N}$ that solves the following system of equations:

$$\sum_{i=1}^{2N} \int_{\Gamma_i} f j^a_{(i)} \, |\beta\rangle = 0 \tag{2.16}$$

for any holomorphic function f on Σ^0. Here $j^a_{(i)}(z), z \in \Gamma_i$, is an element of the loop algebra in the representation H_i.

Generalized conformal blocks form a vector space, and we chose a basis $\{|\beta\rangle\}$ in this space.

STATEMENT 2.2. *The space of conformal blocks is in one-to-one correspondence with the space* Inv *of invariants in* $V_1 \otimes \cdots \otimes V_{2N}$, *i.e., with the space of vectors* $u_\beta \in V_1 \otimes \cdots \otimes V_{2N}$ *invariant with respect to the diagonal action of the algebra* $\mathcal{G}$:

$$\sum_{i=1}^{2N} \tau_{(i)}(t_a) u_\beta = 0. \tag{2.17}$$

IDEA OF THE PROOF. We look for solutions to (2.16) in the form

$$|\beta\rangle = (1 + X^a_{-n,i} j^a_{-n,(i)} + X^{ab}_{-n,-m;ij} j^a_{-n,(i)} j^b_{-m,(j)} + \cdots)|u_\beta\rangle, \tag{2.18}$$

where $|u_\beta\rangle$ is the "ground state" corresponding to the invariant u_β, i.e., for $n > 0$,

$$j^a_{n,(i)} |u_\beta\rangle = 0 \tag{2.19}$$

and $|u_\beta\rangle$ solves equations (2.16) for $f = 1$. □

For general values of k, we can find the coefficients X uniquely. For a very similar construction, see [**FZ**]).

2.2.3. *Definition of "current correlators"*. Below we present a definition of the current correlators, which is just a version of a definition given in [**BPZ**].

DEFINITION. The m-current correlators $|j^{a_1}(z_1), \ldots, j^{a_m}(z_m); \beta\rangle$, $m \geqslant 0$, are defined inductively, as sections of the holomorphic cotangent bundle on $(\Sigma^0)^m$ taking values in $\mathcal{G}^m \otimes H_{\text{open}}$.

Define the function $f_z(w)$ by the condition that it is holomorphic for $w \neq z$ and has a simple pole at $w = z$ with

$$f_z(w) \sim \frac{1}{2\pi i} \frac{1}{z - w} \quad \text{as } w \to z$$

in some local coordinate near the point z. Such functions are defined up to a holomorphic function on the surface Σ^0:

$$\begin{aligned}
&|j^{a_1}(z_1), \ldots, j^{a_{m+1}}(z_{m+1}); \beta\rangle \\
&= \sum_{i=1}^{2N} \int_{\Gamma_i} f_{z_{m+1}} j^{a_{m+1}}_{(i)} |j^{a_1}(z_1), \ldots, j^{a_m}(z_m); \beta\rangle \\
&\quad - \sum_{l=1}^{m} \big(f^{a a_l a_{m+1}} f_{z_{m+1}}(z_l) |j^{a_1}(z_1), \ldots, j^{a}(z_l), \ldots, j^{a_m}(z_m); \beta\rangle \\
&\qquad - \delta_{a_{m+1} a_l} k d f_{z_{m+1}}(z_l) |j^{a_1}(z_1), \ldots, j^{a_{l-1}}(z_{l-1}), \\
&\qquad\qquad j^{a_{l+1}}(z_{l+1}), \ldots, j^{a_m}(z_m); \beta\rangle \big),
\end{aligned} \tag{2.20}$$

and the 0-current correlator is the generalized conformal block $|\beta\rangle$ itself.

REMARK. One can easily check that the dependence of the definition of the current correlators on the local coordinate (through such a dependence of f_z) is consistent with $j(z)$ being a 1-form on the surface.

PROPOSITION 2.1 (properties of current correlators). 1) *Current correlators are symmetric under permutations of points.*

2) *When z_m tends to a point $w \in \Gamma_i$, the m-current correlator tends to the result of the action of the operator $j^{a_m}_{(i)}(w)$ on the $m-1$ current correlator:*

$$\lim_{z_m \to w \in \Gamma_i} |j^{a_1}(z_1), \dots, j^{a_m}(z_m); \beta\rangle = j^{a_m}_{(i)}(w)\, |j^{a_1}(z_1), \dots, j^{a_{m-1}}(z_{m-1}); \beta\rangle. \tag{2.21}$$

The proof is a direct verification by induction in m.

2.2.4. *Generalized conformal blocks with primary fields.* The definition of generalized conformal block can be extended to cover the case of "insertions of primary fields". Namely, let us mark K points P_n, $n = 1, \dots, K$, on the open surface and attach to each such point a finite-dimensional representation $(V_n, \tau_{(n)})$, in addition to the above construction. Then, the *generalized conformal block on the open surface with primary fields*

$$|\Phi_1(P_1) \cdots \Phi_K(P_K); \beta\rangle \in H_1 \otimes \cdots \otimes H_{2N} \otimes V_1 \otimes \cdots \otimes V_K$$

is defined as a solution to the equations

(2.22)
$$\sum_{i=1}^{2N} \int_{\Gamma_i} f j^a_{(i)}\, |\Phi_1(P_1) \cdots \Phi_K(P_K); \beta\rangle = \sum_{n=1}^{K} f(P_n) \tau^a_{(n)}\, |\Phi_1(P_1) \cdots \Phi_K(P_K); \beta\rangle$$

for all holomorphic functions f on an open surface.

After sewing an open surface to a closed one, we can obtain a generalized conformal block for a genus N surface with insertions of primary fields. This construction would be quite similar to the case with no insertions present (see below), so we shall not describe it here.

2.2.5. *Generating function.* In order to show the relationship between generalized conformal blocks and standard definitions, we introduce the generating function $|\bar{A}; \beta\rangle$ for current correlators (the generating parameter is a Lie-algebra-valued $(0,1)$-form $\bar{A}$ smooth on Σ^0):

$$|\bar{A}; \beta\rangle = \sum_{m=1}^{\infty} \frac{1}{m!} \int \cdots \int |j^{a_1}(z_1), \dots, j^{a_m}(z_m); \beta\rangle\, \bar{A}^{a_1}(z_1) \cdots \bar{A}^{a_m}(z_m), \tag{2.23}$$

where each integral is taken as the principal value over Σ^0 (see remark below), and $|\bar{A}; \beta\rangle$ is a formal series in $\bar{A}$ (taking values in $H_1 \otimes \cdots \otimes H_{2N}$).

REMARK. The principal-value integral of the product of a meromorphic function and a smooth function on a complex manifold Σ can be defined as follows. Choose a holomorphic local coordinate z at the pole, and a small number ϵ, and take a disc around $z = 0$: $D(\epsilon) = \{x \in \Sigma \mid |z(x)| < \epsilon\}$. Then define the integral over $\Sigma \setminus D(\epsilon)$. One can show that the limit of the integral as $\epsilon \to 0$ exists and that this limit is independent of the choice of the local coordinate.

PROPOSITION 2.2. *The generating functions $|\bar{A}; \beta\rangle$ satisfy the Lie-algebraic anomaly equation,*

$$c^a(z; k)\, |\bar{A}; \beta\rangle = 0, \tag{2.24}$$

where the differential operator $c^a(z;k)$ *was defined in* (2.9).

The next step in constructing conformal blocks is sewing up a closed surface from the open one. This is done by means of a sewing covector.

2.3. Sewing covector and GCB **on compact surfaces.**

2.3.1. *Construction of the sewing covector.* The ith and the $(N+i)$th boundaries can be sewed together by a smooth bijective (orientation reversing) map s_i,

$$s_i\colon \Gamma_i \to \Gamma_{N+i}, \tag{2.25}$$

such that for any local holomorphic coordinate Z_{N+i} in the neighborhood of Γ_{N+i} the inverse image, $s_i^* Z_{N+i}$, can be holomorphically continued from Γ_i to some neighborhood of Γ_i. The boundaries sewn together will form a cycle A_i on the resulting surface.

If Σ is equipped with a projective structure, and s_i is projective (preserves the restriction of the projective structure to boundary components Γ_i and Γ_{N+i}), then the sewn surface is also equipped with a projective structure. This happens, for example, in the Schottky construction of Riemann surfaces (see Appendix A).

DEFINITION OF THE SEWING COVECTOR. The sewing covector, $\langle S_i|$, is an element in $H_i^* \otimes H_{N+i}^*$ solving the equation

$$\langle S_i|\left(\int_{\Gamma_i} (s_i^*(f) j^a_{(i)}) - \int_{\Gamma_{N+i}} f j^a_{(N+i)}\right) = 0 \tag{2.26}$$

for all smooth functions f on Γ_{N+i}.

The sewing covector exists (for general values of k) and is unique (up to a scalar factor) only if the "ground state representations" (see the definition of Verma modules above in this section) V_i and V_{N+i} for the Verma modules H_i and H_{N+i} are conjugate to each other.

2.3.2. *Twisting.* Constant maps define an embedding of the finite-dimensional Lie algebra $\mathcal{G}$ in the loop algebra; thus, the highest-weight representations of the Lie algebra are decomposed into finite-dimensional irreducible representations of $\mathcal{G}$. The representations H_i are integrable to representations of the finite-dimensional Lie group G. Let us denote by $\hat{\tau}_i(h)$ the action of $h \in G$ in H_i.

DEFINITION. The twisted current correlator, $\langle j^{a_1}(z_1), \dots, j^{a_m}(z_m); g, \beta\rangle$, on a genus N surface with twists $g = (g_1, \dots, g_N)$ is defined by

$$\begin{aligned}&\langle j^{a_1}(z_1), \dots, j^{a_m}(z_m); g, \beta\rangle \\ &\quad = \langle S_1 \hat{\tau}_1(g_1) \cdots S_N \hat{\tau}_N(g_N)\, | j^{a_1}(z_1), \dots, j^{a_m}(z_m); \beta\rangle.\end{aligned} \tag{2.27}$$

If $m = 0$, we call the result of the pairing (2.27) a *generalized conformal block on a compact surface* and denote it by $GCB_\beta(g)$.

The operators $\hat{\tau}_i(g_i)$ in this context will be called *twist operators.*

REMARKS. 1) We cannot construct a sewing covector twisted by an element of the loop group, since the representations H_i do not extend to representations of the loop group.

2) We are a bit sloppy about the convergence of the series for the matrix elements in the above definition. In fact, we can only prove the convergence for sewn Riemann surfaces which are close to degenerate ones (all A-cycles are nearly pinched).

2.3.3. *Properties of the twisted current correlator.*

PROPOSITION 2.3. *The twisted current correlator has the following properties:*

1) $\langle j^{a_1}(z_1), \ldots, j^{a_m}(z_m); g, \beta\rangle$ *(considered as a function of* z_1 *and* a_1*) is a meromorphic section of the holomorphic vector bundle on the surface determined by constant transition functions* g_i *on the cycles* A_i*:*

$$\begin{aligned} &\lim_{z_1 \to_L w \in A_i} \langle j^{a_1}(z_1), \ldots, j^{a_m}(z_m); g, \beta\rangle \\ &\quad = \mathrm{Ad}(g_i)^{a_1 b} \lim_{z_1 \to_R w \in A_i} \langle j^{b}(z_1), \ldots, j^{a_m}(z_m); g, \beta\rangle, \end{aligned} \tag{2.28}$$

where $\to_L$ *(resp.* $\to_R$*) means taking the limit from the left (resp. right) side of the cycle. It has poles only at* $z_1 = z_i$*, and near the pole (say, at* $z_1 = z_2$*) the current correlator has the following form:*

$$\begin{aligned} &\langle j^{a_1}(z_1), j^{a_2}(z_2) \ldots, j^{a_m}(z_m); g, \beta\rangle \\ &\quad = \frac{f^{a_1 a_2 b}}{z_1 - z_2}\langle j^{b}(z_2), \ldots, j^{a_m}(z_m); g, \beta\rangle \\ &\qquad + \frac{k\delta^{a_1 a_2}}{(z_1 - z_2)^2}\langle j^{a_3}(z_3) \ldots, j^{a_m}(z_m); g, \beta\rangle + \text{regular terms.} \end{aligned} \tag{2.29}$$

2) *The* A*-periods of an* m*-current correlator can be expressed as Lie derivatives with respect to the twisting elements of the* $(m-1)$*-current correlators:*

$$\int_{A_i} \langle j^{a_1}(z_1), \ldots, j^{a_m}(z_m); g, \beta\rangle = \mathcal{L}^{i a_m}\langle j^{a_1}(z_1), \ldots, j^{a_{m-1}}(z_{m-1}); g, \beta\rangle. \tag{2.30}$$

Here $\mathcal{L}^{ia}$ *is the derivative with respect to a right-invariant vector field on* G*:*

$$\mathcal{L}^{ia} f(g_1, \ldots, g_N) = \left.\frac{d}{d\varepsilon}\right|_{\varepsilon=0} f(g_1, \ldots, e^{\varepsilon t^a} g_i, \ldots, g_N). \tag{2.31}$$

The proof is a direct verification

COROLLARY 2.1. *Twisted current correlators can be found from generalized conformal blocks* $GCB_\beta(g)$*.*

PROOF. Meromorphic 1-forms with values in the sections of a holomorphic bundle with vanishing Chern class are completely determined by their A-periods and poles. Both are expressed in terms of $(m-1)$- and $(m-2)$-current correlators. □

PROPOSITION 2.4. *For general values of twists and complex structures, the generalized conformal blocks,* $GCB_\beta(g)$*, form a basis in the space of* Ad_G*-invariant holomorphic functions on* G^N*.*

IDEA OF THE PROOF. Consider $\tau_1(g_1) \otimes \cdots \otimes \tau_N(g_N)$ as an element of $V_1 \otimes V_1^* \otimes \cdots \otimes V_N \otimes V_N^*$. We expect that near the point in the moduli space where all the A-cycles shrink to points we have

$$GCB_\beta(g) \sim (\langle \tau_{i_1}(g_1) \cdots \tau_{i_N}(g_N)\, \mathrm{Inv}_\beta(V_1 \otimes V_1^* \otimes \cdots \otimes V_N \otimes V_N^*)\rangle + \text{small terms}). \tag{2.32}$$

Here $\langle \cdot, \cdot \rangle$ stands for the natural pairing between vectors in a representation and its dual, $\mathrm{Inv}_\alpha(V_1 \otimes \cdots \otimes V_m)$ stands for a basis in the space of invariants of the representation $V_1 \otimes \cdots \otimes V_m$, and "small terms" is a series of terms which rapidly vanish as the A-cycles shrink. The first term in (2.32) comes from the first term in (2.18).

Moreover, it follows easily from the Peter–Weyl theorem that the functions

$$\langle \tau_{i_1}(g_1)\cdots\tau_{i_N}(g_N)\,\mathrm{Inv}_\beta(V_1\otimes V_1^*\otimes\cdots\otimes V_N\otimes V_N^*)\rangle \tag{2.33}$$

form a basis in the space of Ad_G-invariant holomorphic functions on G^N—see the appendix in [**Los1**].

Thus we expect that almost everywhere (up to subvariety of complex codimension 1) the generalized conformal blocks form a basis in the space of Ad_G-invariant functions. □

The generating function for current correlators allows us to make contact with the standard ("functional-integral") definition of conformal blocks. First, define the generating function on the compact surface by means of the sewing covectors,

$$GCB_\beta(\bar{A}) = \langle S_1\cdots S_N|\bar{A};\beta\rangle, \tag{2.34}$$

where now $\bar{A}$ is a smooth Lie-algebra-valued $(0,1)$-form on the compact surface. Then this functional of $\bar{A}$ satisfies equation (2.8).

2.3.4. *On the relationship between $GCB_\beta(\bar{A})$ and $GCB_\beta(g)$.* At first sight, we have constructed two seemingly different objects: $GCB_\beta(g)$ and $GCB_\beta(\bar{A})$. The key observation is that differentiating $GCB_\beta(\bar{A})$ with respect to $\bar{A}$ generates current correlators and that the same is true for differentiating $GCB_\beta(g)$ with respect to g_i. This gives rise to the conjecture that these conformal blocks become identical when $\bar{A}$ concentrates to a step-like function on the A-cycles.

To be more precise, define smeared G-valued step functions $g_i(z)$ (not holomorphic) in the vicinity of the A-cycles A_i, respectively. Each of these functions must be equal to $1\in G$ on the left of A_i, to g_i on the right of A_i, and interpolate between these two values in a small neighborhood of A_i. Next, define

$$\bar{A} = g_i^{-1}(z)\bar{\partial}g_i(z) \tag{2.35}$$

in the vicinity of the A-cycles A_i, and $\bar{A}=0$ everywhere else. Then we conjecture the following

THEOREM 2.1.

$$GCB_\beta(g) = GCB_\beta(\bar{A})\exp\Big\{-k\sum_i S_{WZNW}[g_i(z)]\Big\}, \tag{2.36}$$

where $S_{WZNW}[g_i(z)]$ is the WZNW action for the field $g_i(z)$ (with twisted boundary conditions).

SKETCH OF THE PROOF. The proof consists of two parts:

1) First, using the cocycle property (2.5) of the WZNW action, one proves that the right-hand side of (2.36) is invariant with respect to the gauge transformations (2.4) of $\bar{A}$, and, therefore, depends only on g_i.

2) Second, one verifies that the derivatives with respect to g_i of the left-hand side and of the right-hand side are equal. □

REMARKS. 1) In equation (2.36), a proper branch of the multivalued functional S_{WZNW} must be taken. Namely, the continuation of the function $g_i(z)$ inside the surface must be in the same topological class as the constant for $g_i=1$.

2) We ignore the questions of convergence for the formal series in $\bar{A}$. Therefore, we can only give the above theorem the status of a conjecture.

3) The WZNW action $S_{WZNW}[g_i(z)]$ diverges as $g_i(z)$ tends to a step function. This compensates the divergence of $GCB_\beta(\bar{A})$, due to the self-action of the current. We must remark, however, that taking this limit is not required, since the relation (2.36) holds even for smeared functions $g_i(z)$, provided they are defined in nonintersecting domains.

2.4. Construction of the KZB connection on GCB.

Strategy. Consider the family of Riemann surfaces obtained from the open Riemann surface by a family of sewing maps $s_i(t)$, $i = 1, \dots, N$, where t are real parameters on this family. The equations on the sewing covector (2.26) take the form

$$\langle S_i(t)| \Big(\int_{\Gamma_i} (s_i(t)^*(f) j^a_{(i)}) - \int_{\Gamma_{N+i}} f j^a_{(N+i)} \Big) = 0. \tag{2.37}$$

Thus, given a family $\langle S_i(t)|$ of solutions to (2.37), we can define a family of generalized conformal blocks $GCB(g, t, \beta)$ by setting

$$GCB_\beta(g, t) = \langle S_1(t)|\, \hat{\tau}_1(g_1) \cdots S_N(t) \hat{\tau}_N(g_N)\, |\beta\rangle. \tag{2.38}$$

The KZB connection on the space of GCB's over the family of Riemann surfaces is induced from a connection on the space of solutions to (2.37) constructed using the "energy-momentum tensor" (EMT, for short) operators.

We shall see later that for special ("projective") families, the KZB connection will be integrable because:

1) EMT operators represent Virasoro algebra;

2) the central term vanishes for "projective" families.

It is not possible to cover all complex structures by such "projective" families, but it is possible to do this by complexifying the parameter t.

Thus we arrive at the Schottky parametrization of Riemann surfaces.

2.4.1. *Definition of the energy-momentum tensor.* Let Γ be a loop in the complex plane, and An be a thin annulus containing Γ (i.e., a formal neighborhood of Γ). We suppose that the annulus is equipped with a projective structure and, thus, with a complex structure.

Let TAn denote the holomorphic tangent bundle to the annulus An, and $T\Gamma$ its restriction to Γ. Below we shall identify sections of $T\Gamma$ and their holomorphic continuations to the annulus as sections of TAn.

DEFINITION. We call a set of operators $\int_\Gamma Tv$ (labeled by sections v of $T\Gamma$) the *Virasoro algebra* if

$$\Big[\int_\Gamma T v_1, \int_\Gamma T v_2 \Big] = C_{\mathrm{Vir}}(v_1, v_2) + \int_\Gamma T[v_1, v_2]. \tag{2.39}$$

Here $[v_1, v_2]$ denotes the commutator of the holomorphic vector fields v_1 and v_2 on the annulus, C_{Vir} is a 2-cocycle in the algebra of holomorphic vector fields on the annulus,

$$C_{\mathrm{Vir}}(v_1, v_2) = c \int_\Gamma \Big(\frac{\partial v_1^Z}{\partial Z} \Big) d \Big(\frac{\partial v_2^Z}{\partial Z} \Big). \tag{2.40}$$

Here Z denotes a projective coordinate on the annulus, and v^Z stands for the component of the vector field v in this coordinate: $v = v^Z\, \partial/\partial Z$.

The number c in (2.40) is called the *central charge* of the Virasoro algebra.

REMARK. One can show that the cocycle (2.40) does not depend on the choice of projective coordinate.

DEFINITION. We say that the representation H of the loop algebra has an *energy-momentum tensor* (EMT) algebra, if the representation of the universal enveloping algebra associated to H contains a Virasoro algebra $\int_\Gamma Tv$ such that

$$\left[\int_\Gamma Tv, \int_\Gamma jf\right] = \int_\Gamma (\mathcal{L}_v f) j. \tag{2.41}$$

Here and below, $\mathcal{L}_v$ denotes the Lie derivative along the vector field v.

Now we can come back to the problem of finding a connection on the space of solutions to the sewing covector equations (2.37) for a family of Riemann surfaces. Namely, let us consider the N families of sewing maps $s_i(t)\colon \Gamma_i \to \Gamma_{N+i}$ and consider maps $\phi_i(t)\colon \Gamma_i \to \Gamma_i$ such that

$$\phi_i(t) = s_i^{-1}(0) \cdot s_i(t). \tag{2.42}$$

Let $t_{(A)}$ be coordinates parametrizing the family of sewing maps, and let us define the vector fields $v_{(A,i)}$ as derivatives of the maps $\phi_i(t)$ with respect to these coordinates:

$$v_{(A,i)} = \frac{\partial \phi_i}{\partial t_{(A)}}. \tag{2.43}$$

PROPOSITION 2.5. 1) *The equation*

$$\frac{\partial}{\partial t_{(A)}} \langle S_i| - \langle S_i| \int_{\Gamma_i} Tv_{(A,i)} = 0 \tag{2.44}$$

defines a connection on the space of solutions to the sewing covector equations for a family of Riemann surfaces.

2) *The curvature* $F_{(AB),i}$ *of this connection is given by the cocycle*

$$C_{\text{Vir}}(v_{(A,i)}, v_{(B,i)})$$

of the Virasoro algebra.

In particular, the connection is flat if the maps $\phi_i(t)$ are projective, i.e., preserve the projective structure. In this case we call the family of Riemann surfaces a *projective family.*

The proof is straightforward.

REMARK. It is clear that one projective family cannot include all complex structures. Consider the Riemann surface in the Schottky parametrization (see Appendix A). The projective family $s_i(t) = \gamma_i(t)$ corresponds to $\phi_i(t) \in SL(2,R)_{(i)} \subset SL(2,C)$, where $SL(2,R)_{(i)}$ denotes the subgroup of the group of projective transformations that preserves A_i. Nevertheless, for the purposes of the present paper, it is sufficient to prove properties of KZB equations over a projective family, because they form a real submanifold in the Teichmüller space, and KZB equations are holomorphic in t.

2.4.2. *Sugawara form of the energy-momentum tensor.* We define the operator $T_k(z)$ as an operator acting on H [**KZ**]:

$$T_k(z) = \lim_{w \to z} \frac{1}{k+h^*}\left(j^a(z)j^a(w) - k \dim G \, \frac{dz\, dw}{(z-w)^2}\right). \tag{2.45}$$

REMARK. The second term in (2.45) needs a projective structure for its definition.

PROPOSITION 2.6. *The operators* $\int_\Gamma T_k(z)v(z)$ *form the energy-momentum algebra for the loop algebra with central charge*

$$c_k = k\,\frac{\dim G}{k+h^*}. \tag{2.46}$$

DEFINITION. The *Knizhnik–Zamolodchikov–Bernard connection* is the flat connection in the bundle of generalized conformal blocks over the "projective" family of Riemann surfaces constructed by using the energy-momentum tensor algebra (2.45) with the Sugawara form of the energy-momentum tensor.

Previously we mentioned that we can identify the space of GCB's on a closed surface with the space of Ad_G-invariant elements of the space of holomorphic functions on G^N. In this identification, a correlator of n currents becomes an nth order differential operator on G^N taking values in the nth power of the Lie algebra.

In particular, the correlator of the energy-momentum tensor becomes a second-order operator. From the explicit form of the energy-momentum tensor (2.45), we determine its dependence on k,

$$\langle T_k(z)|g,\alpha\rangle = \left(\frac{1}{k+h^*}\,A_2(z) + \frac{k}{k+h^*}\,A_0(z)\right) GCB_\alpha(g,k). \tag{2.47}$$

Thus, for the GCB's over the projective family of Riemann surfaces we obtain

$$\left(\frac{\partial}{\partial t_A} + B_{(0;A)} - \frac{1}{\kappa}\,B_{(2;A)}\right) GCB_\alpha(g,\kappa,t) = 0. \tag{2.48}$$

Here A_n, B_n are nth order differential operators and $\kappa = k + h^*$.

Equation (2.48) is almost what we want—a flat connection with a spectral parameter.

Now we go on and show that there is a bilinear pairing that turns the KZB connection into a GS quadruple.

2.5. Partition function of ***bc*****-system and bilinear pairing on GCB's.**

Strategy. Our strategy in constructing a bilinear pairing respected by the KZB connection is as follows. We shall start by introducing the so-called *bc*-system in a framework parallel to our description of GCB above. Namely, using the chiral Heisenberg algebra, we write down the equations for generalized conformal blocks on a surface with boundary and for the sewing covectors. These equations will uniquely determine the generalized conformal block on an open surface up to a constant factor [**Va**] (recall that in the case of a loop algebra we had a finite-dimensional space of invariants).

Then we show that the $\dim G$ copies of the *bc*-system form a (reducible) representation of the loop algebra $\mathcal{G}_{h^*}$, where h^* is the dual Coxeter number. Moreover, the twist operators $\hat{\tau}^{bc}(g)$ (see 2.3.2) in the $\dim G$ copies of the *bc*-system can be extended to operators $\hat{\tau}^{bc}_{\dim G}(g)$ taking values in holomorphic top forms on G. Thus, sewing the generalized conformal block in the $\dim G$ copies of the *bc*-system, we obtain a generalized conformal block on the closed surface $Z_{bc}(g)$ (which is often called "the chiral determinant"), and this block is naturally a holomorphic top form on G^N. We shall construct the energy-momentum tensor for the bc-systems and thus obtain $Z_{bc}(g,t)$ for a family of Riemann surfaces.

Then we define the bilinear pairing between GCB's (transported by the flat EMT connection) as follows:

$$\langle GCB_\alpha(-\kappa,t), GCB_\beta(\kappa,t)\rangle = \int_{G_c^N} Z_{bc}(g,t) GCB_\alpha(g,-\kappa,t) GCB_\beta(g,\kappa,t), \tag{2.49}$$

where G_c is the maximal compact subgroup of the complex group G.

Our aim is to show that the pairing defined above is independent of t. To do this, we consider the total system, i.e., the system composed of the loop algebra with the central charge $\kappa = k + h^*$, the loop algebra with the central charge $-\kappa$, and $\dim G$ copies of the bc-system.

We shall observe that:

1) the total system contains a very special current J whose "zero mode" $\int_\Gamma J$ is a nilpotent Z_2-odd operator Q ($Q^2 = 0$);
2) the energy-momentum tensor of the total system is Q-exact;
3) the twist operator in the total system is replaced by the operator $\hat{\tau}_{\mathrm{tot},\dim G}$ such that $\{Q, \hat{\tau}_{\mathrm{tot},\dim G}\}$ is a d-exact form on G.

Finally, we shall show how it follows from 1), 2), and 3) that the derivative of the product (2.49) with respect to t equals zero.

2.5.1. *A single bc-system.*

DEFINITION. We define a Heisenberg algebra associated to the loop Γ as a Z_2-graded algebra generated by odd elements $\int_\Gamma f\cdot b$ and $\int_\Gamma h\cdot c$ which satisfy

$$\left\{\int_\Gamma f\cdot b, \int_\Gamma h\cdot c\right\} = \int_\Gamma fh, \tag{2.50}$$

$$\left\{\int_\Gamma f_1\cdot b, \int_\Gamma f_2\cdot b\right\} = \left\{\int_\Gamma h_1\cdot c, \int_\Gamma h_2\cdot c\right\} = 0, \tag{2.51}$$

where f is a function on Γ and h is a 1-form on Γ. As in the case of currents, we define local (fermionic) operators on Γ: a scalar operator $c(w)$ and a 1-form operator $b(z)$ so that

$$\{b(z), c(w)\} = \delta^{(1)}(z-w). \tag{2.52}$$

The analog of the Verma module (see 2.2.1) is the Fock representation H_{bc}, i.e., the irreducible representation containing the vacuum $|0\rangle$:

$$\int_\Gamma bz^n\,|0\rangle = \int_\Gamma cz^n\,dz\,|0\rangle, \qquad n \geqslant 0. \tag{2.53}$$

Now we come back to the open Riemann surface Σ^0 and associate to each boundary component Γ_i the Fock representation H_i^{bc} with vacuum $|0,i\rangle$. The operators acting in H_i^{bc} will be denoted by $b_{(i)}(z)$, $c_{(i)}(z)$, for $z\in\Gamma_i$.

DEFINITION OF A GENERALIZED CONFORMAL BLOCK FOR A SURFACE WITH BOUNDARY. The bc conformal block for a surface with boundary is an element

$$|\rangle \in H^{bc}_{\text{open}} = H^{bc}_1 \otimes \cdots \otimes H^{bc}_{2N}$$

which solves the system

$$\sum_{i=1}^{2N} \int_{\Gamma_i} c_{(i)} h \,|\rangle = 0, \tag{2.54}$$

$$\sum_{i=1}^{2N} \int_{\Gamma_i} b_{(i)} f \,|\rangle = 0 \tag{2.55}$$

for all 1-forms h and functions f holomorphic on Σ^0.

PROPOSITION 2.7. *Such a state is unique up to a numeric factor.*

PROOF. This follows from the Riemann–Roch theorem. □

DEFINITION. Here we inductively define correlators in the bc-system. Using the functions f_z as in the definition of current correlators (2.20), and 1-forms $h_z(w)$ which are holomorphic on the open surface for $w \neq z$ and have a simple pole at $w = z$ with residue equal to 1, we define:

(2.56)
$$\begin{aligned} &|b(z_1), \dots, b(z_{n+1}), c(w_1), \dots, c(w_m)\rangle \\ &\quad = \sum_i \int_{\Gamma_i} f_{z_{n+1}} b_{(i)} |b(z_1), \dots, b(z_n), c(w_1), \dots, c(w_m)\rangle \\ &\qquad - \sum_{k=1}^{m} f_{z_{n+1}}(w_k) |b(z_1), \dots, b(z_n), c(w_1), \dots, c(w_{k-1}), c(w_{k+1}), \dots, c(w_m)\rangle, \end{aligned}$$

(2.57)
$$\begin{aligned} &|b(z_1), \dots, b(z_n), c(w_1), \dots, c(w_{m+1})\rangle \\ &\quad = \sum_i \int_{\Gamma_i} h_{w_{m+1}} c_{(i)} |b(z_1), \dots, b(z_n), c(w_1), \dots, c(w_m)\rangle \\ &\qquad - \sum_{k=1}^{n} h_{w_{m+1}}(z_k) |b(z_1), \dots, b(z_{k-1}), b(z_{k+1}), \dots, b(z_n), c(w_1), \dots, c(w_m)\rangle. \end{aligned}$$

2.5.2. $\dim G$ *copies of the bc-system and the "chiral determinant"* $Z_{bc}(g)$. Now we take $\dim G$ copies of the bc-system, each copy labeled by a subscript; i.e., we shall deal with the operators b_a and c_a, $a = 1, \dots, \dim G$,

$$\{b_a(z), c_b(w)\} = \delta_{ab}\delta(z - w). \tag{2.58}$$

PROPOSITION 2.8. *The space* $(H^{bc})^{\otimes \dim G}$ *forms a (reducible) representation of the loop algebra with the central charge* $2h^*$.

PROOF. Consider the currents

$$j^a(z) = f^{abc} b_b(z) c_c(z). \tag{2.59}$$

□

REMARK. Note that the operators $b_a(z)$ and $c_a(w)$ form an evaluation representation of this algebra (associated to the adjoint representation):

$$\left[\int jf, b_a(z)\right] = f^{abc} f_b(z) b_c(z), \qquad \left[\int jf, c_a(w)\right] = f^{abc} f_b(w) c_c(w). \tag{2.60}$$

Sewing. The twisted sewing covectors are defined as in the case of the loop algebra. We start by defining ordinary sewing covectors $\langle S_i|$ by the equations

$$\begin{aligned} \langle S_i| \Big(\int_{\Gamma_i} (s_i^* f b_{a,(i)}) - \int_{\Gamma_{N+i}} f b_{a,(N+i)} \Big) &= 0, \\ \langle S_i| \Big(\int_{\Gamma_i} (s_i^* h c_{a,(i)}) - \int_{\Gamma_{N+i}} h c_{a,(N+i)} \Big) &= 0 \end{aligned} \tag{2.61}$$

for all functions f and 1-forms h on Γ_{N+i}.

Now we introduce twists.

The convolutions of the current (2.59) with constant functions represent the finite-dimensional Lie algebra $\mathcal{G}$ in $(H_i^{bc})^{\otimes \dim G}$,

$$\hat{\tau}_i(t_a) = \int_{\Gamma_i} j^a_{(i)}. \tag{2.62}$$

One can check that this representation can be integrated to a representation of the group G. This allows us to define the twist operators $\hat{\tau}^{bc}(g_i)$ as operators in $(H_i^{bc})^{\otimes \dim G}$. This construction is completely analogous to that for the loop algebra.

Finally, we define the correlators of b and c fields on a sewn Riemann surface as follows:

$$\begin{aligned} &\langle b(z_1), \ldots, b(z_n), c(w_1), \ldots, c(w_m); g\rangle \\ &\quad = \langle S_1^{bc} \hat{\tau}^{bc}(g_1) \cdots S_N^{bc} \hat{\tau}^{bc}(g_N) | b(z_1), \ldots, b(z_n), c(w_1), \ldots, c(w_m)\rangle. \end{aligned} \tag{2.63}$$

PROPOSITION 2.9. *The correlator*

$$\langle b(z_1), \ldots, b(z_n), c(w_1), \ldots, c(w_m); g\rangle, \tag{2.64}$$

regarded as a function of w_k, is a meromorphic section of the vector bundle $E(g)$ determined by the transition functions g_i on A_i cycles, with the simple poles at points z_l with residues

$$\langle b(z_1), \ldots, b(z_{l-1}), b(z_{l+1}), \ldots, b(z_n), c(w_1), \ldots c(z_{k-1}), c(z_{k+1}), \ldots, c(w_m); g\rangle. \tag{2.65}$$

The same correlator (2.64), *regarded as a function of z_l, is a meromorphic* 1*-form taking values in the vector bundle $E(g)$ with simple poles at the points w_k with the residues given by the correlators* (2.65).

PROPOSITION 2.10. *The correlator* (2.64) *is nonzero only if*

$$n - m = (N-1) \dim G. \tag{2.66}$$

PROOF. This follows from the Riemann–Roch theorem. □

It turns out that in the bc system the twist operators $\hat{\tau}^{bc}(g)$ can be naturally extended from functions to forms on G. This will allow us to construct a measure on the space of twists, that is to say, on G^N.

DEFINITION. For any $p \leqslant \dim G$ we define $\hat{\tau}_p^{bc}(g)$, a p-form on G, taking values on the operators acting on the representation space $(H^{bc})^{\otimes \dim G}$, as follows. Let $\hat{\tau}_0^{bc}(g)$ be a twist operator in the bc-system. Then the pairing between the p-form $\hat{\tau}_p^{bc}(g)$ and left-invariant vector fields $e_{a_1}(g), \dots, e_{a_p}(g)$ on G is given by

$$\hat{\tau}_p^{bc}(g)(e_{a_1}(g), \dots, e_{a_p}(g)) = \int_\Gamma b_{a_1}, \dots, \int_\Gamma b_{a_p} \hat{\tau}_0^{bc}(g). \tag{2.67}$$

Now we can construct the conformal block on a sewed surface (the "chiral determinant") Z_{bc}, which can be used for the pairing, by sewing open conformal block with the insertion of $\prod_{a=1}^{\dim G} c_a(w_0)$, using the "twist" forms $\hat{\tau}_{\dim G}^{bc}$ (one for each handle) described above:

$$Z_{bc}(g, w_0) = \left\langle \prod_{i=1}^{N} S_i \hat{\tau}_{\dim G}^{bc}(g_i) \middle| \prod_{a=1}^{\dim G} c_a(w_0) \right\rangle. \tag{2.68}$$

PROPOSITION 2.11. *The energy-momentum tensor for the bc-system is given by the following expression:*

$$T_{bc}(z) = \lim_{w \to z} \left(b_a(z) \partial c_a(w) - \dim G \frac{dz\, dw}{(z-w)^2} \right). \tag{2.69}$$

We define $Z_{bc}(g, t)$ for a family of Riemann surfaces with the help of this energy-momentum tensor as we did in subsection 2.4 above.

2.5.3. *The total system and invariance of pairing.* Now we take three chiral algebras: the first one is the loop algebra of group G with the central charge k (in this case $\kappa = k + h^*$), the second one again the loop algebra, but now the central charge is $-k - 2h^*$ (i.e. $\kappa' = -\kappa$), and the third algebra is the sum of $\dim G$ copies of the bc-system.

Below we shall label everything concerning the second system by a prime.

The energy-momentum tensor of the total theory is just the sum of the energy-momentum tensors of the above theories:

$$T^{\text{tot}}(z) = T_\kappa(z) + T'_{\kappa'}(z) + T_{bc}(z). \tag{2.70}$$

The main feature of the total system is that it contains the so-called BRST current J:

$$J(z) = c_a(z) \left(j^a(z) + j'^a(z) + \frac{1}{2} j^a_{bc}(z) \right). \tag{2.71}$$

Using the current J, we define the operator Q as follows:

$$Q = \int_\Gamma J. \tag{2.72}$$

PROPOSITION 2.12. *The main properties of Q, established in* [**FF**], *are:*

1) *Q is nilpotent, i.e.,*

$$Q^2 = 0. \tag{2.73}$$

2) *The total current $j^a + j'^a + j^a_{bc}$ is Q-exact, namely*

$$\{Q, b^a\} = j^a + j'^a + j^a_{bc}. \tag{2.74}$$

3) *The total energy-momentum tensor is Q-exact,*

$$T = \{Q, \theta\}, \qquad \theta = \frac{1}{2\kappa} b^a j^a + \frac{1}{2\kappa'} b^a j^a. \tag{2.75}$$

DEFINITION. We define the modified twist operators, $\hat{\tau}_{\mathrm{tot},p}$, acting in $H_{\mathrm{tot}} = H \otimes H' \otimes (H^{bc})^{\otimes \dim G}$ as the product of $\hat{\tau}_p^{bc}$, the "twist" operator $\tau(g)$ in the κ system and the "twist" operator $\tau'(g)$ in the $\kappa' = -\kappa$ system:

$$\hat{\tau}_{\mathrm{tot},p}(g) = \hat{\tau}_p^{bc}(g)\hat{\tau}(g)\hat{\tau}'(g). \tag{2.76}$$

PROPOSITION 2.13. *The supercommutator of Q with $\hat{\tau}_{\mathrm{tot},p}(g)$ (observe that $\hat{\tau}_{\mathrm{tot},p}(g)$ is fermionic for odd p) gives an exact form:*

$$[Q, \hat{\tau}_{\mathrm{tot},p}(g)]_{\mathrm{Super}} = d\hat{\tau}_{\mathrm{tot},p-1}(g), \tag{2.77}$$

where d is the external differential on G.

PROOF. This follows from (2.74). □

THEOREM 2.2. *For a projective family of Riemann surfaces, we have*

$$\frac{d}{dt}\int_{G^N_{\mathrm{comp}}} Z_{bc}(g,t)GCB_\alpha(g,-\kappa,t)GCB_\beta(g,\kappa,t) = 0. \tag{2.78}$$

IDEA OF THE PROOF [**Los1**]. The derivative along t of the integrand in (2.78) is equal to the correlator of the total energy-momentum tensor. Since the total energy-momentum tensor equals $\{Q,\theta\}$ (see (2.75)), the integrand can be represented as a sum of two terms. The first term in the integrand has the following structure:

$$\left\langle \mathrm{sew} \,\middle|\, \theta \sum_{i=1}^{2N} \int_{\Gamma_i} J_{(i)} \,\middle|\, \prod_{a=1}^{\dim G} c_a(w_0), \mathrm{tot} \right\rangle, \tag{2.79}$$

where

$$\left| \prod_{a=1}^{\dim G} c_a(w_0), \mathrm{tot} \right\rangle \tag{2.80}$$

is a conformal block of the total system on an open surface (with insertions of c-fields), and $\langle\mathrm{sew}|$ is the sewing covector multiplied by twist operators. One can check that in the total system the correlator

$$\left| J(z) \prod_{a=1}^{\dim G} c_a(w_0), \mathrm{tot} \right\rangle, \tag{2.81}$$

considered as a function of z, is a holomorphic 1-differential on Σ^0; thus, the first term is equal to zero (as an integral of a holomorphic 1-differential along the boundary).

The second term in the integrand contains the supercommutator of Q with twist operators $\hat{\tau}_{\mathrm{tot},\dim G}$, and is a d-exact form on G_c^N due to (2.77). Thus, the derivative along t changes the integrand by an exact form that does not contribute to the integral over the compact manifold G_c. □

In order to put the KZB equations in the traditional form [**B2**], we turn to modified GCB's.

DEFINITION. The modified generalized conformal blocks are Ad_G-invariant functions on G^N given by

$$GCB_\beta^{\mathrm{mod}}(g,\kappa,t) = \left(\frac{Z_{bc}(g,t)}{\mathrm{Haar}(g)}\right)^{1/2} GCB_\beta(g,\kappa,t), \tag{2.82}$$

where $\mathrm{Haar}(g)$ is the Haar measure on G_c^N holomorphically continued to a holomorphic top form on G^N.

The space of modified invariants is equipped with the bilinear pairing

$$\int_{G_c^N} GCB_\beta^{\mathrm{mod}}(g) GCB_\alpha^{\mathrm{mod}}(g)\,\mathrm{Haar}(g). \tag{2.83}$$

COROLLARY 2.2. *The KZB connection on modified conformal blocks has the form*

$$\frac{\partial}{\partial t_B} - \frac{A_{(2,B)}}{\kappa}, \tag{2.84}$$

where $A_{(2,B)}$ is a second order differential operator on G^N, selfadjoint with respect to the pairing (2.83).

PROOF. This follows from the previous theorem and from the fact that the 0th-order differential operator

$$\frac{\partial}{\partial t_B} \frac{1}{2} \log Z_{bc} + B_{(0,B)}, \tag{2.85}$$

see (2.48), can be antiselfadjoint (with respect to the pairing (2.83)) only if it equals zero. □

3. Explicit description of Bernard operators and direct proof of their properties

In this section we explicitly construct the KZB connection (in the so-called "Bernard parametrization" [**B1, B2**]) and verify its properties. Instead of presenting a consistent derivation of the connection, we write down only the resulting expression for the connection form. The considerations in the previous section should be sufficient to simply translate the "physical" derivation of [**Iv**] into the language of coinvariants.

We shall construct the KZB connection 1-form (the Bernard operators) as

$$A(z) = \Delta(z) + U(z), \tag{3.1}$$

where $\Delta(z)$ is a quadratic differential operator on Ad_G-invariant functions on G^N and $U(z)$ is a scalar ("potential") term (as functions of z they both are holomorphic two-differentials on the compact Riemann surface Σ). We start by introducing a basis of twisted 1-forms on the surface—the building blocks for our constructions. Then we define L-operators and construct the Bernard Laplacian $\Delta(z)$. After some auxiliary constructions (in particular, introducing the function Π, "the square root of the twisted bc partition function Z_{bc}" [**B2, Los1**]), we write down an explicit expression for the "potential term" $U(z)$. Finally, we state and prove the main theorem of this section, describing the properties of the connection.

3.1. Twisted 1-forms. Let Σ be a Riemann surface of genus N, and let the loops $A_1, \ldots, A_N$ be A-generators of $\pi_1(\Sigma)$. Any ordered set of N group elements $g_1, \ldots, g_N$ of G (twisting elements) defines transition functions for bundles of twisted meromorphic 1-forms (for some given locations and orders of poles). We encountered examples of such twisted 1-forms in the previous section, namely, the current correlators and the correlators of b-fields. Below we define several bases of twisted meromorphic 1-forms. For convenience, we mark a point w_0 on the surface and allow the forms to have single poles at w_0. We shall see later that the result of the construction is independent of the choice of w_0.

In Appendix A, we list the explicit expressions for all these twisted 1-forms in Schottky coordinates.

3.1.1. *Twisted 1-forms* $\omega_i(z)$. We start off by defining N elements $\omega_i(z)$ of $\Omega^1(\Sigma \setminus (\bigcup_{i=1}^N A_i \cup \{w_0\})) \otimes \operatorname{End}\mathcal{G}$ by the following properties:

(i) The form $\omega_i(z)$ has a simple pole at $z = w_0$.

(ii) The form $\omega_i(z)$ is a twisted 1-form; i.e., its jumps across the loops A_j are governed by the adjoint action of g_j:

$$\omega_i(z_{j+}) = \operatorname{Ad}_{g_j} \omega_i(z_{j-}), \qquad j = 1, \ldots, N. \tag{3.2}$$

(iii) The integrals of these forms along the loops A_i are

$$\oint_{A_j} \omega_i(z_-) = \delta_{ij}\mathbf{1} \tag{3.3}$$

($\mathbf{1}$ is the unit element of $\operatorname{End}\mathcal{G}$).

PROPOSITION 3.1. *The forms* $\omega_i(z)$ *satisfying the above conditions* (i), (ii), (iii) *exist and are unique.*

3.1.2. *Twisted 1-forms* $\theta(z;w)$. Another twisted 1-form that we shall use is

$$\theta^a_b(z;w) \in \Omega^1\left(\Sigma \setminus \left(\bigcup_{i=1}^N A_i \cup \{w_0, w\}\right)\right) \otimes \operatorname{End}\mathcal{G}.$$

It is defined by the following properties:

(i) The form $\theta(z;w)$ is a twisted 1-form with respect to (a,z):

$$\theta^a_b(z_{j+};w) = (\operatorname{Ad}_{g_j})^a_c \theta^c_b(z_{j-};w). \tag{3.4}$$

(ii) The form $\theta(z;w)$ has first-order poles at $z = w$ and at $z = w_0$. The residue at $z = w$ is

$$\theta(z;w) \sim \frac{\mathbf{1}}{z-w}. \tag{3.5}$$

(iii) The integrals over the loops A_j vanish:

$$\oint_{A_j, z} \theta(z;w) = 0. \tag{3.6}$$

PROPOSITION 3.2. *Such a form* $\theta^a_b(z;w)$ *exists and is unique.*

3.1.3. *Nonabelian prime-form* $\Omega^{ab}(z,w)$. Finally, we introduce a twisted analog of the prime-form $\Omega^{ab}(z,w)$ defined by the following three conditions:

(i) The form $\Omega^{ab}(z,w)$ is a twisted 1-form with respect to both (a,z) and (b,w):

$$\Omega^{ab}(z_{j+},w) = (\operatorname{Ad}_{g_j})^a_c \Omega^{cb}(z_{j-},w), \tag{3.7}$$

$$\Omega^{ab}(z,w_{j+}) = (\operatorname{Ad}_{g_j})^b_c \Omega^{ac}(z,w_{j-}), \tag{3.8}$$

where z_{j+} and z_{j-} are points on the two sides of the loop A_j (and similarly for w_{j+} and w_{j-}).

(ii) The form $\Omega^{ab}(z,w)$ has a second-order pole at $z = w$, and

$$\Omega^{ab}(z,w) \sim \frac{\delta^{ab}}{(z-w)^2}. \tag{3.9}$$

(iii) The integrals over the loops A_j vanish (with respect to both z and w):

$$\oint_{A_j,z} \Omega^{ab}(z,w) = \oint_{A_j,w} \Omega^{ab}(z,w) = 0. \tag{3.10}$$

PROPOSITION 3.3. *Such a form exists, is unique, and is symmetric:*

$$\Omega^{ab}(z,w) = \Omega^{ba}(w,z). \tag{3.11}$$

PROPOSITION 3.4. *The prime-form $\Omega^{ab}(z,w)$ generates the other two forms $\theta^a_b(z;w)$ and $\omega^a_{ib}(z)$:*

$$\theta^a_b(z;w) = \int_{\xi=w_0}^{w} \Omega^{ab}(z,\xi), \tag{3.12}$$

$$\omega^a_{ib}(z) = \int_{\xi=w_0}^{\gamma_i(w_0)} \Omega^{ab}(z,\xi) = \theta^a_b(z;\gamma_i(w_0)), \tag{3.13}$$

where, in the second formula, the integral of the analytic continuation of $\Omega^{ab}(z,\xi)$ is performed over a loop homotopic to B_i (intersecting the A_i-loop).

3.2. L-operators. Let $\{\xi^a\}$ be an orthonormal (with respect to the Killing form) basis of $\mathcal{G}$. Let $\mathcal{L}^{ia}$ be the right-invariant derivatives on G^N:

$$\mathcal{L}^{ia} f(g_1,\dots,g_N) = \frac{d}{d\varepsilon}\bigg|_{\varepsilon=0} f(g_1,\dots,e^{\varepsilon\xi^a}g_i,\dots,g_N). \tag{3.14}$$

Similarly, define the left-invariant derivatives $\mathcal{R}^{ia} = \mathrm{Ad}_{g_i}\mathcal{L}^{ia}$. Construct the L-operator[2] as

$$L^a(z) = \omega^a_{ib}(z)\mathcal{L}^{ib}. \tag{3.15}$$

PROPOSITION 3.5. *The restriction of $L^a(z)$ to the space of Ad_G-invariant functions does not depend on w_0.*

PROOF. (i) The residue of $L^a(z)$ at $z = w_0$ is

$$\operatorname*{Res}_{w_0} L^a(z) = \sum_i (\mathcal{L}^{ia} - \mathcal{R}^{ia}),$$

which is zero on the space of Ad_G-invariant functions. Therefore, $L^a(z)$ is regular at $z = w_0$.

(ii) Then the difference of two L-operators corresponding to different choices of w_0 obeys properties (ii) and (iii) of Proposition 3.1 and is regular at w_0. From the proof of Proposition 3.1 it follows that this difference must vanish. □

3.3. Natural bilinear form. Consider the bilinear form $(\cdot\,,\cdot)$ on holomorphic functions on G^N defined by

$$(f,g) = \int \prod_i dg_i\, fg, \tag{3.16}$$

[2]The invariant meaning of this L-operator is the derivative along the moduli space MH of holomorphic G-bundles. The space of holomorphic twisted 1-forms on Σ is naturally identified with the cotangent space to MH. One easily sees that the map from MH to T^*MH provided by $L(z)$ defined above coincides with the differential on MH.

where the integration is performed with respect to the invariant measure on the compact group G_c. This bilinear form is Ad_G-invariant. Then the differential operator conjugated to $L^a(z)$ is

$$L^{a\dagger}(z) = -\mathcal{L}^{ib}\omega^a_{ib}(z). \tag{3.17}$$

3.4. Important identities. The twisted 1-forms introduced above satisfy several important identities:

PROPOSITION 3.6.

$$\mathcal{L}^{ia}\Omega^{bc}(w,\xi) = -f^{ade}\oint_{A_i} dz\,\Omega^{dc}(z,\xi)\,\theta^b_e(w;z). \tag{3.18}$$

PROOF. The left-hand side and the right-hand side obey the same transformation properties with respect to w and ξ. □

PROPOSITION 3.7. *The following identity holds:*

(3.19)

$$L^a(z)\Omega^{bc}(w,\xi) - L^b(w)\Omega^{ac}(z,\xi) + [\Omega^{ad}(z,\xi)\theta^b_e(w;\xi) - \Omega^{bd}(w,\xi)\theta^a_e(z;\xi)]f^{cde} + \Omega^{ec}(w,\xi)\theta^a_d(z;w)f^{bde} - \Omega^{ec}(z,\xi)\theta^b_d(w;z)f^{ade} = 0.$$

STEPS OF THE PROOF. (i) Check that the left-hand side is a twisted 1-differential with respect to z and a.

(ii) Check that the left-hand side is regular at $z = w$ and $z = \xi$.

(iii) Check that the integrals along A_j-loops vanish (using the previous proposition). □

SIMPLE COROLLARY (commutator of L-operators).

$$[L^a(z), L^b(w)] = f^{acd}\theta^b_c(w;z)L^d(z) - f^{bcd}\theta^a_c(z;w)L^d(w). \tag{3.20}$$

3.5. Bernard Laplacian. Let us construct the quadratic differential operator (Bernard Laplacian):

$$\Delta(z) = -L^{a\dagger}(z)L^a(z). \tag{3.21}$$

PROPOSITION 3.8. *$\Delta(z)$ commutes with the Ad_G action. The restriction of $\Delta(z)$ to the space of Ad_G-invariant functions is regular at w_0.*

To construct the KZB operators, we must introduce the "potential term", which we shall do after some additional definitions.

3.6. Regularization of singularities. The procedure of regularizing singularities plays an important role in our construction. In physical language, this means a definition of the product of local operators at coinciding points. This procedure corresponds to the normal ordering of operators in the Sugawara construction. In this paper we avoid discussing an invariant description of the normal ordering (which is an important issue) and confine ourselves to a simple coordinate-dependent definition.

Suppose we have a projective structure on the surface Σ (i.e., a local coordinate up to a projective transformation). Then we may define regularized versions of 1-forms introduced above:

$$\Omega^{ab}(z,z)_{\mathrm{reg}} = \lim_{w\to z}\left(\Omega^{ab}(z,w) - \frac{\delta^{ab}}{(z-w)^2}\right). \tag{3.22}$$

From now on we fix the projective structure to be the one induced by the Schottky parametrization of the surface.

3.7. Differential on the moduli space of complex structures. An infinitesimal change in the complex structure on the surface can be described by a Beltrami differential $\mu(z,\bar{z})$, which is a $(-1,1)$-differential (or rather a first Dolbeaut cohomology class with values in holomorphic vector fields). Consequently, 1-forms on the moduli space of complex structures $\mathcal{M}_N$ can be identified with 2-differentials on the surface. The pairing with tangent vectors to $\mathcal{M}_N$ is natural:

$$(T,\mu) = \int_\Sigma T(z)\mu(z,\bar{z})\,dz\,d\bar{z}. \tag{3.23}$$

We may view the exterior derivative d_m on $\mathcal{M}_N$ as a holomorphic operator-valued 2-differential on the surface $d_m(z)$:

$$\delta_\mu F = \int_\Sigma \mu(z,\bar{z})\,d_m(z)F\,dz\,d\bar{z} \tag{3.24}$$

for any function F on $\mathcal{M}_N$.

PROPOSITION 3.9.

$$d_m(z)\Omega^{ab}(u,w) = \Omega^{ac}(u,z)\Omega^{cb}(z,w). \tag{3.25}$$

In the above formula, $\Omega^{ab}(u,w)$ should be considered as a function on the moduli space of surfaces with two marked points equipped with tangent vectors. A cotangent vector to such a space can be identified with two-differentials that are allowed to have second-order poles at u and w, the first-order residues being responsible for moving marked points and the second-order residues, for rescaling the local coordinates at the points. Thus, the second-order residues of $d_m(z)\Omega^{ab}(u,w)$ at $z=u$ and $z=w$ must coincide with $\Omega^{ab}(u,w)$; the first-order residues must coincide with $\Omega^{ab}(u,w)/du$ and $\Omega^{ab}(u,w)/dw$, respectively. Together with the requirement of $d_m(z)\Omega^{ab}(u,w)$ being regular at $u=w$, this fixes the right-hand side of (3.25) uniquely.

3.8. *bc* twisted partition function.

PROPOSITION 3.10. *There exists an* Ad_G*-invariant holomorphic function* Π *on* $G^N\times\mathcal{M}_N$ *(the square root of the twisted bc partition function) such that:*

$$\text{(a)}\qquad 2L^a(z)\log\Pi + \mathcal{L}^{ib}\omega^a_{ib} - f^{abc}\theta^b_c(z;z) = 0, \tag{3.26}$$

$$\text{(b)}\qquad d_m(z)\log\Pi = -\frac{1}{2}\,\Omega^{aa}(z,z)_{\mathrm{reg}}. \tag{3.27}$$

PROOF. This function is given explicitly in the Schottky parametrization by

$$\Pi = \prod_{k=1}^{\infty}\prod_{\gamma\in\mathrm{prim}}\det_{\mathrm{adj}}(1-g_\gamma K^k_\gamma), \tag{3.28}$$

where the product is taken over the primitive conjugate classes of the Schottky group (γ and γ^{-1} are elements of the same primitive class), the determinants are computed in the adjoint representation, and K_γ is the multiplier of the transformation γ.

To prove (a), we verify that the right-hand side is a twisted automorphic 1-form in (a,z) and that the integrals over A_j-loops vanish (from the explicit expression for Π). Part (b) is explicitly verified in [**Mar**]. □

3.9. Bernard operators. Define the "potential" $U(z)$ by

$$U(z) = h^* \Omega^{aa}(z,z)_{\text{reg}} - \frac{1}{\Pi}\Delta(z)\Pi. \tag{3.29}$$

It is a holomorphic 2-differential on the surface.

Define the Bernard operators

$$A(z) = \Delta(z) + U(z). \tag{3.30}$$

It is a holomorphic 2-differential on the surface with values in G-invariant second-order differential operators on G^N. Alternatively, it may be viewed as an operator-valued 1-form on $\mathcal{M}_N$ (which has $\dim \mathcal{M}_N$ components).

3.10. Properties of the KZB connection.

MAIN THEOREM 3.1. (i) *All the operators* $A(z)$ *commute*:

$$[A(z), A(w)] = 0. \tag{3.31}$$

(ii) $A(z)$ *is a closed* 1*-form on* $\mathcal{M}_N$.

(iii) $A(z)$ *is self-conjugate with respect to the bilinear form* $(\cdot\,,\cdot)$ *on* $C(G^N)$ *introduced above.*

PROOF. (i) and (ii) have straightforward verifications using Propositions 3.8–3.10 (the proof is sketched in Appendices B and C; for more details see [**Iv**]).

(iii) is obvious. □

3.11. KZB equation on surfaces with marked points. The KZB equation on a surface with marked points may be obtained as the limit of the general form of the KZB equation when some of the handles of the surface degenerate to double points. In other words, the KZB connection on moduli of surfaces with marked points is the restriction of the KZB connection to the boundary of the moduli space of complex structures. Below we give an explicit description of this procedure.

Consider a Riemann surface Σ of genus N_1+N_2+M; by shrinking $M+1$ cycles Σ degenerates into two surfaces Σ_1 and Σ_2 of genera N_1 and N_2 connected by M double points. We expect that under such a degeneration the generalized conformal blocks on the original surface factor into products of generalized conformal blocks on the two surfaces with marked points. We shall see how this happens from the explicit form of the KZB equation.

Let the twisting group elements for the surface Σ and the auxiliary point w_0 be chosen in such a way that the twists $g_1, \ldots, g_M$ are placed to the first M of the double points (i.e., it is the loops $A_1, \ldots, A_M$ that shrink to points) and w_0 is placed to the $(M+1)$th double point. Then one can easily verify that on the degenerate surface, the twisted 1-forms $\omega^a_{bi}(z)$ become

$$\omega^a_{ib}(z)_\Sigma = \begin{cases} \omega^a_{ib}(z)_{\Sigma_1} & \text{for } z \in \Sigma_1, \\ 0 & \text{for } z \in \Sigma_2, \end{cases} \tag{3.32}$$

if the twist i belongs to Σ_1 (and similarly if it belongs to Σ_2), and

$$\omega^a_{ib}(z)_\Sigma = \begin{cases} \theta^a_b(z; w_i)_{\Sigma_1} & \text{for } z \in \Sigma_1, \\ (\mathrm{Ad}_{g_i})^c_b \theta^a_c(z; w_i)_{\Sigma_2} & \text{for } z \in \Sigma_2, \end{cases} \tag{3.33}$$

if $1 \leqslant i \leqslant M$, i.e., if the twist i is at one of the double points w_i.

Therefore, for the L-operator we have

$$L(z) = L_1(z) + L_2(z), \tag{3.34}$$

where

$$L_1(z) = \begin{cases} \omega^a_{ib}(z)_{\Sigma_1}\mathcal{L}^{ib} + \sum_{j=1}^M \theta^a_b(z;w_j)_{\Sigma_1}\mathcal{L}^{jb} & \text{for } z \in \Sigma_1, \\ 0 & \text{for } z \in \Sigma_2, \end{cases} \tag{3.35}$$

$$L_2(z) = \begin{cases} 0 & \text{for } z \in \Sigma_1, \\ \omega^a_{ib}(z)_{\Sigma_2}\mathcal{L}^{ib} + \sum_{j=1}^M \theta^a_b(z;w_j)_{\Sigma_2}\mathcal{R}^{jb} & \text{for } z \in \Sigma_2. \end{cases} \tag{3.36}$$

We note first that $L_1(z)$ and $L_2(z)$ commute and that they commute with the action of the twists at the degenerate points. Therefore, we may restrict these operators to any representations $V_1, \dots, V_M$ placed at marked points of Σ_1 (and $V_1^*, \dots, V_M^*$ at marked points of Σ_2), and call the operator

$$L(z) = \omega^a_{ib}(z)\mathcal{L}^{ib} + \sum_j \theta^a_b(z;w_j)t^b_j \tag{3.37}$$

the *L-operator on a surface with marked points.* This operator acts in the space of invariants $\mathrm{Inv}_G(C(G^N) \otimes V_1 \otimes V_2 \otimes \cdots \otimes V_M)$ (the adjoint action on $C(G^N)$ is assumed); t^b_j is the representation of the Lie algebra in V_j.

From the decomposition of the L-operator, it follows that

$$\Delta(z)_\Sigma = \Delta(z)_{\Sigma_1} + \Delta(z)_{\Sigma_2}. \tag{3.38}$$

It remains to prove a similar decomposition for the "potential" $U(z)$ in order to verify that the KZB equation on the degenerate surface splits into two commuting equations on Σ_1 and Σ_2. From the explicit form of Π in the Schottky parametrization (3.28), it is easy to see that on the degenerate surface Σ we have

$$\Pi_\Sigma = \Pi_{\Sigma_1}\Pi_{\Sigma_2}, \tag{3.39}$$

where Π_{Σ_α} is the function Π on Σ_α, and that Π does not depend on the twists at the double points.

Then for the potential $U(z)$ we get

$$U(z) = -\frac{1}{\Pi_\Sigma}(2h^* + \Delta(z))\Pi_\Sigma = \begin{cases} U(z)_{\Sigma_1} & \text{for } z \in \Sigma_1, \\ U(z)_{\Sigma_2} & \text{for } z \in \Sigma_2. \end{cases} \tag{3.40}$$

Thus we have proved that the KZB equation on a surface with marked points is a limiting case of more general equations on higher-genus surfaces without marked points. The main theorem about the properties of the KZB connection therefore also holds for the KZB connection on a surface with marked points.

3.12. Examples. Finally, we illustrate our discussion by the three very simple examples of the KZB connection. In all three the "potential" term vanishes.

(1) The sphere with marked points. This gives us the well-known KZ equation:

$$A(z) = \sum_{i \neq j} \frac{t^a_i t^a_j}{(z - z_i)(z - z_j)} = d_m(z)\left[\sum_{i \neq j} t^a_i t^a_j \log(z_i - z_j)\right]. \tag{3.41}$$

(2) The torus without marked points. The torus can be parametrized as a quotient of $\mathbf{C}^*$ by the equivalence $z \sim qz$. In this parametrization,

$$A(z) = \frac{1}{z^2}\Delta_G = d_m(z)[\log q\Delta_G], \tag{3.42}$$

where $\Delta_G = \mathcal{L}^a\mathcal{L}^a$ is the Laplacian on the group G.

(3) The Abelian group $G = U(1)$, a surface of arbitrary genus without marked points. Then

$$A(z) = (d_m(z)\tau_{ij})\mathcal{L}^i\mathcal{L}^j, \tag{3.43}$$

where $\tau_{ij} = \int_{B_j} \omega_i(z)\,dz$ are the periods of the surface.

Acknowledgments

We would like to thank A. Alekseev, I. Frenkel, V. Fock, A. Gerasimov, A. Levin, G. Moore, A. Morozov, N. Nekrasov, M. Olshanetsky, A. Rosly, and S. Shatashvili for helpful discussions, and A. Rosly for help in the final stage of preparing the text.

Appendix A. Schottky parametrization of Riemann surfaces

In the Schottky parametrization, the surface is constructed as a quotient of the Riemann sphere (more precisely, of the sphere with the fixed points of the group deleted) by the action of the Schottky group. The Schottky group Γ is the group freely generated by N projective maps γ_i such that one can find $2N$ circles A_i and $A_i' = \gamma_i(A_i)$, $i = 1, \ldots, N$, all external to each other, so that γ_i maps the exterior of A_i onto the interior of A_i'. The region exterior to all the circles A_i and A_i' is a fundamental domain of Γ. The surface is obtained by sewing each circle A_i' to A_i by the action of γ_i. Then the circles A_i become A-cycles on the surface.

For future use, we need two more definitions. The first one is the parametrization of a projective transformation γ by its fixed points u_γ, v_γ (repulsive and attractive) and its multiplier K_γ defined by

$$\frac{\gamma(z) - u_\gamma}{\gamma(z) - v_\gamma} = K_\gamma \frac{z - u_\gamma}{z - v_\gamma}, \qquad |K_\gamma| < 1. \tag{A.1}$$

Finally, we shall need to extend the twists introduced on the handles of the surface to the group homomorphism between Γ and G:

$$g\colon \Gamma \to G, \qquad g_{\gamma_i} = g_i, \qquad g_{\gamma\mu} = g_\gamma g_\mu. \tag{A.2}$$

The global coordinate z on the Riemann sphere defines a projective structure on the surface. It is remarkable that in this family of projective structures the stress-energy tensor of the bc system $\Omega^{aa}(z,z)_{\text{reg}}$ can be integrated to the partition function Π^2 according to (3.27); because of this, in the Schottky projective structure the KZB connection becomes flat (not only projectively flat).

To illustrate the definitions of Section 3 and to facilitate comparison with other papers, we list the Poincaré series for the twisted meromorphic 1-forms $\omega_{ib}^a(z)$, $\theta_b^a(z;w)$ and $\Omega^{ab}(z,w)$:

$$\omega_{ib}^a(z) = \sum_{\gamma\in\Gamma} (g_\gamma^{-1})_b^a \left[\frac{\gamma'(z)}{\gamma(z) - \gamma_i(w_0)} - \frac{\gamma'(z)}{\gamma(z) - w_0}\right] dz, \tag{A.3}$$

$$\theta_b^a(z;w) = \sum_{\gamma\in\Gamma} (g_\gamma^{-1})_b^a \left[\frac{\gamma'(z)}{\gamma(z) - w} - \frac{\gamma'(z)}{\gamma(z) - w_0}\right] dz, \tag{A.4}$$

$$\Omega^{ab}(z,w) = \sum_{\gamma\in\Gamma} (g_\gamma^{-1})_b^a \frac{\gamma'(z)\,dz\,dw}{(\gamma(z) - w)^2}. \tag{A.5}$$

(The sum is over the Schottky group, the matrix elements are taken in the adjoint representation, and indices are raised and lowered by the Killing form.)

Appendix B. The KZB connection form is closed

The Bernard Laplacian is an exact 1-form and can be integrated using (3.25):

$$\Delta(z) = d_m(z)\Big(\mathcal{L}^{ib}\Big[\int_{\xi=w_0}^{\gamma_i(w_0)}\int_{\eta=w_0}^{\gamma_j(w_0)}\Omega^{bc}(\xi,\eta)\Big]\mathcal{L}^{jc}\Big). \tag{B.1}$$

(Since $\Omega^{ab}(z,w)$ has no first-order residue at $z = w$, the double integral is well defined.)

To verify that $U(z)$ represents an exact 1-form, we check that

$$d_m(w)U(z) - d_m(z)U(w) = 0, \tag{B.2}$$

where the operators $d_m(z)$ are understood as in Proposition 3.9.

Since we are interested in the antisymmetric part with respect to the interchange $z \leftrightarrow w$, we shall use the symbol $\simeq$ to indicate that the antisymmetric parts of the two expressions are equal. We have

$$d_m(z)U(w) = h^* d_m(z)\Omega^{aa}(w,w)_{\mathrm{reg}} - \frac{1}{2}\, d_m(z)\Big(\frac{1}{\Pi}\,\Delta(w)\Pi\Big). \tag{B.3}$$

The first summand has zero antisymmetric part, since $\Omega^{aa}(w,w)_{\mathrm{reg}}$ is an exact 1-form on the moduli space (see (3.27)), and $d_m^2 = 0$. Also, since $\Delta(w)$ is exact, by using (3.26) and (3.27) we arrive at

$$\begin{aligned} d_m(z)U(w) &\simeq -\frac{1}{2}\Big[\Big(d_m(z)\frac{1}{\Pi}\Big)\Delta(w)\Pi + \frac{1}{\Pi}\Delta(w)\Big(d_m(z)\Pi\Big)\Big] \\ &= \frac{1}{4}\,[L^b(w)L^b(w)\Omega^{aa}(z,z)_{\mathrm{reg}} + \Xi^b(w)L^b(w)\Omega^{aa}(z,z)_{\mathrm{reg}}], \end{aligned} \tag{B.4}$$

where we have introduced

$$\Xi^a(z) = f^{abc}\theta_c^b(z;z) \tag{B.5}$$

to simplify notation. From (3.19) we deduce that

$$\begin{aligned} &L^b(w)\Omega^{aa}(z,z)_{\mathrm{reg}} \\ &\qquad = L^a(z)\Omega^{ba}(w,z) + [\Omega^{ea}(w,z)\theta_d^a(z;w)f^{bde} + \Omega^{bd}(w,z)\Xi^d(z)]. \end{aligned} \tag{B.6}$$

Using this relation in rewriting the first summand in (B.4), we obtain

$$\begin{aligned} 4d_m(z)U(w) \simeq\ & L^b(w)L^a(z)\Omega^{ab}(z,w) \\ & + L^b(w)[\Omega^{ea}(w,z)\theta_d^a(z;w)f^{bde} + \Omega^{bd}(w,z)\Xi^d(z)] \\ & - \Xi^b(z)L^b(z)\Omega^{aa}(w,w)_{\mathrm{reg}}. \end{aligned} \tag{B.7}$$

Commuting $L^a(z)$ and $L^b(w)$ in the first term according to (3.20), we get

$$\begin{aligned} 4d_m(z)U(w) &\simeq f^{bcd}\theta_c^a(z;w)L^d(w)\Omega^{ab}(z,w) - f^{edb}\theta_d^a(z;w)L^b(w)\Omega^{ae}(z,w) \\ &\quad + \Omega^{ea}(w,z)L^b(w)\theta_d^a(z;w)f^{bde} + \Omega^{bd}(w,z)L^b(w)\Xi^d(z) \\ &\quad - \Xi^b(z)[L^b(z)\Omega^{aa}(w,w)_{\mathrm{reg}} - L^a(w)\Omega^{ba}(z,w)] \\ &= \Omega^{ea}(w,z)L^b(w)\theta_d^a(z;w)f^{bde} + \Omega^{bd}(w,z)L^b(w)\Xi^d(z) \\ &\quad - \Xi^b(z)\Omega^{ea}(z,w)\theta_d^a(w;z)f^{bde} - \Xi^b(z)\Omega^{bd}(z,w)\Xi^d(w). \end{aligned} \tag{B.8}$$

The last term is symmetric; therefore,

$$\begin{aligned}4d_m(z)U(w) &\simeq \Omega^{ea}(w,z)L^b(w)\theta^a_d(z;w)f^{bde} - \Omega^{ea}(w,z)L^a(z)\theta^b_d(w;w)f^{bde}\\ &\quad - \Xi^b(z)\Omega^{ea}(z,w)\theta^a_d(w;z)f^{bde}\\ &= \Omega^{ea}(w,z)\big[\theta^a_f(z;w)\theta^b_g(w;w)_{\mathrm{reg}}f^{dfg}\\ &\qquad + \theta^g_d(w;w)_{\mathrm{reg}}\theta^a_f(z;w)f^{bfg} - \theta^g_d(z;w)\theta^b_f(w;z)f^{afg}\big]\\ &\quad - \Xi^b(z)\Omega^{ea}(z,w)\theta^a_d(w;z)f^{bde}\\ &\simeq \Omega^{ea}(w,z)\theta^a_f(z;w)\theta^b_g(w;w)_{\mathrm{reg}}f^{bde}f^{dfg}\\ &\quad + \Omega^{ea}(w,z)\theta^g_d(w;w)_{\mathrm{reg}}\theta^a_f(z;w)[-f^{bfd}f^{beg} - f^{bef}f^{bdg}]\\ &\quad + \Omega^{ea}(w,z)\Xi^b(w)\theta^a_d(z;w)f^{bde}\\ &= 0.\end{aligned} \tag{B.9}$$

We used the Jacobi identity

$$f^{abc}f^{ade} + f^{abd}f^{aec} + f^{abe}f^{acd} = 0 \tag{B.10}$$

and a simple corollary of (3.19):

$$\begin{aligned}L^a(z)\theta^b_c(w;u) - L^b(w)\theta^a_c(z;u) + \theta^a_d(z;u)\theta^b_e(w;u)f^{cde}\\ + \theta^e_c(w;u)\theta^a_d(z;w)f^{bde} - \theta^e_c(z;u)\theta^b_d(w;z)f^{ade} = 0.\end{aligned} \tag{B.11}$$

Appendix C. Components of the KZB connection form commute

Let us show explicitly that

$$[A(z), A(w)] = 0, \tag{C.1}$$

where

$$A(z) = \Delta(z) + U(z). \tag{C.2}$$

Using the commutation relation (3.20) and its conjugate, we find that

$$[\Delta(z), \Delta(w)] = -L^{a\dagger}(z)Q^{ab}(z,w)L^b(w) + L^{b\dagger}(w)Q^{ab}(z,w)L^a(z), \tag{C.3}$$

where

$$\begin{aligned}Q^{ab}(z,w) = Q^{ba}(w,z) = -L^a(z)L^{b\dagger}(w) - L^{d\dagger}(w)f^{bcd}\theta^a_c(z;w)\\ + f^{adc}f^{bfe}\theta^d_e(z;w)\theta^f_c(w;z)\end{aligned} \tag{C.4}$$

(the derivatives do not act outside $Q^{ab}(z,w)$).

Now it is obvious that the commutator we wish to prove to be zero is a first-order differential operator. Since it is anti-self-conjugate, it is sufficient to prove that its principal symbol vanishes. Collecting everything together, we obtain

$$\begin{aligned}[A(z), A(w)] = Y^d(z,w)L^d(z) - L^{d\dagger}(z)Y^d(z,w)\\ - Y^d(w,z)L^d(w) + L^{d\dagger}(w)Y^d(w,z),\end{aligned} \tag{C.5}$$

where

$$Y^d(z,w) = L^d(z)U(w) - \frac{1}{2}\left(\bar{L}^a(w)Q^{ad}(w,z) - Q^{ab}(w,z)f^{bcd}\theta^a_c(w;z)\right). \tag{C.6}$$

By lengthy explicit computations using (3.19), (3.20), (3.26), and the following simple corollary of (B.11):

$$\begin{aligned} &L^a(z)\theta^b_c(w;w)_{\text{reg}} - L^b(w)\theta^a_c(z;w) \\ &= -\theta^a_d(z;w)\theta^b_e(w;w)_{\text{reg}} f^{cde} - \theta^a_d(z;w)\theta^e_c(w;w)_{\text{reg}} f^{bde} \\ &\quad + \theta^b_d(w;z)\theta^e_c(z;w) f^{ade} + f^{bcd}\Omega^{ad}(z,w), \end{aligned} \tag{C.7}$$

it is straightforward to show that

$$Y^d(z,w) = 0 \tag{C.8}$$

(see [**Iv**] for computational details).

References

[ADW] S. Axelrod, S. Della Pietra, and E. Witten, *Geometric quantization of Chern-Simons theory*, J. Diff. Geom. **33** (1991), 787–902.

[BPZ] A. A. Belavin, A. N. Polyakov, and A. B. Zamolodchikov, *Infinite conformal symmetries in two-dimensional quantum field theory*, Nuclear Phys. B **241** (1984), 333–380.

[B1] D. Bernard, *On the Wess–Zumino–Witten model on the torus*, Nuclear Phys. B **303** (1988), 77–93.

[B2] ———, *On the Wess–Zumino–Witten model on Riemann surfaces*, Nuclear Phys. B **309** (1988), 145–174.

[Dub] B. A. Dubrovin, *Geometry and integrability of topological-antitopological fusion*, Comm. Math. Phys. **152** (1993), 539–564.

[FF] J. M. Figueroa-O'Farrill, *The equivalence between the gauged WZWN and GKO conformal theories*, Stony Brook Preprint, ITP-SB-89-41, June 1989 ITP-SB-89-41 (1989).

[Fe] G. Felder, *The KZB equations on Riemann surfaces*, hep-th 9609153, Symétries Quantiques (Les Houches, 1995), North-Holland, Amsterdam, 1998, pp. 687–725.

[FGK] G. Felder, K. Gawedzki, and A. Kupiainen, *Spectra of Wess–Zumino–Witten models with arbitrary simple groups*, Comm. Math. Phys. **117** (1989), 117–158.

[FS] L. Faddeev and S. Shatashvili, *Algebraic and Hamiltonian methods in the theory of non-abelian anomalies*, Theoret. Math. Fiz. **60** (1984), no. 2, 206–217; English transl., Theoret. and Math. Phys. **60** (1984), 770–778.

[Fad] L. Faddeev, *Operator anomaly for the Gauss law*, Phys. Lett. B **145** (1984), 81–84.

[FZ] I. B. Frenkel and Y. Zhu, *Vertex operator algebras associated to representation of affine and Virasoro algebras*, Duke Math. J. **66** (1992), 123–168.

[G] K. Gawedzki, *SU(2) WZW theory at higher genus*, Comm. Math. Phys. **169** (1995), 329–371.

[Gr] P. A. Griffiths, *Topics in algebraic and analytic geometry*, Princeton University Press, Princeton, 1974.

[GW] D. Gepner and E. Witten, *String theory on group manifolds*, Nuclear Phys. B **278** (1986), 493–549.

[H1] N. Hitchin, *Flat connections and geometric quantization*, Comm. Math. Phys. **131** (1990), 347–380.

[H2] ———, *Stable bundles and integrable systems*, Duke Math. J. **54** (1987), 91–114.

[Iv] D. Ivanov, *Knizhnik-Zamolodchikov-Bernard equations on Riemann surfaces*, Internat. J. Modern Phys. A **10** (1995), 2507–2536.

[K] V. G. Kac, *Infinite dimensional Lie algebras*, Cambridge University Press, 1990

[KM] M. Kontsevich and Yu. Manin, *Gromov–Witten classes, quantum cohomology, and enumerative geometry*, Comm. Math. Phys. **164** (1994), no. 3, 525–562.

[KZ] V. G. Knizhnik and A. B. Zamolodchikov, *Current algebra and Wess–Zumino model in two dimensions*, Nuclear Phys. B **247** (1984), 83–103.

[Los1] A. Losev, *Coset construction and Bernard equations*, CERN Preprint, TH.6215, 1991.

[Los2] ———, *Structures of K. Saito theory of primitive form in topological theories coupled to topological gravity*, Integrable Models and Strings (Espoo, 1993), Lecture Notes Phys., Vol 436, Springer–Verlag, Berlin, 1994, pp. 172–193.

[Mar] E. Martinec, *Conformal field theory on a (super-)Riemann surface*, Nuclear Phys. B **281** (1987), 157–210.

[Mick] J. Mickelsson, *Chiral anomalies in even and odd dimensions*, Comm. Math. Phys. **97** (1985), 361–370.

[MS] G. Moore and N. Seiberg, *Lectures on RCFT*, Physics, Geometry, and Topology (H. C. Lee, ed.), Plenum, New York, 1990, pp. 263–361.

[PS] G. Pressley and G. Segal, *Loop groups*, Oxford University Press, Oxford, 1986.

[Sa1] K. Saito, *Period mapping associated to a prinitive form*, Publ. Res. Inst. Math. Sci. **19** (1983), 1231–1264.

[Sa2] ———, *The higher residue pairing $K_F^(k)$ for a family of hypersurface singular points*, Singularities, Part 2 (Arcata, Calif, 1981), Proc. Symp. Pure Math., Vol. 40, Amer. Math. Soc., Providence, RI, 1983, pp. 441–463.

[V] E. Verlinde, *Fusion rules and modular transformations in 2D conformal field theory*, Nuclear Phys. B **300** (1988), 360–376.

[Va] C. Vafa, *Operator formulation on Riemann surfaces*, Phys. Lett. B **190** (1987), 47–54.

[W] E. Witten, *Non-Abelian bosonization*, Comm. Math. Phys. **92** (1984), 455–472.

[WZ] J. Wess and B. Zumino, *Consequences of anomalous Ward identities*, Phys. Lett. B **37** (1971), 95–97.

Department of Physics, MIT, Cambridge, MA 02139; and Landau Institute for Theoretical Physics, Moscow, Russia

Institute of Theoretical and Experimental Physics, B. Cheremushkinskaya 25, 117259, Moscow, Russia

Amer. Math. Soc. Transl.
(2) Vol. **191**, 1999

Kadomtsev–Petviashvili Hierarchy and Generalized Kontsevich Model

S. Kharchev

ABSTRACT. This survey is devoted to integrability properties of the generalized Kontsevich Model, which is a universal matrix model describing the conformal field theories with $c < 1$. A careful analysis of the model with arbitrary polynomial potential of order $p+1$ is presented. In the case of monomial potential, the partition function is proved to be a τ-function of the p-reduced Kadomtsev–Petviashvili hierarchy satisfying the $\mathbf{L}_{-p}$ Virasoro constraint. It is shown that the deformations of the "monomial" phase to a "polynomial" phase have a natural interpretation in the context of so-called equivalent hierarchies. The dynamical transition between equivalent integrable systems occurs exactly along the flows of the dispersionless Kadomtsev–Petviashvili hierarchy; the coefficients of the potential are shown to be directly related with the flat (quasiclassical) times arising in the $N = 2$ Landau–Ginzburg topological model. It is proved that the partition function of a generic generalized Kontsevich model can be presented as a product of a "quasiclassical" factor and a nondeformed partition function that depends only on the sum of transformed integrable flows and flat times. The Virasoro constraint for the solution with an arbitrary potential is shown to be a standard $\mathbf{L}_{-p}$-constraint of the (equivalent) p-reduced hierarchy with the times additively corrected by the flat coordinates. The rich structure of the model involves almost all aspects of classical integrability. Therefore, the essential details of the fermionic approach to the Kadomtsev–Petviashvili hierarchy, as well as the notions of equivalent integrable systems and their quasiclassical analogs, are collected together in parallel with a step-by-step investigation of the universal matrix model under consideration.

1. Introduction

Recently, matrix models have played an important role in the theory of 2-dimensional gravity, topological models, and statistical physics (see [**1**] and references therein). This paper is devoted to the study of a particular 1-matrix model in an external matrix field which is regarded as the "universal" one. The structure

1991 *Mathematics Subject Classification.* Primary 35Q53, 58F07; Secondary 81T40.

This work was partly supported by grant RFBR 96-02-19085 and by grant 96-15-96455 for Support of Scientific Schools.

of the model is essentially defined by the matrix integral of the typical form

$$Z_N^V[M] \sim \int dX\, e^{-\operatorname{Tr} V(X) + \operatorname{Tr} XV'(M)}, \tag{1.1}$$

where M, X are Hermitian $N \times N$ matrices and $dX \sim \prod_{i,j=1}^N dX_{ij}$. In (1.1) $V(X)$ is an arbitrary potential (see the exact formulation below). The model with $V(X) = \frac{1}{3}X^3$ (the Kontsevich model) was derived in [**2**] as the generating function of the intersection numbers on the moduli spaces, i.e., by purely geometrical arguments, guided by Witten's treatment of 2-dimensional topological gravity [**3**]. Unfortunately, a similar interpretation of the more complicated model with an arbitrary polynomial potential is still lacking. Actually, the same model (though in somewhat implicit form) first appeared in [**4**], inspired by more "physical" arguments [**5, 6**]. The advantage of [**4**] consists in the fact that it starts from the *integrable* properties of the model from the very beginning: it gives a clear interpretation of the Kontsevich partition function as a concrete solution of the 2-reduced Kadomtsev–Petviashvili (KP) hierarchy, that is, the Korteweg-de-Vries one. This allows to generalize the original Kontsevich model immediately. In [**7**], the partition function with an arbitrary potential has been suggested as "the universal" matrix model under the name of Generalized Kontsevich Model (GKM) (independently, the integral (1.1) with monomial potential of finite order has been considered in [**8**]–[**10**]). The universality of the GKM is based on the following facts [**7, 11, 12**]:

(i) For monomial potential it properly describes the (sophisticated) double scaling limit of any multimatrix model.
(ii) The GKM partition function with polynomial potential of order $p+1$ is a τ-function of the Kadomtsev–Petviashvili hierarchy, properly reducible, at the points associated with multi-matrix models, to a solution of the p-reduced hierarchy. Moreover, it satisfies an additional equation, which reduces to the conventional Virasoro constraint (string equation) for multi-matrix models when the potential degenerates to a monomial one.
(iii) It allows the deformations of the potential (probably discontinuous), changing the latter from the form associated with a given multi-matrix to those corresponding to the others.
(iv) The GKM with arbitrary polynomial potential is directly connected with $N=2$ supersymmetric Landau–Ginzburg theories.
(v) The partition function (1.1) with potential $V(X) \sim X^2 + n \log X$ describes the standard 1-matrix model before the double-scaling limit.
(vi) By adding negative powers of X, the model gives a particular solution of the Toda lattice (TL) hierarchy.

Besides, the GKM is a nontrivial (and more or less explicit) example of a solution of the Kadomtsev–Petviashvili hierarchy corresponding to the Riemann surface of infinite genus. As an integrable system with an infinite number of degrees of freedom, it possesses a very rich structure, encoding many features which are absent in finite-dimensional systems: the Virasoro constraints and, more generally, the W-constraints do not exhaust the complexity of the model. It turns out that the GKM is properly designed to describe the quasiclassical (dispersionless) solutions parametrized by the coefficients of the potential $V(X)$, being at the same time the exact solution of the original hierarchy. This makes the study of GKM very promising from the point of view of physical applications, as well as in the context of purely mathematical aspects concerning integrable structures. In this paper we shall deal with the integrablility properties of (1.1) only.

To investigate the GKM model in detail, a long route must be followed. The point is that this model, being an excellent example of the explicit solution of an integrable system (KP or even TL hierarchy), unifies many aspects of the latter. Besides the well elaborated general strategy [**13, 14, 15**] to describe the above hierarchies, some more subtle notions have to be implemented. Originally the author was tempted to dump all the details concerning the standard material into appendices (including the fermionic approach to the τ-function). After some contemplation, it become clear that in this case the paper would contain only the Introduction as the main body, with a lot of appendices, so that the structure would be the same. Therefore, for pedagogical reasons, I decided to arrange things as self-consistently as possible.

The paper is organized as follows. In the first three sections we follow the approach developed in [**13**]; the material here (except of some details) is quite standard. In Section 2 we give the essentials concerning the most important integrable system, namely, the Kadomtsev–Petviashvili hierarchy. We discuss briefly the pseudo-differential calculus and introduce the notion of Baker–Akhiezer function as well as the central object, the τ-function, as a solution of the evolution equations. In Section 3 the fermionic approach to the realization of gl_∞ is presented. This approach is of importance when representing the solutions of the KP hierarchy in "explicit" form in terms of the fermionic correlators.

In Section 4 we represent the τ-function in specific determinant form, using the fermionic approach introduced in the previous section. Such a representation is very natural from the Grassmannian approach to integrable systems [**15**] and, what is more important for our purposes, it yields a simple proof of the integrability of the GKM partition function.

The generalized Kontsevich model is introduced in Section 5. First of all, we simplify the GKM partition function by the standard integration over the angle variables, thus obtaining the integral over the eigenvalues $x_1, \ldots, x_N$ of the matrix X. After this, we are able to write the partition function in determinant form, which is the starting point in the investigation of its integrability properties. We prove that, in the case of the monomial potential $V(X) = X^{p+1}/(p+1)$, the GKM integral is a solution of the p-reduced KP hierarchy. Moreover, it also satisfies the string equation. In turn, these two conditions fix the solution of the KP hierarchy uniquely; this is exactly the GKM partition function with monomial potential.

The case of an arbitrary polynomial potential is more complicated (and richer). It requires the notion of equivalent hierarchies [**24**], which is thoroughly discussed in Section 6. We prove that the solution with polynomial potential can be generated from the corresponding solution of the p-reduced KP hierarchy by the action of the Virasoro group and is represented as another p-reduced τ-function corrected by the exponential factor with some quadratic form.

In order to investigate in detail the nature of the transformations between the equivalent hierarchies, one more notion is required, namely, the notion of the quasiclassical hierarchy, which is described in Section 7 following the approach developed in [**25**]–[**29**]. We show that the quadratic form is related to the quasiclassical τ-function. Moreover, we demonstrate that it is possible to describe the quasiclassical hierarchy directly in terms of the GKM.

The last section contains the complete description of the GKM partition function with an arbitrary polynomial potential. It is proved that after redefinition of times, the partition function of a generic GKM can be presented as the product of a "quasiclassical" factor and a nondeformed partition function, the latter being the solution of the equivalent p-reduced KP hierarchy. We show how to extract the genuine partition function, which depends only on the sum of the transformed integrable flows and flat (quasiclassical) times, and which satisfies the standard $\mathbf{L}_{-p}$-constraint of the (equivalent) p-reduced hierarchy.

2. Kadomtsev–Petviashvili hierarchy

2.1. KP hierarchy: Lax equations. Let $\{T\} = (T_1, T_2, \ldots, T_i, \ldots)$ be an infinite set of variables. Consider the pseudo-differential operator (the Lax operator)

$$L = \partial + \sum_{i=1}^{\infty} u_{i+1}(T)\partial^{-i}, \qquad \partial = \frac{\partial}{\partial T_1}, \tag{2.1}$$

where ∂^{-1} is the formal inverse to ∂, i.e., $\partial^{-1} \circ \partial = \partial \circ \partial^{-1} = 1$; for any function $f(T_1)$ and any $n \geqslant 1$, we have

$$\partial^{-n} \circ f = \sum_{i=0}^{\infty} (-1)^i \frac{(n+i-1)!}{i!\,(n-1)!} \frac{\partial^i f}{\partial T_1^i} \circ \partial^{-n-i}. \tag{2.2}$$

Note that the Lax operator L can be written as

$$L = W \circ \partial \circ W^{-1}, \tag{2.3}$$

where

$$W \equiv 1 + \sum_{i=1}^{\infty} w_i(T)\partial^{-1} \tag{2.4}$$

and the inverse W^{-1} can be calculated term by term by using the Leibniz rule (2.2); an easy exercise gives

(2.5)
$$W^{-1} = 1 - w_1\partial^{-1} + (-w_2 + w_1^2)\partial^{-2} + (-w_3 + 2w_1w_2 - w_1w_1' - w_1^3)\partial^{-3} + \ldots,$$

where $'$ denotes the derivative with respect to T_1. Comparing (2.1) and (2.3), one can find the relationship between the functions $\{u_i\}$ and $\{w_i\}$:

$$\begin{aligned} u_2 &= w_1', \qquad u_3 = -w_2' + w_1w_1', \\ u_4 &= -w_3' + w_1w_2' + w_1'w_2 - w_1^2w_1' - (w_1')^2, \quad \text{etc.} \end{aligned} \tag{2.6}$$

Let L_+^k denote the differential part of the pseudo-differential operator L^k; for example,

$$L_+ = \partial, \quad L_+^2 = \partial^2 + 2u_2, \quad L_+^3 = \partial^3 + 3u_2\partial + 3(u_3 + u_2'), \quad \text{etc.} \tag{2.7}$$

We also use the symbol L_-^k to denote the purely pseudo-differential part of L^k; evidently, $L^k = L_+^k + L_-^k$.

By definition, the dependence of the functions $\{u_i\}$ on the *time* variables $(T_1, T_2, \ldots)$ is determined by the Lax equations

$$\frac{\partial L}{\partial T_k} = [L_+^k, L], \qquad k \geqslant 1. \tag{2.8}$$

It can be shown that this set of equations is equivalent to the zero-curvature condition

$$\frac{\partial L_+^n}{\partial T_k} - \frac{\partial L_+^k}{\partial T_n} = [L_+^k, L_+^n]. \tag{2.9}$$

The set of equations (2.8) (or, equivalently, (2.9)) is called the *Kadomtsev–Petviashvili* (KP) *hierarchy*. Let $n = 2$, $m = 3$ in (2.9). Using the explicit expression for the differential polynomials L_+^2, L_+^3, one can easily obtain the simplest equation of the KP hierarchy, the Kadomtsev–Petviashvili equation

$$\frac{\partial}{\partial T_1}\left(4\frac{\partial u_2}{\partial T_3} - 12u_2\frac{\partial u_2}{\partial T_1} - \frac{\partial^3 u_2}{\partial T_1^3}\right) - 3\frac{\partial^2 u_2}{\partial T_2^2} = 0. \tag{2.10}$$

2.2. Baker–Akhiezer functions. Evolution equations for the KP hierarchy (2.8) or (2.9) are the compatibility conditions of the following equations:

$$L\Psi = z\Psi, \qquad \partial_{T_n}\Psi = L_+^n\Psi. \tag{2.11}$$

The function $\Psi(T, z)$ that satisfies this system is called the *Baker–Akhiezer function.* Introduce the conjugation $\partial^* = -\partial$ and put

$$\begin{aligned} L^* &= -\partial + (-\partial)^{-1}\circ u_2 + (-\partial)^{-2}\circ u_3 + \dots, \\ W^* &= 1 + (-\partial)^{-1}\circ w_1 + (-\partial)^{-2}\circ w_2 + \dots, \end{aligned} \tag{2.12}$$

so that $L^* = -(W^*)^{-1}\circ\partial\circ W^*$. The *adjoint Baker–Akhiezer function* $\Psi^*(T, z)$ satisfies, by definition, the set of equations

$$L^*\Psi^* = z\Psi^*, \qquad \partial_{T_n}\Psi^* = -(L_+^n)^*\Psi^*. \tag{2.13}$$

It can be shown that solutions of systems (2.12), (2.13) can be represented in the form

$$\begin{aligned} \Psi(T, z) &= W(T, \partial)e^{\xi(T,z)} \equiv e^{\xi(T,z)}\sum_{i=0}^{\infty} w_i(T)z^{-i}, \\ \Psi^*(T, z) &= W^*(T, \partial)^{-1}e^{-\xi(T,z)}, \end{aligned} \tag{2.14}$$

where

$$\xi(T, z) \equiv \sum_{k=1}^{\infty} T_k z^k. \tag{2.15}$$

In [**13**] the following fundamental theorem was proved. Let $\Psi(T, z)$, $\Psi^*(T, z)$ be the Baker–Akhiezer functions of the KP hierarchy. There exists a function $\tau(T)$ such that

$$\Psi(T, z) = \frac{\tau(T_k - k^{-1}z^{-k})}{\tau(T_k)}\, e^{\xi(t,z)}, \qquad \Psi^*(T, z) = \frac{\tau(T_k + k^{-1}z^{-k})}{\tau(T_k)}\, e^{-\xi(t,z)}. \tag{2.16}$$

It is not hard to see that all the functions $u_i(T)$, $i \geqslant 2$, can be represented in terms of τ. For example,

$$\begin{aligned} u_2 = \partial_{T_1}^2 \log\tau, \qquad u_3 &= \frac{1}{2}(\partial_{T_1}^3 + \partial_{T_1}\partial_{T_3})\log\tau, \\ u_4 = \frac{1}{6}(\partial_{T_1}^4 - 3\partial_{T_1}^2\partial_{T_2} &+ 2\partial_{T_1}\partial_{T_2})\log\tau - (\partial_{T_1}^2\log\tau)^2, \qquad \text{etc.} \end{aligned} \tag{2.17}$$

Substitution of the first relation into (2.17) gives the representation of the KP equation in bilinear form:

$$\frac{1}{12}\tau\left(\frac{\partial^4\tau}{\partial T_1^4}-4\frac{\partial^2\tau}{\partial T_1\partial T_3}+3\frac{\partial^2\tau}{\partial T_2^2}\right)-\frac{1}{3}\frac{\partial\tau}{\partial T_1}\left(\frac{\partial^3\tau}{\partial T_1^3}-\frac{\partial\tau}{\partial T_3}\right) \tag{2.18}$$

$$+\frac{1}{4}\left(\frac{\partial^2\tau}{\partial T_1^2}+\frac{\partial\tau}{\partial T_2}\right)\left(\frac{\partial^2\tau}{\partial T_1^2}-\frac{\partial\tau}{\partial T_2}\right)=0. \tag{2.19}$$

As it turns out, it is possible to rewrite all the nonlinear equations of the KP hierarchy as an infinite set of *bilinear* equations for the τ-function [**13**] in more or less compact form by using the Hirota symbols.

One should note that it is possible to consider a more general integrable system, namely the Toda lattice (TL) hierarchy [**14**], which can be thought of as a specific "gluing" of the two KP hierarchies. In this case the solutions depend on two infinite sets of times, $\{T_k\}$ and $\{\overline{T}_k\}$, parametrizing the KP parts as well on the discrete time n that mixes the KP evolutions. The τ-function of the TL hierarchy $\tau_n(T,\overline{T})$ also satisfies an infinite set of bilinear equations [**14**]; the simplest evolution is described by the famous Toda equation

$$\tau_n\frac{\partial^2\tau_n}{\partial T_1\partial\overline{T}_1}-\frac{\partial\tau_n}{\partial T_1}\frac{\partial\tau_n}{\partial\overline{T}_1}=-\tau_{n+1}\tau_{n-1}. \tag{2.20}$$

The main problem is to describe the generic solutions of these hierarchies. This will be done in Section 4.

2.3. Reduction. The KP hierarchy is called *p-reduced* if for some natural $p\geqslant 2$ the operator L^p has only the differential part, i.e.,

$$(L^p)_-=0. \tag{2.21}$$

In this case $L_+^{np}=L^{np}$ for any $n\geqslant 1$, and from (2.11), (2.16) it follows that

$$\frac{\partial}{\partial T_{np}}\frac{\tau(T_k-k^{-1}z^{-k})}{\tau(T_k)}=0. \tag{2.22}$$

From the last relation it is clear that, on the level of the τ-function, the condition of p-reduction reads

$$\frac{\partial\tau(T)}{\partial T_{np}}=\mathrm{Const}\cdot\tau(T),\qquad n=1,2,\dots. \tag{2.23}$$

Equivalently, relations (2.23) themselves can be taken as a definition of p-reduced hierarchy.

3. Free field realization of gl_∞

3.1. Free fermions and vacuum states. Let us consider the infinite set of fermionic modes ψ_i, ψ_i^*, $i\in\mathbb{Z}$, which satisfy the usual anticommutation relations

$$\{\psi_i,\psi_j^*\}=\delta_{ij},\quad \{\psi_i,\psi_j\}=\{\psi_i^*,\psi_j^*\}=0,\qquad i,j\in\mathbb{Z}. \tag{3.1}$$

Totally empty (true) *vacuum* $|+\infty\rangle$ is determined by the relations

$$\psi_i\,|+\infty\rangle=0,\qquad i\in\mathbb{Z}. \tag{3.2}$$

Then the nth "*vacuum*" *state* $|n\rangle$ is defined as follows:

$$|n\rangle=\psi_n^*\psi_{n+1}^*\cdots|+\infty\rangle, \tag{3.3}$$

thus satisfying the conditions (which themselves can be taken as the definition of this state)

(3.4) $$\psi_k^* |n\rangle = 0, \quad k \geqslant n, \qquad \psi_k |n\rangle = 0, \quad k < n.$$

Similarly, the *left (dual) nth vacuum* $\langle n|$ is defined by the conditions

(3.5) $$\langle n| \psi_k^* = 0, \quad k < n, \qquad \langle n| \psi_k = 0, \quad k \geqslant n.$$

One can select a particular state, for example, $|0\rangle$, and consider the normal ordering of the fermions with respect to this preferred vacuum. In this case the annihilation operators will be ψ_i, $i < 0$, and ψ_i^*, $i \geqslant 0$, and, therefore, the normal ordering is defined as follows:

(3.6) $$\psi_i \psi_j^* = :\psi_i \psi_j^*: + \theta(-i-1)\delta_{ij}.$$

3.2. Boson-fermion correspondence. It is convenient to introduce the free fermionic fields

(3.7) $$\psi(z) \equiv \sum_{i\in\mathbb{Z}} \psi_i z^i, \qquad \psi^*(z) \equiv \sum_{i\in\mathbb{Z}} \psi_i^* z^{-i},$$

which, in turn, can be expressed in terms of the free *bosonic field* $\varphi(z)$

(3.8) $$\varphi(z) = q - ip\log z + i\sum_{k\in\mathbb{Z}} \frac{J_k}{k} z^{-k}, \quad [q,p] = i, \qquad [J_m, J_n] = m\delta_{m+n,0},$$

according to the well-known formulas

(3.9) $$\psi(z) = \, :e^{i\varphi(z)}: \, \equiv e^{iq} e^{p\log z} \exp\Big(\sum_{k=1}^{\infty} \frac{J_{-k}}{k} z^k\Big) \exp\Big(-\sum_{k=1}^{\infty} \frac{J_k}{k} z^{-k}\Big),$$

(3.10)
$$\psi^*(z) = z :e^{-i\varphi(z)}: \, \equiv z e^{-iq} e^{-p\log z} \exp\Big(-\sum_{k=1}^{\infty} \frac{J_{-k}}{k} z^k\Big) \exp\Big(\sum_{k=1}^{\infty} \frac{J_k}{k} z^{-k}\Big).$$

Note that under the formal Hermitian conjugation

(3.11) $$(z)^\dagger = z^{-1}, \quad (J_k)^\dagger = J_{-k}, \qquad (q)^\dagger = q, \quad (p)^\dagger = p,$$

we have the involution

(3.12) $$(\psi(z))^\dagger = \psi^*(z), \qquad (\psi^*(z))^\dagger = \psi(z).$$

It can be shown that the vacua $|n\rangle$ are eigenfunctions of the operator p,

(3.13) $$p\,|n\rangle = n\,|n\rangle, \qquad \langle n|\,p = n\,\langle n|$$

and the zero bosonic mode shifts the vacua, i.e., changes its charge

(3.14) $$\begin{cases} e^{imq}\,|n\rangle = |n+m\rangle, \\ \langle n|\,e^{imq} = \langle n-m|, \end{cases} \qquad m \in \mathbb{Z}.$$

Using the definition (3.8), one can show that

(3.15) $$:e^{i\alpha\varphi(z)}::e^{i\beta\varphi(w)}: = (z-w)^{\alpha\beta} :e^{i\alpha\phi(z)+i\beta\phi(w)}:$$

and, therefore,

(3.16) $$\psi(z)\psi^*(w) = \frac{w}{z-w} :e^{i\varphi(z)-i\varphi(w)}: \, \equiv \, :\psi(z)\psi^*(w): + \frac{w}{z-w}.$$

The last expression, being expanded near the point $w \sim z$, enables one to rewrite the bosonic field via the fermionic fields:

$$i\partial_z \varphi(z) = \frac{1}{z} :\psi(z)\psi^*(z): = \sum_{k\in\mathbb{Z}} J_k z^{-k-1}, \tag{3.17}$$

or, equivalently, the bosonic currents can be represented as a bilinear combinations of the fermionic modes:

$$J_k = \sum_{i\in\mathbb{Z}} :\psi_i \psi^*_{i+k}:, \qquad k \in \mathbb{Z}. \tag{3.18}$$

Obviously, the normal ordering in (3.18) is essential only for $J_0 \equiv p$. Using (3.18), it is easy to see that

$$\begin{cases} J_k \left|n\right\rangle \equiv 0, \\ \left\langle n\right| J_{-k} \equiv 0, \end{cases} \qquad k > 0,\ n \in \mathbb{Z}. \tag{3.19}$$

One should mention that not only the bosonic currents can be expressed as bilinear combinations of the free fermions. Actually, this is true for the whole family of gl_∞ generators (sometimes called $W_{1+\infty}$-generators); for example, one can derive a similar boson-fermion correspondence for the Virasoro generators:

$$\mathbf{L}_k \equiv \frac{1}{2} \sum_{i\in\mathbb{Z}} :J_i J_{k-i}: = \sum_{i\in\mathbb{Z}} \left(i + \frac{k+1}{2} \right) :\psi_i \psi^*_{i+k}:. \tag{3.20}$$

The bosonization formulas are a very useful tool for calculating different correlators containing the fermionic operators.

4. τ-functions in free field representation

In this section solutions of the KP hierarchy (more generally, the Toda hierarchy) are represented in the form of fermionic correlators parametrized by an infinite set of continuous variables. The fermionic language is very convenient for integrable systems, since it enables us to represent an arbitrary solution in specific determinant form. This, in turn, allows us to identify the GKM partition function with the appropriate solution of the hierarchy.

4.1. Fermionic correlators, Wick theorem, and solution of the KP (TL) hierarchy. Let us introduce the "Hamiltonians"

$$H(T) \equiv \sum_{k=1}^{\infty} T_k J_k, \qquad \overline{H}(\overline{T}) \equiv \sum_{k=1}^{\infty} \overline{T}_k J_{-k}, \tag{4.1}$$

where $\{T_k\}$ and $\{T_k\}$ are infinite sets of parameters (sometimes called *positive* and *negative times* respectively). We define the fermionic correlators (τ-functions) with the following parameterization by these times:

$$\tau_n(T, \overline{T}|g) = \left\langle n\right| e^{H(T)} g e^{-\overline{H}(\overline{T})} \left|n\right\rangle \equiv \left\langle n\right| g(T, \overline{T}) \left|n\right\rangle, \tag{4.2}$$

where

$$g = :\exp\left\{ \sum_{i,j\in\mathbb{Z}} A_{ij} \psi_i \psi^*_j \right\}: \tag{4.3}$$

with $\|A_{ij}\| \in gl_\infty$. In most cases we shall write $\tau_n(T, \overline{T})$ for brevity. We assume that the (infinite) matrix $\|A_{ij}\|$ satisfies requirements such that the correlator (4.2) is well defined. As an example, the matrix with almost all zero entries is suitable.

A wide class of suitable matrices is that of the Jacobian ones: $A_{ij} = 0$ for $|i-j| \gg 1$. More general conditions can be found in [**15**]. One should mention that the normal ordering in (4.3) is taken with respect to the zero vacuum state $|0\rangle$ (see (3.6)); it is equivalent to (3.16).

Note also that every element of type (4.3) rotates the fermionic modes:

$$(4.4) \qquad g\psi_i g^{-1} = R_{ki}\psi_k, \qquad g\psi_i^* g^{-1} = R_{ik}^{-1}\psi_k^*$$

with some (infinite) matrix $\|R\| \in GL_\infty$. As an example, the exponentials containing the Hamiltonians give the transformations

(4.5)

$$\begin{aligned} e^{H(T)}\psi(z)e^{-H(T)} &= e^{\xi(T,z)}\psi(z), & e^{H(T)}\psi^*(z)e^{-H(T)} &= e^{-\xi(T,z)}\psi^*(z), \\ e^{\overline{H}(\overline{T})}\psi(z)e^{-\overline{H}(\overline{T})} &= e^{\xi(\overline{T},z^{-1})}\psi(z), & e^{\overline{H}(\overline{T})}\psi^*(z)e^{-\overline{H}(\overline{T})} &= e^{-\xi(\overline{T},z^{-1})}\psi^*(z), \end{aligned}$$

because of the commutator relations $[J_k, \psi(z)] = z^k\psi(z)$, $[J_k, \psi^*(z)] = z^{-k}\psi^*(z)$ (the latter are simple consequences of the fermionic representation (3.19)).

The fermionic correlators introduced above have a very specific dependence on the infinite sets of times $\{T_k\}$, $\{\overline{T}_k\}$. The main statement is that the correlators (4.2) solve the Toda lattice hierarchy; in particular, as a function of the positive times $\{T_k\}$, these correlators are solutions of the KP hierarchy: each particular solution is parametrized by the given matrix $\|A_{ij}\|$. This can be proved in full generality by using the so-called bilinear identity [**13**]. For our present purposes it suffices, however, to show that the simplest equations of the above-mentioned hierarchies are satisfied. It is possible to deduce them starting directly from the fermionic correlators. Note that we shall only deal with the KP hierarchy in what follows. Nevertheless, as an instructive example, let us derive the 2-dimensional Toda equation, which is the first equation of the Toda hierarchy. This example shows the natural appearance of the determinant representations in the context of integrable systems; besides, a similar technique will be used below quite extensively.

All the correlators similar to (4.2) are expressed in terms of the free fields, and so the Wick theorem is applicable; as an example,

$$(4.6) \qquad \frac{\langle n|\, \psi_{i_1} \cdots \psi_{i_k} g(T,\overline{T})\psi_{j_1}^* \cdots \psi_{j_k}^* \,|n\rangle}{\langle n|\, g(T,\overline{T}) \,|n\rangle} = \det \left. \frac{\langle n|\, \psi_{i_a} g(T,\overline{T})\psi_{j_b}^* \,|n\rangle}{\langle n|\, g(T,\overline{T}) \,|n\rangle} \right|_{a,b=1}^{k}.$$

This key observation gives an easy way to prove that the τ-function (4.2) satisfies the standard Toda equation. Indeed, using the fermionic representation (3.18) of the currents J_k together with the definition of the vacuum states (3.4), (3.5), one obtains

$$(4.7) \quad \partial_{T_1}\partial_{\overline{T}_1}\tau_n = -\langle n|\, J_1 e^{H(T)} g e^{-\overline{H}(\overline{T})} J_{-1} \,|n\rangle = -\langle n|\, \psi_{n-1}\psi_n^* g(T,\overline{T})\psi_n\psi_{n-1}^* \,|n\rangle.$$

Using the Wick theorem, this expression can be written in the form

$$(4.8) \quad \begin{aligned} \partial_{T_1}\partial_{\overline{T}_1}\tau_n = -\frac{1}{\tau_n}\big\{ &\langle n|\, \psi_{n-1}\psi_n^* g(T,\overline{T}) \,|n\rangle \langle n|_n\, g(T,\overline{T})\psi_n\psi_{n-1}^* \,|n\rangle \\ &+ \langle n|\, \psi_{n-1} g(T,\overline{T})\psi_{n-1}^* \,|n\rangle \langle n|\, \psi_n^* g(T,\overline{T})\psi_n \,|n\rangle \big\}. \end{aligned}$$

Recalling the definitions again, one can rewrite every term in the last formula in terms of the τ-functions and their derivatives, namely,

$$\begin{aligned}\langle n|\,\psi_{n-1}g(T,\overline{T})\psi^*_{n-1}\,|n\rangle &= \tau_{n-1}, & \langle n|\,\psi^*_n g(T,\overline{T})\psi_n\,|n\rangle &= \tau_{n+1},\\ \langle n|\,\psi_{n-1}\psi^*_n g(T,\overline{T})\,|n\rangle &= \partial_{T_1}\tau_n, & \langle n|\,g(T,\overline{T})\psi_n\psi^*_{n-1}\,|n\rangle &= -\partial_{\overline{T}_1}\tau_n,\end{aligned} \tag{4.9}$$

and, therefore, (4.8) reduces to the Toda equation

$$\partial_{T_1}\partial_{\overline{T}_1}\log = -\frac{\tau_{n+1}\tau_{n-1}}{\tau_n^2}, \tag{4.10}$$

which is equivalent to (2.20). Similar (though more involved) calculations show that τ_n as a function of the positive times T_1, T_2, T_3 satisfies the Kadomtsev–Petviashvili equation (2.19) for any fixed n. Let us stress again that the complete list of bilinear equations for the τ-functions is presented in [**13, 14**].

4.2. Determinant representation of τ-functions. Here we represent an arbitrary solution of the KP hierarchy in determinant form; this is crucial in what follows. Let us calculate the fermionic correlator $\langle n+N|\,\psi(\mu_N)\cdots\psi(\mu_1)g\,|n\rangle$ in two different ways. First of all, using the definition of the vacua and applying the Wick theorem, the correlator can be written in determinant form:

$$\begin{aligned}\langle n+N|\,\psi(\mu_N)\cdots\psi(\mu_1)g\,|n\rangle &= \langle n|\,\psi^*_n\cdots\psi^*_{n+N-1}\psi(\mu_N)\cdots\psi(\mu_1)g\,|n\rangle\\ &= \langle n|\,g\,|n\rangle\det\frac{\langle n|\,\psi^*_{n+i-1}\psi(\mu_j)g\,|n\rangle}{\langle n|\,g\,|n\rangle}.\end{aligned} \tag{4.11}$$

On the other hand, using the boson-fermion correspondence (3.9), the normal ordering (3.15), and the formulas (3.13), (3.14), (3.19) describing the action of the different operators on the vacuum state $\langle N|$, one can write

$$\begin{aligned}\langle n+N|\,\psi(\mu_N)\cdots\psi(\mu_1)g\,|n\rangle &\equiv \Delta(\mu)\langle n+N|:\exp\left\{i\sum_{j=1}^N\phi(\mu_j)\right\}:g\,|n\rangle\\ &= \Delta(\mu)\prod_{j=1}^N\mu_j^n\,\langle n|\exp\left\{\sum_{k=1}^\infty T_kJ_k\right\}g\,|n\rangle,\end{aligned} \tag{4.12}$$

where in the right-hand side the τ-function appears with the specific parametrization of the positive times

$$T_k \equiv -\frac{1}{k}\sum_{j=1}^N\mu_j^{-k}. \tag{4.13}$$

The parametrization (4.13) was introduced in [**16**]. We shall call such a representation of times the *Miwa parametrization* (accordingly, the set $\{\mu_i\}$ is called the set of *Miwa variables*). Note that for N finite only the first N times $T_1,\dots,T_N$ are functionally independent. Equivalently, only the first N equations of the KP hierarchy have a nontrivial meaning (all higher equations are functionally dependent on the first N ones). We shall deal with such a restricted hierarchy in what follows. Comparing the relations (4.11), (4.12), one arrives at the following statement. For any finite N, the τ-functions of the KP hierarchy written in the Miwa variables (4.13) can be represented in determinant form

$$\tau_n(T) = \langle n|\,g\,|n\rangle\frac{\det\phi_i^{(\mathrm{can})}(\mu_j)|_{i,j=1}^N}{\Delta(\mu)}, \tag{4.14}$$

where the *canonical basis vectors*

$$\phi_i^{(\mathrm{can})}(\mu) = \mu^{-n}\,\frac{\langle n|\,\psi^*_{n+i-1}\psi(\mu)g\,|n\rangle}{\langle n|g|n\rangle}, \qquad i = 1, 2, \ldots, \tag{4.15}$$

have the following asymptotics:

$$\phi_i^{(\mathrm{can})}(\mu) = \mu^{i-1} + O(\mu^{-1}), \qquad \mu \to \infty. \tag{4.16}$$

Moreover, the converse statement is true. Namely, any functions $\tau(\mu_1, \ldots, \mu_N)$ of the form

$$\tau(T) = \frac{\det \phi_i(\mu_j)}{\Delta(\mu)}, \qquad T_k \equiv -\frac{1}{k}\sum_{j=1}^{N} \mu_j^{-k}, \tag{4.17}$$

whose basis vectors $\phi_i(\mu)$, $i = 1, 2, \ldots$, have the asymptotics

$$\phi_i(\mu) = \mu^{i-1}(1 + O(\mu^{-1})), \qquad \mu \to \infty, \tag{4.18}$$

solve the KP hierarchy. The set $\{\phi_i(\mu)\}$ satisfying the asymptotics (4.18) is naturally identified with the projective coordinates of a point of the Grassmannian [**15**]. More precisely, the vectors $\{\phi_i(\mu)\}$ can be transformed to canonical ones by taking appropriate linear combinations (clearly, such a transformation does not change the determinant in (4.17)). Then there exists an element (4.3) of the Grassmannian such that the transformed basis vectors can be written as fermionic correlators (4.15) (for some fixed n) and, consequently, $\tau(\mu_1, \ldots, \mu_N)$ has the form (4.2) in the Miwa parametrization (4.13). To summarize, any infinite set of vectors (4.18) describes a particular solution of the KP hierarchy in determinant form (4.17).

4.3. Time derivatives. Let us find the expression of the time derivatives $\partial\tau/\partial T_k$ for the τ-function written in determinant form (4.14). As in (4.13), we consider a finite number N of Miwa variables. Hence, only the first N times T_k are functionally independent, and all formulas below make sense for $\partial\tau/\partial T_1, \ldots, \partial\tau/\partial T_N$ only. From (4.12) we obtain

$$\begin{aligned} \frac{\partial\tau_n}{\partial T_k} &= \frac{\prod \mu_i^{-n}}{\Delta(\mu)}\,\langle n+N|\,\psi(\mu_N)\cdots\psi(\mu_1)J_k g\,|n\rangle \\ &\equiv \frac{\prod \mu_i^{-n}}{\Delta(\mu)}\Big\{\langle n+N|\,J_k\psi(\mu_N)\cdots\psi(\mu_1)g|n\rangle \\ &\qquad + \sum_{i=1}^{N}\langle n+N|\,\psi(\mu_N)\cdots[\psi(\mu_i), J_k]\cdots\psi(\mu_1)g\,|n\rangle\Big\}. \end{aligned} \tag{4.19}$$

Since the currents $J_k = \sum_{j\in\mathbb{Z}} \psi_j\psi^*_{j+k}$ satisfy the commutation relations $[J_k, \psi(\mu)] = \mu^k\psi(\mu)$, the last expression can be written in the form

$$\begin{aligned} \frac{\partial\tau_n}{\partial T_k} = \frac{\prod \mu_i^{-n}}{\Delta(\mu)}\,\langle n|\,\psi^*_n\cdots\psi^*_{n+N-1}\Big\{\sum_{j=n+N-k}^{n+N-1}\psi_j\psi^*_{j+k}\Big\}\psi(\mu_N)\cdots\psi(\mu_1)g\,|n\rangle \\ - \tau_n(x)\sum_{i=1}^{N}\mu_i^k, \end{aligned} \tag{4.20}$$

where, according to the definition of the vacua (3.5), the action of J_k on the state $\langle n+N|$ reduces to the action of a finite number of fermionic modes with $n+N-k \leqslant j \leqslant n+N-1$. This fact allows us to represent the expression (4.20) in compact determinant form. Indeed, since $j \geqslant n+N-k$ and $k \leqslant N$ (i.e., $j \geqslant n$), it is clear that $\langle n|\,\psi_j\psi^*_{j+k} = 0$, and moving the operator $\sum_{j=n+N-k}^{n+N-1}\psi_j\psi^*_{j+k}$ to the left state results in appropriate shifts of the modes $\psi^*_n,\ldots,\psi^*_{n+N-1}$. For example, for $k=1$, one obtains the sole correlator $\langle n|\,\psi^*_n\cdots\psi^*_{n+N-2}\psi^*_{n+N}g\psi(\mu_n)\cdots\psi(\mu_1)g\,|n\rangle$, and therefore the first term in (4.20) has a determinant form similar to (4.14) (with the last row shifted $\phi^{(\mathrm{can})}_N \to \phi^{(\mathrm{can})}_{N+1}$). It is evident that for arbitrary $k \leqslant N$ the first term in (4.20) can be represented as the sum of shifted determinants

$$\frac{\langle n|\,g\,|n\rangle}{\Delta(\mu)}\sum_{m=1}^{N}\begin{vmatrix} \phi^{(\mathrm{can})}_1(\mu_1) & \ldots & \phi^{(\mathrm{can})}_1(\mu_N) \\ \cdots & \cdots & \cdots \\ \phi^{(\mathrm{can})}_{m-1}(\mu_1) & \ldots & \phi^{(\mathrm{can})}_{m-1}(\mu_N) \\ \phi^{(\mathrm{can})}_{m+k}(\mu_1) & \ldots & \phi^{(\mathrm{can})}_{m+k}(\mu_N) \\ \phi^{(\mathrm{can})}_{m+1}(\mu_1) & \ldots & \phi^{(\mathrm{can})}_{m+1}(\mu_N) \\ \cdots & \cdots & \cdots \\ \phi^{(\mathrm{can})}_N(\mu_1) & \ldots & \phi^{(\mathrm{can})}_N(\mu_N) \end{vmatrix}. \tag{4.21}$$

Hence, one arrives at the following formula:

$$\frac{\partial}{\partial T_k}\left(\frac{\det\phi^{(\mathrm{can})}_i(\mu_j)}{\Delta(\mu)}\right) \tag{4.22}$$

$$= \frac{1}{\Delta(\mu)}\sum_{m=1}^{N}\begin{vmatrix} \phi^{(\mathrm{can})}_1(\mu_1) & \ldots & \phi^{(\mathrm{can})}_1(\mu_N) \\ \cdots & \cdots & \cdots \\ \phi^{(\mathrm{can})}_{m-1}(\mu_1) & \ldots & \phi^{(\mathrm{can})}_{m-1}(\mu_N) \\ \phi^{(\mathrm{can})}_{m+k}(\mu_1) - \mu_1^k\phi^{(\mathrm{can})}_m(\mu_1) & \ldots & \phi^{(\mathrm{can})}_{m+k}(\mu_N) - \mu_N^k\phi^{(\mathrm{can})}_m(\mu_N) \\ \phi^{(\mathrm{can})}_{m+1}(\mu_1) & \ldots & \phi^{(\mathrm{can})}_{m+1}(\mu_N) \\ \cdots & \cdots & \cdots \\ \phi^{(\mathrm{can})}_N(\mu_1) & \ldots & \phi^{(\mathrm{can})}_N(\mu_N) \end{vmatrix}.$$

Introducing the formal operator which shifts the indices of the canonical basis vectors

$$B(\mu)\phi^{(\mathrm{can})}_i(\mu) \equiv \phi^{(\mathrm{can})}_{i+1}(\mu), \tag{4.23}$$

one can write the final answer in a more compact notation:

$$\frac{\partial}{\partial T_k}\left(\frac{\det\phi^{(\mathrm{can})}_i(\mu_j)}{\Delta(\mu)}\right) = \frac{1}{\Delta(\mu)}\sum_{m=1}^{N}(B^k(\mu_m) - \mu_m^k)\det\phi^{(\mathrm{can})}_i(\mu_j). \tag{4.24}$$

This is the first important formula that we need in what follows. As an immediate application, one can consider the translation of the notion of p-reduced KP hierarchy to the language of the Grassmannian. Suppose that for some natural $p > 1$ the quantity $\mu^p\phi^{(\mathrm{can})}_m(\mu)$ can be expanded in the canonical basis vectors, i.e., for any $m \geqslant 1$,

$$\mu^p\phi^{(\mathrm{can})}_m(\mu) \subset \mathrm{Span}\{\phi^{(\mathrm{can})}(\mu)\}. \tag{4.25}$$

Writing

$$\phi^{(\mathrm{can})}_m(\mu) \equiv \mu^{m-1} + \sum_{j=1}^{\infty}\alpha_{mj}\mu^{-j}, \tag{4.26}$$

it is easy to see that for any $n \geqslant 1$ the following expansion holds:

$$\mu^{np}\phi_m^{(\mathrm{can})}(\mu) = \phi_{m+np}^{(\mathrm{can})}(\mu) + \sum_{j=1}^{np} \alpha_{mj}\phi_{np-j+1}^{(\mathrm{can})}. \tag{4.27}$$

Due to the determinant structure in (4.22), every row containing the terms $\phi_{m+np}^{(\mathrm{can})} - \mu^{np}\phi_m^{(\mathrm{can})}$ gives a nontrivial contribution $-\alpha_{m,np-m+1}\phi_m^{(\mathrm{can})}$, $1 \leqslant m \leqslant np$ (provided $np \leqslant N$); hence,

$$\frac{\partial \tau(T)}{\partial T_{np}} = -\tau(T) \sum_{m=1}^{np} \alpha_{m,np-m+1}, \qquad np \leqslant N, \tag{4.28}$$

assuming that (4.25) holds. In the limit as $N \to \infty$, this is exactly the case of the p-reduced KP hierarchy. Hence, conditions (4.25) and (4.28) are equivalent [**15**].

4.4. Action of the Virasoro generators. Literally the same calculation can be performed for any W-generators. Consider, for example, the Virasoro generators

$$\mathbf{L}_k(T) = \frac{1}{2} \sum_{a+b=-k} abT_aT_b + \sum_{a-b=-k} aT_a \frac{\partial}{\partial T_b} + \frac{1}{2} \sum_{a+b=k} \frac{\partial^2}{\partial T_a \partial T_b}; \tag{4.29}$$

then, evidently,

$$\mathbf{L}_k(T)\tau_n(T) = \langle n|\, e^{H(T)}\mathbf{L}_k(J)\, g\, |n\rangle, \tag{4.30}$$

where the fermionic Virasoro generators $\mathbf{L}_k(J)$ (3.20) satisfy the commutations relations

$$[\mathbf{L}_k(J), \psi(\mu)] = \left(\mu^{k+1}\frac{\partial}{\partial \mu} + \frac{k+1}{2}\mu^k\right)\psi(\mu) \equiv A_k(\mu)\psi(\mu). \tag{4.31}$$

Consider the subset $\{\mathbf{L}_{-k}(J),\ k > 0\}$. Since $\langle n + N|\, \mathbf{L}_{-k}(J) = 0$ for $k > 0$, instead of (4.19) one obtains

$$\begin{aligned} &\mathbf{L}_{-k}(T)\tau_n(T) \\ &\quad = -\frac{\prod \mu_i^{-n}}{\Delta(\mu)} \sum_{m=1}^{N} \langle n + N|\, \psi(\mu_N) \cdots [\psi(\mu_m), L_{-k}(J)] \cdots \psi(\mu_1) g\, |n\rangle \\ &\quad = -\frac{\prod \mu_i^{-n}}{\Delta(\mu)} \sum_{m=1}^{N} A_{-k}(\mu_m)\, \langle n + N|\, \psi(\mu_N) \cdots \psi(\mu_1) g\, |n\rangle \\ &\quad = nkT_k\tau_n(T) - \frac{\langle n|\, g\, |n\rangle}{\Delta(\mu)} \sum_{m=1}^{N} A_{-k}(\mu_m) \det \phi_i^{(\mathrm{can})}(\mu_j). \end{aligned} \tag{4.32}$$

In particular, the standard τ-function of the KP hierarchy $\tau_{n=0}(T) \equiv \tau(T)$ satisfies the relations

$$\begin{gathered} \mathbf{L}_{-k}(T)\left(\frac{\det \phi_i^{(\mathrm{can})}(\mu_j)}{\Delta(\mu)}\right) = -\frac{1}{\Delta(\mu)} \sum_{m=1}^{N} A_{-k}(\mu_m) \det \phi_i^{(\mathrm{can})}(\mu_j), \\ A_{-k}(\mu) = \mu^{1-k}\frac{\partial}{\partial \mu} + \frac{1-k}{2}\mu^{-k} \end{gathered} \tag{4.33}$$

(we shall see below that the GKM partition function corresponds exactly to the choice of the 0-vacuum state).

Similarly to (4.25), consider the case in which, for some $q > 1$, we have

$$A_{-q}(\mu)\phi_i^{(\mathrm{can})}(\mu) \subset \mathrm{Span}\{\phi^{(\mathrm{can})}(\mu)\}. \tag{4.34}$$

From (4.33) it follows that the solution of the KP hierarchy is invariant with respect to the action of the corresponding Virasoro generator:

$$\mathbf{L}_{-q}(T)\tau(T) = 0. \tag{4.35}$$

In the next section it will be shown that the GKM partition function satisfies conditions quite similar to (4.28) and (4.35).

Relations (4.24) and (4.33) are the simplest examples of W-generators acting on τ-functions in the Miwa parametrization. Using the fermionic representation, it is possible to write out similar expressions for the higher generators.

5. Generalized Kontsevich model: preliminary investigation

5.1. GKM: definition. Recall that the standard Hermitian one-matrix model is defined as a multiple integral over the $n \times n$ Hermitian matrix X:

$$Z_n[t] = \int e^{-\operatorname{Tr} S(X,t)}\, dX, \tag{5.1}$$

where the action $S(X,t)$ depends on infinitely many coupling constants ("the times")

$$S(X,t) = \sum_{k=1}^{\infty} t_k X^k \tag{5.2}$$

and the measure

$$dX = \prod_{i=1}^{n} dX_{ii} \prod_{i<j} 2\, d(\operatorname{Re} X_{ij})\, d(\operatorname{Im} X_{ij}) \tag{5.3}$$

is chosen in such a way that the following normalization condition is satisfied:

$$\int e^{-\frac{1}{2}\operatorname{Tr} X^2}\, dX = (2\pi)^{n^2/2}. \tag{5.4}$$

After integration over the angle variables [**21**], the partition function (5.1) reduces to an n-tuple integral over the eigenvalues $x_1, \dots, x_n$ of the matrix X:

$$Z_n[t] = \frac{(2\pi)^{n(n-1)/2}}{\prod_{k=1}^{n} k!} \int \Delta^2(x) \prod_{i=1}^{n} e^{-S(x_i,t)}\, dx_i, \tag{5.5}$$

where

$$\Delta(x) \equiv \prod_{i>j} (x_i - x_j) \tag{5.6}$$

is the Vandermonde determinant and

$$U_n \equiv \frac{(2\pi)^{n(n-1)/2}}{\prod_{k=1}^{n} k!} \tag{5.7}$$

is the volume of the group $SU(n)$. The partition function (5.5) possesses a remarkable integrability property: as a function of the times $\{t_k\}$ and of the discrete variable n (the size of the matrix) it is a solution of the so-called Toda chain hierarchy. In particular, the function

$$\tau_n(t) \equiv \frac{1}{n!\,U_n} Z_n[t] \tag{5.8}$$

satisfies the famous Toda equation

$$\frac{\partial^2 \log \tau_n}{\partial t_1^2} = \frac{\tau_{n+1}\tau_{n-1}}{\tau_n^2}. \tag{5.9}$$

The main object we shall discuss below is a completely different one-matrix integral depending on the external $N \times N$ Hermitian matrix M:

$$Z_N^V[M] = \frac{\int e^{-S(M,Y)}\, dY}{\int e^{-S_2(M,Y)}\, dY}, \tag{5.10}$$

where the measure is the same as in (5.3) (with n replaced by N). The explicit dependence on the matrix M comes from the action $S(M,Y)$ and its quadratic part $S_2(M,Y)$; for any Taylor series $V(Y)$ we set, by definition,

$$S(M,Y) = \operatorname{Tr}\left[V(Y+M) - V'(M)Y - V(M)\right] \tag{5.11}$$

so that this action does not contain constant and linear terms in Y. The denominator in (5.10) is interpreted as a natural normalization factor and is nothing but the Gaussian integral determined by the quadratic part of the original action:

$$S_2(M,Y) = \lim_{\varepsilon\to 0} \frac{1}{\varepsilon^2} S(M,\varepsilon Y). \tag{5.12}$$

It is clear that the integral (5.10) depends only on the eigenvalues $\mu_1, \dots, \mu_N$ of the external matrix M. It is more reasonable, however, to use another parametrization of the partition function Z_N^V, treating it as a function of the *time variables* T_k defined by the relations

$$T_k = -\frac{1}{k} \sum_{i=1}^{N} \mu_i^{-k}; \tag{5.13}$$

these relations are appropriate analogs of times entering in the definition of the standard matrix model (5.1).[1] The appearance of such variables is very natural for certain reasons to be discussed below.

The matrix model (5.10) is called the *generalized Kontsevich model.* The reason for this is that for the special choice of potential

$$V(Y) = Y^3/3 \tag{5.14}$$

the integral (5.10) becomes the partition function of the original Kontsevich model [**2**]:

$$Z_N^{(2)}[M] = \frac{\int dY\, e^{-\frac{1}{3}\operatorname{Tr} Y^3 - \operatorname{Tr} MY^2}}{\int dY\, e^{-\operatorname{Tr} MY^2}}. \tag{5.15}$$

The expression (5.15) was derived in [**2**] as a representation of the generating functional of intersection numbers of the stable cohomology classes on the universal moduli space, i.e., it is defined as the partition function of Witten's $2d$ topological gravity [**3**]. In [**18**] (see also [**20, 19**] for alternative derivations) it was shown

[1] Nevertheless, to write the explicit expression of the partition function in the times $\{T_k\}$ requires some additional work.

that as $N \to \infty$ the partition function $Z_\infty^{(2)}$, considered as a function of the time variables (5.15), satisfies the set of Virasoro constraints

$$\mathbf{L}_n^{(2)} Z_\infty^{(2)} = 0, \qquad n \geqslant -1, \tag{5.16}$$

$$\begin{aligned} \mathbf{L}_n^{(2)} = \frac{1}{2} \sum_{k \text{ odd}} kT_k \frac{\partial}{\partial T_{k+2n}} + \frac{1}{4} \sum_{\substack{a+b=2n \\ a,\, b \text{ odd and } > 0}} \frac{\partial^2}{\partial T_a \, \partial T_b} \\ + \frac{1}{4} \sum_{\substack{a+b=-2n \\ a,\, b \text{ odd and } > 0}} aT_a bT_b + \frac{1}{16} \delta_{n,0} - \frac{\partial}{\partial T_{3+2n}}. \end{aligned} \tag{5.17}$$

Constraints (5.16) are exactly the equations from [**5, 6**], imposed on the square root of the partition function (5.1) in the double-scaling limit.

5.2. GKM in determinant form. After the shift of the integration variable

$$X = Y + M \tag{5.18}$$

the numerator in (5.10) can be written in the form

$$\int e^{-S(Y,M)} \, dY = e^{\operatorname{Tr}[V(M) - MV'(M)]} \, F[V'(M)], \tag{5.19}$$

where

$$F[\Lambda] = \int e^{-\operatorname{Tr} V(X) + \operatorname{Tr} \Lambda X} \, dX, \qquad \Lambda \equiv V'(M). \tag{5.20}$$

Using integration over the angular variables of the matrix X according to [**22, 23**], one obtains

$$F[\Lambda] = (2\pi)^{N(N-1)/2} \frac{1}{\Delta(\lambda)} \int \Delta(x) \prod_{i=1}^{N} e^{-V(x_i) + \lambda_i x_i} \, dx_i, \tag{5.21}$$

where $\{\lambda_i\}$ and $\{x_i\}$ are eigenvalues of the matrices Λ and X respectively. Therefore, the function $F[V'(M)]$ in (5.20) can be represented as

$$F[V'(M)] \sim \frac{1}{\Delta(V'(\mu))} \det \left\{ \int x^{j-1} e^{-V(x) + V'(\mu_i)x} \, dx \right\} \Bigg|_{i,j=1}^{N}, \tag{5.22}$$

where $\Delta(V'(\mu)) \equiv \prod_{i>j} (V'(\mu_i) - V'(\mu_j))$ in accordance with the definition (5.6); the inessential constant factor is omitted.

We proceed now to the denominator of (5.10):

$$D_N^V[M] \equiv \int dY \, e^{-S_2(M,Y)}. \tag{5.23}$$

Making use of the $SU(N)$-invariance of the measure dY, one can easily diagonalize M in (5.23). Of course, this does not imply any integration over angular variables and produces no factors like $\Delta(Y)$. Then to evaluate (5.23), it remains to use the obvious rule of Gaussian integration,

$$\int dY e^{-\sum_{i,j}^{N} S_{ij}(M) Y_{ij} Y_{ji}} \sim \prod_{i,j}^{N} S_{ij}^{-1/2}(M) \tag{5.24}$$

(a constant factor is omitted again), and substitute the explicit expression for $U_{ij}(M)$. If the potential is represented as a formal series,

$$V(Y) = \sum_{k=1}^{\infty} \frac{v_k}{k} Y^k \tag{5.25}$$

(and thus is assumed analytic in Y at $Y = 0$), the definition (5.12) implies that

$$S_2(M, Y) = \frac{1}{2} \sum_{k=2}^{\infty} v_k \Big\{ \sum_{a+b=k-2} \operatorname{Tr} M^a Y M^b Y \Big\} \tag{5.26}$$

and, consequently,

$$S_{ij} = \sum_{k=2}^{\infty} v_k \Big\{ \sum_{a+b=k-2} \mu_i^a \mu_j^b \Big\} = \sum_{n=0}^{\infty} V_k \frac{\mu_i^k - \mu_j^k}{\mu_i - \mu_j} = \frac{V'(\mu_i) - V'(\mu_j)}{\mu_i - \mu_j}. \tag{5.27}$$

Hence,

$$\int e^{-S_2(M,Y)}\, dY = \frac{\Delta(\mu)}{\Delta(V'(\mu))} \prod_{i=1}^{N} [V''(\mu_i)]^{-1/2}; \tag{5.28}$$

the substitution of (5.19), (5.22) and (5.28) into (5.10) yields the following representation of the GKM partition function:

$$\begin{aligned} Z_N^V[M] &= \frac{\Delta(V'(\mu))}{\Delta(\mu)} \prod_{i=1}^{N} \{[V''(\mu_i)]^{-1/2} e^{V(\mu_i) - \mu_i V'(\mu_i)}\} F[V'(M)] \\ &\equiv \frac{\det \Phi_i^V(\mu_j)|_{i,j=1}^N}{\Delta(\mu)}, \end{aligned} \tag{5.29}$$

where

$$\Phi_i^V(\mu) = [V''(\mu)]^{1/2} e^{V(\mu) - \mu V'(\mu)} \int x^{i-1} e^{-V(x) + x V'(\mu)}\, dx. \tag{5.30}$$

5.3. Functional relations. The vectors (5.30) form a linearly independent infinite set. In the generic situation, the basis vectors determining the τ-function are functionally independent since they are parametrized by an arbitrary gl_∞ matrix. On the contrary, in the GKM case, the solution is parametrized, loosely speaking, by a vector (the coefficient of the potential $V(x)$). In this sense the solution (5.29) is degenerate; the degeneration results in functional relations (the constraints) on the level of the basis vectors, which, in turn, can be considered as a definition of the GKM from the Grassmannian point of view [**15**].

Consider the model parametrized by an arbitrary polynomial potential of degree $p + 1$ ($p \geqslant 2$):

$$V(x) = \sum_{k=1}^{p+1} \frac{v_k}{k} x^k. \tag{5.31}$$

First of all, after multiplying (5.30) by $V'(\mu)$, integration by parts gives (assuming vanishing boundary conditions):

(5.32)

$$V'(\mu)\Phi_i^V(\mu) = [V''(\mu)]^{1/2}e^{V(\mu)-\mu V'(\mu)}\int x^{i-1}e^{-V(x)}\frac{\partial}{\partial x}e^{xV'(\mu)}\,dx$$
$$= [V''(\mu)]^{1/2}e^{V(\mu)-\mu V'(\mu)}\int \{x^{i-1}V'(x) - (i-1)x^{i-2}\}e^{-V(x)+xV'(\mu)}\,dx$$

i.e.,

$$V'(\mu)\Phi_i^V(\mu) = \sum_{k=1}^{p+1} v_k\Phi_{i+k-1}^V(\mu) - (i-1)\Phi_{i-1}^V(\mu), \qquad i = 1, 2, \dots . \tag{5.33}$$

This relation generalizes the notion of p-reduced KP hierarchy; for the monomial potential, one obtains condition (4.25) exactly. We shall show below (Section 6) that the general constraint (5.33) has a natural interpretation in terms of equivalent hierarchies.

There is another type of constraint generalizing (4.34). Indeed,

(5.34)

$$\Phi_i^V(\mu) = [V''(\mu)]^{1/2}e^{V(\mu)-\mu V'(\mu)}\frac{1}{V''(\mu)}\frac{\partial}{\partial\mu}\int e^{-V(x)+xV'(\mu)}\,dx \equiv A^V(\mu)\Phi_{i-1}^V(\mu),$$

where $A^V(\mu)$ is the first-order differential operator of the following special form:

$$A^V(\mu) = \frac{e^{V(\mu)-\mu V'(\mu)}}{[V''(\mu)]^{1/2}}\frac{\partial}{\partial\mu}\frac{e^{-V(\mu)+\mu V'(\mu)}}{[V''(\mu)]^{1/2}} = \frac{1}{V''(\mu)}\frac{\partial}{\partial\mu} + \mu - \frac{V'''(\mu)}{2[V''(\mu)]^2}. \tag{5.35}$$

Thus, we have the functional relation

$$\Phi_{i+1}^V(\mu) = A^V(\mu)\Phi_i^V(\mu), \tag{5.36}$$

which leads to a kind of string equation similar to (4.35). To obtain the differential (with respect to the time variables) constraint on the GKM partition function resulting from (5.36), the notion of quasiclassical hierarchy is required (Section 7).

5.4. GKM as a solution of the KP hierarchy. We proved that the GKM partition function (5.10) can be represented in determinant form

$$Z_N^V[M] = \frac{\det \Phi_i^V(\mu_j)|_{i,j=1}^N}{\Delta(\mu)}, \tag{5.37}$$

where the vectors $\Phi_i^V(\mu)$ are defined by (5.30). Moreover, using the steepest descent method, it is not hard to find the following asymptotics of the GKM basis vectors:

$$\Phi_i^V(\mu) = \mu^{i-1}(1 + O(\mu^{-p-1})), \qquad \mu \to \infty; \tag{5.38}$$

compare with (4.17), (4.18). From the above consideration it follows that the partition function (5.37), being written in terms of the Miwa times (4.13), solves the KP hierarchy, i.e.,

$$Z[T] \sim \tau_n(T) \tag{5.39}$$

with some (yet unknown) value of the vacuum state n (see the definition (4.2)).

Before we proceed further, an important remark concerning the dependence on N in the formula (5.37) is worth mentioning. The entire set $\{\Phi_i^V(\mu)\}$ is certainly N-independent and infinite. It is evident that the Φ_i^V's are linear independent.

The right-hand side of (5.37) naturally represents the τ-function for an *infinitely large* matrix M. In order to return to the case of finite N, it is enough to require that all the eigenvalues of M, except $\mu_1, \ldots, \mu_N$, tend to infinity. In this sense the partition function $Z_N^V[M]$ is independent of N; the entire dependence on N comes from the argument M, the number N being the number of finite eigenvalues of M. As a simple check of consistency, let us additionally carry μ_N to infinity in (5.37); then, according to (5.38),

$$\det{}_N \Phi_i^V(\mu_j) = (\mu_N)^{N-1} \det{}_{N-1} \Phi_i^V(\mu_j)(1 + O(1/\mu_N)), \tag{5.40}$$

$$\Delta_N(\mu) \sim \mu_N^{N-1} \Delta_{N-1}(\mu)(1 + O(1/\mu_N)). \tag{5.41}$$

Therefore,

$$Z_N^V[M] \underset{\mu_N \to \infty}{\sim} Z_{N-1}^V[M](1 + O(1/\mu_N)). \tag{5.42}$$

This is the exact statement about the N-dependence of the GKM partition function. In this sense one can claim that the GKM partition function is independent of N. Therefore, we often omit the subscript N in what follows.

As the solution of the KP hierarchy, the partition function (5.37) is parametrized by the coefficients of the polynomial V. Since the latter depends only on a finite number of parameters, the original matrix integral describes a very particular τ-function. Therefore, the question arises whether it possible to write out some kind of constraint which would naturally select this specific solution from the huge set of typical τ-functions parametrized by the gl_∞ matrix $\|A_{ij}\|$ (4.3). It turns out that the GKM τ-function satisfies the subset of $W_{1+\infty}$ constraints; indeed, one can find a number of differential (in KP times $\{T_k\}$) operators that annihilate the function (5.37). This gives an invariant description of the model in the spirit of [**5**]. The problem is to describe the action of these operators on the τ-function, which is essentially written in the Miwa variables. Of course, from [**4, 17**], it is well known how to reformulate all the constraints on the level of the basis vectors: complete information concerning the invariant properties of the τ-functions can be deciphered from relations similar to (5.36), (5.33) and vice versa; this has been demonstrated explicitly in subsections 4.3 and 4.4. In the case of monomial potential, the invariant properties of the basis vectors indeed give complete information (see below). It is important, however, that the relations mentioned above are not enough to describe the nontrivial evolution of the GKM partition function with respect to deformations of the potential V (say, from the monomial to an arbitrary polynomial of the same degree). The account of such deformations results in a highly involved mixture of standard KP flows and so-called quasiclassical (or dispersionless) ones. In order to interpret the latter evolution, one needs to know the action of the operators that do not annihilate the τ-function of GKM. The noninvariant actions cannot be reformulated in terms of the basis vectors; explicit formulas on the level of τ-functions are required.

5.5. GKM with monomial potential. p-reduced KP hierarchy and $\mathbf{L}_{-p}$ constraint. Consider the GKM partition function in the simplest case of the monomial potential V of the form $V(X) = X^{p+1}/(p+1)$:

$$Z^{(p)}[M] = \frac{e^{-\frac{p}{p+1}\operatorname{Tr} M^{p+1}} \int dX e^{\operatorname{Tr}[-\frac{1}{p+1}X^{p+1} + M^p X]}}{\int dX\, e^{-\frac{1}{2}\operatorname{Tr}[\sum_{a+b=p-2} M^a X M^b X]}} = \frac{\det \Phi_i^{(p)}(\mu_j)}{\Delta(\mu)}. \tag{5.43}$$

The basis vectors

$$\Phi_i^{(p)}(\mu) \equiv \sqrt{p\mu^{p-1}}\, e^{-\frac{p}{p+1}\mu^{p+1}} \int x^{i-1} e^{-\frac{1}{p+1}x^{p+1}+x\mu^p}\, dx \tag{5.44}$$

satisfy the obvious relations

$$\mu^p \Phi_i^{(p)}(\mu) = \Phi_{i+p}^{(p)}(\mu) - (i-1)\Phi_{i-1}^{(p)}(\mu), \tag{5.45}$$

$$A^{(p)}(\mu)\Phi_i^{(p)}(\mu) = \Phi_{i+1}^{(p)}(\mu), \tag{5.46}$$

where

$$A^{(p)}(\mu) \equiv \frac{1}{p\mu^{p-1}}\frac{\partial}{\partial\mu} - \frac{p-1}{2p\mu^p} + \mu \tag{5.47}$$

is the Kac–Schwarz operator [**4**]. Note that up to a linear term this operator is proportional to the Virasoro operator A_{-p} defined in (4.33).

We have already seen that the partition function (5.43) is a τ-function of the KP hierarchy. Now more concrete statements can be made. First of all, the GKM τ-function is a solution of the p-reduced KP hierarchy. Moreover, $Z^{(p)}[T]$ is independent of the times T_{np}:

$$\frac{\partial Z^{(p)}[T]}{\partial T_{np}} = 0, \qquad n = 1, 2, \ldots. \tag{5.48}$$

In addition, the partition function (5.43) satisfies the $\mathbf{L}_{-p}$ constraint:

$$\frac{1}{p}\mathbf{L}_{-p} Z^{(p)}[T] + \frac{\partial Z^{(p)}[T]}{\partial T_1} = 0. \tag{5.49}$$

Let us comment on relation (5.48). One should note that due to (5.38), the first $p+1$ vectors $\Phi_1^{(p)}(\mu), \ldots, \Phi_{p+1}^{(p)}(\mu)$ have a canonical structure (4.16). Therefore, for $k = p$, formula (4.22) holds if one substitutes $\Phi_i^{(p)}$, $i = 1, \ldots, N$, for the canonical GKM vectors $\Phi_i^{(\mathrm{can})}$ (see a more careful discussion of this point in 5.6 below). Moreover, (5.45) implies that the combination $\Phi_{i+p}^{(p)} - \mu^p \Phi_i^{(p)}$ does not contain the vector $\Phi_i^{(p)}$. Hence, from (4.28) we obtain

$$\frac{\partial Z^{(p)}[T]}{\partial T_p} = 0. \tag{5.50}$$

From the general KP theory one can deduce that the constraint (5.50) implies all higher relations of the form $\partial_{T_{np}} Z^{(p)}[T] = \mathrm{Const} \cdot Z^{(p)}[T]$. Actually, this follows from the relations (5.45) and from the discussion in 4.3. Thus, $Z^{(p)}[T]$ is, indeed, the τ-function of the p-reduced KP hierarchy; i.e., the corresponding Lax operator satisfies the constraint

$$L^p = (L^p)_+. \tag{5.51}$$

Unfortunately, no *simple* proof of the stronger statement (5.48), namely, of the complete independence from the times T_{np}, $n \geqslant 1$, exists (see [**7**] and, especially, [**10**] for details).

To derive the constraint (5.49), one needs the canonical structure of the GKM vectors $\Phi_1^{(p)}(\mu), \ldots, \Phi_{p+1}^{(p)}(\mu)$ again. It is important that, because of this fact, relations (4.24) (with $k = 1$) and (4.33) (with $k = p$) can be written in terms of $\{\Phi_i^{(p)}\}$. The Kac–Schwarz operator (5.47) coincides with the formal shift operator

$B(\mu)$ (4.23) due to (5.46), while $A_{-p}(\mu)$ in (4.33) is represented as $p(A^{(p)}(\mu)-\mu)$. Hence, one arrives at the relations

$$\frac{\partial Z^{(p)}}{\partial T_1}=\frac{1}{\Delta(\mu)}\sum_{m=1}^{N}(A^{(p)}(\mu_m)-\mu_m)\det\Phi_i^{(p)}(\mu_j), \tag{5.52}$$

$$\mathbf{L}_{-p}Z^{(p)}=-p\,\frac{1}{\Delta(\mu)}\sum_{m=1}^{N}(A^{(p)}(\mu_m)-\mu_m)\det\Phi_i^{(p)}(\mu_j), \tag{5.53}$$

thus getting the constraint (5.48). Note that the latter can be written in the form

$$\frac{1}{2p}\sum_{k=1}^{p-1}k(p-k)T_kT_{p-k}+\frac{1}{p}\sum_{k=1}^{\infty}(k+p)\left(T_{k+p}+\frac{p}{p+1}\,\delta_{k,1}\right)\frac{\partial\log Z^{(p)}}{\partial T_k}=0. \tag{5.54}$$

To conclude, the GKM τ-function with monomial potential satisfies the usual $\mathbf{L}_{-p}$-constraint (the integrated version of the string equation) with the shifted times

$$T_k\to T_k+\frac{p}{p+1}\,\delta_{k,p+1}. \tag{5.55}$$

5.6. General case: V'-reduction and transformation of times. In the general situation of an arbitrary polynomial of degree p, one has the following matrix model:

$$Z^V[T]=\frac{e^{\mathrm{Tr}\,[V(M)-MV'(M)]}}{\int e^{-S_2(X,M)}\,dX}\int e^{\mathrm{Tr}\,[-V(X)+XV'(M)]}\,dX. \tag{5.56}$$

The partition function (5.56) can be represented in standard determinant form:

$$Z^V[T]=\frac{\det\Phi_i^V(\mu_j)}{\Delta(\mu)}. \tag{5.57}$$

Therefore, $Z^V[T]$ is a τ-function of the KP hierarchy. Its basis vectors

$$\Phi_i^V(\mu)=[V''(\mu)]^{1/2}e^{V(\mu)-\mu V'(\mu)}\int x^{i-1}e^{-V(x)+xV'(\mu)}\,dx \tag{5.58}$$

satisfy the relations

$$V'(\mu)\Phi_i^V(\mu)=\sum_{k=1}^{p+1}v_k\Phi_{i+k-1}^V(\mu)-(i-1)\Phi_{i-1}^V(\mu),\qquad i=1,2,\dots, \tag{5.59}$$

$$\Phi_{i+1}^V(\mu)=A^V(\mu)\Phi_i^V(\mu), \tag{5.60}$$

where $A_V(\mu)$ is the first-order differential operator

$$A^V(\mu)=\frac{1}{V''(\mu)}\frac{\partial}{\partial\mu}-\frac{V'''(\mu)}{2[V''(\mu)]^2}+\mu. \tag{5.61}$$

As before, these relations impose severe restrictions on the hierarchy. It can be shown [**7**] that $Z^V[T]$ satisfies the generalized Virasoro constraint

$$\mathbf{L}^VZ^V[T]=0, \tag{5.62}$$

where

$$\begin{aligned}\mathbf{L}^V = &\sum_{n\geqslant 1} \operatorname{Tr}\left[\frac{1}{V''(M)M^{n+1}}\right]\frac{\partial}{\partial T_k} \\ &-\frac{1}{2}\sum_{i,j}\frac{1}{V''(\mu_i)V''(\mu_j)}\frac{V''(\mu_i)-V''(\mu_j)}{\mu_i-\mu_j}+\frac{\partial}{\partial T_1}.\end{aligned} \tag{5.63}$$

For the monomial potential, this constraint is reduced to (5.54), while, in general, it is impossible to write a compact expression of (5.63) in the original times (4.13). Nevertheless, one can construct a set of new times $\{\widetilde{T}_k\}$ as linear combinations of the "old" ones, $\{T_k\}$, in such a way that the operator (5.63) is transformed to the standard one expressed in the $\widetilde{T}_k$'s. The way to find the appropriate linear combinations is as follows. From (5.59) one sees that the GKM basis vectors determine an invariant point of the Grassmannian such that

$$\mathcal{P}(\mu)\Phi_i^V(\mu) \subset \operatorname{Span}\{\Phi^V(\mu)\}, \qquad \mathcal{P}(\mu) \equiv V'(\mu). \tag{5.64}$$

This condition is a natural generalization of the standard p-reduction and is called a V'-reduction. The general ideology [**15**] tells us that the pseudo-differential Lax operator

$$L\Psi = \mu\Psi, \qquad L = \partial + u_2\partial^{-1} + u_3\partial^{-2} + \dots, \tag{5.65}$$

corresponding to this point possesses the property

$$[\mathcal{P}(L)]_- = 0; \tag{5.66}$$

i.e., $V'(L)$ is a differential operator of order p. Therefore, there exists a Lax operator for the KP hierarchy

$$\widetilde{L}\Psi = \widetilde{\mu}\Psi, \qquad \widetilde{L} = \partial + \widetilde{u}_2\partial^{-1} + \widetilde{u}_3\partial^{-2} + \dots, \tag{5.67}$$

such that

$$\widetilde{L}^p = \mathcal{P}(L) \tag{5.68}$$

and, certainly, the relation between the spectral parameters of the corresponding hierarchies is

$$\widetilde{\mu} = \mathcal{P}^{1/p}(\mu). \tag{5.69}$$

Now it is clear that the relevant spectral parameter is $\widetilde{\mu}$ rather than μ. Therefore, the times appropriate for describing the V'-reduced KP hierarchy should be determined by the relations

$$\widetilde{T}_k = -\frac{1}{k}\sum_i \widetilde{\mu}_i^{-k} \equiv -\frac{1}{k}\sum_i \mathcal{P}^{-k/p}(\mu_i). \tag{5.70}$$

In order to find the relation between $\{T_k\}$ and $\{\widetilde{T}_k\}$, one introduces the notion of residue operation Res. For any Laurent series $F(\lambda) = \sum_k F_k\lambda^k$, we put

$$\operatorname{Res} F(\lambda)\, d\lambda = F_{-1}. \tag{5.71}$$

It is easy to see that this operation possesses the properties

$$\begin{gathered}\operatorname{Res}\frac{dF(\lambda)}{d\lambda}d\lambda = 0, \qquad \operatorname{Res} F d_\lambda G = -\operatorname{Res} G d_\lambda F, \\ \operatorname{Res} F d_\lambda G = \operatorname{Res} F_+ d_\lambda G_- + \operatorname{Res} F_- d_\lambda G_+\end{gathered} \tag{5.72}$$

for any two Laurent series $F(\lambda) \equiv F_+(\lambda) + F_-(\lambda)$ and $G(\lambda) \equiv G_+(\lambda) + G_-(\lambda)$, where F_+ (F_-) are the parts of the corresponding Laurent series containing only nonnegative (negative) powers in λ.

Using the properties of Res, one finds the relations

$$\widetilde{T}_k = \frac{1}{k} \sum_{m=k}^{\infty} mT_m \operatorname{Res} \lambda^{m-1} \mathcal{P}^{-k/p}(\lambda)\, d\lambda, \tag{5.73}$$

$$T_k = \sum_{m=k}^{\infty} \widetilde{T}_m \operatorname{Res} \lambda^{-k-1} \mathcal{P}^{m/p}(\lambda)\, d\lambda. \tag{5.74}$$

Now let us prove that for an arbitrary polynomial potential the GKM partition function (5.56) is independent of the time $\widetilde{T}_p$:

$$\frac{dZ^V[T(\widetilde{T})]}{\partial \widetilde{T}_p} = 0; \tag{5.75}$$

i.e., the V'-reduced KP hierarchy resembles the standard p-reduction, while involving evolution along new integrable flows $\widetilde{T}_k$.

Actually, we shall derive more general formulas for the derivatives of Z^V with respect to the first p times $\widetilde{T}_1, \dots, \widetilde{T}_p$. To do so, we must calculate the derivatives with respect to the old times. Let us apply formula (4.22) to the GKM partition function. Immediately a problem arises. Indeed, the GKM vectors (5.58) have nice integral representations, but these are not the canonical ones because of the asymptotics (5.38). On the other hand, formula (4.22) is valid only for canonical basis vectors. No compact integral representation for $\Phi_i^{(\mathrm{can})}(\mu)$ exists for the GKM. Therefore, it is impossible to find the matrix integral representation for the derivatives $\partial_{T_k} Z^V$ which would be valid for *all* times $k \geqslant 1$. Fortunately, this problem disappears when we consider the derivatives with respect to the first p times $T_1, \dots, T_p$. The key point is that the first $p+1$ basis vectors $\Phi_1^V(\mu), \dots, \Phi_{p+1}^V(\mu)$ already are in canonical form (see (5.38)). As a corollary, one can directly use (4.22) with the simple substitution $\Phi^{(\mathrm{can})} \to \Phi^V(\mu)$ (i.e., without any modification) for derivatives with respect to these times. The derivatives with respect to higher times do not allow such replacements. To illustrate this statement, one can check the "marginal" derivative $\partial_{T_p} Z^V$ which contains, for example, the particular term (see (4.22) with $k = p$)

$$\frac{1}{\Delta(\mu)} \begin{vmatrix} \Phi_1^{(\mathrm{can})}(\mu_1) & \dots & \Phi_1^{(\mathrm{can})}(\mu_N) \\ \dots & \dots & \dots \\ \Phi_{N-1}^{(\mathrm{can})}(\mu_1) & \dots & \Phi_{N-1}^{(\mathrm{can})}(\mu_N) \\ \Phi_{N+p}^{(\mathrm{can})}(\mu_1) & \dots & \Phi_{N+p}^{(\mathrm{can})}(\mu_N) \end{vmatrix}. \tag{5.76}$$

Obviously, the first $N-1$ rows in this expression can be written in terms of the GKM vectors (5.58). The only trouble can come from the last row. But due to the asymptotics (5.38), the canonical vectors can be represented in the GKM basis as

$$\Phi_{N+p}^{(\mathrm{can})} = \Phi_{N+p}^V + (\alpha_{N+p} \Phi_{N-1}^V + \text{lower terms})$$

with some constant α_{N+p}, and the row with entries

$$\alpha_{N+p} \Phi_{N-1}^V(\mu_1), \dots, \alpha_{N+p} \Phi_{N-1}^V(\mu_N)$$

(as well as the rows with lower terms) does not contribute to the determinant (5.76). This conclusion is true for all other determinants resulting to $\partial_{T_p} Z^V$. Hence, formula (4.22) with $k = p$ remains unchanged if one simply substitutes Φ_i^V for $\Phi_i^{(\mathrm{can})}$. The same reasoning can certainly be applied to all derivatives $\partial_{T_k} Z^V$ with $k \leqslant p$. On the contrary, the derivative $\partial_{T_{p+1}} Z$ contains a determinant similar to (5.76) with $\Phi_{N+p+1}^{(\mathrm{can})}(\mu_1), \ldots, \Phi_{N+p+1}^{(\mathrm{can})}(\mu_N)$ in the last row. In this case the transformation

$$\Phi_{N+p+1}^{(\mathrm{can})} = \Phi_{N+p+1}^V + (\alpha_{N+p+1}\Phi_N^V + \text{lower terms})$$

yields an additional term proportional to Z^V. Evidently, the higher derivatives become more and more involved when we express them in terms of the noncanonical vectors (5.58).

For the reasons described above, only the first p derivatives have simple integral representations. In this case the formula (4.22) gives for $1 \leqslant k \leqslant p$

$$\frac{\partial Z^V[T]}{\partial T_k} = \frac{e^{\mathrm{Tr}[V(M) - MV'(M)]}}{\int e^{-S_2(X,M)} dX} \int \mathrm{Tr}\left[X^k - M^k\right] e^{\mathrm{Tr}[-V(X) + XV'(M)]}\, dX \tag{5.77}$$

or, in compact notation,

$$\frac{\partial}{\partial T_k} \log Z^V[T] = \langle \mathrm{Tr}\, X^k - \mathrm{Tr}\, M^k \rangle, \qquad 1 \leqslant k \leqslant p, \tag{5.78}$$

where

$$\langle \mathcal{F}(X) \rangle \equiv \frac{\int \mathcal{F}(X) e^{\mathrm{Tr}[-V(X) + XV'(M)]}\, dX}{\int e^{\mathrm{Tr}[-V(X) + XV'(M)]}\, dX}. \tag{5.79}$$

Indeed, the calculation of $\langle \mathrm{Tr}\, X^k \rangle$ does not differ in technical details from those resulting in (5.29), and gives exactly the right-hand side of (4.22) with $\Phi^{(\mathrm{can})}$ replaced by Φ^V.[2] Using relation (5.74) between the old and new times, it is easy to find the formulas we need:

$$\frac{\partial}{\partial \widetilde{T}_k} \log Z^V[T(\widetilde{T})] = \langle \mathrm{Tr}\, [\mathcal{P}^{k/p}(X)]_+ - \mathrm{Tr}\, [\mathcal{P}^{k/p}(M)]_+ \rangle, \qquad 1 \leqslant k \leqslant p. \tag{5.80}$$

Note that the right-hand side of (5.80) is expressed in terms of the eigenvalues of the transformed matrix $\widetilde{M}$; i.e., M must be replaced by the solution of the equation

$$\mathcal{P}(M) = \widetilde{M}^p. \tag{5.81}$$

Relation (5.75) can be readily proved now. Indeed,

$$\frac{\partial}{\partial \widetilde{T}_p} \log Z^V[T(\widetilde{T})] = \langle \mathrm{Tr}\, V'(X) - \mathrm{Tr}\, V'(M) \rangle, \tag{5.82}$$

and the right-hand side vanishes since the expression under the integral is a total derivative. We have shown that the partition function (5.56), written in the times $\{\widetilde{T}_k\}$, has something to do with the solution of the p-reduced KP hierarchy. Therefore, it is natural to expect that the complicated Virasoro constraint (5.62), (5.63), being represented as a differential operator with respect to $\{\widetilde{T}_k\}$, can be

[2]The fermionic approach together with the above arguments allows us to write the more complicated derivatives quite explicitly. Without proof, we present the formula

$$\frac{\partial^2 \log Z_N^V}{\partial T_k\, \partial T_m} = \langle (\mathrm{Tr}\, X^k - \mathrm{Tr}\, M^k)(\mathrm{Tr}\, X^m - \mathrm{Tr}\, M^m) \rangle, \qquad 1 \leqslant k + m \leqslant p.$$

The generalization is evident.

simplified also. This expectation is *almost* true but slightly premature: the point is that the partition function $Z^V[T(\widetilde{T})]$ is not a τ-function in general. Indeed, to express the partition function (5.56) in the new times (5.70) means to replace the spectral parameters $\{\mu_i\}$ in (5.57) by the (formal) solution of (5.69). Evidently, the transformation

$$\mu = \widetilde{\mu}(1 + O(\widetilde{\mu}^{-1})) \tag{5.83}$$

destroys the structure of the Vandermonde determinant, and hence the function $Z^V[M(\widetilde{M})]$ is not in standard form. Nevertheless, the situation can be repaired: one can extract the genuine τ-function of the p-reduced KP hierarchy from $Z^V[M(\widetilde{M})]$. To describe this procedure, we must elaborate the notion of equivalent hierarchies.

6. Equivalent hierarchies

6.1. Definition. Consider the spectral problem $L\Psi = \mu\Psi$, where the operator L defining the KP hierarchy has the standard form (2.1). For any given function f,

$$f(\mu) = \sum_{i=-\infty}^{0} f_i \mu^{i+1}, \qquad f_0 = 1, \tag{6.1}$$

with time independent coefficients, one can construct the new L-operator

$$\widetilde{L} = f(L), \tag{6.2}$$

which has the same structure as the original one. The spectral problem is now

$$\widetilde{L}\Psi = \widetilde{\mu}\Psi, \tag{6.3}$$

where

$$\widetilde{\mu} \equiv f(\mu). \tag{6.4}$$

The new operator $\widetilde{L}$ (6.2) determines a KP hierarchy, which is called *equivalent* to the original one [**24**]. Introducing the differential operators $\widetilde{B}_k \equiv (\widetilde{L}^k)_+$, one can construct the evolution equations

$$\frac{d\widetilde{L}}{\partial \widetilde{T}_k} = [\widetilde{B}_k, \widetilde{L}], \tag{6.5}$$

which may be regarded as a definition of the times $\{\widetilde{T}_i\}$. Obviously,

$$\widetilde{B}_m = B_k \frac{\partial T_k}{\partial \widetilde{T}_m}. \tag{6.6}$$

The question is, what is the relationship between the solutions of the equivalent hierarchies determined by the operators L and $\widetilde{L}$? First of all, one needs to establish the explicit relationship between $\{T_i\}$ and $\{\widetilde{T}_i\}$. The second step is to find the τ-function of the "deformed" $\widetilde{L}$-hierarchy which corresponds to an arbitrary given function $\tau(T)$ of the original L-hierarchy. This gives a specific mapping between the equivalent hierarchies.

6.2. Variation of the spectral parameter. It is evident that relation (6.4) can be regarded as a transformation of the original spectral parameter μ under the action of the Virasoro generators. Let

$$\sum_{k=1}^{\infty} a_k A_{-k}(\mu) \equiv \frac{1}{W'(\mu)}\frac{\partial}{\partial\mu} + \frac{1}{2}\left(\frac{1}{W'(\mu)}\right)' \equiv A(\mu), \tag{6.7}$$

where the differential operators $A_k(\mu)$ are determined in (4.31). The function $W(\mu)$ has the asymptotic behavior

$$W'(\mu) = \frac{\mu^{s-1}}{a_s}(1 + O(\mu^{-1})), \qquad \mu \to \infty, \tag{6.8}$$

where a_s is the first nonzero coefficient in the sum (6.7). The exponential operator $\exp A(\mu)$ can be disentangled as

$$\begin{aligned} &\exp\left\{\frac{1}{W'(\mu)}\frac{\partial}{\partial\mu} + \frac{1}{2}\left(\frac{1}{W'(\mu)}\right)'\right\} \\ &\quad = \{\partial_\mu(W^{-1}(W(\mu)+1))\}^{1/2} \exp\left(\frac{1}{W'(\mu)}\frac{\partial}{\partial\mu}\right), \end{aligned} \tag{6.9}$$

where W^{-1} is the function inverse to W. It is convenient to introduce the function

$$f(\mu) = W^{-1}(W(\mu)+1) \equiv \widetilde{\mu}, \tag{6.10}$$

which has the Laurent expansion

$$f(\mu) = \mu(1 + O(\mu^{-1})), \qquad \mu \to \infty, \tag{6.11}$$

due to (6.8). We describe the transformation of the spectral parameter by the formula

$$\exp\left(\frac{1}{W'(\mu)}\frac{\partial}{\partial\mu}\right)\mu = W^{-1}(W(\mu)+1) \equiv f(\mu). \tag{6.12}$$

We have seen that the action of the operator $e^{A(\mu)}$ is expressed in terms of the function f rather than W. Therefore, from here on we shall denote the function $A(\mu)$, entering in the definition (6.7) by $A_f(\mu)$, keeping in mind the relation (6.10) between the functions $f(\mu)$ and $W(\mu)$. Note that the relations between the coefficients a_k and f_i are rather complicated.

Introduce the two sets of times

$$T_k = -\frac{1}{k}\sum_i \mu_i^{-k}, \qquad \widetilde{T}_k = -\frac{1}{k}\sum_i \widetilde{\mu}_i^{-k}, \tag{6.13}$$

where $\widetilde{\mu} = f(\mu)$. It is easy to find relations between these times using the residue operation,

$$\widetilde{T}_k = \frac{1}{k}\sum_{m=k}^{\infty} mT_m \operatorname{Res} \lambda^{m-1} f^{-k}(\lambda)\, d\lambda, \tag{6.14}$$

$$T_k = \sum_{m=k}^{\infty} \widetilde{T}_m \operatorname{Res} \lambda^{-k-1} f^m(\lambda)\, d\lambda, \tag{6.15}$$

in complete analogy with (5.73), (5.74), where $f(\lambda) = \mathcal{P}^{1/p}(\lambda)$. Note that in operator form

$$\widetilde{T}_k(\{\widetilde{\mu}\}) \equiv -\frac{1}{k}\sum_i \widetilde{\mu}_i^{-k} = \prod_i \exp\left(\frac{1}{W'(\mu_i)}\frac{\partial}{\partial\mu_i}\right)T_k(\{\mu\}). \tag{6.16}$$

Since

$$W'(\widetilde{\mu})\,d\widetilde{\mu} = W'(\mu)\,d\mu, \tag{6.17}$$

the transformation (6.15) can be written as

$$T_k(\{\mu\}) \equiv \prod_i \exp\left(-\frac{1}{W'(\widetilde{\mu}_i)}\frac{\partial}{\partial\widetilde{\mu}_i}\right)\widetilde{T}_k(\{\widetilde{\mu}\}) = -\frac{1}{k}\sum_i (f^{-1}(\widetilde{\mu}_i))^{-k}, \tag{6.18}$$

where f^{-1} is the function inverse to f.[3]

6.3. τ-functions of equivalent hierarchies. Consider the correspondence between the τ-functions of equivalent hierarchies. Let $\tau(T)$ be a solution of the original hierarchy. Using (6.15), one can regard it as a function of the times $\{\widetilde{T}_k\}$; i.e., one can deal with $\tau[T(\widetilde{T})]$. It is reasonable to assume that the latter object has something to do with the equivalent hierarchy determined by the operator $\widetilde{L}$ (6.2). Actually, $\tau[T(\widetilde{T})]$ is not a solution of the equivalent hierarchy. Nevertheless, when corrected by an appropriate factor, this function does determine the solution we need. Namely, one can show that the expression

$$e^{\frac{1}{2}A_{km}\widetilde{T}_k\widetilde{T}_m}\tau[T(\widetilde{T})] \equiv \widetilde{\tau}(\widetilde{T}) \tag{6.19}$$

with some definite matrix A_{km} is a τ-function of the equivalent hierarchy. The easiest way to prove this statement is to consider the τ-functions in determinant form (4.17). It is clear that the transformation of times $T \to \widetilde{T}$ corresponds to the transformation of the Miwa variables $\mu_i = f^{-1}(\widetilde{\mu}_i)$. In terms of $\widetilde{\mu}_i$, the original function $\tau[T(\widetilde{T})]$ is not the ratio of two determinants (since $\Delta(\mu)|_{\mu=f^{-1}(\widetilde{\mu})} \equiv \Delta(\mu(\widetilde{\mu}))$ is not the Vandermonde determinant in terms of $\{\widetilde{\mu}_i\}$) and, therefore, does not correspond to any τ-function. Nevertheless, the τ-function of the equivalent hierarchy can be easily extracted. Indeed, consider the identical transformation:[4]

$$\tau[T(\widetilde{T})] \equiv \left\{\frac{\Delta(\widetilde{\mu})}{\Delta(\mu(\widetilde{\mu}))}\prod_i [f'(\mu(\widetilde{\mu}))]^{1/2}\right\}\widetilde{\tau}(\widetilde{T}), \tag{6.20}$$

where $\widetilde{\tau}(\widetilde{T})$ as a function of the times (6.16) is in the determinant form (4.17), i.e.,

$$\widetilde{\tau}(\widetilde{T}) = \frac{\det\widetilde{\phi}_i(\widetilde{\mu}_j)}{\Delta(\widetilde{\mu})} \tag{6.21}$$

with the basis vectors

$$\widetilde{\phi}_i(\widetilde{\mu}) = [f'(\mu(\widetilde{\mu}))]^{-1/2}\phi_i(\mu(\widetilde{\mu})). \tag{6.22}$$

By direct calculation one can show that the prefactor in the right-hand side of (6.20) may be represented in the form

$$\frac{\Delta(\widetilde{\mu})}{\Delta(\mu(\widetilde{\mu}))}\prod_i [f'(\mu(\widetilde{\mu}))]^{1/2} = e^{-\frac{1}{2}A_{km}\widetilde{T}_k\widetilde{T}_m}, \tag{6.23}$$

---

[3] Note that $f^{-1}(\widetilde{\mu}) = W^{-1}(W(\widetilde{\mu}) - 1)$, and compare with (6.10).

[4] By $f'(\mu(\widetilde{\mu}))$ we mean the function $\partial_\mu f(\mu)$ calculated at the point $\mu = f^{-1}(\widetilde{\mu})$.

where

$$A_{km} = \operatorname{Res} f^k(\lambda)\, d_\lambda (f^m(\lambda))_+. \tag{6.24}$$

Thus, one arrives at relation (6.19). We omit the brute force derivation of the identity (6.23) because the technical details are not instructive here; instead, we present below the "physical" proof of (6.19) on the level of the fermionic correlators. Such an approach has two advantages: first of all, it explains very clearly the meaning of the identical redefinition in (6.20)–(6.22); also, it describes the explicit transformation of the point of the Grassmannian while passing to the equivalent hierarchy.

6.4. Proof of the equivalence formula. Let us consider the identity

$$\tau(T|g) \equiv \prod_i \exp\left(-\frac{1}{W'(\tilde{\mu}_i)}\frac{\partial}{\partial \tilde{\mu}_i}\right)\tau(\widetilde{T}|g) \tag{6.25}$$

(see (6.18)). The last (rather trivial) relation can be reformulated in terms of the fermionic correlators as follows. Using the correspondence (4.12) between the fermionic correlators and the τ-functions, one can write[5]

$$\begin{aligned} \langle 0|\, e^{H(T)} g\,|0\rangle &= \frac{\langle N|\,\psi(\mu_N)\cdots\psi(\mu_1) g\,|0\rangle}{\langle N|\,\psi(\mu_N)\cdots\psi(\mu_1)\,|0\rangle} \\ &= \frac{\prod_i \exp(-\frac{1}{W'(\tilde{\mu}_i)}\frac{\partial}{\partial\tilde{\mu}_i})\langle N|\,\psi(\tilde{\mu}_N)\cdots\psi(\tilde{\mu}_1) g\,|0\rangle}{\prod_i \exp(-\frac{1}{W'(\tilde{\mu}_i)}\frac{\partial}{\partial\tilde{\mu}_i})\langle N|\,\psi(\tilde{\mu}_N)\cdots\psi(\tilde{\mu}_1)\,|0\rangle} \\ &\equiv \frac{\langle N|\,\psi(\tilde{\mu}_N)\cdots\psi(\tilde{\mu}_1) e^{\mathbf{L}_f(J)} g\,|0\rangle}{\langle N|\,\psi(\tilde{\mu}_N)\cdots\psi(\tilde{\mu}_1) e^{\mathbf{L}_f(J)}\,|0\rangle} = \frac{\langle 0|\, e^{H(\widetilde{T})} e^{\mathbf{L}_f(J)} g\,|0\rangle}{\langle 0|\, e^{H(\widetilde{T})} e^{\mathbf{L}_f(J)}\,|0\rangle}, \end{aligned} \tag{6.26}$$

where

$$\mathbf{L}_f(J) = \sum_{k=1}^{\infty} a_k \mathbf{L}_{-k}(J) \tag{6.27}$$

similarly to (6.7). Thus we have proved that the transition to the equivalent hierarchy results in the following identity between the τ-functions of the corresponding hierarchies:

$$\tau(T|g) = \frac{\tau(\widetilde{T}|e^{\mathbf{L}_f(J)} g)}{\tau(\widetilde{T}|e^{\mathbf{L}_f(J)})}; \tag{6.28}$$

i.e., one needs to redefine the times together with the appropriate change of the point of the Grassmannian

$$g \to e^{\mathbf{L}_f(J)} g \equiv g_f \tag{6.29}$$

---

[5] From (4.31), (6.7), (6.9)–(6.12) it is clear that

$$e^{\mathbf{L}_f(J)}\psi(\mu) e^{-\mathbf{L}_f(J)} = (\partial_\mu f(\mu))^{1/2}\psi(f(\mu));$$

i.e., the fermions are transformed as $\frac{1}{2}$-differentials. The inverse transformation is

$$e^{-\mathbf{L}_f(J)}\psi(\tilde{\mu}) e^{\mathbf{L}_f(J)} = (\partial_{\tilde{\mu}} f^{-1}(\tilde{\mu}))^{1/2}\psi(f^{-1}(\tilde{\mu})) \sim \exp\left(-\frac{1}{W'(\tilde{\mu}_i)}\frac{\partial}{\partial\tilde{\mu}_i}\right)\psi(\tilde{\mu}).$$

During the calculations in (6.26) we used the latter of these two formulas. In fact, the only things that we need are $\psi(\mu) \sim e^{-\mathbf{L}_f(J)}\psi(\tilde{\mu}) e^{\mathbf{L}_f(J)}$ and $\langle N|\,\mathbf{L}_f = 0$.

and simultaneously renormalize the τ-function. Formula (6.28) coincides with (6.20). Indeed, the numerator is a τ-function that can be written in determinant form (6.21) with basis vectors

$$\widetilde{\phi}_i(\widetilde{\mu}) = \frac{\langle 0|\, \psi^*_{i-1}\psi(\widetilde{\mu})e^{\mathbf{L}_f(J)}g\,|0\rangle}{\langle 0|\, g\,|0\rangle}, \qquad i = 1, 2, \dots . \tag{6.30}$$

These vectors coincide with the previously defined ones (6.22). To show this, let us move the Virasoro element to the left state $\langle 0|$. One can discard the adjoint action of $e^{\mathbf{L}_f}$ on ψ^*_i, since

$$[\mathbf{L}_{-k}(J), \psi^*_i] = \left(\frac{k-1}{2} - i\right)\psi^*_{i-k} \tag{6.31}$$

and, by definition, $\mathbf{L}_f$ contains only the Virasoro generators $\mathbf{L}_{-k}$ with $k > 0$. Therefore, for any $i \geqslant 1$

$$e^{\mathbf{L}_f(J)}\psi^*_{i-1}e^{-\mathbf{L}_f(J)} = \psi^*_{i-1} + \text{lower modes}. \tag{6.32}$$

But the negative modes annihilate the left state $\langle 0|$ while the positive lower modes generate the lower basis vectors, which do not contribute to $\det \widetilde{\phi}_i(\widetilde{\mu}_j)$. Therefore, without loss of generality one may assume that

$$\begin{aligned} \widetilde{\phi}_i(\widetilde{\mu}) &= \frac{\langle 0|\, \psi^*_{i-1}e^{-\mathbf{L}_f(J)}\psi(\widetilde{\mu})e^{\mathbf{L}_f(J)}g\,|0\rangle}{\langle 0|\, g\,|0\rangle} \\ &= (\partial_{\widetilde{\mu}} f^{-1}(\widetilde{\mu}))^{1/2}\, \frac{\langle 0|\, \psi^*_{i-1}\exp(-\frac{1}{W'(\widetilde{\mu}_i)}\frac{\partial}{\partial \widetilde{\mu}_i})\psi(\widetilde{\mu})g\,|0\rangle}{\langle 0|\, g\,|0\rangle} \\ &\equiv (\partial_\mu f(\mu(\widetilde{\mu})))^{-1/2}\phi_i(\mu(\widetilde{\mu})), \end{aligned} \tag{6.33}$$

where in the last equality the original basis vectors $\phi_i(\mu)$ are expressed in terms of the deformed spectral parameter $\widetilde{\mu}$ via the substitution $\mu = f^{-1}(\widetilde{\mu})$. We obtain the basis vectors (6.22). Note that the appearance of the normalization factor in definition (6.22) is very natural, since the basis vectors are transformed as $\frac{1}{2}$-differentials under the action of the Virasoro group (see footnote 5).

Thus, the only problem is to calculate the trivial τ-function

$$\tau(\widetilde{T}|e^{\mathbf{L}_f}) = \langle 0|\, e^{H(\widetilde{T})}e^{\mathbf{L}_f(J)}\,|0\rangle = e^{\mathbf{L}_f(T)} \cdot 1, \tag{6.34}$$

which corresponds to the point of Grassmannian

$$g_0 = e^{\mathbf{L}_f(J)}. \tag{6.35}$$

It is evident that this function is the exponential of quadratic combinations of $\widetilde{T}$. To find the explicit expression, let us consider its derivative with respect to the arbitrary time $\widetilde{T}_k$:

$$\partial_{\widetilde{T}_k}\tau(\widetilde{T}|e^{\mathbf{L}_f}) = \langle 0|\, e^{H(\widetilde{T})}J_k e^{\mathbf{L}_f(J)}\,|0\rangle. \tag{6.36}$$

To calculate the right-hand side of (6.36) one can use the following trick. The motion of the current J_k through $e^{\mathbf{L}_f(J)}$ to the right state $|0\rangle$ results in the appearance of all lower modes J_i with $i \leqslant k$, since

$$[\mathbf{L}_{-i}, J_k] = -kJ_{k-i}. \tag{6.37}$$

The positive modes annihilate the right state and do not contribute to (6.36). Then let us move the negative modes through $e^{\mathbf{L}_f(J)}$ back to the left. In this commutation no positive modes arise (due to (6.37) again). The commutation of negative modes with e^H yields a linear combination of times because of the commutation relations $[J_m, J_n] = m\delta_{n+m,0}$. Finally, all negative modes annihilate the left state $\langle 0|$, and the final result is $\tau(\widetilde{T}|e^{\mathbf{L}_f})$ multiplied by the linear combination of times. The explicit calculation is as follows. From (6.37) we can write

$$(6.38)\qquad [\mathbf{L}_f, J(\mu)] = \partial_\mu\left(\frac{1}{W'(\mu)}\, J(\mu)\right)$$

and, therefore, exponentiation gives the differential operator

$$\exp\left(\frac{1}{W'(\mu)}\,\partial_\mu + \left(\frac{1}{W'(\mu)}\right)'\right),$$

which can be disentangled similarly to (6.9). Thus,

$$(6.39)\qquad e^{\mathbf{L}_f} J(\mu) e^{-\mathbf{L}_f} = \partial_\mu f(\mu)\, J(f(\mu)).$$

The inverse transformation is

$$(6.40)\qquad e^{-\mathbf{L}_f} J(\mu) e^{\mathbf{L}_f} = \partial_\mu f^{-1}(\mu)\, J(f^{-1}(\mu)).$$

Multiplying (6.40) by μ^k and taking the residue, one obtains

$$(6.41)\qquad J_k e^{\mathbf{L}_f} = e^{\mathbf{L}_f} \operatorname{Res} f^k(\lambda) J(\lambda)\, d\lambda,$$

and the action of this operator on the right state reduces to

$$(6.42)\qquad J_k e^{\mathbf{L}_f} |0\rangle = e^{\mathbf{L}_f} \operatorname{Res} f^k(\lambda) J^{(-)}(\lambda)\, d\lambda\, |0\rangle,$$

where $J^{(-)}(\lambda)$ denotes the linear combination of the negative current modes in the expansion

$$(6.43)\qquad J(\lambda) \equiv \sum_{k=0}^{\infty} J_k \lambda^{-k-1} + \sum_{k=-\infty}^{-1} J_k \lambda^{-k-1} \equiv J^{(+)}(\lambda) + J^{(-)}(\lambda).$$

Note that $J^{(-)}(\lambda)$ contains only nonnegative degrees of the spectral parameter λ. Recall that we denote by $(F(\lambda))_+$ the part of the Laurent series $F(\lambda)$ containing nonnegative degrees of λ. From (6.39)

$$(6.44)\qquad e^{\mathbf{L}_f} J^{(-)}(\lambda) = (\partial_\lambda f(\lambda)\, J^{(-)}(f(\lambda)))_+ e^{\mathbf{L}_f}.$$

We should stress that the positive modes $J^{(+)}(f(\lambda))$ do not contribute to the right-hand side of the last formula, since $\partial_\lambda f(\lambda) = 1 + O(\lambda^{-1})$ and $J^{(+)}(f(\lambda))$ contains only the negative degrees of λ. Combining (6.42) and (6.44), one obtains

$$(6.45)\qquad \begin{aligned} J_k e^{\mathbf{L}_f} |0\rangle &= \operatorname{Res} f^k(\lambda)(\partial_\lambda f(\lambda)\, J^{(-)}(f(\lambda)))_+\, d\lambda\, e^{\mathbf{L}_f} |0\rangle \\ &\equiv \sum_{m=1}^{\infty} \frac{1}{m} \operatorname{Res} f^k(\lambda)\, d_\lambda (f^m(\lambda))_+\, J_{-m} e^{\mathbf{L}_f} |0\rangle. \end{aligned}$$

After substitution of (6.45) into (6.36) and taking into account the commutation relations

$$(6.46)\qquad e^{H(\widetilde{T})} J_{-m} e^{-H(\widetilde{T})} = J_{-m} + m\widetilde{T}_m,$$

one arrives at the equation

$$(6.47)\qquad \partial_{\widetilde{T}_k} \tau(\widetilde{T}|e^{\mathbf{L}_f}) = \tau(\widetilde{T}|e^{\mathbf{L}_f}) \sum_{m=1}^{\infty} A_{km} \widetilde{T}_m,$$

where

$$A_{km} = \operatorname{Res} f^k(\lambda)\, d_\lambda (f^m(\lambda))_+. \tag{6.48}$$

It is easy to show that $A_{km} = A_{mk}$. From (6.47) the final answer is

$$\langle 0|\, e^{H(\widetilde{T})} e^{\mathbf{L}_f}\, |0\rangle = \exp\left\{\frac{1}{2}\sum_{k,m=1}^{\infty} A_{km}\widetilde{T}_k\widetilde{T}_m\right\}, \tag{6.49}$$

and relation (6.28) between the τ-functions of the equivalent hierarchies takes the form

$$\tau[T(\widetilde{T})|g] = e^{-\frac{1}{2}A_{km}\widetilde{T}_k\widetilde{T}_m}\, \tau(\widetilde{T}|e^{\mathbf{L}_f} g). \tag{6.50}$$

6.5. Equivalent GKM. In the GKM context, equivalent hierarchies are naturally described by the function

$$\widetilde{\mu} = \mathcal{P}^{1/p}(\mu), \qquad \mathcal{P}(\mu) \equiv V'(\mu). \tag{6.51}$$

Applying the general formula (6.20), one can represent the original partition function (written in the new times $\widetilde{T}$) as follows:

$$Z^V[T(\widetilde{T})] = \frac{\Delta(\widetilde{\mu})}{\Delta(\mu)}\prod_i \left(\frac{V''(\mu_i)}{p\widetilde{\mu}_i^{p-1}}\right)^{1/2} \widetilde{Z}^V[\widetilde{T}] = e^{-\frac{1}{2}A_{ij}\widetilde{T}_i\widetilde{T}_j}\, \widetilde{Z}^V[\widetilde{T}], \tag{6.52}$$

where

$$A_{ij} = \operatorname{Res} \mathcal{P}^{i/p}(\lambda)\, d_\lambda(\mathcal{P}^{j/p}(\lambda))_+ \tag{6.53}$$

and the τ-function of the equivalent hierarchy is described by the matrix integral

$$\widetilde{Z}^V[\widetilde{T}] = \frac{\Delta(\widetilde{\mu}^p)}{\Delta(\widetilde{\mu})}\prod_i (p\widetilde{\mu}_i^{p-1})^{1/2}\, e^{\operatorname{Tr}[V(M) - M\widetilde{M}^p]}\int e^{\operatorname{Tr}[-V(X)+X\widetilde{M}^p]}\, dX. \tag{6.54}$$

The right-hand side of (6.54) must be expressed in terms of the matrix $\widetilde{M}$. Using relation (6.51) it is easy to find that

$$\mu = \frac{1}{p}\sum_{k=-\infty}^{p+1} kt_k\widetilde{\mu}^{k-p}, \tag{6.55}$$

$$V(\mu) - \mu V'(\mu) = -\sum_{k=-\infty}^{p+1} t_k\widetilde{\mu}^k, \tag{6.56}$$

where

$$t_k \equiv -\frac{p}{k(p-k)}\operatorname{Res} \mathcal{P}^{(p-k)/p}(\lambda)\, d\lambda. \tag{6.57}$$

Note that the parameters $t_1, \dots, t_{p+1}$ are independent; they are related to the coefficients of the potential V and can be interpreted as the times generating some integrable evolution. Indeed, (6.57) can be regarded as the set of equations that determine the coefficients of the potential as functions of these additional times. Equations (6.57) naturally arise in the dispersionless KP hierarchy (see below). Note also that all higher positive times are zero because of the polynomiality of $\mathcal{P}$, while the negative "times" $\{t_k,\, k < 0\}$ are complicated functions of the independent positive times.

The substitution of (6.56) into (6.54) results in a matrix integral depending on two sets of times:

$$\widetilde{Z}^V[\widetilde{T},t] = \exp\left(\sum_{k=1}^{\infty} kt_{-k}\widetilde{T}_k\right)\left\{\frac{\Delta(\widetilde{\mu}^p)}{\Delta(\widetilde{\mu})}\prod_i (p\widetilde{\mu}_i^{p-1})^{1/2} \times \exp\left(-\sum_{k=1}^{p+1} t_k \operatorname{Tr}\widetilde{M}^k\right)\int e^{\operatorname{Tr}[-V(X)+X\widetilde{M}^p]}\,dX\right\}, \tag{6.58}$$

where the coefficients of $V(X)$ are functions of the quasiclassical times $t_1, \dots, t_{p+1}$ given by (6.57). It follows from (6.53) that $A_{i,np} = A_{np,i} = 0$. Moreover, $t_{-p} = 0$ and, due to (5.75),

$$\frac{\partial \widetilde{Z}^V[\widetilde{T}]}{\partial \widetilde{T}_p} = 0. \tag{6.59}$$

Thus, we have extracted the τ-function of the p-reduced KP hierarchy from the general matrix integral (5.56). The last logical step to reveal the genuine integrable object hidden in (5.56) is to consider the part of the partition function (6.58) without the exponential prefactor with linear $\widetilde{T}$-dependence, i.e., the matrix integral

$$\tau^V[\widetilde{T},t] = \frac{\Delta(\widetilde{\mu}^p)}{\Delta(\widetilde{\mu})}\prod_i (p\widetilde{\mu}_i^{p-1})^{1/2} \exp\left(-\sum_{k=1}^{p+1} t_k \operatorname{Tr}\widetilde{M}^k\right)\int e^{\operatorname{Tr}[-V(X)+X\widetilde{M}^p]}\,dX. \tag{6.60}$$

This is exactly the object we need. First of all, this τ-function has the standard determinant form

$$\tau^V[\widetilde{T},t] = \frac{\det \phi_i^V(\widetilde{\mu}_j)}{\Delta(\widetilde{\mu})} \tag{6.61}$$

with the basis vectors

$$\phi_i^V(\widetilde{\mu}) = \sqrt{p\widetilde{\mu}^{p-1}}\exp\left(-\sum_{k=1}^{\infty} t_k\widetilde{\mu}^k\right)\int x^{i-1}e^{-V(x)+\widetilde{\mu}^p x}\,dx \tag{6.62}$$

satisfying the p-reduction condition

$$\widetilde{\mu}^p\phi_i^V = \sum_{j=1}^{p+1} v_j\phi_{i+j-1}^V - (i-1)\phi_{i-1}^V \tag{6.63}$$

as well as a Virasoro-type constraint

$$A(\widetilde{\mu})\phi_i^V = \phi_{i+1}^V, \tag{6.64}$$

where

$$A(\widetilde{\mu}) \equiv \frac{1}{p\widetilde{\mu}^{p-1}}\frac{\partial}{\partial\widetilde{\mu}} - \frac{p-1}{2p\widetilde{\mu}^p} + \frac{1}{p}\sum_{k=1}^{p+1} kt_k\widetilde{\mu}^{k-p}. \tag{6.65}$$

The partition function (6.60) possesses remarkable properties. First of all, it is a solution of the p-reduced KP hierarchy, i.e.,

$$\frac{\partial\tau^V[\widetilde{T},t]}{\partial\widetilde{T}_p} = 0; \tag{6.66}$$

this is a corollary of (6.58). Further, relation (6.64) implies that (6.60) satisfies the standard $\mathbf{L}_{-p}$-constraint

$$\mathbf{L}^V_{-p}\tau^V[\widetilde{T},t] = 0, \tag{6.67}$$

(6.68)

$$\mathbf{L}^V_{-p} = \frac{1}{2p}\sum_{k=1}^{p-1} k(p-k)(\widetilde{T}_k + t_k)(\widetilde{T}_{p-k} + t_{p-k}) + \frac{1}{p}\sum_{k=1}^{\infty}(k+p)(\widetilde{T}_{k+p} + t_{p+k})\frac{\partial}{\partial \widetilde{T}_k},$$

where the KP times are naturally shifted by the corresponding quasiclassical ones. We shall give a direct proof of this statement in subsection 8.1 below.

Moreover, the form of the $\mathbf{L}_{-p}$-operator (6.68) gives us a hint that an object depending only on the sum $T_k + t_k$ should exist. This is the case indeed. Consider the product

$$\mathcal{Z}^V[\widetilde{T},t] \equiv \tau^V[\widetilde{T},t]\,\tau_0(t), \tag{6.69}$$

where $\tau_0(t)$ is a τ-function of the *quasiclassical p-reduced KP hierarchy.* We shall show in 8.2 below that $\mathcal{Z}^V[\widetilde{T},t]$ depends only on the sum of the KP times and the quasiclassical times:

$$\left(\frac{\partial}{\partial \widetilde{T}_k} - \frac{\partial}{\partial t_k}\right)\mathcal{Z}^V[\widetilde{T},t] = 0, \qquad k = 1,\dots p. \tag{6.70}$$

Of course, $\mathcal{Z}^V$ also satisfies the constraint (6.67).

To prove the above statements, some essentials concerning quasiclassical hierarchies are required. At this point one sees that the GKM includes almost all the fundamental notions of integrability theory, thus mixing these ingredients together.

7. The quasiclassical KP hierarchy

7.1. Basic definitions. A general treatment of the quasiclassical limit in the theory of integrable systems can be found in [**25**]–[**27**] and references therein. Here we outline the description of the so-called quasiclassical or dispersionless KP hierarchy [**28**], which is the appropriate limit of the standard KP hierarchy (a careful investigation of this limit was given in [**29**]). Consider the quasiclassical version of the L-operator[6]

$$\mathcal{L} = \lambda + \sum_{i=1}^{\infty} u_{i+1}\lambda^{-i}, \tag{7.1}$$

where the functions u_i depend on the infinite set of time variables $(t_1, t_2, t_3, \dots) = \{t\}$ and the evolution along these times is determined by the Lax equations

$$\frac{\partial \mathcal{L}}{\partial t_i} = \{\mathcal{L}^i_+, \mathcal{L}\}, \qquad i = 1, 2, \dots, \tag{7.2}$$

where the functions $\mathcal{L}^i_+(\{t\},\lambda)$ are polynomials in λ, and, in complete analogy with the standard KP theory, are defined as the nonnegative parts of the corresponding degrees of the $\mathcal{L}$-operator:

$$\mathcal{L}^i_+ \equiv \mathcal{L}^i - \mathcal{L}^i_-. \tag{7.3}$$

[6]In what follows we shall use the term "operator" in order to maintain the resemblance with the ordinary KP-terminology, but one should perceive that we are certainly dealing with functions, not with genuine operators.

In (7.2) the Poisson bracket $\{\cdot,\cdot\}$ is the quasiclassical analog of the commutator; for any functions $F(t_1,\lambda)$, $G(t_1,\lambda)$

$$\{F,G\} = \frac{\partial F}{\partial \lambda}\frac{\partial G}{\partial t_1} - \frac{\partial F}{\partial t_1}\frac{\partial G}{\partial \lambda}. \tag{7.4}$$

It is useful to introduce the additional operator

$$\mathcal{M} = \sum_{n=1}^{\infty} n t_n \mathcal{L}^{n-1} + \sum_{i=1}^{\infty} h_{i+1}\mathcal{L}^{-i-1} \equiv \sum_{i\in\mathbb{Z}} i\, t_i \mathcal{L}^{i-1}, \tag{7.5}$$

which satisfies the following equations:[7]

$$\frac{\partial \mathcal{M}}{\partial t_i} = \{\mathcal{L}^i_+, \mathcal{M}\}, \qquad i = 1,2,\ldots, \tag{7.6}$$

$$\{\mathcal{L},\mathcal{M}\} = 1. \tag{7.7}$$

Originally, the differential prototype of (7.5) for the KP hierarchy was introduced in [**30**] in order to describe the symmetries of the evolution equations.

In [**25**], [**28**] it was proved that there exists a function $S(\{t\},\lambda)$ whose total derivative is given by

$$dS = \sum_{i=1}^{\infty} \mathcal{L}^i_+ \, dt_i + \mathcal{M}\, d_\lambda \mathcal{L} \tag{7.8}$$

and, consequently,

$$\left(\frac{\partial S}{\partial t_i}\right)_{\mathcal{L}} = \mathcal{L}^i_+, \qquad d_\lambda S = \mathcal{M}\, d_\lambda \mathcal{L}. \tag{7.9}$$

The function S is the direct quasiclassical analog of the logarithm of the Baker–Akhiezer function; the solution to (7.9) can be represented in the form [**28**]

$$S = \sum_{n=1}^{\infty} t_n \mathcal{L}^n - \sum_{j=1}^{\infty} \frac{1}{j} h_{j+1}\mathcal{L}^{-j} \equiv \sum_{j\in\mathbb{Z}} t_j \mathcal{L}^j. \tag{7.10}$$

7.2. Quasiclassical τ-function and p-reduction. The notion of the quasiclassical τ-function can be introduced as follows. In [**28**] it was proved that

$$\frac{\partial h_{i+1}}{\partial t_j} = \frac{\partial h_{j+1}}{\partial t_i} = \operatorname{Res} \mathcal{L}^i \, d_\lambda \mathcal{L}^j_+, \qquad i,j \geqslant 1, \tag{7.11}$$

where the residue operation is defined in (5.71), (5.72). Therefore, there exists a function whose derivatives with respect to t_i coincide with h_{i+1}. By definition, the quasiclassical τ-function is defined by the relations

$$h_{i+1} = \frac{\partial \log \tau}{\partial t_i}, \qquad i \geqslant 1. \tag{7.12}$$

In [**29**] it was shown that the τ-function defined above satisfies some dispersionless variant of the bilinear Hirota equations, so the definition (7.12) is reasonable.

Let us consider the p-reduced quasiclassical KP hierarchy; this means that for some natural p the function $\mathcal{P} \equiv \mathcal{L}^p$ is a polynomial in λ, i.e.,

$$\mathcal{P}_- = 0. \tag{7.13}$$

---

[7] One should stress that "the negative times" t_{-i}, $i > 0$ are functions of the independent set $\{t_i,\, i > 0\}$ determined by the evolution equations (7.6). They have nothing to do with the actual negative times of the Toda lattice hierarchy.

One can construct the "dual" function

$$\mathcal{Q} = \frac{1}{p}\mathcal{M}\mathcal{L}^{1-p} \equiv \frac{1}{p}\sum_{j\in\mathbb{Z}} jt_j\mathcal{L}^{j-p}, \tag{7.14}$$

which satisfies the equation

$$\{\mathcal{P}, \mathcal{Q}\} = 1 \tag{7.15}$$

as a corollary of (7.7). In [**25**], [**28**], a particular case of the p-reduced hierarchy was discussed, namely, the case in which the function $\mathcal{Q}(\{t\}, \lambda)$ is also polynomial in λ:

$$\mathcal{Q}_- = 0. \tag{7.16}$$

This constraint restricts the possible solutions of the p-reduced quasiclassical KP hierarchy to a very specific subset; as Krichever has shown [**26**], the τ-function satisfies an infinite set of quasiclassical W-constraints when (7.16) holds.[8] In particular, the τ-function satisfies the $\mathbf{L}_{-1}$-constraint

$$\frac{1}{2}\sum_{i=1}^{p-1} i(p-i)t_it_{p-i} + \sum_{i=1}^{\infty}(p+i)t_{p+i}\frac{\partial \log \tau}{\partial t_i} = 0 \tag{7.17}$$

(see the proof below).

7.3. Quasiclassical times and the structure of solutions. When the constraints (7.13), (7.16) are satisfied, one can construct solutions of the hierarchy as follows [**26**]. Note that, evidently,

$$\operatorname{Res}\mathcal{L}^{i-1}d_\lambda\mathcal{L} = \delta_{i,0}. \tag{7.18}$$

Multiplying (7.14) by $\mathcal{L}^{p-i-1}d_\lambda\mathcal{L}$ and taking the residue with the help (5.71), it is easy to show that

$$t_i = -\frac{p}{i(p-i)}\operatorname{Res}\mathcal{L}^{p-i}d_\lambda\mathcal{Q}, \qquad i\in\mathbb{Z}. \tag{7.19}$$

From these equations for $i > 0$ one can determine (at least, in principle) the coefficients of $\mathcal{Q} = \sum_{i=0}^{\infty} q_i\lambda^i$ and $\mathcal{L}$ as functions of the times $t_1, t_2, \dots$, while the same equations for $i < 0$ then give a parametrization of the "negative times" $t_{-i} = -i^{-1}h_{i+1}$:

$$t_{-i} = -\frac{1}{i}\frac{\partial \log \tau}{\partial t_i} \tag{7.20}$$

in terms of $t_1, t_2, \dots$. Consider the simplest situation when $\mathcal{Q}(\lambda)$ is a polynomial of the first order. From (7.14), (7.16) it is easily seen that such a condition is equivalent to switching off all the times with $i > p+1$: $t_{p+2} = t_{p+3} = \dots = 0$. In this case

$$\mathcal{Q}(\{t\}, \lambda) = \frac{p+1}{p}t_{p+1}\lambda + t_p, \tag{7.21}$$

and equations (7.19) are reduced to

$$t_i = -\frac{(p+1)t_{p+1}}{i(p-i)}\operatorname{Res}\mathcal{P}^{(p-i)/p}(\lambda)\,d\lambda, \qquad i \leqslant p+1. \tag{7.22}$$

---

[8] Equations (7.13), (7.16), and (7.15) are the analogs of the Douglas equations [**31**], which, in turn, are equivalent [**5**] to the W-constraints in the KP hierarchy.

Equations (7.22) determine the coefficients of the polynomial $\mathcal{P}$ as functions of the first $p+1$ times $t_1, t_2, \ldots, t_{p+1}$.

It is easy to see that the first time t_1 is contained (linearly) only in the λ-independent term of $\mathcal{P}(t,\lambda)$. Therefore,

$$(7.23)\qquad \frac{\partial \mathcal{P}}{\partial t_1} = -\frac{p+1}{p}\, t_{p+1}, \qquad \frac{\partial \mathcal{L}_+^i}{\partial t_1} = 0 \quad (i = 1, \ldots, p).$$

The Lax equations (7.2) are now reduced to the form

$$(7.24)\qquad \frac{\partial \mathcal{P}}{\partial t_i} = -\frac{\partial \mathcal{P}_+^{i/p}}{\partial \lambda}\, \frac{p}{(p+1)t_{p+1}}, \qquad i = 1, \ldots, p.$$

The remaining equations (7.22) determine the functions $t_{-i}(t_1, \ldots, t_{p+1})$, $i \geqslant 1$. It is possible to find the explicit time dependence in a straightforward way. For example, using the equation of motion (7.24), one obtains

$$(7.25)\qquad -j\,\frac{\partial t_{-j}}{\partial t_i} = \operatorname{Res} \mathcal{P}^{j/p}\, d_\lambda \mathcal{P}_+^{i/p} = \operatorname{Res} \mathcal{P}^{i/p}\, d_\lambda \mathcal{P}_+^{j/p},$$

where one uses the properties (5.72) of the residue operation. In particular, the differentiation of t_{-1} leads to the simple relation (since $d_\lambda \mathcal{P}_+^{1/p} \equiv d\lambda$)

$$(7.26)\qquad \frac{\partial t_{-1}}{\partial t_i} = -\operatorname{Res} \mathcal{P}^{i/p}\, d\lambda = \frac{i(p-i)}{p+1}\, \frac{t_{p-i}}{t_{p+1}}.$$

After integrating these equations, one arrives at the relation

$$(7.27)\qquad t_{-1} = -\frac{\partial \log \tau}{\partial t_1} = \frac{1}{2(p+1)t_{p+1}} \sum_{i=1}^{p-1} i(p-i) t_i t_{p-i},$$

which is equivalent to the $\mathbf{L}_{-1}$-constraint (7.17) with $t_{p+2} = \cdots = 0$.

Note that without loss of generality one can choose

$$(7.28)\qquad t_{p+1} = \frac{p}{p+1}$$

by rescaling the lower times, and therefore the main equations (7.22), (7.24) acquire the standard form[9]

$$(7.29)\qquad t_i = -\frac{p}{i(p-i)} \operatorname{Res} \mathcal{P}^{(p-i)/p}(\lambda)\, d\lambda, \qquad i \leqslant p+1,$$

$$(7.30)\qquad \frac{\partial \mathcal{P}}{\partial t_i} = -\frac{\partial \mathcal{P}_+^{i/p}}{\partial \lambda}, \qquad i = 1, \ldots, p.$$

7.4. Comparison with GKM. The structure of the quasiclassical hierarchy has a nice interpretation in the GKM framework. First of all, the prepotential $V'(\mu) \equiv \mathcal{P}(\mu)$ of GKM generates the solution of the quasiclassical KP hierarchy subjected the constraints

$$(7.31)\qquad \mathcal{P}_-(\mu) = 0, \qquad \mathcal{Q}(\mu) \sim \mu.$$

The easiest way to see this is to note that the definitions (6.57) and (7.29) are the same. Moreover, *all* the quasiclassical ingredients are naturally reproduced.

[9] Note also that the function $\mathcal{P}$ does not contain any term proportional to λ^p, due to the structure of the $\mathcal{L}$-operator (7.1). Hence, $\partial\mathcal{P}/\partial t_p = 0$.

Consider the first basis vector from the set $\{\phi_i^V(\mu)\}$ defined by (6.62). Neglecting the exponential prefactor, one can easily see that the object

$$\Psi(t,\mu) = \sqrt{p\mu^{p-1}} \int e^{-V(x)+x\mu^p}\,dx \tag{7.32}$$

is a Baker–Akhiezer function of the p-reduced quasiclassical KP hierarchy[10] (recall that the coefficients of V are parametrized by the quasiclassical times according to (6.57)). It is evident that $\Psi(t,\mu)$ has the usual asymptotics

$$\Psi(t,\mu) \underset{\mu\to\infty}{\longrightarrow} \exp\left(\sum_{k=1}^{\infty} t_k\mu^k\right)(1+O(\mu^{-1})). \tag{7.33}$$

Using the equations of motion for the quasiclassical KP hierarchy, we can write

$$\frac{\partial V}{\partial t_k} = -\mathcal{P}_+^{k/p} \tag{7.34}$$

(this is a consequence of equations (7.30) or, *equivalently*, the corollary of the parametrization (7.29)). One can easily show that the Baker–Akhiezer function (7.32) satisfies the usual equations of the p-reduced KP hierarchy:

$$[\mathcal{P}(\partial_{t_1})+t_1]\,\Psi(t,\mu) = \mu^p\Psi(t,\mu), \qquad \frac{\partial\Psi(t,\mu)}{\partial t_i} = \mathcal{P}_+^{k/p}(\partial_{t_1})\,\Psi(t,\mu), \tag{7.35}$$

where the polynomials $\mathcal{P}_+^{k/p}(\mu)$ are functions of the times $t_1,\dots,t_p$. Hence, the function (7.32) gives an explicit example of an exact solution. On the other hand, it is important that $\mathcal{P}_+^{k/p}(\mu)$ does not depend on t_1 for $k<p$, and therefore in the corresponding equations (7.35) we can treat $\partial/\partial t_1$ as a formal *parameter*, not as an operator; i.e., it is a particular case of a quasiclassical system. Thus, we see that "the quasiclassical limit" can be naturally treated in the (pseudo-differential) context of the standard hierarchy, and quasiclassical solutions are *exact* solutions of the full p-reduced KP hierarchy restricted to the "small phase space". The exact Baker–Akhiezer function (7.32) gives an explicit solution of the quasiclassical evolution equations along the first p flows, since the following standard relation holds:

$$\Psi(t,\mu) = \exp\left(\sum_{k=1}^{p+1} t_k\mu^k\right)\frac{\tau(t_k - k^{-1}\mu^{-k})}{\tau(t_k)}. \tag{7.36}$$

Evaluating the Baker–Akhiezer function (7.32) by the steepest descent method, it is possible to find all the derivatives of the τ-function appearing in the right-hand side of (7.36). To conclude, we can say that the quasiclassical hierarchy is completely determined by the GKM integrals.

As a consequence of the above reasoning, one can see that the upper $p\times p$ diagonal minor of the matrix A_{ij} (6.53) (which appears here for the first time in the context of equivalent hierarchies) can be written as the second derivative of the quasiclassical τ-function. Indeed, in the p-reduced case formulas (7.11) and (7.12) read[11]

$$\frac{\partial^2 \log\tau_0(t)}{\partial t_i\,\partial t_j} = \operatorname{Res}\mathcal{P}^{i/p}\,d_\lambda\mathcal{P}_+^{j/p}, \qquad i,j=1,\dots,p, \tag{7.37}$$

[10] The use of μ instead of $\widetilde{\mu}$ should not lead to confusion.

[11] We denote the quasiclassical τ-function by τ_0 in what follows.

and the right-hand side is nothing but (6.53). Further, the "negative" times entering in the partition function (6.58) are also represented with the help of τ_0 because of (7.20),

$$kt_{-k} = -\frac{\partial \log \tau_0(t)}{\partial t_k}, \qquad k = 1, \dots, p. \tag{7.38}$$

Before returning to the GKM τ-function, we must prove some useful statements concerning the homogeneity of the quasiclassical τ-function.

7.5. Homogeneity property.

LEMMA 7.1. *Conditions* (7.13), (7.16) *imply*

$$S_- = 0. \tag{7.39}$$

PROOF. Recall that $d_\lambda S = \mathcal{M} d_\lambda \mathcal{L}$. Therefore,

$$\operatorname{Res} \mathcal{L}_+^i d_\lambda S = \operatorname{Res} \mathcal{L}_+^i \mathcal{M}\, d_\lambda \mathcal{L} \equiv p \operatorname{Res} \mathcal{L}_+^i \left(\frac{1}{p} \mathcal{M} \mathcal{L}^{1-p} \right) \mathcal{L}^{p-1}\, d_\lambda \mathcal{L} \equiv \operatorname{Res} \mathcal{L}_+^i \mathcal{Q}\, d_\lambda \mathcal{P}. \tag{7.40}$$

Since, by definition, $\mathcal{P}_- = 0$, one obtains further

$$\operatorname{Res} \mathcal{L}_+^i \mathcal{Q}\, d_\lambda \mathcal{P} = \operatorname{Res} (\mathcal{L}_+^i \mathcal{Q})_-\, d_\lambda \mathcal{P} = \operatorname{Res} (\mathcal{L}_+^i \mathcal{Q}_-)_-\, d_\lambda \mathcal{P}. \tag{7.41}$$

On the other hand, $\operatorname{Res} \mathcal{L}_+^i d_\lambda S = \operatorname{Res} \mathcal{L}_+^i d_\lambda S_-$. Thus, finally,

$$\operatorname{Res} \mathcal{L}_+^i\, d_\lambda S_- = \operatorname{Res} (\mathcal{L}_+^i \mathcal{Q}_-)_-\, d_\lambda \mathcal{P}, \qquad i = 1, 2, \dots . \tag{7.42}$$

Therefore, if $\mathcal{Q}_- = 0$, then $\operatorname{Res} \mathcal{L}_+^i (d_\lambda S)_- = 0$ for any $i = 1, 2, \dots$. The last equality is equivalent to

$$\operatorname{Res} \lambda^i\, d_\lambda S_- = 0, \tag{7.43}$$

and, consequently, (7.39) holds. □

LEMMA 7.2. *The constraint* (7.39) *is equivalent to the homogeneity condition*

$$\sum_{n=1}^{\infty} t_n \frac{\partial t_{-i}}{\partial t_n} = t_{-i}. \tag{7.44}$$

PROOF. Using the explicit representation of S, one obtains

$$\operatorname{Res} S\, d_\lambda \mathcal{L}_+^i = \sum_{n=1}^{\infty} t_n \operatorname{Res} \mathcal{L}^n\, d_\lambda \mathcal{L}_+^i - \sum_{j+1}^{\infty} \frac{1}{j} h_{j+1} \operatorname{Res} \mathcal{L}^{-j}\, d_\lambda \mathcal{L}_+^i. \tag{7.45}$$

But for $j > 0$

$$\operatorname{Res} \mathcal{L}^{-j}\, d_\lambda \mathcal{L}_+^i \equiv \operatorname{Res} \mathcal{L}^{-j}\, d_\lambda (\mathcal{L}^i - \mathcal{L}_-^i) = \operatorname{Res} \mathcal{L}^{-j}\, d_\lambda \mathcal{L}^i = i \operatorname{Res} \mathcal{L}^{i-j-1} d_\lambda \mathcal{L} = i \delta_{ij}$$

due to (7.18); using (7.11), we obtain

$$\operatorname{Res} S d_\lambda \mathcal{L}_+^i = \sum_{n=1}^{\infty} t_n \frac{\partial h_{i+1}}{\partial t_n} - h_{i+1} \tag{7.46}$$

or, equivalently,

$$\operatorname{Res} \mathcal{L}_+^i\, d_\lambda S_- = h_{i+1} - \sum_{n=1}^{\infty} t_n \frac{\partial h_{i+1}}{\partial t_n}, \tag{7.47}$$

and in the case $S_- = 0$ one arrives at (7.44) by using the identification $h_{i+1} = -it_{-i}$. □

Note that, expressed in terms of the τ-function (7.12), the homogeneity condition (7.44) has the form

$$\sum_{n=1}^{\infty} t_n \frac{\partial \log \tau_0}{\partial t_n} = 2\log \tau_0. \tag{7.48}$$

8. Polynomial GKM: synthesis

8.1. $\mathbf{L}_{-p}$-constraint. Here we present a proof of the Virasoro constraint (6.67), (6.68).

Let $\phi_i^{(\mathrm{can})}$ be the canonical basis vectors corresponding to the GKM vectors and let ϕ_i^V be defined by (6.62). From the general formula (4.22) one gets the following expression for the derivative with respect to the first time $\widetilde{T}_1$:

$$\frac{\partial \tau^V}{\partial \widetilde{T}_1} = \frac{1}{\Delta(\widetilde{\mu})} \begin{vmatrix} \phi_1^{(\mathrm{can})}(\widetilde{\mu}_1) & \dots & \phi_1^{(\mathrm{can})}(\widetilde{\mu}_N) \\ \dots & \dots & \dots \\ \phi_{N-1}^{(\mathrm{can})}(\widetilde{\mu}_1) & \dots & \phi_{N-1}^{(\mathrm{can})}(\widetilde{\mu}_N) \\ \phi_{N+1}^{(\mathrm{can})}(\widetilde{\mu}_1) & \dots & \phi_{N+1}^{(\mathrm{can})}(\widetilde{\mu}_N) \end{vmatrix} - \tau^V \sum_{m=1}^{N} \widetilde{\mu}_m. \tag{8.1}$$

Now consider the action of the Virasoro generator $\mathbf{L}_{-p}$ given by (4.33). The operator $A_{-p}(\widetilde{\mu})$, being expressed via the operator $A(\widetilde{\mu})$ (6.65),

$$A_{-p}(\widetilde{\mu}) = pA(\widetilde{\mu}) - \sum_{k=1}^{p-1} k t_k \widetilde{\mu}^{k-p} - p t_p - p\widetilde{\mu}, \tag{8.2}$$

results in the relation

$$\begin{aligned} \frac{1}{p} L_{-p}(\widetilde{T})\tau^V &= -\frac{1}{\Delta(\widetilde{\mu})} \sum_{k=1}^{N} A(\widetilde{\mu}_m) \det \phi_i^{(\mathrm{can})}(\widetilde{\mu}_j) \\ &\quad - \sum_{k=1}^{p-1} k(p-k)\widetilde{T}_{p-k} t_k + \tau^V \left(N t_p + \sum_{m=1}^{N} \widetilde{\mu}_m \right), \end{aligned} \tag{8.3}$$

where the KP times $\widetilde{T}_k$ are expressed through the Miwa variables $\widetilde{\mu}_i$ according to (5.70). In order to prove the analog of constraint (5.54), one must calculate the action of the operator $A(\widetilde{\mu})$ on the canonical basis vectors starting from (6.64). The vectors (6.62) are not in canonical form; let

$$\phi_i^V(\widetilde{\mu}) = \widetilde{\mu}^{i-1} + \alpha_i \widetilde{\mu}^{i-2} + \dots, \qquad i = 1, 2, \dots \tag{8.4}$$

(see the exact expression for α_i below). Now $\phi_i^{(\mathrm{can})} = \phi_i^V - \alpha_i \phi_{i-1}^V + \dots$ and, therefore,

$$A(\widetilde{\mu})\phi_i^{(\mathrm{can})} = \phi_{i+1}^{(\mathrm{can})} + (\alpha_{i+1} - \alpha_i)\phi_i^{(\mathrm{can})} + \dots, \qquad \alpha_1 = 0. \tag{8.5}$$

It is evident that

(8.6)

$$\frac{1}{\Delta(\widetilde{\mu})} \sum_{k=1}^{N} A(\widetilde{\mu}_m) \det \phi_i^{(\mathrm{can})}(\widetilde{\mu}_j) = \frac{1}{\Delta(\widetilde{\mu})} \begin{vmatrix} \phi_1^{(\mathrm{can})}(\widetilde{\mu}_1) & \dots & \phi_1^{(\mathrm{can})}(\widetilde{\mu}_N) \\ \dots & \dots & \dots \\ \phi_{N-1}^{(\mathrm{can})}(\widetilde{\mu}_1) & \dots & \phi_{N-1}^{(\mathrm{can})}(\widetilde{\mu}_N) \\ \phi_{N+1}^{(\mathrm{can})}(\widetilde{\mu}_1) & \dots & \phi_{N+1}^{(\mathrm{can})}(\widetilde{\mu}_N) \end{vmatrix} + \alpha_{N+1}\tau^V,$$

and, after substituting (8.6), (8.1) into (8.3), one arrives at the relation

$$\left(\frac{1}{p}\mathbf{L}_{-p}+\frac{\partial}{\partial\widetilde{T}_1}\right)\tau^V=-\sum_{k=1}^{p-1}k(p-k)\widetilde{T}_{p-k}t_k+(Nt_p-\alpha_{N+1})\tau^V. \tag{8.7}$$

The last step is to calculate the coefficients α_i in the expansion (8.4). Recall that the vectors ϕ_i^V (6.62) are related with the original ones (5.58) as follows (here we take into account relation (6.56)):

$$\begin{aligned}\phi_i^V(\widetilde{\mu})&=\sqrt{\frac{p\widetilde{\mu}^{p-1}}{V''(\mu)}}\exp\left(\sum_{k=-\infty}^{-1}t_k\widetilde{\mu}^k\right)\Phi_i^V(\mu)\\&=\sqrt{\frac{p\widetilde{\mu}^{p-1}}{V''(\mu)}}e^{t_{-1}\widetilde{\mu}^{-1}}\mu^{i-1}(1+O(\mu^{-2}));\end{aligned} \tag{8.8}$$

see the asymptotics (5.38). Using the residue technique, it is easy to find that

$$\frac{p\widetilde{\mu}^{p-1}}{V''(\mu)}=\sum_{i=-\infty}^{1}\frac{i(p+i)}{p+1}\frac{t_{p+i}}{t_{p+1}}\widetilde{\mu}^{i-1}=1+O(\widetilde{\mu}^{-2}); \tag{8.9}$$

i.e., this term does not contribute to α_i. Therefore, from (8.8) and (6.55) we can write

$$\phi_i^V(\widetilde{\mu})=\widetilde{\mu}^{i-1}+(t_{-1}+(i-1)t_p)\widetilde{\mu}^{i-2}+\dots. \tag{8.10}$$

Hence, $\alpha_i=(i-1)t_p+t_{-1}$ and, consequently,

$$Nt_p-\alpha_{N+1}=-t_{-1}, \tag{8.11}$$

where t_{-1} is just the corresponding derivative of the quasiclassical τ-function defined by (7.27); using the convention (7.28), it now reads

$$t_{-1}=\frac{1}{2p}\sum_{k=1}^{p-1}k(p-k)t_kt_{p-k}. \tag{8.12}$$

Hence, equation (8.6) acquires the form

(8.13)

$$\frac{1}{2p}\sum_{k=1}^{p-1}k(p-k)(\widetilde{T}_k+t_k)(\widetilde{T}_{p-k}+t_{p-k})+\frac{1}{p}\sum_{k=1}^{\infty}(k+p)(\widetilde{T}_{k+p}+t_{p+k})\frac{\partial\log\tau^V}{\partial\widetilde{T}_k}=0,$$

and the $\mathbf{L}_{-p}$-constraint (6.67), (6.68) is established.

8.2. Complete description of time dependence. Let us bring all the essential facts together.

(i) We have rewritten the original matrix integral (5.56) in terms of the new times (5.74) as follows:

$$Z^V[T(\widetilde{T})]\equiv\frac{\Delta(\widetilde{\mu})}{\Delta(\mu)}\prod_i\left(\frac{V''(\mu_i)}{p\widetilde{\mu}_i^{p-1}}\right)^{1/2}\widetilde{Z}^V[\widetilde{T}]=e^{-\frac{1}{2}A_{ij}\widetilde{T}_i\widetilde{T}_j}\widetilde{Z}^V[\widetilde{T}]. \tag{8.14}$$

(ii) The matrix $\|A_{ij}\|$ depends on the quasiclassical times $\{t_i\}$ related to the coefficients of the polynomial $\mathcal{P}(\lambda)\equiv V'(\lambda)$ by formula (7.29), and can be compactly written in the form

$$A_{ij}(t)=\operatorname{Res}\mathcal{P}^{i/p}(\lambda)\,d_\lambda\mathcal{P}_+^{j/p}(\lambda). \tag{8.15}$$

Moreover, for $1 \leqslant i, j \leqslant p$,

$$A_{ij}(t) = \frac{\partial^2 \log \tau_0(t)}{\partial t_i\, \partial t_j}, \tag{8.16}$$

and τ_0 is the τ-function of the p-reduced quasiclassical KP hierarchy.

(iii) The "preliminary" p-reduced τ-function of the GKM is a solution of the equivalent hierarchy; it can be presented as

$$\widetilde{Z}^V[\widetilde{T}] = \frac{\Delta(\widetilde{\mu}^p)}{\Delta(\widetilde{\mu})} \prod_i (p\widetilde{\mu}_i^{p-1})^{1/2}\, e^{\mathrm{Tr}[V(M)-MV'(M)]} \int e^{\mathrm{Tr}[-V(X)+XV'(M)]}\, dX, \tag{8.17}$$

where $\widetilde{\mu}^p \equiv V'(\mu)$. Note that the right-hand side of (8.17) is expressed in terms of the matrix $\widetilde{M}$. In order to stress this dependence, we denote the corresponding terms by $M|_{\widetilde{M}}$, etc. below.

(iv) The time derivatives of the partition function (8.14) can also be written as matrix integrals:

$$\frac{\partial}{\partial \widetilde{T}_k} \log Z^V[T(\widetilde{T})] = \langle \mathrm{Tr}\,[\mathcal{P}^{k/p}(X)]_+ - \mathrm{Tr}[\mathcal{P}^{k/p}(M)]_+ \rangle, \qquad 1 \leqslant k \leqslant p. \tag{8.18}$$

Let us differentiate $\widetilde{Z}^V$ with respect to the quasiclassical times t_k, keeping $V'(M) = \widetilde{M}^p$ fixed. It is clear that

$$\frac{\partial}{\partial t_k} \mathrm{Tr}\, V(X) = -\,\mathrm{Tr}\,[\mathcal{P}^{k/p}(X)]_+, \tag{8.19}$$

because of the quasiclassical evolution equations. To calculate the derivative of $\{V(M) - MV'(M)\}|_{\widetilde{M}}$ one must take into account the fact that, besides the coefficient of V, the elements of the matrix M also depend on the times $\{t_k\}$ as functions of $\widetilde{M}$. Therefore,

$$\begin{aligned} &\frac{\partial}{\partial t_k} \mathrm{Tr}\,[V(M)|_{\widetilde{M}} - M|_{\widetilde{M}} \widetilde{M}^p] \\ &\qquad \equiv \sum_{i=1}^{p+1} \frac{1}{i} \frac{\partial v_i(t)}{\partial t_k} \mathrm{Tr}\, M^k + \mathrm{Tr}\left[\frac{\partial V(M)}{\partial M}\frac{\partial M}{\partial t_k} - \frac{\partial M}{\partial t_k}\widetilde{M}^p\right]_{\widetilde{M}}. \end{aligned} \tag{8.20}$$

The second term in this expression vanishes identically; thus,

$$\frac{\partial}{\partial t_k} \mathrm{Tr}\,[V(M) - M\widetilde{M}^p)] = -\,\mathrm{Tr}\,[\mathcal{P}_+^{k/p}(M)|_{\widetilde{M}}] \tag{8.21}$$

and, finally,

$$\frac{\partial}{\partial t_k} \log \widetilde{Z}^V[\widetilde{T}] = \langle \mathrm{Tr}\,[\mathcal{P}^{k/p}(X)]_+ - \mathrm{Tr}[\mathcal{P}^{k/p}(M)]_+|_{\widetilde{M}} \rangle, \qquad 1 \leqslant k \leqslant p. \tag{8.22}$$

Therefore, comparing the last relation with (8.18), we see that

$$\frac{\partial}{\partial T_k} \log Z^V[T(\widetilde{T})] = \frac{\partial}{\partial t_k} \log \widetilde{Z}^V[\widetilde{T}]. \tag{8.23}$$

From (8.14) and (8.23) we obtain

$$\left(\frac{\partial}{\partial \widetilde{T}_k} - \frac{\partial}{\partial t_k}\right) \log Z^V[\widetilde{T}, t] = \sum_{i=1}^{\infty} A_{ki}\widetilde{T}_i, \qquad 1 \leqslant k \leqslant p. \tag{8.24}$$

Introducing the τ-function τ^V in agreement with (6.58), (6.60), we can write

$$\widetilde{Z}^V[\widetilde{T},t] = \exp\Big(\sum_{i=1}^{\infty} it_{-i}\widetilde{T}_i\Big)\tau^V[\widetilde{T},t], \tag{8.25}$$

where the negative times t_{-i}, $i = 1, 2, \dots$, satisfy the equations

$$-i\,\frac{\partial t_{-i}}{\partial t_k} = A_{ki}, \qquad k = 1, \dots, p, \tag{8.26}$$

in accordance with (7.25). The substitution of τ^V into (8.24) results in the equation

$$\begin{aligned}\Big(\frac{\partial}{\partial \widetilde{T}_k} - \frac{\partial}{\partial t_k}\Big)\log \tau^V[\widetilde{T},t] &= -kt_{-k} + \sum_{i=1}^{\infty}\Big(A_{ki} + i\,\frac{\partial t_{-i}}{\partial t_k}\Big)\widetilde{T}_i \\ &= -kt_{-k} \equiv \frac{\partial \log \tau_0(t)}{\partial t_k},\end{aligned} \tag{8.27}$$

because of relations (8.26). Thus, the partition function

$$\mathcal{Z}^V[\widetilde{T},t] \equiv \tau^V[\widetilde{T},t]\,\tau_0(t) \tag{8.28}$$

depends only on the sum $\widetilde{T}_k + t_k$:

$$\frac{\partial \mathcal{Z}^V}{\partial \widetilde{T}_k} = \frac{\partial \mathcal{Z}^V}{\partial t_k}. \tag{8.29}$$

Let us find the relationship between the partition functions Z^V and $\mathcal{Z}^V$. Due to (8.14), (8.25), and (8.28), these functions are proportional up to the exponential with

$$-\frac{1}{2}\,A_{ij}\widetilde{T}_i\widetilde{T}_j + it_{-i}\widetilde{T}_i - \log \tau_0(t). \tag{8.30}$$

Using the homogeneity relation (7.44) we can rewrite (8.30) as

$$-\frac{1}{2}\sum_{ij} A_{ij}(t)\,(\widetilde{T}_i + t_i)(\widetilde{T}_j + t_j); \tag{8.31}$$

hence,

$$Z^V[T(\widetilde{T})] = \mathcal{Z}^V(\widetilde{T} + t)\exp\Big\{-\frac{1}{2}\sum_{ij} A_{ij}(t)(\widetilde{T}_i + t_i)(\widetilde{T}_j + t_j)\Big\}. \tag{8.32}$$

This formula gives a complete description of the polynomial GKM with respect to the "quantum" ($\{\widetilde{T}_k\}$) and quasiclassical ($\{t_k\}$) times.

Acknowledgments

The author is deeply indebted to his coauthors of [**7**], [**11**] and [**12**]—A. Marshakov, A. Mironov, and A. Morozov—for stimulating discussions.

References

[1] P. Di Francesco, P. Ginsparg, and J. Zinn-Justin, *2D gravity and random matrices*, Phys. Rep. **254** (1995), 1–133; `hep-th/9306153`; A. Morozov, *Integrability and matrix models*, Phys. Usp. **37** (1994), 1–55; `hep-th/9303139`.

[2] M. Kontsevich, *Intersection theory on the module space of curves and the matrix Airy functions*, Comm. Math. Phys. **147** (1992), 1–23.

[3] E. Witten, *On the structure of the topological phase of two-dimensional gravity*, Nuclear Phys. B **340** (1990), 281–332.

[4] V. Kac and A. Schwarz, *Geometric interpretation of the partition function of 2D gravity*, Phys. Lett. **257** (1991), 329–334.

[5] M. Fukuma, H. Kawai, and R. Nakayama, *Continuum Schwinger–Dyson equations and universal structures in two-dimensional quantum gravity*, Internat. J. Modern Phys. A **6** (1991), 1385–1406.

[6] R. Dijkgraaf, E. Verlinde, and H. Verlinde, *Loop equations and Virasoro constraints in non perturbative two-dimensional quantum gravity*, Nuclear Phys. B **348** (1991), 435–456.

[7] S. Kharchev, A. Marshakov, A. Mironov, A. Morozov, and A. Zabrodin, *Towards unified theory of 2d gravity*, Nuclear Phys. B **380** (1992), 181–240.

[8] M. Adler and P. van Moerbeke, *A matrix integral solution to two-dimensional gravity*, Comm. Math. Phys. **147** (1992), 25–56.

[9] D. J. Gross and M. Newman, *Unitary and Hermitian matrices in an external field II: the Kontsevich model and continuum Virasoro constraints*, Nuclear Phys. B **380** (1992), 168–180.

[10] C. Itzykson and J.-B. Zuber, *Combinatorics of the modular group II. The Kontsevich integrals*, Internat. J. Modern Phys. **7** (1992), 5661–5706.

[11] S. Kharchev, A. Marshakov, A. Mironov, and A. Morozov, *Landau–Ginzburg topological theories in the framework of GKM and equivalent hierarchies*, Modern Phys. Lett. A **8** (1993), 1047–1061.

[12] ———, *Generalized Kontsevich model versus Toda hierarchy and discrete matrix models*, Nuclear Phys. B **397** (1993), 339–378.

[13] E. Date, M. Jimbo, M. Kashiwara, and T. Miwa, *Transformation groups for soliton equations*, Proc. RIMS symp. nonlinear integrable systems—classical theory and quantum theory, M. Jimbo and T. Miwa (eds.), World Scientific, Singapore, 1983, pp. 39–119.

[14] K. Ueno and K. Takasaki, *Toda lattice hierarchy*, Group Representations and Systems of Differential Equations (Tokyo, 1982; K. Okamoto, ed.), Adv. Stud. Pure Math., vol. 4, North-Holland, Amsterdam, 1984, pp. 1–95.

[15] G. Segal and G. Wilson, *Loop groups and equations of KdV type*, Inst. Hautes Études Sci. Publ. Math. **61** (1985), 5–65; A. Pressley and G. Segal, *Loop groups*, Clarendon Press, Oxford, 1986.

[16] T. Miwa, *On Hirota's difference equations*, Proc. Japan Acad. Ser. A **58** (1982), 9–12.

[17] M. Fukuma, H. Kawai, and R. Nakayama, *Infinite-dimensional Grassmannian structure of two-dimensional quantum gravity*, Comm. Math. Phys. **143** (1992), 371–403.

[18] A. Marshakov, A. Mironov, and A. Morozov, *On equivalence of topological and quantum 2d gravity*, Phys. Lett. B **274** (1992), 280.

[19] E. Witten, *On the Kontsevich model and other models of two-dimensional gravity*, Proc. XX Internat. Conf. Differential Geometric Methods in Theoretical Physics (New York, 1991), Vol. 1, World Sci. Publ., Singapore, 1992, pp. 176–216.

[20] Yu. Makeenko and G. Semenoff, *Properties of Hermitian matrix model in an external field*, Modern Phys. Lett. A **6** (1991), 3455–3466.

[21] M. L. Mehta, *A method of integration over matrix variables*, Comm. Math. Phys. **79** (1981), 327–340.

[22] C. Itzykson and J.-B. Zuber, *The planar approximation* II, J. Math. Phys. **21** (1980), 411–421.

[23] S. Chadha, G. Mahoux, and M. L. Mehta, *A method of integration over matrix variables*, 2, J. Phys. A **14** (1981), 579–586.

[24] T. Shiota, *Characterization of Jacobian varieties in terms of soliton equations*, Invent. Math. **83** (1986), 333–382.

[25] I. Krichever, *The dispersionless Lax equations in topological minimal models*, Comm. Math. Phys. **143** (1992), 415–429.

[26] ———, *The τ-function of the universal Whitham hierarchy, matrix models and topological field theories*, Comm. Pure Appl. Math. **47** (1994), 437–475; hep-th/9205110.

[27] B. Dubrovin, *Geometry of 2D topological field theories*, hep-th/9407018, Integrable Systems and Quantum Groups, Lecture Notes Math., Vol. 1620, Springer-Verlag, Berlin, 1996, pp. 120–348.

[28] K. Takasaki and T. Takebe, *Sdiff(2) KP hierarchy*, Infinite Analysis (Kyoto, 1991) Adv. Ser. Math. Phys Vol. 16, World Sci., River Edge, NJ, 1992, pp. 888–922.

[29] ———, *Integrable hierarchies and dispersionless limit*, Rev. Math. Phys. **7** (1995), 743–808; hep-th/9405096.

[30] A. Orlov and E. Shulman, *Additional symmetries for integrable equations and conformal algebra representations*, Lett. Math. Phys. **12** (1986), 171–179; P. Grinevich and A. Orlov, *Flag spaces in KP theory and Virasoro action on* $\det \bar{\partial}_j$ *and Segal–Wilson τ-function*, Preprint Cornell Univ., September, 1989.

[31] M. Douglas, *Strings in less than one dimension and the generalized KdV hierarchies*, Phys. Lett. B **238** (1990), 176–180.

Institute of Theoretical and Experimental Physics, Bol. Cheremushkinskaya st., 25, Moscow, 117 259, Russia

E-mail address: kharchev@vitep5.itep.ru

Amer. Math. Soc. Transl.
(2) Vol. **191**, 1999

Yangian Algebras and Classical Riemann Problems

S. Khoroshkin, D. Lebedev, and S. Pakuliak

ABSTRACT. We investigate different Hopf algebras associated with Yang's solution of the quantum Yang–Baxter equation. It is shown that for the precise definition of the algebra one needs commutation relations for the deformed algebra of formal currents and a specialization of the Riemann problem for the currents. We consider two different types of Riemann problems. They lead to the centrally extended Yangian double associated with $\mathfrak{sl}_2$ and to the degeneration of the scaling limit of the elliptic affine algebra. Although the defining relations for the generating functions of the two algebras coincide, their properties and the theory of infinite-dimensional representations are quite different. We discuss also the Riemann problems for twisted algebras and for scaled elliptic algebras.

1. Introduction

The Yangian $Y(\mathfrak{g})$, where $\mathfrak{g}$ is a simple Lie algebra, was introduced by Drinfeld [**1**] as a Hopf algebra such that a quantization of the Yang rational solution of the classical Yang–Baxter equation can be done in the tensor category of finite-dimensional representations of $Y(\mathfrak{g})$. Later, the algebraic structure of the quantum double $DY(\mathfrak{g})$ of the Yangian was studied in [**2**] and the corresponding universal R-matrix was calculated explicitly. The Yangian double admits a central extension $\widehat{DY(\mathfrak{g})}$ [**3**] (see also [**4**]), and the intertwining operators for its infinite-dimensional representations can be used for the calculation of form-factors of local operators in the $SU(2)$-invariant Thirring model [**5**].

Recently, a detailed analysis of the scaling limit $\mathcal{A}_{\hbar,\eta}(\widehat{\mathfrak{sl}_2})$ of the elliptic algebra was carried out [**6**]. The corresponding intertwining operators can be used for the

---

1991 *Mathematics Subject Classification.* Primary 17B65, 81R10; Secondary 17B37.

The work of the first author was supported in part by grants RFBR-96-01-00814a, INTAS-93-0166-Ext, No. 96-15-96455 for support of scientific schools, and by Award No. RM2-150 of the U. S. Civilian Research & Development Foundation (CRDF) for the Independent States of the Former Soviet Union.

The work of the second author was supported in part by grants RFBR-95-01-01101, INTAS-93-0166-Ext, French MENESRIP/DAEIF, No. 96-15-96455 for support of scientific schools, and by Award No. RM2-150 of the U.S. Civilian Research & Development Foundation (CRDF) for the Independent States of the Former Soviet Union.

The work of the third author was supported in part by grants RFBR-95-01-01106, No. 96-15-96455 for support of scientific schools, and by Award No. RM2-150 of the U.S. Civilian Research & Development Foundation (CRDF) for the Independent States of the Former Soviet Union.

calculations of the form-factors in the XXZ model in the gapless regime [**7**] and in the sine-Gordon model [**8**]. It turns out that the rational degeneration $\mathcal{A}_\hbar(\widehat{\mathfrak{sl}_2})$ of the algebra $\mathcal{A}_{\hbar,\eta}(\widehat{\mathfrak{sl}_2})$ ($\eta \to 0$) can be described on the level of generating functions (L-operators) by the same set of relations as the centrally extended Yangian double $\widehat{DY(\mathfrak{sl}_2)}$, while the structure of the algebras and their infinite-dimensional representations look very different.

To investigate this phenomenon, we go back to the original ideology of the classical inverse scattering method, where the Riemann problem of factoring a matrix-valued function into the product of functions analytic in certain domains plays the crucial role [**9**]. We claim that the complete structure of the quantum algebra, including its coalgebraic structure, can be fixed by the following data: a formal algebra of (deformed) currents and a specific Riemann problem. For the algebras related to the $\mathfrak{sl}_2$ case, which we study here, this means that they are given by the formal relations for the total currents $e(u)$, $f(u)$, and $h^\pm(u)$ (here u is a spectral parameter) and the decompositions $e(u) = e^+(u) - e^-(u)$, $f(u) = f^+(u) - f^-(u)$. The operator-valued generating functions $e^\pm(u)$ and $f^\pm(u)$ of the spectral parameter are defined as certain integral transforms of the total currents $e(u)$ and $f(u)$, which can be uniquely determined by the prescribed analytic properties of $e^\pm(u)$ and $f^\pm(u)$ (see Section 3). The specification of the Riemann problem uniquely defines the precise expressions of the currents as generating functions of the elements of the algebra. It converts the relationship between the currents into relations between the generators of the algebra. Moreover, it defines the comultiplication structure of the quantum algebra. Assuming here that $e^\pm(u)$, $f^\pm(u)$, and $h^\pm(u)$ are the Gauss coordinates of certain L-operators, we can use the universal comultiplication formulas for Gauss coordinates (see Section 7).

For the algebras $\widehat{DY(\mathfrak{sl}_2)}$ and $\mathcal{A}_\hbar(\widehat{\mathfrak{sl}_2})$, the commutation relations between the currents can be computed by a standard procedure called the Ding–Frenkel isomorphism [**10**]. Actually these calculations coincide for both algebras, since they are formal algebraical manipulations which only use the structure of the Yang (quantum) R-matrix $R(u) = 1 + \hbar P/u$, where P is a flip. This is done in Section 2.

The Riemann problem for $\widehat{DY(\mathfrak{sl}_2)}$ involves the decomposition of a function with a finite number of singularities into the sum of functions analytic in the vicinity of zero and infinity. The solution is defined by the Cauchy integrals

$$e^\pm(u) = \oint \frac{e(v)\,dv}{2\pi i(u-v)},$$

where the closed contour encircling zero goes from the left (from the right) of the point u. The elements of the corresponding algebra are Taylor coefficients of $e^\pm(u)$, $f^\pm(u)$, and $h^\pm(u)$. They generate precisely $\widehat{DY(\mathfrak{sl}_2)}$. The Riemann problem that corresponds to the algebra $\mathcal{A}_\hbar(\widehat{\mathfrak{sl}_2})$ involves the decomposition of a function vanishing at infinity into the sum of functions analytic in some half-planes. The solution is given by the Cauchy integral with the same kernel over a contour that is a straight line parallel to the real axis going under or over u. The generators of the algebra, indexed by real numbers, are the coefficients of the inverse Laplace transforms of the currents $e^\pm(u)$, $f^\pm(u)$, and of $h^\pm(u)$ (Section 3). An analysis of the natural completions of the two algebras and of their properties show that they cannot be transformed one into another by means of projective transforms, because the corresponding Riemann problems are essentially different due to different asymptotic conditions for the functions involved (Section 4).

These two Riemann problems may be modified by imposing other asymptotic behaviors on $e^{\pm}(u)$, $f^{\pm}(u)$. Thus we get the twisted versions of these two algebras and, as a limit, the popular algebras with a simple comultiplication introduced by Drinfeld (new realization) (see Section 7). We demonstrate also in that section that the further trigonometric generalization of the Yangian current algebra essentially leads to the algebra defined in [**6**], and an application of the Riemann problem to a strip produces the scaled elliptic algebra $\mathcal{A}_{\hbar,\eta}(\widehat{\mathfrak{sl}_2})$.

It is worth mentioning that the algebra $\mathcal{A}_{\hbar}(\widehat{\mathfrak{sl}_2})$ is much more appropriate for applications to quantum integrable field theories. The main advantage is that the algebra $\mathcal{A}_{\hbar}(\widehat{\mathfrak{sl}_2})$ is graded, while $\widehat{DY(\mathfrak{sl}_2)}$ is a filtered algebra. Moreover, the grading operator d can be diagonalized in infinite-dimensional representations of $\mathcal{A}_{\hbar}(\widehat{\mathfrak{sl}_2})$ in such a way that the trace over these representations is well defined, contrary to the case of $\widehat{DY(\mathfrak{sl}_2)}$ (see Section 6). The application of the universal R-matrix to finite-dimensional representations gives integral forms of the corresponding R-matrices (Section 5). One may consider the universal R-matrix for $\mathcal{A}_{\hbar}(\widehat{\mathfrak{sl}_2})$ as another quantization of the classical rational solution of the Yang–Baxter equation.

2. The algebra of L-operators

2.1. The aim of this section is to develop the well-known technique of L-operators, starting from Yang's solution of the Yang–Baxter equation, without referring to the precise meaning of L-operators and to their analytic properties. The result can be thought of as a formal algebra, which turns out to be a genuine Hopf algebra when one defines the L-operators to be explicit generating functions of its elements.

Let $\overline{R}(u)$ be a rational solution of the quantum Yang–Baxter equation:

$$\overline{R}(u) = \frac{u - i\hbar P}{u - i\hbar}, \tag{2.1}$$

where $P \in \operatorname{End} \mathbb{C}^2 \otimes \mathbb{C}^2$,

$$P = \begin{pmatrix} 1 & 0 & 0 & 0 \\ 0 & 0 & 1 & 0 \\ 0 & 1 & 0 & 0 \\ 0 & 0 & 0 & 1 \end{pmatrix},$$

is a permutation operator and $\hbar$ is the deformation parameter. Note that the algebras that we discuss below can be defined for arbitrary complex values of the deformation parameter. But representation theory, especially the infinite-dimensional representation theory, depends on the specific values of this parameter. So we fix $\hbar \in \mathbb{R}$ and $\hbar > 0$. Moreover, this choice is in accordance with applications to massive field theory, where one should put $\hbar = \pi$ [**8, 11**].

Following the formalism of Faddeev–Reshetikhin–Takhtadjan [**12**], one can use $\overline{R}(u)$ for the construction of the bialgebra whose generators are gathered into the matrix elements of the quantum L-operator $L(u)$,

$$L(u)=\begin{pmatrix} L_{11}(u) & L_{12}(u)\\ L_{21}(u) & L_{22}(u)\end{pmatrix}, \tag{2.2}$$

that satisfies the Yang–Baxter relation

$$\overline{R}(u-v)L_1(u)L_2(v)=L_2(v)L_1(u)\overline{R}(u-v),$$

where $L_1(u)=L(u)\otimes 1$ and $L_2(u)=1\otimes L(u)$.

More precisely, we would like to consider the Hopf algebras that are quantum doubles and have a family of $(2\ell+1)$-dimensional representations $\pi_z^{(\ell)}$ parametrized by the parameter $z\in\mathbb{C}$. Here ℓ is the spin of the representations. For the simplest nontrivial two-dimensional representation, we require the L-operator to be proportional to the R-matrix (2.1):

$$\pi_z^{(1/2)}(L(u))\sim \overline{R}(u-z). \tag{2.3}$$

As shown in [**12**], such an algebra can be defined via two generating matrix-valued functions $L^{\pm}(u)$ that satisfy the relations

$$\begin{aligned}\overline{R}(u_1-u_2)L_1^{+}(u_1)L_2^{-}(u_2)&=L_2^{-}(u_2)L_1^{+}(u_1)\overline{R}(u_1-u_2),\\ \overline{R}(u_1-u_2)L_1^{\pm}(u_1)L_2^{\pm}(u_2)&=L_2^{\pm}(u_2)L_1^{\pm}(u_1)\overline{R}(u_1-u_2),\end{aligned} \tag{2.4}$$

$$\text{q-det}\,L^{\pm}(u)=L_{11}^{\pm}(u-i\hbar)L_{22}^{\pm}(u)-L_{12}^{\pm}(u-i\hbar)L_{21}^{\pm}(u)=1. \tag{2.5}$$

As usual, the relation (2.5) on the q-determinant factors the algebra over the primitive central elements that appear due to degeneracy of the R-matrix at the critical points $u=\pm i\hbar$. This ensures the existence of the antipode and turns the bialgebra into a Hopf algebra. The two L-operators $L^{\pm}(u)$ generate two dual Hopf subalgebras of the quantum double.

It was shown in [**2**] that the relations (2.4), (2.5) can be interpreted as the defining relations for the quantum double of the Yangian associated with $\mathfrak{sl}_2$ [**13**].

The two-dimensional representation (2.3) is

$$\pi_z^{(1/2)}(L^{\pm}(u))=\rho^{\pm}(u-z)\overline{R}(u-z), \tag{2.6}$$

where the functions $\rho^{\pm}(u)$ satisfy the equation

$$\rho^{\pm}(u)\rho^{\pm}(u-i\hbar)=\frac{u-i\hbar}{u},$$

which follows from the q-determinant condition (2.5) and can be chosen, in particular, as follows:

$$\rho^{\pm}(u)=\left[\frac{\Gamma^2(\frac12\mp\frac{u}{2i\hbar})}{\Gamma(1\mp\frac{u}{2i\hbar})\Gamma(\mp\frac{u}{2i\hbar})}\right]^{\mp1}. \tag{2.7}$$

Moreover, the solution (2.7) is unique under certain analyticity conditions on the L-operators $L^{\pm}(u)$, which we discuss later. Nevertheless, the formal current algebras that we define below do not depend on the choice of $\rho^{\pm}(u)$.

The latter algebra of L-operators appears to have only finite-dimensional representations. In order to construct infinite-dimensional representations, we must perform the central extension of this formal algebra. This can be done as follows. The algebra of L-operators (2.4), (2.5) admits a family of shifting automorphisms $T_z L^{\pm}(u) = L^{\pm}(u-z)$. Then the extension of the quantum double by the infinitesimal shift operator d and the element c dual to it [**14, 3**] leads to the following commutation relations [**15, 16**]:

$$\begin{aligned} &[L(u), c] = 0, \qquad e^{\mathrm{ad}} L^{\pm}(u) = L^{\pm}(u+a) e^{\mathrm{ad}}, \\ &R^{+}(u_1 - u_2 + ic\hbar/2) L_1^{+}(u_1) L_2^{-}(u_2) = L_2^{-}(u_2) L_1^{+}(u_1) R^{+}(u_1 - u_2 - ic\hbar/2), \\ &R^{\pm}(u_1 - u_2) L_1^{\pm}(u_1) L_2^{\pm}(u_2) = L_2^{\pm}(u_2) L_1^{\pm}(u_1) R^{\pm}(u_1 - u_2), \\ &\text{q-det}\, L^{\pm}(u) = L_{11}^{\pm}(u - i\hbar) L_{22}^{\pm}(u) - L_{12}^{\pm}(u - i\hbar) L_{21}^{\pm}(u) = 1, \end{aligned} \tag{2.8}$$

where

$$R^{\pm}(u) = \rho^{\pm}(u) \overline{R}(u).$$

We call this algebra the (*centrally extended*) *algebra of L-operators.* Let us stress that this is still a formal algebra, since we did not specify the meaning of its generating functions $L^{\pm}(u)$.

We can go further, proceeding in algebraic manipulations with the formal algebra of L-operators. First, it is natural to factor the quantum determinant explicitly and have three generating functions instead of four, as in the classical passage from $\mathfrak{gl}_2$ to $\mathfrak{sl}_2$. This can be done by means of the Gauss decomposition of the L-operators.

Let

$$L^{\pm}(u) = \begin{pmatrix} 1 & f^{\pm}(u) \\ 0 & 1 \end{pmatrix} \begin{pmatrix} k_1^{\pm}(u) & 0 \\ 0 & k_2^{\pm}(u) \end{pmatrix} \begin{pmatrix} 1 & 0 \\ e^{\pm}(u) & 1 \end{pmatrix} \tag{2.9}$$

be the Gauss decomposition of the L-operators. Condition (2.5) implies that the product of the entries of diagonal matrix in right-hand side of (2.9) is equal to one: $k_1^{\pm}(u) k_2^{\pm}(u + i\hbar) = 1$, therefore they are invertible and

$$k_1^{\pm}(u) = (k_2^{\pm}(u + i\hbar))^{-1}.$$

Let

$$h^{\pm}(u) = (k_2^{\pm}(u + i\hbar))^{-1} (k_2^{\pm}(u))^{-1}.$$

We call the operator-valued generating functions $e^{\pm}(u)$, $f^{\pm}(u)$, and $h^{\pm}(u)$ the *Gauss coordinates* of the L-operators. We have the following:

PROPOSITION 2.1. *The Gauss coordinates* $e^{\pm}(u)$, $f^{\pm}(u)$, *and* $h^{\pm}(u)$ *satisfy the following commutation relations*:

$$h^{\pm}(u)h^{\pm}(v) = h^{\pm}(v)h^{\pm}(u),$$

$$[e^{\pm}(u), f^{\pm}(v)] = i\hbar \frac{h^{\pm}(u) - h^{\pm}(v)}{u - v},$$

$$[h^{\pm}(u), e^{\pm}(v)] = i\hbar \frac{\{h^{\pm}(u), e^{\pm}(u) - e^{\pm}(v)\}}{u - v},$$

$$[h^{\pm}(u), f^{\pm}(v)] = -i\hbar \frac{\{h^{\pm}(u), f^{\pm}(u) - f^{\pm}(v)\}}{u - v},$$

$$[e^{\pm}(u), e^{\pm}(v)] = i\hbar \frac{(e^{\pm}(u) - e^{\pm}(v))^2}{u - v},$$

$$[f^{\pm}(u), f^{\pm}(v)] = -i\hbar \frac{(f^{\pm}(u) - f^{\pm}(v))^2}{u - v},$$

$$h^{+}(u)h^{-}(v) = \frac{(u - v - i\hbar(1 + c/2))(u - v + i\hbar(1 + c/2))}{(u - v + i\hbar(1 - c/2))(u - v - i\hbar(1 - c/2))} h^{-}(v)h^{+}(u),$$

$$[e^{\pm}(u), f^{\mp}(v)] = i\hbar \frac{h^{\pm}(u)}{u - v \pm ic\hbar/2} - i\hbar \frac{h^{\mp}(v)}{u - v \mp ic\hbar/2},$$

$$[h^{\pm}(u), e^{\mp}(v)] = i\hbar \frac{\{h^{\pm}(u), e^{\pm}(u) - e^{\mp}(v)\}}{u - v \mp ic\hbar/2},$$

$$[h^{\pm}(u), f^{\mp}(v)] = -i\hbar \frac{\{h^{\pm}(u), f^{\pm}(u) - f^{\mp}(v)\}}{u - v \pm ic\hbar/2},$$

$$[e^{\pm}(u), e^{\mp}(v)] = i\hbar \frac{(e^{\pm}(u) - e^{\mp}(v))^2}{u - v \mp ic\hbar/2},$$

$$[f^{\pm}(u), f^{\mp}(v)] = -i\hbar \frac{(f^{\pm}(u) - f^{\pm}(v))^2}{u - v \pm ic\hbar/2}.$$

PROOF. The proof is a direct substitution of the Gauss decomposition of the L-operators (2.9) into (2.8). The particular case of (2.8) for $u_1 = u_2 - i\hbar$,

$$k_2^{\pm}(u)e^{\pm}(u) = e^{\pm}(u - i\hbar)k_2^{\pm}(u), \qquad k_2^{\pm}(u)f^{\pm}(u - ih) = f^{\pm}(u)k_2^{\pm}(u),$$

is very useful for this algebraic exercise. For a similar treatment of quantum affine algebras, see [**10**].

One can see that the algebra of Gauss coordinates does not depend on the choice of the factor $\rho^{\pm}(u)$ in the definition of $R^{\pm}(u)$. It refers only to the original Yang R-matrix. □

2.2. Hopf structure of the algebra of L-operators. The L-operator language is convenient for the description of the coalgebraic structure. The comultiplication map for the algebra (2.8) of formal L-operators is given by the formulas

$$\Delta c = c^{(1)} + c^{(2)} = c \otimes 1 + 1 \otimes c, \qquad \Delta d = d \otimes 1 + 1 \otimes d,$$

(2.10) $$\Delta' L^{\pm}(u) = L(u \pm i\hbar c^{(2)}/4) \dot{\otimes} L(u \mp i\hbar c^{(1)}/4)$$

or in components

(2.11) $$\Delta L_{ij}^{\pm}(u) = \sum_{k=1}^{2} L_{kj}^{\pm}(u \mp i\hbar c^{(2)}/4) \otimes L_{ik}(u \pm i\hbar c^{(1)}/4).$$

The antipode and counit are

$$S(L^{\pm}(u)) = (L^{\pm}(u))^{-1}, \tag{2.12}$$

$$\epsilon(L^{\pm}_{ij}(u)) = \delta_{ij}. \tag{2.13}$$

The comultiplications of the Gauss coordinates $e^{\pm}(u)$, $f^{\pm}(u)$, and $h^{\pm}(u)$ read as follows:

$$\Delta e^{\pm}(u) = e^{\pm}(u') \otimes 1 + \sum_{p=0}^{\infty}(-1)^p (f^{\pm}(u' - i\hbar))^p h^{\pm}(u') \otimes (e^{\pm}(u''))^{p+1}, \tag{2.14}$$

$$\Delta f^{\pm}(u) = 1 \otimes f^{\pm}(u'') + \sum_{p=0}^{\infty}(-1)^p (f^{\pm}(u'))^{p+1} \otimes h^{\pm}(u'')(e^{\pm}(u'' - i\hbar))^p, \tag{2.15}$$

$$\Delta h^{\pm}(u) = \sum_{p=0}^{\infty}(-1)^p (p+1)(f^{\pm}(u' - i\hbar))^p h^{\pm}(u') \otimes h^{\pm}(u'')(e^{\pm}(u'' - i\hbar))^p, \tag{2.16}$$

where $u' = u \mp i\hbar c^{(2)}/4$ and $u'' = u \pm i\hbar c^{(1)}/4$. The proof of these formulas can be found in [**6**].

2.3. Current realization of the algebra of L-operators. To construct the infinite-dimensional representation theory of the centrally extended L-operator algebra the Gauss coordinates are inconvenient, and it is natural to introduce the total currents. Let

$$\begin{aligned} e(u) &= e^{+}\left(u + \frac{ic\hbar}{4}\right) - e^{-}\left(u - \frac{ic\hbar}{4}\right), \\ f(u) &= f^{+}\left(u - \frac{ic\hbar}{4}\right) - f^{-}\left(u + \frac{ic\hbar}{4}\right) \end{aligned} \tag{2.17}$$

be the combination of the Gauss coordinates of the L-operators which we call the total currents. One can verify that the commutation relations between the total currents and $h^{\pm}(u)$ close:

$$\begin{gathered} [d, e(u)] = \frac{d}{du}\, e(u), \qquad [d, f(u)] = \frac{d}{du}\, f(u), \\ h^{\pm}(u) e(v) = \frac{(u - v - i\hbar(1 \pm c/4))}{(u - v + i\hbar(1 \mp c/4))}\, e(v) h^{\pm}(u), \\ h^{\pm}(u) f(v) = \frac{(u - v + i\hbar(1 \pm c/4))}{(u - v - i\hbar(1 \mp c/4))}\, f(v) h^{\pm}(u), \\ e(u)e(v) = \frac{(u - v - i\hbar)}{(u - v + i\hbar)}\, e(v)e(u), \qquad f(u)f(v) = \frac{(u - v + i\hbar)}{(u - v - i\hbar)}\, f(v)f(u), \\ [e(u), f(v)] = -i\hbar\left[\delta\left(u - v + \frac{ic\hbar}{2}\right) h^{+}\left(u + \frac{ic\hbar}{4}\right) \right. \\ \left. - \delta\left(u - v - \frac{ic\hbar}{2}\right) h^{-}\left(v + \frac{ic\hbar}{4}\right)\right], \end{gathered} \tag{2.18}$$

where the δ-function is defined by the formal equality

$$\delta(u - v) = \frac{1}{u - v} - \frac{1}{u - v}$$

and satisfies the relation

$$g(u)\delta(u - v) = g(v)\delta(u - v).$$

In the next section we develop a technique for extracting the correctly defined (Hopf) algebra from the above current algebra, provided certain analytic conditions are assumed.

3. Factorization and the Riemann problems

3.1. Until now we did not specify the analytic properties of L-operators. Let us fix them in the following manner:

$$L^+(u) \text{ is analytic in some neighborhood of } u=\infty,$$
$$L^-(u) \text{ is analytic in some neighborhood of } u=0.$$

To simplify the considerations, we first set $c=0$ and restore the central element only in the final formulas. We would like to invert the relation

$$e(u)=e^+(u)-e^-(u), \tag{3.1}$$

i.e., to express the generating functions $e^{\pm}(u)$ through the total current $e(u)$ by means of some integral transforms. More precisely, let us suppose that

(i) the function $e(u)$ is an analytic function having only isolated singularities on the compactified complex plane;
(ii) $e^+(u)$ is analytic at infinity and $e^-(u)$ is analytic at zero.

We call the solution of (3.1) with conditions (i) and (ii) the *classical Riemann problem for the circle.* Let us first fix a closed counterclockwise-oriented contour Γ around 0 (for instance, $\Gamma=\{|u|=1\}$). Then the Cauchy type integrals

$$\tilde{e}^{\pm}(u)=\oint_\Gamma \frac{dv}{2\pi i}\,\frac{e(v)}{u-v},$$

where u is outside or inside Γ, give two functions $\tilde{e}^{\pm}(u)$, analytic outside and inside of Γ, such that the relation (3.1) is valid for the points $u\in\Gamma$. The limiting values of these functions as u tends to the contour are given by the Sokhotsky–Plemelj relations:

$$\tilde{e}^{\pm}(u)=\frac{1}{2}\left[\text{V.P.}\oint_\Gamma \frac{dv}{\pi i}\,\frac{e(v)}{u-v}\pm e(u)\right],\qquad u\in\Gamma. \tag{3.2}$$

Suppose also that the function $e(u)$ is analytic at the points of Γ, so that we have the identity

$$e(u)=\pm\int_{C^{\pm}_{u,\epsilon}} \frac{dv}{\pi i}\,\frac{e(v)}{u-v}, \tag{3.3}$$

where $C^{\pm}_{u,\epsilon}$ are small semicircles around the point u of radius ϵ drawn in opposite directions as shown in Figure 1. Using (3.3), we can obtain the solution of (3.1) in the domain of analyticity of $e(u)$ near the contour Γ:

$$e^{\pm}(u)=\oint_{\Gamma_\pm} \frac{dv}{2\pi i}\,\frac{e(v)}{u-v}=\oint_{|v|\lessgtr|u|} \frac{dv}{2\pi i}\,\frac{e(v)}{u-v}, \tag{3.4}$$

where the contours $\Gamma_\pm=\Gamma\oplus C^{\pm}_{u,\epsilon}$ are shown in Figure 2. The distinction between the generating functions $\tilde{e}^{\pm}(u)$ and $e^{\pm}(u)$ is that the former are related to the Riemann problem with fixed contour, while the latter have to do with the problem in which the contours $\Gamma_\pm$ are not strictly fixed.

Now we forget the initially fixed contour Γ and use the integral transforms (3.4) as the solution of (3.1) satisfying the condition (ii), which is valid due to (i). The

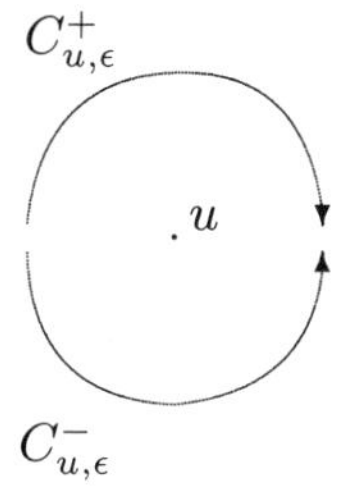

FIGURE 1

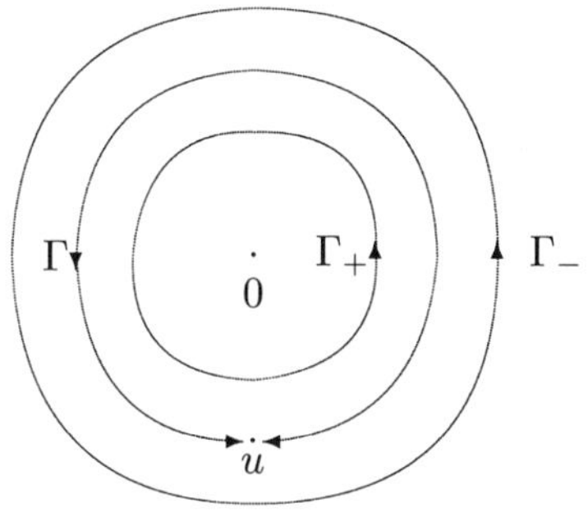

FIGURE 2

contours Γ_+ (Γ_-) should be chosen close to the point u, which means that they are located in the regions $|u| - \epsilon < |v| < |u|$ ($|u| < |v| < |u| + \epsilon$) of analyticity of the functions $e(v)$.

Restoring the dependence on the central element in (2.17), we obtain similarly

$$\begin{aligned} e^+(u) &= \oint_{|v|<|u-ic\hbar/4|} \frac{dv}{2\pi i} \frac{e(v)}{(u-v-ic\hbar/4)}, \\ f^+(u) &= \oint_{|v|<|u+ic\hbar/4|} \frac{dv}{2\pi i} \frac{f(v)}{(u-v+ic\hbar/4)}, \\ e^-(u) &= \oint_{|v|>|u+ic\hbar/4|} \frac{dv}{2\pi i} \frac{e(v)}{(u-v+ic\hbar/4)}, \\ f^-(u) &= \oint_{|v|>|u-ic\hbar/4|} \frac{dv}{2\pi i} \frac{f(v)}{(u-v-ic\hbar/4)}, \end{aligned} \tag{3.5}$$

where all contours in the integrals are counterclockwise circles around the point $v = 0$ such that the corresponding points $u \pm i\hbar c/4$ are either outside the contours for the generating functions $e^+(u)$, $f^+(u)$ or inside the contours for the generating functions $e^-(u)$, $f^-(u)$.

The integral transformations (3.5) are the solutions of the Riemann problems for the factorization of the entire function into the sum of functions analytic in the neighborhood of the given point ($u = 0$ and $u = \infty$, respectively).

The δ-function in the commutation relation of the total currents $e(u)$ and $f(v)$ can be also presented in terms of the same Riemann problem:

$$\delta(u-v) = \frac{1}{u-v}\bigg|_{|u|>|v|} - \frac{1}{u-v}\bigg|_{|u|<|v|} = \sum_{n+m=-1} u^n v^m. \tag{3.6}$$

The relations (3.5) dictate the precise sense of the operator-valued functions $e(u)$, $f(u)$, $e^{\pm}(u)$, and $f^{\pm}(u)$ as the generating series of the elements of the algebra $\widehat{DY(\mathfrak{sl}_2)}$ (centrally extended Yangian double). If we decompose the currents $e^{\pm}(u)$ into Taylor series at the points of their regularity (∞ and 0)

$$e^{\pm}(u) = \mp i\hbar \sum_{\substack{k\geqslant 0\\ k<0}} e_k (u \mp ic\hbar/4)^{-k-1}, \qquad f^{\pm}(u) = \mp i\hbar \sum_{\substack{k\geqslant 0\\ k<0}} f_k (u \pm ic\hbar/4)^{-k-1},$$

then, due to (3.1), we have the presentation

$$e(u) = -i\hbar \sum_{n\in\mathbb{Z}} e_n u^{-n-1}, \qquad f(u) = -i\hbar \sum_{n\in\mathbb{Z}} f_n u^{-n-1}. \tag{3.7}$$

We see that the decompositions (2.17) modulo technical shifts are just the decompositions of a formal power series into the parts with positive and negative powers, and the coefficients of the series are given as

$$\begin{aligned} e_n &= \pm\frac{1}{2\pi\hbar}\oint_{C_{\pm}} v^n e^{\pm}(v \mp ic\hbar/4)\,dv,\\ f_n &= \pm\frac{1}{2\pi\hbar}\oint_{C_{\pm}} v^n f^{\pm}(v \pm ic\hbar/4)\,dv, \end{aligned} \tag{3.8}$$

or, equivalently (this follows from the definition (3.4)),

$$e_n = \frac{1}{2\pi\hbar}\oint_{C_{\pm}} v^n e(v), dv, \qquad f_n = \frac{1}{2\pi\hbar}\oint_{C_{\pm}} v^n f(v)\,dv, \tag{3.9}$$

where C_+ is a contour surrounding infinity (clockwise, like Γ_+) and is taken for $n \geqslant 0$, while C_- is a contour surrounding zero (counterclockwise, like Γ_-) and is taken for $n < 0$. Note that in the following we use (3.8) as the basic definitions, since for the functions $e^{\pm}(u)$ and $f^{\pm}(u)$ of distribution type, the integrals in (3.8) make precise sense, unlike the integrals in (3.9). It is natural also to assume that the currents $h^{\pm}(u)$ satisfy the same analyticity conditions as $e^{\pm}(u)$ and $f^{\pm}(u)$, and put

$$h^{\pm}(u) = 1 \mp i\hbar \sum_{\substack{k\geqslant 0\\ k<0}} h_k u^{-k-1}, \quad \text{where } h_n = \pm\frac{1}{2\pi\hbar}\int_{C_{\pm}} (h^{\pm}(u)-1)u^n\,du, \tag{3.10}$$

with the same rule for the signs. The unit in the above formula is introduced because of the group-like nature of the Cartan generating series.

3.2. There is another choice of the analytic data. Let us fix the following analytic behavior of the operators $L^{\pm}(u)$:

$$L^{+}(u) \text{ is analytic in } \mathbb{C} \text{ for } \operatorname{Im} u < -A,$$
$$L^{-}(u) \text{ is analytic in } \mathbb{C} \text{ for } \operatorname{Im} u > A,$$

for some positive A. In a similar way, we see that for a meromorphic function $e(u)$ decreasing as $\operatorname{Re} u \to \pm\infty$, the Cauchy type integrals

$$\tilde{e}^{\pm}(u) = \int_{-\infty}^{\infty} \frac{dv}{2\pi i} \frac{e(v)}{u-v}$$

are analytic in the closed half-planes $\operatorname{Im} u \leqslant 0$ and $\operatorname{Im} u \geqslant 0$; they satisfy the relation (3.1) for real u and can be used for the following presentations of the functions $e^{\pm}(u)$ satisfying (3.1) for all u and analytic in the half-planes $\operatorname{Im} u < -A$ and $\operatorname{Im} u > A$ for some positive A:

$$e^{\pm}(u) = \int_{\operatorname{Im} v \gtrless \operatorname{Im} u} \frac{dv}{2\pi i} \frac{e(v)}{u-v}. \tag{3.11}$$

The contours of the integrations in (3.11) are close to the point u, which means that there are no singularities of the function $e(z)$ in the strips $\operatorname{Im} u \lesseqgtr \operatorname{Im} z \lesseqgtr \operatorname{Im} v$.

Restoring the central elements, we obtain

$$\begin{aligned} e^{+}(u) &= \int_{\operatorname{Im}(u-v)<c\hbar/4} \frac{dv}{2\pi i} \frac{e(v)}{(u-v-ic\hbar/4)}, \\ f^{+}(u) &= \int_{\operatorname{Im}(u-v)<-c\hbar/4} \frac{dv}{2\pi i} \frac{f(v)}{(u-v+ic\hbar/4)}, \\ e^{-}(u) &= \int_{\operatorname{Im}(u-v)>-c\hbar/4} \frac{dv}{2\pi i} \frac{e(v)}{(u-v+ic\hbar/4)}, \\ f^{-}(u) &= \int_{\operatorname{Im}(u-v)>c\hbar/4} \frac{dv}{2\pi i} \frac{f(v)}{(u-v-ic\hbar/4)}. \end{aligned} \tag{3.12}$$

The corresponding integral transforms are related to the Riemann problem of factoring the decreasing meromorphic function into a sum of functions analytic in certain upper and lower half-planes of the complex plane.

The δ-function in the commutation relation of the total currents $e(u)$ and $f(v)$ should also be the solution of the same Riemann problem, and is given by the integral

$$\delta(u-v) = \lim_{\epsilon \to 0} \left[\frac{1}{u-v-i\epsilon} - \frac{1}{u-v+i\epsilon} \right] = i \int_{-\infty}^{\infty} d\lambda\, e^{-i\lambda(u-v)}. \tag{3.13}$$

As in the case of the Riemann problem for the circle, the integral relations (3.12) dictate the precise sense of the operator-valued functions $e(u)$, $f(u)$, $e^{\pm}(u)$, and $f^{\pm}(u)$ as the generating integrals of the elements of the algebra $\mathcal{A}_{\hbar}(\widehat{\mathfrak{sl}_2})$ (degeneration of $\mathcal{A}_{\hbar,\eta}(\widehat{\mathfrak{sl}_2})$ as $\eta \to 0$, [**6**]). Namely, the functions $e_{+}(u)$, $f_{+}(u)$ and $e_{-}(u)$, $f_{-}(u)$, analytic in the half-planes $\operatorname{Im} u < -A$ and $\operatorname{Im} u > A$, can be represented via Laplace integrals:

$$\begin{aligned} e^{\pm}(u) &= \hbar \int_{0}^{\pm\infty} d\lambda\, e^{-i\lambda u} \hat{e}_{\lambda} e^{-c\hbar|\lambda|/4}, \\ f^{\pm}(u) &= \hbar \int_{0}^{\pm\infty} d\lambda\, e^{-i\lambda u} \hat{f}_{\lambda} e^{c\hbar|\lambda|/4}. \end{aligned} \tag{3.14}$$

Then, due to (2.17), we have the representation of the total currents

$$e(u) = \hbar \int_{-\infty}^{\infty} d\lambda\, e^{-i\lambda u} \hat{e}_{\lambda}, \qquad f(u) = \hbar \int_{-\infty}^{\infty} d\lambda\, e^{-i\lambda u} \hat{f}_{\lambda} \tag{3.15}$$

and relation (2.17) is the decomposition of the formal Fourier integral into the sum of two Laplace transforms.

Due to the definition and the invertion formulas for the Laplace transforms, the Fourier modes $\hat{e}_\lambda$ and $\hat{f}_\lambda$ are the following integrals:

(3.16)
$$\hat{e}_\lambda = \pm\frac{e^{c\hbar|\lambda|/4}}{2\pi\hbar}\int_{C_\pm} du\, e^{i\lambda u}e^{\pm}(u), \qquad \hat{e}_0 = \frac{1}{2\pi\hbar}\left(\int_{C_+} du\, e^{+}(u) - \int_{C_-} du\, e^{-}(u)\right),$$

(3.17)
$$\hat{f}_\lambda = \pm\frac{e^{-c\hbar|\lambda|/4}}{2\pi\hbar}\int_{C_\pm} du\, e^{i\lambda u}f^{\pm}(u), \qquad \hat{f}_0 = \frac{1}{2\pi\hbar}\left(\int_{C_+} du\, f^{+}(u) - \int_{C_-} du\, f^{-}(u)\right),$$

where C_+ is a line parallel to the real axis and belonging to the half-plane $\operatorname{Im} u < -A$ for $\lambda > 0$, and C_- is also a line parallel to the real axis but in the other half-plane $\operatorname{Im} u > A$ for $\lambda < 0$. All the integrals are principal value integrals. Again we can use total currents $e(u)$ and $f(u)$ in (3.17):

$$\hat{e}_\lambda = \frac{1}{2\pi\hbar}\int_{C_\pm} du\, e^{i\lambda u}e(u), \qquad \hat{e}_0 = \frac{1}{2\pi\hbar}\left(\int_{C_+} du\, e(u) + \int_{C_-} du\, e(u)\right),$$
$$\hat{f}_\lambda = \frac{1}{2\pi\hbar}\int_{C_\pm} du\, e^{i\lambda u}f(u), \qquad \hat{f}_0 = \frac{1}{2\pi\hbar}\left(\int_{C_+} du\, f(u) + \int_{C_-} du\, f(u)\right).$$

It is also natural to suppose that

$$h^{\pm}(u) = 1 + \hbar\int_0^{\pm\infty} d\lambda\; e^{-i\lambda u}\; \hat{h}_\lambda e^{-c\hbar|\lambda|/4}, \tag{3.18}$$

where $\hat{h}_\lambda$ are given by the following principal value integrals:

$$\hat{h}_\lambda = \pm\frac{e^{c\hbar|\lambda|/4}}{2\pi\hbar}\int_{C_+} du\, e^{i\lambda u}(h^{+}(u) - 1) \quad \text{for } \lambda \neq 0, \tag{3.19}$$

$$\hat{h}_0 = \frac{1}{2\pi\hbar}\left(\int_{C_+} du\, (h^{+}(u) - 1) - \int_{C_-} du\, (h^{-}(u) - 1)\right). \tag{3.20}$$

We see that two different Riemann problems correspond to two different algebras. We show in the next sections that they are the centrally extended Yangian double and the rational degeneration of $\mathcal{A}_{\hbar,\eta}(\widehat{\mathfrak{sl}_2})$. Let us note once more that these Hopf algebras can be completely defined by the formal algebra of total currents $e(u)$, $f(u)$, $h^{\pm}(u)$ and by the type of the Riemann problem. We shall return to this point in the last section.

4. Algebras $\widehat{DY(\mathfrak{sl}_2)}$ and $\mathcal{A}_\hbar(\widehat{\mathfrak{sl}_2})$

4.1. The algebra $\widehat{DY(\mathfrak{sl}_2)}$. Let us consider the above algebra specialized by the Riemann problem (3.5) for the circle in more detail. For the reader's convenience we identify $-i\hbar = \nu$. As we have seen before, this algebra (we denote it for now by the letter D) can be defined as the algebra generated by the elements

$c, d, e_n, f_n, h_n, n \in \mathbb{Z}$, gathered into the generating functions

$$
\begin{gathered}
e(u) = \nu \sum_{k\in\mathbb{Z}} e_k u^{-k-1}, \qquad f(u) = \nu \sum_{k\in\mathbb{Z}} f_k u^{-k-1}, \\
h^{\pm}(u) = 1 \pm \nu \sum_{\substack{k\geqslant 0 \\ k<0}} h_k u^{-k-1}
\end{gathered} \tag{4.1}
$$

which satisfy the relations

$$
\begin{gathered}
[d, e(u)] = \frac{\mathrm{d}}{\mathrm{d}u} e(u), \qquad [d, f(u)] = \frac{\mathrm{d}}{\mathrm{d}u} f(u), \\
h^+(u)h^-(v) = \frac{(u-v+\nu(1+c/2))(u-v-\nu(1+c/2))}{(u-v-\nu(1-c/2))(u-v+\nu(1-c/2))} h^-(v)h^+(u), \\
h^{\pm}(u)e(v) = \frac{(u-v+\nu(1\pm c/4))}{(u-v-\nu(1\mp c/4))} e(v)h^{\pm}(u), \\
h^{\pm}(u)f(v) = \frac{(u-v-\nu(1\pm c/4))}{(u-v+\nu(1\mp c/4))} f(v)h^{\pm}(u), \\
e(u)e(v) = \frac{(u-v+\nu)}{(u-v-\nu)} e(v)e(u), \qquad f(u)f(v) = \frac{(u-v-\nu)}{(u-v+\nu)} f(v)f(u), \\
[e(u), f(v)] = \nu \left[\delta\left(u-v-\frac{c\nu}{2}\right)h^+\left(u-\frac{c\nu}{4}\right) - \delta\left(u-v+\frac{c\nu}{2}\right)h^-\left(v-\frac{c\nu}{4}\right)\right].
\end{gathered} \tag{4.2}
$$

The total currents $e(u)$ and $f(u)$ decompose into the sums

$$
e(u) = e^+(u - c\nu/4) - e^-(u + c\nu/4), \qquad f(u) = f^+(u + c\nu/4) - f^-(u - c\nu/4),
$$

where

$$
e^{\pm}(u) = \pm\nu \sum_{\substack{k\geqslant 0 \\ k<0}} e_k (u \mp c\nu/4)^{-k-1}, \qquad f^{\pm}(u) = \pm\nu \sum_{\substack{k\geqslant 0 \\ k<0}} f_k (u \pm c\nu/4)^{-k-1},
$$

due to relations (3.5), and the generators of the algebra e_k, f_k, h_k are given by the inversion formulas (3.8), (3.10). The coalgebraic structure is given by relations (2.14)–(2.16) for the generating functions $h^{\pm}(u)$ and $e^{\pm}(u)$, $f^{\pm}(u)$.

Let us translate these data into the language of the generators e_k, f_k, h_k. Due to the definition of these generators, the translation should be done in two steps. In the first step, following the definitions (3.5), from relations (4.2) we obtain the relations for the currents $e^{\pm}(u)$, $f^{\pm}(u)$, $h^{\pm}(u)$. These relations will be precisely those given by Proposition 2.1. Then we put the spectral parameters into the domains of analyticity of these currents and use the definitions (3.8) and (3.10) of the generators e_n, f_n, and h_n. An example of such a calculation will be given for the Riemann problem on a line. Note that for the Riemann problem for the circle, the result coincides with the formal substitution of the power series (4.1), (3.6) into relations (4.2).

At $c = 0$ the full set of the commutation relations is

$$[d, e_n] = -ne_{n-1}, \quad [d, f_n] = -nf_{n-1}, \quad [d, h_n] = -nh_{n-1}, \tag{4.3}$$

$$[h_0, e_n] = 2e_n, \quad [h_0, f_n] = -2f_n, \tag{4.4}$$

$$[h_k, h_n] = 0, \qquad [e_k, f_n] = h_{k+n}, \tag{4.5}$$

$$[h_{k+1}, e_n] - [h_k, e_{n+1}] = \nu\{h_k, e_n\}, \quad [h_{k+1}, f_n] - [h_k, f_{n+1}] = -\nu\{h_k, f_n\}, \tag{4.6}$$

$$[e_{k+1}, e_n] - [e_k, e_{n+1}] = \nu\{e_k, e_n\}, \quad [f_{k+1}, f_n] - [f_k, f_{n+1}] = -\nu\{f_k, f_n\}. \tag{4.7}$$

When $c \neq 0$ the commutation relations of the type $[h_k, e_n]$, $[h_k, f_n]$, $[h_k, h_n]$, and $[e_k, f_n]$ become more complicated while the rest are unchanged:

$$[h_{k+1}, e_n] - [h_k, e_{n+1}] + c\nu\vartheta(k)[h_k, e_n]/4 = \nu\{h_k, e_n\}, \tag{4.8}$$

$$[h_{k+1}, f_n] - [h_k, f_{n+1}] + c\nu\vartheta(k)[h_k, f_n]/4 = -\nu\{h_k, f_n\}, \tag{4.9}$$

$$[h_0, h_k] = 0, \qquad k \in \mathbb{Z},$$

$$\begin{aligned}&[h^+_{k+2}, h^-_{-n-1}] - 2[h^+_{k+1}, h^-_{-n}] + [h^+_{k+2}, h^-_{-n-1}] \\ &\quad = \eta^2(1 + c^2/4)[h^+_k, h^-_{-n-1}] - 2c\eta^2\{h^+_k, h^-_{-n-1}\}, \qquad k, n \geqslant 0,\end{aligned} \tag{4.10}$$

$$[e_n, f_p] = \sum_{k=0}^{n+p} h_{n+p-k}(-c\nu/4)^k B^k_{n,p}, \tag{4.11}$$

$$[e_{-n-1}, f_{-p-1}] = \sum_{k\geqslant 0} h_{-n-p-k-2}(-c\nu/4)^k D^k_{n,p}, \tag{4.12}$$

$$[e_n, f_{-p-1}] = \nu^{-1} \sum_{k\geqslant 0} [(-1)^k h^+_{n-p-1-k} - h^-_{n-p-1-k}](-c\nu/4)^k A^k_{n,p}, \tag{4.13}$$

$$[e_{-p-1}, f_n] = \nu^{-1} \sum_{k\geqslant 0} [h^+_{n-p-1-k} - (-1)^k h^-_{n-p-1-k}](-c\nu/4)^k A^k_{n,p}, \tag{4.14}$$

where in the last four relations $n, p \geqslant 0$,

$$A^k_{n,p} = \sum_{k'=0}^{k} C^{k'}_n C^p_{k+p-k'}, \qquad B^k_{n,p} = \sum_{k'=0}^{k} (-1)^{k'} C^{k'}_n C^{k-k'}_p,$$

$$D^k_{n,p} = \sum_{k'=0}^{k} (-1)^{k'} C^n_{n+k'} C^p_{p+k-k'},$$

$C^{k'}_k$ are binomial coefficients,

$$\vartheta(k) = \begin{cases} 1, & k \geqslant 0, \\ -1, & k < 0, \end{cases}$$

and in order to write out the commutation relations between the Cartan generators h_k and the commutation relations (4.13) and (4.14), we introduce the shorter notation $h^+_{-1} = 1$, $h^+_k = \nu h_k$, for $k > -1$ and $h^+_k = 0$ for $k < -1$. Similarly, $h^-_{-1} = 1 - \nu h_{-1}$, $h^-_k = -\nu h_k$ for $k < -1$ and $h^-_k = 0$ for $k > -1$.

One can see that the commutation relations (4.8), (4.14) coincide with the commutation relations for the generators of the centrally extended Yangian double $\widehat{DY(\mathfrak{sl}_2)}$ (as well as comultiplication rules). More precisely, the algebra D defined by the relations (4.14) admits the filtration

$$\cdots \subset D_{-n} \subset \cdots \subset D_{-1} \subset D_0 \subset D_1 \cdots \subset D_n \cdots \subset D \tag{4.15}$$

defined by the conditions $\deg e_k = \deg f_k = \deg h_k = k$; $\deg\{x \in D_m\} \leqslant m$. Then the formal completion of D over this filtration is a Hopf algebra which coincides with $\widehat{DY(\mathfrak{sl}_2)}$.

4.2. The algebra $\mathcal{A}_\hbar(\widehat{\mathfrak{sl}_2})$. Let us turn our attention to the Riemann problem for the half-planes (3.5). As we have seen in the previous section, the corresponding algebra $\mathcal{A}$ is generated by the elements $\hat{e}_\lambda, \hat{f}_\lambda, \hat{h}_\lambda, \lambda \in \mathbb{R}$, c and $\tilde{d}$ gathered into generating integrals,

$$e(u) = \hbar \int_{-\infty}^{+\infty} d\lambda\, e^{-i\lambda u} \hat{e}_\lambda, \qquad f(u) = \hbar \int_{-\infty}^{+\infty} d\lambda\, e^{-i\lambda u} \hat{f}_\lambda,$$
$$h^{\pm}(u) = 1 + \hbar \int_0^{\pm\infty} d\lambda\, e^{-i\lambda u}\, \hat{h}_\lambda e^{-c\hbar|\lambda|/4}.$$

They satisfy the commutation relations (2.18) and the comultiplication rules (2.14)–(2.16) for the generating integrals $h^{\pm}(u)$ and $e^{\pm}(u)$, $f^{\pm}(u)$, which, due to (3.12), have the form

$$\begin{aligned} e^{\pm}(u) &= \hbar \int_0^{\pm\infty} d\lambda\, e^{-i\lambda u}\, \hat{e}_\lambda e^{-c\hbar|\lambda|/4}, \\ f^{\pm}(u) &= \hbar \int_0^{\pm\infty} d\lambda\, e^{-i\lambda u}\, \hat{f}_\lambda e^{c\hbar|\lambda|/4}. \end{aligned} \tag{4.16}$$

The elements $\hat{e}_\lambda$, $\hat{f}_\lambda$, $\hat{h}_\lambda$ are given by the inversion formulas (3.16), (3.17), (3.19), (3.20). For the description of the algebraical structure of the algebra $\mathcal{A}_\hbar(\widehat{\mathfrak{sl}_2})$ it is more convenient to use the generators $\hat{\kappa}$ that are the Fourier modes of the logarithm of the currents $h^{\pm}(u)$:

$$\kappa^{\pm}(u) = \log h^{\pm}(u) = \hbar \int_0^{\pm\infty} d\lambda\, e^{-i\lambda u} \hat{\kappa}_\lambda.$$

The generators $\hat{\kappa}_\lambda$ and $\hat{h}_\lambda$ are related as follows:

$$\hat{h}_\lambda = e^{c\hbar|\lambda|/4}\Bigg(\hat{\kappa}_\lambda + \sum_{n\geqslant 2} \frac{(\hbar)^{n-1}}{n!} \int_0^\lambda d\lambda_1 \cdots \int_0^{\lambda_{n-2}} d\lambda_{n-1}\, \hat{\kappa}_{\lambda_1} \cdots \hat{\kappa}_{\lambda_{n-1}} \hat{\kappa}_{\lambda - \sum_{i=1}^{n-1}\lambda_i}\Bigg). \tag{4.17}$$

In particular, $\hat{h}_0 = \hat{\kappa}_0$. The relation inverse to (4.17) is

$$\hat{\kappa}_\lambda = e^{-c\hbar|\lambda|/4}\Bigg(\hat{h}_\lambda + \sum_{n\geqslant 2} \frac{(-\hbar)^{n-1}}{n} \int_0^\lambda d\lambda_1 \cdots \int_0^{\lambda_{n-2}} d\lambda_{n-1}\, \hat{h}_{\lambda_1} \cdots \hat{h}_{\lambda_{n-1}} \hat{h}_{\lambda - \sum_{i=1}^{n-1}\lambda_i}\Bigg).$$

The commutation relations given in Proposition 2.1 and written in terms of the generators $\hat{e}_\lambda$, $\hat{f}_\lambda, \hat{\kappa}_\lambda$, $\lambda \in \mathbb{R}$, c and $\tilde{d}$ take the form

$$[c, \text{everything}] = 0, \quad [\tilde{d}, \hat{e}_\lambda] = -\lambda \hat{e}_\lambda, \quad [\tilde{d}, \hat{f}_\lambda] = -\lambda \hat{f}_\lambda, \quad [\tilde{d}, \hat{\kappa}_\lambda] = -\lambda \hat{\kappa}_\lambda, \tag{4.18}$$

$$[\hat{\kappa}_\lambda, \hat{e}_\mu] = \frac{2\sinh \hbar\lambda}{\hbar\lambda} e^{-\hbar c|\lambda|/4} \hat{e}_{\lambda+\mu}, \qquad [\hat{\kappa}_\lambda, \hat{f}_\mu] = -\frac{2\sinh \hbar\lambda}{\hbar\lambda} e^{\hbar c|\lambda|/4} \hat{f}_{\lambda+\mu}, \tag{4.19}$$

$$[\hat{e}_\lambda, \hat{e}_\mu] = \hbar \int_\mu^\lambda d\tau \, [\theta(\tau-\mu) - \theta(\tau-\lambda)] \hat{e}_\tau \hat{e}_{\lambda+\mu-\tau}, \tag{4.20}$$

$$[\hat{f}_\lambda, \hat{f}_\mu] = -\hbar \int_\mu^\lambda d\tau \, [\theta(\tau-\mu) - \theta(\tau-\lambda)] \, \hat{f}_\tau \hat{f}_{\lambda+\mu-\tau}, \tag{4.21}$$

$$[\hat{\kappa}_\lambda, \hat{\kappa}_\mu] = \frac{4}{\hbar^2\lambda} \sinh(\hbar\lambda) \sinh(\hbar\lambda c/2) \, \delta(\lambda+\mu), \tag{4.22}$$

$$\begin{aligned}[\hat{e}_\lambda, \hat{f}_\mu] &= 2\hbar^{-1} \sinh(\lambda\hbar c/2)\delta(\lambda+\mu) \\ &\quad + [e^{(\lambda-\mu)\hbar c/4}\theta(\lambda+\mu) + e^{(\mu-\lambda)\hbar c/4}\theta(-\lambda-\mu)]\hat{h}_{\lambda+\mu},\end{aligned} \tag{4.23}$$

where the step-function $\theta(\lambda)$ is defined as

$$\theta(\lambda) = \begin{cases} 1 & \text{for } \lambda > 0, \\ 1/2 & \text{for } \lambda = 0, \\ 0 & \text{for } \lambda < 0. \end{cases}$$

Note the similarity of the relations (4.19), (4.22), and (4.23) with some of the relations for the quantum affine algebra $U_q(\widehat{sl_2})$ in the new realization [**14**].

Let us obtain formula (4.20) from the commutation relation in terms of the total current

$$[e(u), e(v)] = -i\hbar \frac{\{e(u), e(v)\}}{u-v}.$$

The first step is to obtain the commutation relations between the currents $e^{\pm}(u)$ using the Riemann problem (3.12). For simplicity, we obtain the commutation relations between currents $e^+(u)$ and $e^+(v)$, and in the calculations we set $c = 0$. Other cases, as well as the reconstruction of the central element, can easily be considered. We have

$$\begin{aligned}[e^+(u), e^+(v)] &= -i\hbar \int_{C_+} \frac{d\tilde{u}}{2\pi i} \int_{C_+} \frac{d\tilde{v}}{2\pi i} \frac{\{e(\tilde{u}), e(\tilde{v})\}}{(\tilde{u}-\tilde{v})(u-\tilde{u})(v-\tilde{v})} \\ &= -i\hbar \int_{C_+} \frac{d\tilde{u}}{2\pi i} \int_{C_+} \frac{d\tilde{v}}{2\pi i} \frac{e(\tilde{u})e(\tilde{v})}{\tilde{u}-\tilde{v}} \left[\frac{1}{(u-\tilde{u})(v-\tilde{v})} - \frac{1}{(u-\tilde{v})(v-\tilde{u})} \right] \\ &= i\hbar \int_{C_+} \frac{d\tilde{u}}{2\pi i} \int_{C_+} \frac{d\tilde{v}}{2\pi i} \frac{e(\tilde{u})e(\tilde{v})}{u-v} \left[\frac{1}{u-\tilde{u}} - \frac{1}{v-\tilde{u}} \right] \left[\frac{1}{u-\tilde{v}} - \frac{1}{v-\tilde{v}} \right] \\ &= i\hbar \frac{(e^+(u) - e^+(v))^2}{u-v},\end{aligned}$$

where the contour C_+ goes above the points u, v; in the second line of the computation we used the fact that $e^2(u) = 0$, which follows from the commutation relation (2.18), in order to interchange the integration contours. Due to this condition, the pole of the integrand when $\tilde{u} = \tilde{v}$ is superficial. Let us stress that the condition that the square of the total current vanishes is a special property of deformed infinite-dimensional algebras (like the centrally extended Yangian double, quantum affine algebra, etc.) and has no classical analogs.

The next step is to use the commutation relation

$$[e^+(u), e^+(v)] = i\hbar\,\frac{(e^+(u) - e^+(v))^2}{u - v}$$

to obtain relation (4.20) for $\lambda, \mu > 0$. Others cases can be obtained from the commutation relations $[e^\pm(u), e^\mp(v)]$ and $[e^-(u), e^-(v)]$. This can be done by using the convolution property of the inverse Laplace transform and the following lemma (recall that we consider the case of $c = 0$ for simplicity).

LEMMA 4.1.

$$\hat{g}(\lambda, \mu) = i\int_{C_+}\frac{du}{2\pi}\int_{C_+}\frac{dv}{2\pi}e^{i\lambda u}e^{i\mu v}\,\frac{e^+(u) - e^+(v)}{u - v} = \begin{cases} \hat{e}_{\lambda+\mu} & \textit{for } \lambda + \mu > 0, \\ \hat{e}_\lambda/2 & \textit{for } \lambda > 0,\ \mu = 0, \\ \hat{e}_\mu/2 & \textit{for } \mu > 0,\ \lambda = 0, \\ 0 & \textit{otherwise.} \end{cases}$$

Now we have

(4.24)

$$\begin{aligned} [\hat{e}_\lambda, \hat{e}_\mu] &= i\hbar\int_{C_+}\frac{du}{2\pi}\int_{C_+}\frac{dv}{2\pi}e^{i\lambda u}e^{i\mu v}\left[e^+(u)\,\frac{e^+(u) - e^+(v)}{u - v} - e^+(v)\,\frac{e^+(u) - e^+(v)}{u - v}\right] \\ &= \hbar\int_{-\infty}^{\infty}d\tau\int_{-\infty}^{\infty}d\nu\,[\hat{e}_\tau\theta(\tau)\delta(\nu) - \hat{e}_\nu\theta(\nu)\delta(\tau)]\,\hat{g}(\lambda - \tau, \mu - \nu) \\ &= \hbar\int_{-\infty}^{\infty}d\tau\,[\theta(\tau - \mu) - \theta(\tau - \lambda)]\,\hat{e}_\tau\hat{e}_{\lambda+\mu-\tau}. \end{aligned}$$

Note that the kernel of the integrand in (4.24) is such that the integral is supported on a finite domain of the real axis. One can verify that for $c \neq 0$ the calculation (4.24) will be modified a little bit, but will yield the same result (4.20).

In the terms of the generators $\hat{h}_\lambda$, the commutation relations (4.19) also can be written in convolution form:

$$[\hat{h}_\lambda, \hat{e}_\mu] = 2\hat{e}_{\lambda+\mu} + \hbar\int_0^\lambda d\tau\,[\theta(\tau) - \theta(\tau - \lambda)]\{\hat{h}_\tau, \hat{e}_{\lambda+\mu-\tau}\}, \tag{4.25}$$

$$[\tilde{\hat{h}}_\lambda, \hat{f}_\mu] = -2\hat{f}_{\lambda+\mu} - \hbar\int_0^\lambda d\tau\,[\theta(\tau) - \theta(\tau - \lambda)]\{\tilde{\hat{h}}_\tau, \hat{f}_{\lambda+\mu-\tau}\}, \tag{4.26}$$

where $\tilde{\hat{h}}_\lambda = \hat{h}_\lambda e^{-\hbar c|\lambda|/2}$.

The comultiplication rules (2.14)–(2.16) can be rewritten for the generators $\hat{e}_\lambda$, $\hat{f}_\lambda$, and $\hat{h}_\lambda$ in terms of multiple convolution integrals. For example, at $c = 0$ we have

$$\Delta\hat{e}_\lambda = \hat{e}_\lambda \otimes 1 + 1 \otimes \hat{e}_\lambda + \hbar\int_0^\lambda d\tau\,\hat{h}_{\lambda-\tau} \otimes \hat{e}_\tau + o(\hbar^2).$$

The precise definition of the algebra $\mathcal{A}_\hbar(\widehat{\mathfrak{sl}_2})$ means that one should consider the proper completion of the free tensor topological algebra generated by $\hat{e}_\lambda$, $\hat{f}_\lambda$, $\hat{h}_\lambda$, $\lambda \in \mathbb{R}$, c and $\tilde{d}$ over the ideal generated by relations (4.21). This means in particular that the completed algebra is generated by the formal integrals

$$\int_{-\infty}^{+\infty}\hat{e}_\lambda g(\lambda)d\lambda, \quad \int_{-\infty}^{+\infty}\hat{f}_\lambda g'(\lambda)d\lambda, \quad \int_{-\infty}^{+\infty}\hat{\kappa}_\lambda g''(\lambda)d\lambda, \tag{4.27}$$

where $g(\lambda)$, $g'(\lambda)$, and $g''(\lambda)$ are integrable functions decreasing faster then $e^{-a|\lambda|}$ for some positive a (see the details in [**6**]). The completion $\widehat{\mathcal{A}}$ is a Hopf algebra which coincides with the rational degeneration $\mathcal{A}_\hbar(\widehat{\mathfrak{sl}_2})$ of the scaled elliptic algebra $\mathcal{A}_{\hbar,\eta}(\widehat{\mathfrak{sl}_2})$ [**6**]. As well as the Yangian double, it is the quantum double of its Hopf subalgebra, generated by the positive Fourier harmonics of the currents and one of the elements c and d.

4.3. The comparison. Let us compare the two Yangian type algebras obtained from the Yang R-matrix and the two Riemann problems. The commutation relations for their generators look very different, although they were obtained from identical equations for the L-operators. In order to visualize their common nature, one should look at relations (4.4), (4.6)–(4.10) as being difference equations (4.6)–(4.10) with initial conditions (4.4). One can solve these equations. For instance, the solution of (4.7) for the generators e_n has the form

$$[e_k, e_l] = \nu \sum_{p=l}^{k-1} e_p e_{k+l-p-1}, \qquad k > l, \tag{4.28}$$

which is the discrete analog of relation (4.20). Note that relations (4.20), (4.21), (4.25), and (4.26), written in terms of convolutions, allow us to order the quadratic expressions over the generators in terms of iterated integrals (see details in [**6**]). On the other hand, one can differentiate the relations (4.20), (4.21), (4.25), and (4.26) over the parameters λ and μ and get the continuous analogs of the relations (4.8)–(4.14):

$$[\hat{e}'_\lambda, \hat{e}_\mu] - [\hat{e}_\lambda, \hat{e}'_\mu] = \hbar\{\hat{e}_\lambda, \hat{e}_\mu\}. \tag{4.29}$$

Nevertheless, only the integral relations (4.21), (4.25), and (4.26) have a precise meaning in the completed algebra and admit further generalizations [**6**], contrary to their differential consequences (4.29).

The following natural question then arises: why are the algebras $\widehat{DY(\mathfrak{sl}_2)}$ and $\mathcal{A}_\hbar(\widehat{\mathfrak{sl}_2})$ essentially different? The main difference concerns properties of the elements d for $\widehat{DY(\mathfrak{sl}_2)}$ and of $\tilde{d}$ for the algebra $\mathcal{A}_\hbar(\widehat{\mathfrak{sl}_2})$, and the different types of completions.

In the discrete case of the Yangian double, the operator $[d, \cdot\,] = d/du$ acts in $\widehat{DY(\mathfrak{sl}_2)}$ as degree -1 operator preserving the filtration (4.15) and has no nonzero eigenvalues in the completed algebra. It would encounter the divergence of the partition functions of infinite-dimensional representations (see Section 6).

In contrast to this, the operator $[\tilde{d}, \cdot\,] = -i\, d/du$ acts as a degree zero operator in the algebra $\mathcal{A}_\hbar(\widehat{\mathfrak{sl}_2})$ preserving the grading. The elements $\hat{e}_\lambda$, $\hat{f}_\lambda$, $\hat{h}_\lambda$ are its eigenvectors with eigenvalue λ, $\lambda \in \mathbb{R}$. As a consequence, the characters of the integrable finite-dimensional modules are well defined (see Section 6).

The algebra $\mathcal{A}_\hbar(\widehat{\mathfrak{sl}_2})$ possesses the Cartan anti-involution θ (which can be treated as an involutive anti-isomorphism of the algebra $\mathcal{A}_\hbar(\widehat{\mathfrak{sl}_2})$ over the ring $\mathbb{C}[\hbar]$ or as involutive anti-isomorphism from $\mathcal{A}_\hbar(\widehat{\mathfrak{sl}_2})$ to $\mathcal{A}_{-\hbar}(\widehat{\mathfrak{sl}_2})$). On the level of generating functions, it has the form

$$\begin{gathered} \theta e(u) = f(-u), \qquad \theta f(u) = e(-u), \qquad \theta h^{\pm}(u) = \pm h^{\mp}(-u), \\ \theta d = d, \qquad \theta c = c, \qquad \theta\hbar = -\hbar, \end{gathered} \tag{4.30}$$

and, for the generators,

$$\theta\hat{e}_\lambda = -\hat{f}_{-\lambda}, \qquad \theta\hat{f}_\lambda = -\hat{e}_{-\lambda}, \qquad \theta\hat{h}_\lambda = \hat{h}_{-\lambda}\,.$$

The anti-involution θ exchanges the two subalgebras of $\mathcal{A}_\hbar(\widehat{\mathfrak{sl}_2})$ generated by the Fourier modes of $e^{\pm}(u)$, $f^{\pm}(u)$, $h^{\pm}(u)$; thus the algebra $\mathcal{A}_\hbar(\widehat{\mathfrak{sl}_2})$ can be treated as a contragredient algebra. On the contrary, there is no analog of the Cartan anti-involution for the Yangian double, since it reverses the filtration (4.15) and should exchange Laurent series from a subalgebra generated by negative Fourier harmonics with Laurent polynomials over positive Fourier harmonics. As a result we cannot expect a good notion of restricted dual for highest weight modules of the extended Yangian double.

The two algebras differ on the classical level as well. The classical limit of the centrally extended Yangian double $\widehat{DY(\mathfrak{sl}_2)}$ can be identified with the semi-direct sum of the central extension of meromorphic $\mathfrak{sl}_2$-valued functions and of the one-dimensional Lie algebra generated by the derivative d/dz. The commutation relations for the generators are standard:

$$\begin{gathered}[h_n, e_m] = 2e_{n+m}, \qquad [h_n, f_m] = -2f_{n+m}, \qquad [e_n, e_m] = [f_n, f_m] = 0,\\ [e_n, f_m] = h_{n+m} + n\delta_{n,-m}c, \qquad [h_n, h_m] = 2n\delta_{n,-m}c,\\ [d, e_n] = -ne_{n-1}, \quad [d, f_n] = -nf_{n-1}, \quad [d, h_n] = -nh_{n-1}.\end{gathered}$$

The bialgebra structure can be defined via the decomposition of the algebra into the subalgebras of $\mathfrak{sl}_2$-functions regular at zero and at infinity:

$$(\mathfrak{sl}_2 \otimes \mathbb{C}[z] \oplus \mathbb{C}c) \oplus (\mathfrak{sl}_2 \otimes \mathbb{C}[[z^{-1}]] \oplus \mathbb{C}d).$$

Again we have no symmetry between the two subalgebras and the (quasi)nilpotent action of the operator d.

The classical limit of the Hopf algebra $\mathcal{A}_\hbar(\widehat{\mathfrak{sl}_2})$ can be identified with the completion of the centrally extended algebra of meromorphic $\mathfrak{sl}_2$-valued functions vanishing at infinity (again with the added element $\tilde{d} = -i\,d/dz$). The commutation relations for the generators are now

$$\begin{gathered}[\hat{h}_\lambda, \hat{e}_\mu] = 2\hat{e}_{\lambda+\mu}, \qquad [\hat{h}_\lambda, \hat{f}_\mu] = -2\hat{f}_{\lambda+\mu}, \qquad [\hat{e}_\lambda, \hat{e}_\mu] = [\hat{f}_\lambda, \hat{f}_\mu] = 0,\\ [\hat{e}_\lambda, \hat{f}_\mu] = \hat{h}_{\lambda+\mu} + \lambda\delta(\lambda+\mu)c, \qquad [\hat{h}_\lambda, \hat{h}_\mu] = 2\lambda\delta(\lambda+\mu)c,\\ [\tilde{d}, \hat{e}_\lambda] = -\lambda\hat{e}_\lambda, \qquad [\tilde{d}, \hat{h}_\lambda] = -\lambda\hat{h}_\lambda, \qquad [\tilde{d}, \hat{f}_\lambda] = -\lambda\hat{f}_\lambda.\end{gathered}$$

The bialgebra structure is given by the decomposition of the algebra into the direct sum of subalgebras of $\mathfrak{sl}_2$-valued functions regular in the upper or lower half-plane, which corresponds to taking positive or negative Fourier modes. Again we have the contragredient structure and the diagonalization of the operator $\tilde{d}$.

Note also that the different types of completions used for the definitions of the two algebras actually follow from different positions of the marked singular points in two Riemann problems. In the case of the Yangian double, the two separated points zero and infinity define Laurent series at infinity and a Laurent polynomial at zero, while in the case of $\mathcal{A}_\hbar(\widehat{\mathfrak{sl}_2})$ the singular point near the contour (infinity) produces a continuous family of Fourier harmonics that diagonalize the operator $\tilde{d}$. Thus these two algebras cannot be connected by a projective transform of the complex plane. The same is true for the corresponding Riemann problems.

Let us note that the algebras $\widehat{DY(\mathfrak{sl}_2)}$ and $\mathcal{A}_\hbar(\widehat{\mathfrak{sl}_2})$ given by the commutation relations (4.8)–(4.14) and (4.19)–(4.23) respectively and associated with the simple Lie algebra $\mathfrak{sl}_2$ can be generalized to arbitrary simply-laced Lie algberas. See, for example, [**3**].

5. Finite-dimensional representations and R-matrices

5.1. Finite-dimensional representations. Finite-dimensional representations make sense for the subquotions of the algebras $\widehat{DY(\mathfrak{sl}_2)}$ and $\mathcal{A}_\hbar(\widehat{\mathfrak{sl}_2})$ defined by the condition $c = 0$ with dropped elements d and $\tilde{d}$. Thus their structures are identical for both algebras. Due to the classification theorem (see [**17**]), the irreducible finite-dimensional representations of the algebras $\widehat{DY(\mathfrak{sl}_2)}$ and $\mathcal{A}_\hbar(\widehat{\mathfrak{sl}_2})$ are certain tensor products of the evaluation representations. The evaluation representations can be defined via the evaluation homomorphism $\mathcal{E}v_z\colon \widehat{DY(\mathfrak{sl}_2)} \to U(\mathfrak{sl}_2)$, $\mathcal{A}_\hbar(\widehat{\mathfrak{sl}_2}) \to U(\mathfrak{sl}_2)$, $z \in \mathbb{C}$. For the Yangian double, this homomorphism is of the form

$$(5.1)\qquad \begin{aligned} \mathcal{E}v_z(e(u)) &= \nu\delta\left(u - z - \frac{h-1}{2}\nu\right)\cdot e,\\ \mathcal{E}v_z(f(u)) &= \nu f\cdot\delta\left(u - z - \frac{h-1}{2}\nu\right), \end{aligned}$$

$$(5.2)\qquad \mathcal{E}v_z(h^\pm(u)) = 1 + \frac{\nu}{u - z - \nu(h-1)/2}\,ef - \frac{\nu}{u - z - \nu(h+1)/2}\,fe,$$

and similarly for the algebra $\mathcal{A}_\hbar(\widehat{\mathfrak{sl}_2})$ with the replacement $\nu = -i\hbar$. Here e, f, and h are the generators of the Lie algebra $\mathfrak{sl}_2$:

$$[h,e] = 2e, \qquad [h,f] = -2f, \qquad [e,f] = h,$$

and the right-hand sides of (5.2) are taken for $\operatorname{Im} z \gtrless \operatorname{Im} u$ for $\mathcal{A}_\hbar(\widehat{\mathfrak{sl}_2})$ and $|z| \gtrless |u|$ for $\widehat{DY(\mathfrak{sl}_2)}$ respectively for "$\pm$" generating functions. For instance, for the simplest two-dimensional representation $\pi_z^{(1/2)}$, the action of the generators of the algebra $\widehat{DY(\mathfrak{sl}_2)}$ and of the algebra $\mathcal{A}_\hbar(\widehat{\mathfrak{sl}_2})$ has the form

$$(5.3)\qquad e_k = z^k e_{1,2}, \qquad f_k = z^k e_{2,1}, \qquad h_k = z^k(e_{1,1} - e_{2,2}),$$

$$(5.4)\qquad \begin{aligned} &\hat{e}_\lambda = e^{i\lambda z} e_{1,2}, \qquad \hat{f}_\lambda = e^{i\lambda z} e_{2,1}, \qquad \hat{h}_\lambda = e^{i\lambda z}(e_{1,1} - e_{2,2}),\\ &\hat{\kappa}_\lambda = \frac{e^{i\lambda z}}{\hbar\lambda}((1 - e^{-\lambda\hbar})e_{1,1} + (1 - e^{\lambda\hbar})e_{2,2}), \end{aligned}$$

where $(e_{i,j})_{k,l} = \delta_{ik}\delta_{jl}$ are unit matrices in $\operatorname{End}\mathbb{C}^2 \otimes \mathbb{C}^2$.

5.2. Universal $\mathcal{R}$-matrices. The universal $\mathcal{R}$-matrix for the algebra $\widehat{DY(\mathfrak{sl}_2)}$ was obtained in [**2, 3**] from the analysis of the canonical Hopf pairing of the two Hopf subalgebras $\widehat{DY(\mathfrak{sl}_2)}^\pm$ of $\widehat{DY(\mathfrak{sl}_2)}$, generated by Fourier coefficients of the currents $e^\pm(u)$, $f^\pm(u)$, $h^\pm(u)$. The pairing of these fields is

$$(5.5)\qquad \langle e^+(u), f^-(v)\rangle = \langle f^+(u), e^-(v)\rangle = \frac{\nu}{u - v},$$

$$(5.6)\qquad \langle h^+(u), h^-(v)\rangle = \frac{u - v + \nu}{u - v - \nu},$$

or, in terms of the generators,

$$\langle c,d\rangle=-\nu^{-1},\qquad \langle e_k,f_{-l-1}\rangle=\langle f_k,e_{-l-1}\rangle=-\nu^{-1}\delta_{k,l},$$

$$\langle h_k,h_{-l-1}\rangle=\begin{cases}2(\nu)^{k-l-1}\dfrac{k!}{l!(k-l)!}, & k\geqslant 0,\ 0\leqslant l\leqslant k,\\ 0, & \text{otherwise.}\end{cases}$$

The full pairing is described by the universal R-matrix, which has the form

$$\mathcal{R}=\mathcal{R}_+\cdot\mathcal{C}\cdot\mathcal{R}_0\cdot\mathcal{C}\cdot\mathcal{R}_-, \tag{5.7}$$

where

$$\mathcal{C}=-\frac{\nu}{4}\,(c\otimes d+d\otimes c),$$

$$\mathcal{R}_+=\prod_{k\geqslant 0}^{\rightarrow}\exp(-\nu e_k\otimes f_{-k-1})=\exp(-\nu e_0\otimes f_{-1})\exp(-\nu e_1\otimes f_{-2})\cdots, \tag{5.8}$$

$$\mathcal{R}_-=\prod_{k\geqslant 0}^{\leftarrow}\exp(-\nu f_k\otimes e_{-k-1})=\cdots\exp(-\nu f_1\otimes e_{-2})\exp(-\nu f_0\otimes e_{-1}), \tag{5.9}$$

$$\mathcal{R}_0=\prod_{n\geqslant 0}\exp\Big(-\operatorname*{Res}_{u=v}\Big[\frac{d}{du}\ln h^+(u)\otimes\ln h^-(v-(2n+1)\nu)\Big]\Big), \tag{5.10}$$

and the residue operation Res is defined as follows:

$$\operatorname*{Res}_{u=v}\Big(\sum_{i\geqslant 0}a_iu^{-i-1}\otimes\sum_{k\geqslant 0}b_kv^k\Big)=\sum_{i\geqslant 0}a_i\otimes b_i.$$

As usual, the L-operators $L^{\pm}$ are given by substituting the two-dimensional representation into one tensor component of $\mathcal{R}$ [**16**]:

$$\begin{aligned}L^-(z)&=(\pi_z^{(1/2)}\otimes\mathrm{id})\,\mathcal{C}^{-1}\cdot\mathcal{R}\cdot\mathcal{C}^{-1},\\ L^+(z)&=(\pi_z^{(1/2)}\otimes\mathrm{id})\,\mathcal{C}\cdot(\mathcal{R}^{21})^{-1}\cdot\mathcal{C}.\end{aligned} \tag{5.11}$$

The decomposition (5.7) produces the Gauss decomposition of the L-operators (2.9).

Similarly, the universal R-matrix for the algebra $\mathcal{A}_\hbar(\widehat{\mathfrak{sl}_2})$ can be described by following the same scheme of [**2**]. Let us recall that the main arguments in [**2**] are: the triangular decomposition of the Yangian double with respect to a Hopf pairing, the basic pairing (5.5), (5.6), and the expression of the tensor of the pairing to the subalgebras generated by the Cartan currents as the exponent of the pairing between logarithms of Cartan fields. The last calculation uses the shift automorphisms of $\widehat{DY(\mathfrak{sl}_2)}$. All these arguments remain unchanged for the algebra $\mathcal{A}_\hbar(\widehat{\mathfrak{sl}_2})$ (in the basic pairing (5.5), (5.6) we use $\nu=-i\hbar$ as usual).

This means that the universal R-matrix for $\mathcal{A}_\hbar(\widehat{\mathfrak{sl}_2})$ admits the decomposition (5.7), where the factor $\mathcal{C}$ is unchanged, the ordered products of exponents in the factors $\mathcal{R}_\pm$ become the ordered exponential in integral form, and the factor $\mathcal{R}_0$ is as before

$$\mathcal{R}_0=\exp\Omega,$$

where Ω is the tensor of the pairing of the fields $k^{\pm}(u) = \log h^{\pm}(u)$. The main difference in the case of $\mathcal{A}_\hbar(\widehat{\mathfrak{sl}_2})$ is that the pairing

$$\langle \kappa^+(u), \kappa^-(v)\rangle = \log \frac{u-v-i\hbar}{u-v+i\hbar}$$

can be explicitly and uniquely diagonalized in the generators $\hat\kappa_\lambda$:

$$\langle \hat\kappa_\lambda, \hat\kappa_\mu\rangle = -2\,\frac{\sinh \hbar\lambda}{\hbar^2\lambda}\,\delta(\lambda+\mu).$$

Summarizing the calculation, we have the following.

The universal R-matrix for $\mathcal{A}_\hbar(\widehat{\mathfrak{sl}_2})$ admits the decomposition (5.7), where $\mathcal{C} = \exp(-\hbar(\tilde d\otimes c + c\otimes \tilde d)/4)$,

(5.12)

$$\mathcal{R}_+ = \overrightarrow{P}\exp\Big(-\hbar\int_0^{+\infty} d\lambda\,\hat e_\lambda\otimes \hat f_{-\lambda}\Big), \qquad \mathcal{R}_- = \overleftarrow{P}\exp\Big(-\hbar\int_0^{+\infty} d\lambda\,\hat f_\lambda\otimes \hat e_{-\lambda}\Big),$$

$$\mathcal{R}_0 = \exp\Big(-\int_0^{+\infty} d\lambda\,\frac{\hbar^2\lambda}{2\sinh\hbar\lambda}\,\hat\kappa_\lambda\otimes\hat\kappa_{-\lambda}\Big). \tag{5.13}$$

It is interesting to compare the evaluation of the two expressions of the universal R-matrix to the tensor product $\pi^{(1/2)}_{z_1}\otimes\pi^{(1/2)}_{z_2}$ of two-dimensional representations. The formulas (5.8), (5.9), and (5.10) yield the following decomposition of the four by four R-matrix $R(z)$:

(5.14)

$$R^-(z) = \begin{pmatrix}1&0&0&0\\0&1&\frac{\nu}{z}&0\\0&0&1&0\\0&0&0&1\end{pmatrix}\begin{pmatrix}\rho^-(z)&0&0&0\\0&\frac{z-\nu}{z}\rho^-(z)&0&0\\0&0&\frac{z}{z+\nu}\rho^-(z)&0\\0&0&0&\rho^-(z)\end{pmatrix}\begin{pmatrix}1&0&0&0\\0&1&0&0\\0&\frac{\nu}{z}&1&0\\0&0&0&1\end{pmatrix}$$

or $R^-(z) = \rho^-(z)(z+\nu P)/(z+\nu)$, where P is a permutation of tensor components and

$$\rho^-(z) = \prod_{n\geqslant 0}\frac{(z-2n\nu)(z-(2n+2)\nu)}{(z-(2n+1)\nu)^2}.$$

The infinite product converges for proper z and is equal to a ratio of Γ-functions, namely

$$\rho^-(z) = \frac{\Gamma^2(\frac12-\frac{z}{2\nu})}{\Gamma(1-\frac{z}{2\nu})\Gamma(-\frac{z}{2\nu})}.$$

An application of (5.12) and (5.13) gives $R^-(z)$ in the form (5.14) but with a scalar factor $\rho^-(z)$ presented in integral form

$$\rho^-(z) = \exp\Big(-2\int_0^{+\infty} d\lambda\,\frac{\sinh^2(\hbar\lambda/2)}{\lambda\sinh\hbar\lambda}\,e^{i\lambda z}\Big), \qquad \operatorname{Im} z>0.$$

We see that the two universal R-matrices presented in this section give two different quantizations of the Yang rational solution of the classical YB equation. The solution via the algebra $\mathcal{A}_\hbar(\widehat{\mathfrak{sl}_2})$ has the advantage of having an integral presentation and being uniquely determined by the asymptotics of the scalar factor. On the contrary, there is no definite choice for such a solution for the double of the Yangian (the answer has no definite asymptotics as $|z|$ tends to infinity). Moreover, as we shall see, further similar representations for divergent infinite products automatically appear in the representation theory of $\mathcal{A}_\hbar(\widehat{\mathfrak{sl}_2})$ in regularized form, whereas the representation theory of the Yangian double has no instruments for such a regularization.

6. Basic infinite-dimensional representations

The goal of this section is to construct examples of infinite-dimensional representations of the algebras $\widehat{DY(\mathfrak{sl}_2)}$ and $\mathcal{A}_\hbar(\widehat{\mathfrak{sl}_2})$. This construction demonstrates the difference between these algebras. We start from the discrete algebra $\widehat{DY(\mathfrak{sl}_2)}$ and for simplicity consider only the case of the central element $c = 1$.

6.1. Basic representations of the algebra $\widehat{DY(\mathfrak{sl}_2)}$. Let $\mathcal{H}$ be the Heisenberg algebra generated by the free bosons $a_{\pm n}$, $n = 1, 2, \dots$, with zero modes a_0, p and commutation relations

$$[a_n, a_m] = n\delta_{n+m,0}, \qquad [p, a_0] = 2.$$

Let

$$a_+(z) = \sum_{n\geqslant 1} \frac{a_n}{n} z^{-n} - p\log z, \quad a_-(z) = \sum_{n\geqslant 1} \frac{a_{-n}}{n} z^n + \frac{a_0}{2}, \quad \phi_\pm(z) = \exp a_\pm(z)$$

be the generating functions of the elements of the algebra $\mathcal{H}$. Let $\bar{e}(u)$, $\bar{f}(u)$, $\bar{h}^+(u)$, and $\bar{h}^-(u)$ be following generating series acting in the Fock space $\mathcal{H}$:

$$\bar{e}(u) = \nu\phi_-(u-\nu)\phi_-(u)\phi_+^{-1}(u), \qquad \bar{f}(u) = \nu\phi_-^{-1}(u+\nu)\phi_-^{-1}(u)\phi_+(u),$$
$$\bar{h}^+(u) = \phi_+(u-\nu)\phi_+^{-1}(u), \qquad \bar{h}^-(u) = \phi_-(u-\nu)\phi_-^{-1}(u+\nu).$$

We have the following:

PROPOSITION 6.1 ([**3**]). *The $\mathcal{H}$-valued generating functions (fields)*

$$e(u) = \bar{e}(u+c\nu/4), \quad f(u) = \bar{f}(u-c\nu/4), \quad h^+(u) = \bar{h}^+(u+c\nu/2), \quad h^-(u) = \bar{h}^-(u)$$

satisfy the commutation relations (4.2) *with* $c = 1$.

Let V_α be the formal power series extensions of the Fock spaces

$$V_i = \mathbb{C}[[a_{-1}, \dots, a_{-n}, \dots]] \otimes \Big(\bigoplus_{n\in\mathbb{Z}} \mathbb{C}e^{(n+\alpha)a_0}\Big), \qquad 0 \leqslant \alpha < 1, \tag{6.1}$$

with the action of bosons on these spaces

$$a_n = \begin{cases} \text{the left multiplication by } a_n \otimes 1 & \text{for } n < 0, \\ [a_n, \cdot\,] \otimes 1 & \text{for } n > 0, \end{cases}$$
$$e^{n_1 a_0}(a_{-j_k}\cdots a_{-j_1} \otimes e^{n_2 a_0}) = a_{-j_k}\cdots a_{-j_1} \otimes e^{(n_1+n_2)a_0},$$
$$u^p(a_{-j_k}\cdots a_{-j_1} \otimes e^{na_0}) = u^{2n} a_{-j_k}\cdots a_{-j_1} \otimes e^{na_0}. \tag{6.2}$$

It is clear from (6.2) that the Fock spaces V_α becomes the irreducible representations of the algebra $\widehat{DY(\mathfrak{sl}_2)}$ at level 1 ($c = 1$) for $\alpha = 0$ or $1/2$ with vacuum vectors $1\otimes 1$ and $1\otimes e^{a_0/2}$. These representations are highest weight representations with respect to the Fourier components $\bar{e}_n$, $\bar{f}_n$ of the generating currents $\bar{e}(u) = \sum \bar{e}_n u^{-n-1}$ and

$\bar{f}(u) = \sum \bar{f}_n u^{-n-1}$. The latter generators are related to e_n and f_n by the triangular transformations due to the relations given in Proposition 6.1:

$$\bar{e}_n(1 \otimes 1) = \bar{f}_n(1 \otimes 1) = 0, \quad n < 0,$$
$$\bar{e}_n(1 \otimes e^{a_0/2}) = 0, \quad n < -1, \qquad \bar{f}_n(1 \otimes e^{a_0/2}) = 0, \quad n < 1.$$

Elements of the monomial basis (6.1) are *not* eigenvalues of the filtration operator d, since

$$[d, a_n] = -na_{n-1}, \qquad n \leqslant -1,\ n \geqslant 2,$$
$$[d, a_1] = -p, \qquad [d, p] = 0, \qquad [d, a_0/2] = a_{-1}.$$

They cannot be used, for example, to calculate the character of the Fock space V_α using the operator d. The usual expression $\operatorname{tr}_{V_\alpha}(e^{pd})$ is divergent, and only the ratio $\operatorname{tr}_{V_\alpha}(e^{pd}O)/\operatorname{tr}_{V_\alpha}(e^{pd})$ of such traces can be made finite for certain operators O in the Fock space $\mathcal{H}$ [**18**].

6.2. Realization of the algebra $\mathcal{A}_\hbar(\widehat{\mathfrak{sl}_2})$ by continuous fields. Let us define bosons a_λ, $\lambda \in \mathbb{R}$, which satisfy the commutation relations [**8, 22**]:

$$[a_\lambda, a_\mu] = \frac{4}{\hbar^2} \frac{\sinh(\hbar\lambda)\sinh(\hbar\lambda/2)}{\lambda} \delta(\lambda + \mu) = a(\lambda)\delta(\lambda + \mu). \tag{6.3}$$

Consider the generating functions

$$e(u) = \hbar e^{\gamma} {:}\exp\left(-\hbar \int_{-\infty}^{\infty} d\lambda\, e^{-i\lambda u} \frac{a_\lambda e^{-\hbar|\lambda|/4}}{2\sinh(\hbar\lambda/2)}\right){:}, \tag{6.4}$$

$$f(u) = \hbar e^{\gamma} {:}\exp\left(\hbar \int_{-\infty}^{\infty} d\lambda\, e^{-i\lambda u} \frac{a_\lambda e^{\hbar|\lambda|/4}}{2\sinh(\hbar\lambda/2)}\right){:}, \tag{6.5}$$

$$h^{\pm}(u) = (\hbar e^{\gamma})^{-2} : e\left(u \mp \frac{i\hbar}{4}\right) f\left(u \pm \frac{i\hbar}{4}\right) : = \exp\left(\hbar \int_0^{\pm\infty} d\lambda\, e^{-i\lambda u} a_\lambda\right), \tag{6.6}$$

where γ is the Euler constant. The notion of normal ordered operator becomes more involved in the case of continuous bosons. It requires some kind of ultraviolet regularization to be included in the definition of normal ordered operators such that the product of these operators satisfy the rules [**7**]:

$$\begin{aligned} &{:}\exp\left(\int_{-\infty}^{\infty} d\lambda\, g_1(\lambda)\, a_\lambda\right){:} \cdot {:}\exp\left(\int_{-\infty}^{\infty} d\mu\, g_2(\mu)\, a_\mu\right){:} \\ &= \exp\left(\int_{\widetilde{C}} \frac{d\lambda\, \ln(-\lambda)}{2\pi i} a(\lambda) g_1(\lambda) g_2(-\lambda)\right) {:}\exp\left(\int_{-\infty}^{\infty} d\lambda\, (g_1(\lambda) + g_2(\lambda))\, a_\lambda\right){:}. \end{aligned} \tag{6.7}$$

The contour $\widetilde{C}$ is shown in Figure 3.

With this definition of the normal ordered exponents, we can prove

PROPOSITION 6.2. *The generating functions* (6.4)–(6.6) *satisfy the commutation relations* (2.18).

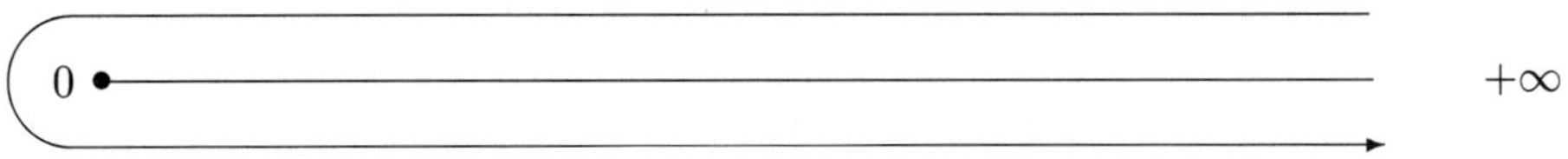

FIGURE 3

The main formula used to prove Proposition 6.2 is

$$\exp\left(\int_{\tilde{C}} \frac{d\lambda\,\ln(-\lambda)}{2\pi i\lambda}\,e^{-x\lambda}\right) = e^{-\gamma} x^{-1}, \qquad \operatorname{Re} x > 0.$$

For the description of the infinite-dimensional representations of the algebra $\mathcal{A}_\hbar(\widehat{\mathfrak{sl}_2})$ in terms of continuous bosons, we need a definition of a Fock space generated by the continuous family of free bosons. We borrow the construction below from [**6**].

Let $a(\lambda)$ be a meromorphic function, regular for $\lambda \in \mathbb{R}$ and satisfying the following conditions:

$$a(\lambda) = -a(-\lambda), \qquad a(\lambda) \sim a_0\lambda, \quad \lambda \to 0, \qquad a(\lambda) \sim e^{a'|\lambda|}, \quad \lambda \to \pm\infty.$$

Let a_λ, $\lambda \in \mathbb{R}$, $\lambda \neq 0$, be free bosons which satisfy the commutation relations $[a_\lambda, a_\mu] = a(\lambda)\delta(\lambda+\mu)$. We define a (right) Fock space $\mathcal{H}_{a(\lambda)}$ as follows. The space $\mathcal{H}_{a(\lambda)}$ is generated as a vector space by the expressions

$$\int_{-\infty}^{0} f_n(\lambda_n) a_{\lambda_n}\, d\lambda_n \cdots \int_{-\infty}^{0} f_1(\lambda_1) a_{\lambda_1}\, d\lambda_1\, |\mathrm{vac}\rangle,$$

where the functions $f_i(\lambda)$ satisfy the condition

$$f_i(\lambda) < C\, e^{(a'/2+\epsilon)\lambda}, \qquad \lambda \to -\infty,$$

for some $\epsilon > 0$ and the functions $f_i(\lambda)$ are analytic in a neighborhood of $\mathbb{R}_-$ except at $\lambda = 0$, where they have a simple pole.

The left Fock space $\mathcal{H}^*_{a(\lambda)}$ is generated by the expressions

$$\langle\mathrm{vac}| \int_{0}^{+\infty} g_1(\lambda_1) a_{\lambda_1}\, d\lambda_1 \cdots \int_{0}^{+\infty} g_n(\lambda_n) a_{\lambda_n}\, d\lambda_n,$$

where the functions $g_i(\lambda)$ satisfy the conditions

$$g_i(\lambda) < C\, e^{-(a'/2+\epsilon)\lambda}, \qquad \lambda \to +\infty,$$

for some $\epsilon > 0$ and the $g_i(\lambda)$ are analytic functions in a neighborhood of $\mathbb{R}_+$ except at $\lambda = 0$, where they also have a simple pole.

The pairing $(\ ,\ )\colon \mathcal{H}^*_{a(\lambda)} \otimes \mathcal{H}_{a(\lambda)} \to \mathbb{C}$ is uniquely defined by the following prescriptions:

(i)
$$(\langle\mathrm{vac}|, |\mathrm{vac}\rangle) = 1,$$

(ii)
$$\left(\langle\mathrm{vac}| \int_{0}^{+\infty} d\lambda\, g(\lambda) a_\lambda, \int_{-\infty}^{0} d\mu\, f(\mu) a_\mu\, |\mathrm{vac}\rangle\right) = \int_{\tilde{C}} \frac{d\lambda\,\ln(-\lambda)}{2\pi i}\, g(\lambda) f(-\lambda) a(\lambda),$$

(iii) the Wick theorem.

Let the vacuums $\langle\mathrm{vac}|$ and $|\mathrm{vac}\rangle$ satisfy the conditions

$$a_\lambda|\mathrm{vac}\rangle = 0, \quad \lambda > 0, \qquad \langle\mathrm{vac}|\, a_\lambda = 0, \quad \lambda < 0,$$

and let $f(\lambda)$ be a function analytic in some neighborhood of the real line, with possibly simple pole at $\lambda = 0$, and with the following asymptotic behavior:

$$f(\lambda) < Ce^{-(a'/2+\epsilon)|\lambda|}, \qquad \lambda \to \pm\infty,$$

for some $\epsilon > 0$. Then, by definition, the operator

$$F = :\exp\Big(\int_{-\infty}^{+\infty} d\lambda\, f(\lambda) a_\lambda\Big):$$

acts on the right Fock space $\mathcal{H}_{a(\lambda)}$ as follows. We have $F = F_- F_+$, where

$$F_- = \exp\Big(\int_{-\infty}^{0} d\lambda\, f(\lambda) a_\lambda\Big), \qquad F_+ = \lim_{\epsilon\to+0} e^{\epsilon \ln \epsilon f(\epsilon) a_\epsilon} \exp\Big(\int_{\epsilon}^{\infty} d\lambda\, f(\lambda) a_\lambda\Big).$$

The action of operator F on the left Fock space $\mathcal{H}^*_{a(\lambda)}$ is defined via another decomposition: $F = \widetilde{F}_- \widetilde{F}_+$, where

$$\widetilde{F}_+ = \exp\Big(\int_{0}^{+\infty} d\lambda\, f(\lambda) a_\lambda\Big), \qquad \widetilde{F}_- = \lim_{\epsilon\to+0} e^{\epsilon \ln \epsilon f(-\epsilon) a_{-\epsilon}} \exp\Big(\int_{-\infty}^{-\epsilon} d\lambda\, f(\lambda) a_\lambda\Big).$$

These definitions imply the following statement:

PROPOSITION 6.3. (i) *The above-defined actions of the operator*

$$F = :\exp\Big(\int_{-\infty}^{+\infty} d\lambda\, f(\lambda) a_\lambda\Big):$$

on the Fock spaces $\mathcal{H}$ and $\mathcal{H}^*$ are adjoint.

(ii) *The product of the normally ordered operators satisfies property* (6.7).

Returning to the level one representation of $\mathcal{A}_\hbar(\widehat{\mathfrak{sl}_2})$, we choose $\mathcal{H} = \mathcal{H}_{a(\lambda)}$ for $a(\lambda)$ defined in (6.3):

$$a(\lambda) = \frac{4}{\hbar^2}\,\frac{\sinh(\hbar\lambda)\sinh(\hbar\lambda/2)}{\lambda}.$$

From the definition of the Fock space $\mathcal{H}$ and from Proposition 6.4, we immediately obtain the construction of a representation of $\mathcal{A}_\hbar(\widehat{\mathfrak{sl}_2})$.

PROPOSITION 6.4. *The relations* (6.4)–(6.6) *define a highest weight right representation of the algebra $\mathcal{A}_\hbar(\widehat{\mathfrak{sl}_2})$ in the Fock space $\mathcal{H}$ and lowest weight left representation in the dual Fock space $\mathcal{H}^*$:*

(6.8)

$$\hat{e}_\lambda\,|\mathrm{vac}\rangle = 0, \quad \hat{f}_\lambda\,|\mathrm{vac}\rangle = 0, \quad \lambda \geqslant 0, \quad \textit{and} \quad \langle\mathrm{vac}|\,\hat{e}_\lambda = 0, \quad \langle\mathrm{vac}|\,\hat{f}_\lambda = 0, \quad \lambda \leqslant 0.$$

The highest weight property (6.8) means that all the matrix elements of the corresponding operators that do not vanishing identically posses this property. Let us demonstrate that $\langle v \mid \hat{e}_\lambda\,|\mathrm{vac}\rangle = 0$ for $\lambda > 0$ and certain $v \in \mathcal{H}^*$. Fix $\hbar > 0$. It is clear that any such matrix element has the form

(6.9)

$$\langle\mathrm{vac}|\prod_{i=1}^{n} f(v_i)\prod_{j=1}^{n-1} e(u_j)\hat{e}_\lambda|\mathrm{vac}\rangle = \frac{1}{2\pi}\int_{-\infty}^{\infty} du_n\, e^{i\lambda u_n}\langle\mathrm{vac}|\prod_{i=1}^{n} f(v_i)\prod_{j=1}^{n} e(u_j)|\mathrm{vac}\rangle$$
$$= G(v;u)\int_{-\infty}^{\infty} du_n\, e^{i\lambda u_n}\frac{\prod_{j=1}^{n-1}(u_n - u_j)(u_n - u_j + i\hbar)}{\prod_{i=1}^{n}(u_n - v_i + i\hbar/2)(u_n - v_i - i\hbar/2)},$$

where $G(v;u)$ is some factor which does not depend on the variable u_n; in order to calculate (6.9) using normal ordering relations for total currents (see [**6**] for details), we must require that $\operatorname{Im} v_i < -\hbar/2$, $i = 1,\dots,n$, and $\operatorname{Im} u_j < 0$, $j = 1,\dots,n-1$. Note that the integrand decreases as the function u_n^{-2} as $u_n \to \pm\infty$, so the integral is convergent. To calculate it, we can close the contour of integration either along a big semicircle in the upper half-plane of u_n for $\lambda > 0$, or along a big semicircle in the lower half-plane. But all the poles of the integrand are in the lower half-plane, so for $\lambda > 0$ the nonvanishing matrix elements of the operator $\hat{e}_\lambda$ become equal to zero and the property $\hat{e}_\lambda|\mathrm{vac}\rangle = 0$ for $\lambda > 0$ is proved. When $\lambda = 0$, relation (6.8) follows from continuity arguments.

In contrast to the case of the discrete algebra $\widehat{DY(\mathfrak{sl}_2)}$, the trace function $\mathrm{tr}_{\mathcal{H}}(e^{pd})$ is well defined now and can be calculated using the grading property of the operator d:

$$[d, a_\lambda] = -\lambda a_\lambda.$$

By definition, this trace is equal to

(6.10)

$$\begin{aligned}\mathrm{tr}_{\mathcal{H}}(e^{pd}) &= \sum_{n=0}^{\infty} \int\!\!\cdots\!\!\int_{0\leqslant\lambda_1<\cdots<\lambda_n<\infty} d\lambda_1\cdots d\lambda_n \, \frac{\langle \mathrm{vac}|a_{\lambda_1}\cdots a_{\lambda_n} e^{pd} a_{-\lambda_n}\cdots a_{-\lambda_1}|\mathrm{vac}\rangle}{\langle \mathrm{vac}|a_{\lambda_1}\cdots a_{\lambda_n} a_{-\lambda_n}\cdots a_{-\lambda_1}|\mathrm{vac}\rangle} \\ &= \sum_{n=0}^{\infty} \frac{1}{n!}\left(\int_0^\infty d\lambda\, e^{p\lambda}\right)^n = e^{1/p}, \qquad \mathrm{Re}\, p < 0.\end{aligned}$$

For the interpretation the generalized form-factors in quantum integrable field theory as the traces (6.10), one must usually put $\mathrm{Re}\, p = 0$ ($p = 2\pi i$). In this case we should understand the result (6.10) as analytically continued from the domain $\mathrm{Re}\, p < 0$. This result can be compared with the asymptotic expansion of the partition function $\prod_{n=1}^{\infty}(1-q^n)^{-1}$. Indeed, the trace of the operator e^{pd} can be presented as the integral over eigenvalues of this operator $e^{p\lambda}$:

$$e^{1/p} = 1 + \int_0^\infty d\lambda\, e^{p\lambda}\mathfrak{p}(\lambda), \quad \text{where } \mathfrak{p}(\lambda) = \sum_{n\geqslant 0} \frac{1}{n!\,(n+1)!}\lambda^n. \tag{6.11}$$

On the other hand, the coefficients $\mathfrak{p}_n$ of the partition function

$$\prod_{n>0} \frac{1}{(1-q^n)} = \sum_{n\geqslant 0} \mathfrak{p}_n q^n$$

have the following asymptotic expansion:

$$\mathfrak{p}_n \sim \sum_k \frac{1}{k!\,(k-1)!} n^k \tag{6.12}$$

in the region as $n \to \infty$. The explicit comparison of (6.11) and (6.12) demonstrates their similarity.

7. Quantized current algebras

7.1. As we have seen in the previous section, the total current algebra is apparently more suitable for constructing infinite-dimensional representations than the standard RLL-formalism that uses the Gauss coordinates of L-operators. On the other hand, it is difficult to define the Hopf structure in terms of total currents, while in the L-operator formalism the Hopf structure $\Delta' L = L \otimes L$ is quite natural because of the RLL-relations and the YB equation for the R-matrix. In this section we intend to discuss the following point of view. It is possible to assign a Hopf algebra structure to the algebra of total currents by supplementing the commutation relations (2.18) with information on the Riemann problem and also some additional information, which we shall discuss below.

Let us start with the formal total current algebra for the currents $e(u)$, $f(u)$, and $h^{\pm}(u)$, given for example by the commutation relations (2.18). This algebra is formal, since in the commutation relations of total currents $[e(u), f(v)]$ the δ-function is defined formally as well as the Cartan "half" currents $h^{\pm}(u)$.

The first step is to fix the Riemann problem serving to divide the total currents $e(u)$ and $f(u)$ into "half"-currents $e^{\pm}(u)$ and $f^{\pm}(u)$. This fixes the notion of δ-function in the commutation relations $[e(u), f(v)]$, the contents of the generating functions $h^{\pm}(u)$ and also the full set of commutation relations between all the generating functions $e^{\pm}(u)$, $f^{\pm}(u)$, and $h^{\pm}(u)$ (see Proposition 2.1). On the other hand, fixing first the analytic properties of δ-functions in this relation leaves freedom for the factorization problem. This freedom is related to the possible twist in the Riemann problem, and will also be discussed below.

Let us assume now that the generating functions $e^{\pm}(u)$, $f^{\pm}(u)$, and $h^{\pm}(u)$ are Gauss coordinates of some L-operator given by (2.9). It is natural to guess that this L-operator satisfies the RLL-relations with some R-matrix.

We claim that there exist universal comultiplication rules for the Gauss coordinates

$$e^{\pm}(u), \quad f^{\pm}(u), \quad h^{\pm}(u) = k_1^{\pm}(u)(k_2^{\pm}(u))^{-1}, \quad \tilde{h}^{\pm}(u) = (k_2^{\pm}(u))^{-1}k_1^{\pm}(u)$$

($h^{\pm}(u)$ is not necessarily equal to $\tilde{h}^{\pm}(u)$ in the general case), which are based on the sole assumption that the R-matrix at a critical point is given by a rank one operator. Let, for instance, $\overline{R}(i\hbar)$ be proportional to $1 - P$, where P is a flip:

$$\overline{R}(i\hbar) = \begin{pmatrix} 0 & 0 & 0 & 0 \\ 0 & 1 & -1 & 0 \\ 0 & -1 & 1 & 0 \\ 0 & 0 & 0 & 0 \end{pmatrix}.$$

This takes place for the Yang R-matrix, for Baxter's elliptic R-matrix, and for the sine-Gordon R-matrix. Then the commutation relations for the L-operators at the critical point of the R-matrix basically reduce to the following:

$$\begin{aligned} k_1^{\pm}(u)f^{\pm}(u+i\hbar) &= f^{\pm}(u)k_1^{\pm}(u), & k_2^{\pm}(u)f^{\pm}(u-i\hbar) &= f^{\pm}(u)k_2^{\pm}(u), \\ k_1^{\pm}(u)e^{\pm}(u) &= e^{\pm}(u+i\hbar)k_1^{\pm}(u), & k_2^{\pm}(u)e^{\pm}(u) &= e^{\pm}(u-i\hbar)k_2^{\pm}(u). \end{aligned} \tag{7.1}$$

The natural map $\Delta' L = L \dot{\otimes} L$ implies the following universal comultiplication rules for the Gauss coordinates (see (2.10)):

$$\begin{aligned} \Delta e^{\pm}(u) &= e^{\pm}(u') \otimes 1 + \sum_{p=0}^{\infty}(-1)^p (f^{\pm}(u'-i\hbar))^p h^{\pm}(u') \otimes (e^{\pm}(u''))^{p+1}, \\ \Delta f^{\pm}(u) &= 1 \otimes f^{\pm}(u'') + \sum_{p=0}^{\infty}(-1)^p (f^{\pm}(u'))^{p+1} \otimes \tilde{h}^{\pm}(u'')(e^{\pm}(u''-i\hbar))^p, \\ \Delta h^{\pm}(u) &= \sum_{p,p'=0}^{\infty} (-1)^{p+p'}((f^{\pm}(u'))^p \otimes (e^{\pm}(u''+i\hbar))^p) \\ &\quad \times (h^{\pm}(u') \otimes h^{\pm}(u''))((f^{\pm}(u'))^{p'} \otimes (e^{\pm}(u''-i\hbar))^{p'}). \end{aligned} \tag{7.2}$$

We see from (7.2) that the choice of the Riemann problem for the algebra of formal currents also enables one to reconstruct the comultiplication structure of the algebra. We shall consider this ideology for some examples.

7.2. The twisting of the Yangian algebras. Besides the comultiplication for the Gauss coordinates given by the formulas (2.14)–(2.16) and related to the natural comultiplication in terms of L-operators, there exists another comultiplication first introduced in the paper [**14**] for quantized affine algebras:

$$(7.3)\quad \begin{gathered} \Delta e(u) = e(u') \otimes 1 + h^+(u') \otimes e(u''), \\ \Delta f(u) = 1 \otimes f(u'') + f(u') \otimes h^-(u''), \qquad \Delta h^\pm(u) = h^\pm(u') \otimes h^\pm(u''). \end{gathered}$$

This comultiplication describes the coproduct of the total currents $e(u)$ and $f(u)$ and at first sight is not related to the L-operator formulation of the corresponding deformed algebra. It was shown in the papers [**2, 19**] that the Hopf algebra corresponding to (7.3) can be obtained as a twisted Hopf algebra, which is equivalent to taking the infinite limit of the shifting automorphism in the quantum Weyl group. We would like to explain this twisting procedure on the example of the algebra $\mathcal{A}_\hbar(\widehat{\mathfrak{sl}_2})$. The twisting of the algebra $\widehat{DY(\mathfrak{sl}_2)}$ can be considered similarly. The essential part of the construction will be a change of the Riemann problems and of the comultiplication formulas.

Fix the parameter $a \in \mathbb{R}$. Consider the automorphism

$$(7.4)\quad \omega_a(e(u)) = e^{iau}e(u), \quad \omega_a(f(u)) = e^{-iau}f(u), \quad \omega_a(h^\pm(u)) = e^{\pm c\hbar a/2}h^\pm(u),$$

which obviously preserves the commutation relations (2.18). The automorphism (7.4) translated to the formal generators $\hat{e}_\lambda$, $\hat{f}_\lambda$, and $\hat{h}_\lambda$ takes the form

$$(7.5)\quad \begin{gathered} \omega_a(\hat{e}_\lambda) = \hat{e}_{\lambda+a}, \qquad \omega_a(\hat{f}_\lambda) = \hat{f}_{\lambda-a}, \\ \omega_a(\hat{h}_\lambda) = \hat{h}_\lambda[e^{c\hbar a/2}\theta(\lambda) + e^{-c\hbar a/2}\theta(-\lambda)] + 4\hbar^{-1}\sinh(c\hbar a/2)\delta(\lambda). \end{gathered}$$

The appearance of the δ-function in (7.5) is possible, since we consider the elements of the algebra $\mathcal{A}_\hbar(\widehat{\mathfrak{sl}_2})$ as integrals of formal generators with exponentially decreasing weight functions.

The R-matrix will change, and its value at the critical point yield the following modification of the commutation relations (7.1):

$$(7.6)\quad \begin{gathered} e^{a\hbar}k_1^\pm(u)f^\pm(u+i\hbar) = f^\pm(u)k_1^\pm(u), \qquad e^{-a\hbar}k_2^\pm(u)f^\pm(u-i\hbar) = f^\pm(u)k_2^\pm(u), \\ k_1^\pm(u)e^\pm(u) = e^{-a\hbar}e^\pm(u+i\hbar)k_1^\pm(u), \qquad k_2^\pm(u)e^\pm(u) = e^{a\hbar}e^\pm(u-i\hbar)k_2^\pm(u). \end{gathered}$$

The comultiplication in the twisted algebra, by (7.6), now reads

$$(7.7)\quad \begin{aligned} \Delta e^\pm(u) &= e^\pm(u') \otimes 1 + \sum_{p=0}^{\infty}(-1)^p(e^{-a\hbar}f^\pm(u'-i\hbar))^p h^\pm(u') \otimes (e^\pm(u''))^{p+1}, \\ \Delta f^\pm(u) &= 1 \otimes f^\pm(u'') + \sum_{p=0}^{\infty}(-1)^p(f^\pm(u'))^{p+1} \otimes h^\pm(u'')(e^{a\hbar}e^\pm(u''-i\hbar))^p, \\ \Delta h^\pm(u) &= \sum_{p,p'=0}^{\infty}(-1)^{p+p'}((f^\pm(u'))^p \otimes (e^{-a\hbar}e^\pm(u''+i\hbar))^p) \\ &\quad \times (h^\pm(u') \otimes h^\pm(u''))((f^\pm(u'))^{p'} \otimes (e^{a\hbar}e^\pm(u''-i\hbar))^{p'}). \end{aligned}$$

It is obvious that the automorphism (7.4) changes the asymptotic properties of the currents $e^{\pm}(u)$ and $f^{\pm}(u)$. These generating functions defined by the Riemann problems (3.12) are analytic in the corresponding domains of the complex plane u and decrease like u^{-1} as $u \to \mp\infty$, respectively. The transformed currents $\omega_a(e^{\pm}(u))$ and $\omega_a(f^{\pm}(u))$ will have changed asymptotics

$$\omega_a(e^{\pm}(u)) \sim e^{\pm a|\mathrm{Im}\, u|}, \quad \omega_a(f^{\pm}(u)) \sim e^{\mp a|\mathrm{Im}\, u|}, \qquad \mathrm{Im}\, u \to \mp\infty. \tag{7.8}$$

Now consider the limit of the twisted algebra as the twisting parameter a tends to $+\infty$. From (7.8) it is clear that

$$\lim_{a\to\infty} \omega_a(e^-(u)) = 0, \qquad \lim_{a\to\infty} \omega_a(f^+(u)) = 0.$$

Because of the Ding–Frenkel relations (2.17), we have

$$\lim_{a\to\infty} e^+(u) = e(u - ic\hbar/4), \qquad \lim_{a\to\infty} f^-(u) = -f(u - ic\hbar/4).$$

So we conclude that the limit of the twisted algebra is an algebra with the commutation relation given by (2.18) and the following Riemann problem:

$$e^-(u) = f^+(u) = 0, \qquad e(u) = e^+(u + ic\hbar/4), \qquad f(u) = -f^-(u + ic\hbar/4). \tag{7.9}$$

The comultiplication (7.3) follows from the general formulas (7.2). As a result, we can assert that Drinfeld's new realization is the deformed current algebra with Riemann problem given in (7.9).

7.3. The algebra $\mathcal{A}_{\hbar,\eta}(\widehat{\mathfrak{sl}_2})$ [6]. Let us consider the following generalization of the total current algebra given by the commutation relations (2.18):

$$\begin{aligned} H^+(u)H^-(v) &= \frac{\sinh \pi\eta(u - v - i\hbar(1 + c/2))}{\sinh \pi\eta(u - v + i\hbar(1 - c/2))} \\ &\quad \times \frac{\sinh \pi\eta'(u - v + i\hbar(1 + c/2))}{\sinh \pi\eta'(u - v - i\hbar(1 - c/2))} H^-(v)H^+(u), \end{aligned} \tag{7.10}$$

$$H^{\pm}(u)H^{\pm}(v) = \frac{\sinh \pi\eta(u - v - i\hbar) \sinh \pi\eta'(u - v + i\hbar)}{\sinh \pi\eta(u - v + i\hbar) \sinh \pi\eta'(u - v - i\hbar)} H^{\pm}(v)H^{\pm}(u), \tag{7.11}$$

$$H^{\pm}(u)E(v) = \frac{\sinh \pi\eta(u - v - i\hbar(1 \pm c/4))}{\sinh \pi\eta(u - v + i\hbar(1 \mp c/4))} E(v)H^{\pm}(u), \tag{7.12}$$

$$H^{\pm}(u)F(v) = \frac{\sinh \pi\eta'(u - v + i\hbar(1 \pm c/4))}{\sinh \pi\eta'(u - v - i\hbar(1 \mp c/4))} F(v)H^{\pm}(u), \tag{7.13}$$

$$E(u)E(v) = \frac{\sinh \pi\eta(u - v - i\hbar)}{\sinh \pi\eta(u - v + i\hbar)} E(v)E(u), \tag{7.14}$$

$$F(u)F(v) = \frac{\sinh \pi\eta'(u - v + i\hbar)}{\sinh \pi\eta'(u - v - i\hbar)} F(v)F(u), \tag{7.15}$$

$$\begin{aligned} [E(u), F(v)] = \hbar \Bigg[&\delta\left(u - v + \frac{ic\hbar}{2}\right) H^+\left(u + \frac{ic\hbar}{4}\right) \\ &- \delta\left(u - v - \frac{ic\hbar}{2}\right) H^-\left(v + \frac{ic\hbar}{4}\right)\Bigg], \end{aligned} \tag{7.16}$$

where the δ-function is defined by formula (3.13) and the periods of the trigonometric functions η and η' are related as follows:

$$\frac{1}{\eta'} - \frac{1}{\eta} = -\hbar c. \tag{7.17}$$

The last relation is necessary in order to make the commutation relations given by (7.10)–(7.16) self-consistent. The "rational" commutation relations (2.18) can be obtained from (7.10)–(7.16) by the degeneration $\eta \to 0$. Note that due to (7.17) we also have $\eta' \to 0$ in this limit. Note that formulas (7.10)–(7.16) differ from the formulas given in [**6**] in that the sign of the central element is reversed. The algebra given by the commutation relations (7.10)–(7.16) is associated with the simple Lie algebra $\mathfrak{sl}_2$, but can be constructed for an arbitrary simply-laced Lie algebra; see for example [**23**].

One can see that the most unusual feature of these relations is the presence of two periods η and η' in the trigonometric functions playing the role of structure constants. Let us comment on the appearance of this phenomenon. If we put $c = 0$, then the periods coincide and the relations are analogous to the ones for $c = 0$ in the quantum affine algebra with $q = e^{i\pi\eta\hbar}$ in the variables $z = e^{\pi\eta u}$, $w = e^{\pi\eta v}$. The question is how to input the central charge so as to have a nontrivial representation theory. First, let us look at the classical picture. In the limit $\hbar \to 0$, $c = 0$, the commutation relations (7.10)–(7.16) are the relations for the Lie algebra of $\mathfrak{sl}_2$-valued (generalized) functions over z vanishing as $|\mathrm{Re}\, z| \to \pm\infty$ (the Cartan currents tend to ± 1) [**20**]:

$$\begin{aligned}[h^{\pm}(u), e(u)] &= 2\pi i\eta \coth \pi\eta(u-v)e(v),\\ [h^{\pm}(u), f(v)] &= -2\pi i\eta \coth \pi\eta(u-v)e(v),\\ [e(u), f(v)] &= \delta(u-v)(h^{+}(u) - h^{-}(v)).\end{aligned} \tag{7.18}$$

The application of the Fourier transform to the generating functions

$$e(u) = e \otimes 2\pi\delta(u-z), \qquad f(u) = f \otimes 2\pi\delta(u-z),$$
$$h^{+}(u) = h \otimes i\pi\eta \coth \pi\eta(u-z), \qquad h^{-}(u) = h \otimes i\pi\eta \coth \pi\eta(u - z - i/\eta),$$

which satisfy (7.18), i.e.

$$e(u) = \int_{-\infty}^{+\infty} d\lambda\, e^{-i\lambda u} \hat{e}_{\lambda}, \qquad h^{\pm}(u) = \int_{-\infty}^{+\infty} d\lambda\, e^{-i\lambda u} \frac{\hat{h}_{\lambda}}{1 - e^{\pm\lambda/\eta}},$$
$$f(u) = \int_{-\infty}^{+\infty} d\lambda\, e^{-i\lambda u} \hat{f}_{\lambda},$$

turns these commutation relations to the standard form

$$[\hat{h}_{\lambda}, \hat{e}_{\mu}] = 2\hat{e}_{\lambda+\mu}, \qquad [\hat{h}_{\lambda}, \hat{f}_{\mu}] = -2\hat{f}_{\lambda+\mu}, \qquad [\hat{e}_{\lambda}, \hat{f}_{\mu}] = \hat{h}_{\lambda+\mu},$$

which admits the standard central extension

$$[\hat{e}_{\lambda}, \hat{f}_{\mu}] = \hat{h}_{\lambda+\mu} + \delta(\lambda+\mu)c, \qquad [\hat{h}_{\lambda}, \hat{h}_{\mu}] = 2\delta(\lambda+\mu)c.$$

Let us look at the value of the corresponding cocycle on the fields $h^{+}(u_i)$ [**20**]:

$$B(h^{+}(u_1), h^{+}(u_2)) = 2i\pi\eta^2 \left(\frac{\pi\eta u}{\sinh^2 \pi\eta u} - \coth \pi\eta u \right). \tag{7.19}$$

The first term in right-hand side of (7.19) is no longer a periodic function with period $1/\eta$. Moreover, the cocycle can be written in terms of the integral over the boundary of the strip Π: $0 < \mathrm{Im}\, z < 1/\eta$ and derivatives over the period $1/\eta$:

$$B(x \otimes \varphi(z), y \otimes \psi(z)) = \frac{\eta^2}{4\pi} \int_{\partial\Pi} dz \left(\frac{d\psi(z)}{d\eta} \varphi(z) - \psi(z) \frac{d\varphi(z)}{d\eta} \right) \langle x, y \rangle, \tag{7.20}$$

where $\langle\ ,\ \rangle$ is the Killing form and $x, y \in \mathfrak{sl}_2$. These arguments signify that the central extension for the quantum algebra should be achieved via a finite shift of the periods of the trigonometric functions in the defining relations (7.10)–(7.16).

To the formal current algebra (7.10)–(7.16) we attach a Riemann problem of the following type.

Given a meromorphic function $g(u)$, we would like to find two functions $g^{\pm}(u)$ satisfying the following conditions:

(i) $g(u) = g^{+}(u) - g^{-}(u)$;
(ii) $g^{\pm}(u)$ are piecewise analytic functions; more precisely, $g^{\pm}(u)$ are analytic on the complement to some collection of horizontal lines in the complex plane u;
(iii) $g^{+}(u)$ $(g^{+}(u))$ has a boundary value on the lower (upper) boundaries of the corresponding strips;
(iv) the following relations hold in any strip of analyticity of $g^{\pm}(u)$:

$$g^{-}(u) = -g^{+}(u - i\eta), \qquad g^{+}(u) = -g^{-}(u + i\eta),$$

where $g^{+}(u-i\eta)$ and $g^{-}(u+i\eta)$ are the analytic continuations of $g^{+}(u)$ and $g^{-}(u)$.

The solution to this Riemann problem is given by the following integrals over horizontal lines close to the point u:

$$g^{\pm}(u) = \pi\eta \int_{\operatorname{Im} v \lessgtr \operatorname{Im} u} \frac{dv}{2\pi i} \frac{g(v)}{\sinh \pi\eta(u - v)}.$$

More precisely, the Riemann problems for the currents $E(u)$ and $F(u)$ are chosen with different periods $1/\eta$, $1/\eta'$ and certain shifts of the spectral parameter:

$$e^{\pm}(u) = \hbar^{-1} \sin \pi\eta\hbar \int_{\operatorname{Im}(u-v) \lessgtr \pm c\hbar/4} \frac{dv}{2\pi i} \frac{E(v)}{\sinh \pi\eta(u - v \mp ic\hbar/4)}, \tag{7.21}$$

$$f^{\pm}(u) = \hbar^{-1} \sin \pi\eta'\hbar \int_{\operatorname{Im}(u-v) \lessgtr \pm c\hbar/4} \frac{dv}{2\pi i} \frac{F(v)}{\sinh \pi\eta'(u - v \pm ic\hbar/4)}. \tag{7.22}$$

The coefficients in front of the integrals in (7.21) and (7.22) are chosen for technical reasons; in this case condition (i) has the form

$$\begin{aligned} e^{+}(u + ic\hbar/4) - e^{-}(u - ic\hbar/4) &= \frac{\sin \pi\eta\hbar}{\pi\eta\hbar} E(u), \\ f^{+}(u - ic\hbar/4) - f^{-}(u + ic\hbar/4) &= \frac{\sin \pi\eta'\hbar}{\pi\eta'\hbar} F(u). \end{aligned} \tag{7.23}$$

These Riemann problems are in agreement with the commutation relations (7.10)–(7.16) in the sense that they yield currents $e^{\pm}(u)$ and $f^{\pm}(u)$ satisfying the commutation relations without the integral terms. As in the case of the Yangian algebras, the Riemann problem is not formulated for the Cartan currents, and we set

$$\tilde{h}^{\pm}(u) = \frac{\sin \pi\eta'\hbar}{\pi\eta'\hbar} H^{\pm}(u), \qquad h^{\pm}(u) = \frac{\sin \pi\eta\hbar}{\pi\eta\hbar} H^{\pm}(u).$$

The commutation relations between currents $e^{+}(u)$, $f^{+}(u)$, and $h^{+}(u)$ are given by the relations (7.24)-(7.29).

$$e^{+}(u_1) f^{+}(u_2) - f^{+}(u_2) e(u_1) = \frac{\sinh i\pi\eta'\hbar}{\sinh \pi\eta' u} h^{+}(u_1) - \frac{\sinh i\pi\eta\hbar}{\sinh \pi\eta u} \tilde{h}^{+}(u_2), \tag{7.24}$$

$$\sinh \pi\eta(u+i\hbar)h^+(u_1)e^+(u_2) - \sinh \pi\eta(u-i\hbar)e^+(u_2)h^+(u_1) = \sinh(i\pi\eta\hbar)\{h^+(u_1), e^+(u_1)\}, \tag{7.25}$$

$$\sinh \pi\eta'(u-i\hbar)h^+(u_1)f^+(u_2) - \sinh \pi\eta'(u+i\hbar)f^+(u_2)h^+(u_1) = -\sinh(i\pi\eta'\hbar)\{h^+(u_1), f^+(u_1)\}, \tag{7.26}$$

$$\sinh \pi\eta(u+i\hbar)e^+(u_1)e^+(u_2) - \sinh \pi\eta(u-i\hbar)e^+(u_2)e^+(u_1) = \sinh(i\pi\eta\hbar)(e^+(u_1)^2 + e^+(u_2)^2), \tag{7.27}$$

$$\sinh \pi\eta'(u-i\hbar)f^+(u_1)f^+(u_2) - \sinh \pi\eta'(u+i\hbar)f^+(u_2)f^+(u_1) = -\sinh(i\pi\eta'\hbar)(f^+(u_1)^2 + f^+(u_2)^2), \tag{7.28}$$

$$h^+(u_1)h^+(u_2) = \frac{\sinh \pi\eta'(u+i\hbar)\sinh \pi\eta(u-i\hbar)}{\sinh \pi\eta(u+i\hbar)\sinh \pi\eta'(u-i\hbar)}\, h^+(u_2)h^+(u_1). \tag{7.29}$$

The rest of the relations follow from condition (iv) for the analytic continuations, which in this case reads as

$$e^-(u) = -e^+(u - i/\eta''), \qquad f^-(u) = -f^+(u - i/\eta''), \qquad \eta'' = \frac{2\eta\eta'}{\eta+\eta'}.$$

We also impose a similar relation on the Cartan currents:

$$h^-(u) = h^+(u - i/\eta'').$$

The integral formulas (7.21) and (7.22) dictate the following presentation of the currents $e^{\pm}(u)$ and $f^{\pm}(u)$:

$$e^{\pm}(u) = \pm\frac{\sin \pi\eta\hbar}{\pi\eta}\int_{-\infty}^{\infty} d\lambda\, e^{-i\lambda u}\,\frac{\hat{e}_\lambda e^{\mp c\hbar\lambda/4}}{1+e^{\mp\lambda/\eta}}, \tag{7.30}$$

$$f^{\pm}(u) = \pm\frac{\sin \pi\eta'\hbar}{\pi\eta'}\int_{-\infty}^{\infty} d\lambda\, e^{-i\lambda u}\,\frac{\hat{f}_\lambda e^{\pm c\hbar\lambda/4}}{1+e^{\mp\lambda/\eta'}}, \tag{7.31}$$

and because of formulas (7.23), the total currents are given by the Fourier transform:

$$E(u) = \hbar\int_{-\infty}^{\infty} d\lambda\, e^{-i\lambda u}\hat{e}_\lambda, \qquad F(u) = \hbar\int_{-\infty}^{\infty} d\lambda\, e^{-i\lambda u}\hat{f}_\lambda.$$

Note that in the limit as $\eta \to 0$ formulas (7.30) and (7.31) become (4.16). It is natural to guess that the Cartan generating functions are given also by the Fourier integrals of the generators $\hat{t}_\lambda$:

$$h^{\pm}(u) = \frac{\sin \pi\eta\hbar}{2\pi\eta}\int_{-\infty}^{\infty} d\lambda\, e^{-i\lambda u}\,\hat{t}_\lambda e^{\pm\lambda/2\eta''}.$$

The generators $\hat{e}_\lambda$ and $\hat{f}_\lambda$ can be defined by means of the inversion of the Fourier integrals in (7.30), (7.31) for the currents $e^+(u)$ and $f^+(u)$ respectively. This can be done in the domains of analyticity of the corresponding currents. These domains depend on the choice of the representation of the algebra $\mathcal{A}_{\hbar,\eta}(\widehat{\mathfrak{sl}_2})$. For instance, the spectral parameter z of the evaluation representation shifts these domains. To obtain the commutation relations between the formal generators $\hat{e}_\lambda$, $\hat{f}_\lambda$, and $\hat{t}_\lambda$, one can use the strip $\{-1/\eta - \hbar c/4 < \operatorname{Im} u < -\hbar c/4\}$, as was done in [**6**]. As in the case of the algebra $\mathcal{A}_\hbar(\widehat{\mathfrak{sl}_2})$, the elements of the algebra $\mathcal{A}_{\hbar,\eta}(\widehat{\mathfrak{sl}_2})$ are formal integrals similar to (4.27) (for the details, see [**6**]).

Considering the generating functions $e^{\pm}(u)$, $f^{\pm}(u)$, $h^{\pm}(u) = k_1^{\mp}(u)(k_2^{\mp}(u))^{-1}$ as the Gauss coordinates of the L-operators and using (7.1), we can obtain from the natural comultiplication in terms of L-operator $\Delta' L(u) = L(u) \dot{\otimes} L(u)$ the comultiplication map for these generating functions. The verification of the agreement of these comultiplication rules with the commutation relations (7.24)–(7.29) brings us to the following phenomenon. Let us demonstrate it on the first term in the comultiplication formulas for the Cartan generating functions $h^{\pm}(u)$ (7.2): $\Delta h^{\pm}(u) = h^{\pm}(u) \otimes h^{\pm}(u)$. The other terms can be considered similarly.

The commutation relation (7.29) can be rewritten in the form

$$g(u_1 - u_2, \xi - \hbar c)h^{\pm}(u_1, \xi)h^{\pm}(u_2, \xi) = h^{\pm}(u_1, \xi)h^{\pm}(u_2, \xi)g(u_1 - u_2, \xi), \tag{7.32}$$

where

$$g(u, \xi) = \frac{\sinh \pi\eta(u - i\hbar)}{\sinh \pi\eta(u + i\hbar)}, \qquad \xi = \eta^{-1}.$$

Due to the fact that $\Delta c = c \otimes 1 + 1 \otimes c = c^{(1)} + c^{(2)}$, we conclude that the commutation relation

$$\begin{aligned} &g(u_1 - u_2, \xi - \hbar(c^{(1)} + c^{(2)}))\Delta h^{\pm}(u_1, \xi)\Delta h^{\pm}(u_2, \xi) \\ &\quad = \Delta h^{\pm}(u_1, \xi)\Delta h^{\pm}(u_2, \xi)g(u_1 - u_2, \xi) \end{aligned}$$

will follow from (7.32) if and only if the comultiplication for the generating function is defined as follows:

$$\Delta h^{\pm}(u, \xi) = h^{\pm}(u, \xi - \hbar c^{(2)}) \otimes h^{\pm}(u, \xi) + \cdots . \tag{7.33}$$

By the dots we denote other terms in the series (7.2). Actually, a similar phenomenon was observed in the paper [**8**] on the theory of the quantum sine-Gordon model and implicitly used for the definition of the intertwining operators in the elliptic algebra $\mathcal{A}_{q,p}(\widehat{\mathfrak{sl}_2})$, which served as the dynamical symmetry algebra of the eight-vertex model [**21**]. The algebra $\mathcal{A}_{\hbar,\eta}(\widehat{\mathfrak{sl}_2})$, which is the scaling limit of $\mathcal{A}_{q,p}(\widehat{\mathfrak{sl}_2})$, was investigated in [**6**]. In the L-operator formalism, the algebra $\mathcal{A}_{q,p}(\widehat{\mathfrak{sl}_2})$ as well as $\mathcal{A}_{\hbar,\eta}(\widehat{\mathfrak{sl}_2})$ is given by the commutation relation

$$R^+(u_1 - u_2, \xi - \hbar c)L(u_1, \eta)L_2(u_2, \eta) = L_2(u_2, \eta)L_1(u_1, \eta)R^+(u_1 - u_2, \xi).$$

Although the comultiplication map moves the algebra into the tensor product of the algebra with parameter shifted by the central element, this unusual Hopf structure allows us to construct the intertwining operators for the infinite-dimensional representations of the algebra $\mathcal{A}_{\hbar,\eta}(\widehat{\mathfrak{sl}_2})$ and interpret them as Zamolodchikov–Faddeev operators in the quantum sine-Gordon model [**6**].

To conclude, we would like to mention the papers [**24**] and [**25**], where the quasi-Hopf twisting of the quantum groups was considered. These papers are the realization of the idea, due to C. Frønsdal [**26**], that the elliptic deformations of quantum groups can be obtained by twisting the corresponding Hopf algebras and belong to the category of Drinfeld's quasi-Hopf algebras [**27**]. In particular, this means that the algebra $\mathcal{A}_{\hbar,\eta}(\widehat{\mathfrak{sl}_2})$ can be obtained by quasi-Hopf twisting of the algebra $\mathcal{A}_{\hbar}(\widehat{\mathfrak{sl}_2})$, and it also explains the comultiplication formulas like (7.33), first considered in [**6**]. Also, the result of these papers can be regarded as the proof that all elliptic deformations of the Hopf algebras are actually isomorphic for generic values of the elliptic parameter, a fact that was observed in [**20**] on the level of the classical limit of the algebra $\mathcal{A}_{\hbar,\eta}(\widehat{\mathfrak{sl}_2})$.

Acknowledgements

S. Khoroshkin and S. Pakuliak are grateful for the warm hospitality of the University Roma 1, where part of this work was done. This visit was arranged in the framework of INFN–ITEP and INFN–JINR exchange programs. S. Khoroshkin also acknowledges the hospitality of the Max Planck Institut für Mathematik. D. Lebedev is grateful to the Institute Girard Desargues, University Claude Bernard Lyon 1, where part of this work was done, for their warm hospitality.

References

[1] V. G. Drinfeld, *Hopf algebras and quantum Yang–Baxter equation*, Dokl. Akad. Nauk SSSR **283** (1985), 1060–1064; Englsih transl. in Soviet Math. Dokl. **32** (1985).

[2] S. Khoroshkin and V. Tolstoy, *Yangian double*, Lett. Math. Phys. **36** (1996), 373–402.

[3] S. Khoroshkin, *Central Extension of the Yangian Double*, in: Actes du Septième Contact Franco-Belge, Reims, June 1995, Algèbre Noncommutative, Groupes Quantique et Invariants, Société Mathématique de France, Collection Séminaires et Congrès, Numéro 2, 1977, pp. 119–135.

[4] K. Iohara and M. Kohno, *A central extension of Yangian double and its vertex representations*, Preprint q-alg/9603032, Lett. Math. Phys. **37** (1996), 319–328.

[5] S. Khoroshkin, D. Lebedev, and S. Pakuliak, *Traces of intertwining operators for the Yangian double*, Lett. Math. Phys. **41** (1997), 31–47.

[6] ———, *Elliptic algebra $\mathcal{A}_{q,p}(\widehat{\mathfrak{sl}_2})$ in the scaling limit*, Preprint ITEP-TH-51/96, q-alg/9702002; Comm. Math. Phys. **190** (1998), 597–627.

[7] M. Jimbo and T. Miwa, *Quantum KZ equation with $|q| = 1$ and correlation functions of the XXZ model in the gapless regime*, J. Phys. A **29** (1996), 2923–2958.

[8] S. Lukyanov, *Free field representation for massive integrable models*, Comm. Math. Phys. **167** (1995), 183–226.

[9] L. D. Faddeev and L. A. Takhtajan, *Hamiltonian method to the theory of solitons*, Springer–Verlag, New York, 1987.

[10] J. Ding and I. B. Frenkel, *Isomorphism of two realizations of quantum affine algebras $U_q(gl(n))$*, Comm. Math. Phys. **156** (1993), 277–300.

[11] F. A. Smirnov, *Form factors in completely integrable field theories*, World Scientific, Singapore, 1992.

[12] L. D. Faddeev, N. Yu. Reshetikhin, and L. A. Takhtajan, *Quantization of Lie groups and Lie algebras*, Algebra i Analiz **1** (1989), no. 1, 178–206; English transl., Leningrad Math. J. **1** (1990), no. 1, 193–225.

[13] V. G. Drinfeld, *Quantum groups*, Proc. Intern. Congress Math. (Berkeley, 1986), Vol. 1, Amer. Math. Soc., Providence, RI, 1987, pp. 798–820.

[14] ———, *A new realization of Yangians and quantum affine algebras*, Dokl. Akad. Nauk SSSR **296** (1987), 13–17; English transl., Soviet Math. Dokl. **36** (1988), 212–216.

[15] N. Yu. Reshetikhin and M. A. Semenov-Tyan-Shansky, *Central extensions of quantum current groups*, Lett. Math. Phys. **19** (1990), 178–142.

[16] I. B. Frenkel and N. Yu. Reshetikhin, *Quantum affine algebras and holonomic difference equations*, Comm. Math. Phys. **146** (1992), 1–60.

[17] A. Chari and A. N. Pressley, *A guide to quantum groups*, Cambridge University Press, Cambridge, 1994.

[18] A. Chervov, *Traces of creating-annihilating operators and Fredholm's formulas*, Preprint, 1997, pp. 1–20, q-alg/9703017; Funktsional. Anal. i Prilozhen. **32** (1998), no. 1, 90–95; English transl., Functional. Anal. Appl. **32** (1998), 71–74.

[19] S. Khoroshkin and V. Tolstoy, *Twisting of quantum (super)algebras. Connection of Drinfeld's and Cartan–Weyl realizations for quantum affine algebras*, Preprint MPI-94/23, hep-th/9404036.

[20] S. Khoroshkin, D. Lebedev, S. Pakuliak, A. Stolin, and V. Tolstoy, *Classical limit of the scaled elliptic algebra $\mathcal{A}_{\hbar,\eta}(\widehat{\mathfrak{sl}_2})$*, Preprint ITEP-TH-1/97, RIMS-1139, q-alg/9703043.

[21] O. Foda, K. Iohara, M. Jimbo, R. Kedem, T. Miwa, and H. Yan, *An elliptic quantum algebra for* $\widehat{\mathfrak{sl}}_2$, Lett. Math. Phys. **32** (1994), 259–268; *Notes on highest weight modules of the elliptic algebra* $\mathcal{A}_{q,p}(\widehat{\mathfrak{sl}_2})$, Progr. Theoret. Phys., Supplement, **118** (1995), 1–34.
[22] M. Jimbo, H. Konno, and T. Miwa, *Massless XXZ model and degeneration of the elliptic algebra* $\mathcal{A}_{q,p}(\widehat{\mathfrak{sl}_2})$, `hep-th/9610079`, Math. Phys. Stud., Vol. 20, Deformation Theory and Symplectic Geometry (Ascona, 1996), Kluwer, Dordrecht, 1997, pp. 117–138.
[23] B. Y. Hou, L. Zhao, and X.-M. Ding, *The algebra* $\mathcal{A}_{\hbar,\eta}(\hat{\mathfrak{g}})$ *and infinite Hopf family of algebras*, Preprint `q-alg/9703046`; J. Geom. Phys. **27** (1998), 249–266.
[24] M. Jimbo, H. Konno, S. Odake, and J. Shiraishi, *Quasi-Hopf twistors for elliptic quantum groups*, Preprint `q-alg/9712029`.
[25] D. Arnaudon, E. Buffenoir, E. Ragoucy, and E. Roche, *Universal solutions of quantum dynamical Yang–Baxter equations*, Preprint `q-alg/9712037`; Lett. Math. Phys. **44** (1998), 201–214.
[26] C. Frønsdal, *Quasi Hopf deformations of quantum groups*, Lett. Math. Phys. **40** (1997), 117–134.
[27] V. G. Drinfeld, *Quasi-Hopf algebras*, Algebra i Analiz **1** (1989), no. 6, 114–148; English transl., Leningrad Math. J. **1** (1990), 1419–1457; *On quasitriangular quasi-Hopf algebras and a group closely connected with* $\mathrm{Gal}(\overline{\mathbb{Q}}/\mathbb{Q})$, Algebra i Analiz **2** (1990), no. 4, pp. 149–181; English transl., Leningrad Math. J. **2** (1991), 829–860.

INSTITUTE OF THEORETICAL & EXPERIMENTAL PHYSICS, 117259 MOSCOW, RUSSIA
E-mail address: `khoroshkin@vitep1.itep.ru`
E-mail address: `dlebedev@vitep5.itep.ru`

BOGOLYUBOV LABORATORY OF THEORETICAL PHYSICS, JINR, 141980 DUBNA, RUSSIA
E-mail address: `pakuliak@thsun1.jinr.ru`

Amer. Math. Soc. Transl.
(2) Vol. **191**, 1999

Vacuum Curves of Elliptic L-Operators and Representations of the Sklyanin Algebra

I. Krichever and A. Zabrodin

ABSTRACT. An algebro-geometric approach to representations of the Sklyanin algebra is proposed. To each 2×2 quantum L-operator an algebraic curve parametrizing its possible vacuum states is associated. This curve is called the vacuum curve of the L-operator. An explicit description of the vacuum curve for quantum L-operators of the integrable spin chain of XYZ type with arbitrary spin ℓ is given. The curve is highly reducible. For a half-integer ℓ it splits into $\ell + 1/2$ components isomorphic to an elliptic curve. For an integer ℓ it splits into ℓ elliptic components and one rational component. The action of elements of the L-operator on functions on the vacuum curve leads to a new realization of the Sklyanin algebra by difference operators in two variables restricted to an invariant functional subspace.

1. Introduction

The Yang–Baxter equation

$$R^{23}(u-v)R^{13}(u)R^{12}(v) = R^{12}(v)R^{13}(u)R^{23}(u-v) \tag{1.1}$$

is a key relation of the theory of quantum integrable models.[1] Each solution of equation (1.1) generates a hierarchy of integrable models. The commutation relations for elements of quantum L-operators of this hierarchy are given by the "intertwining" equation

$$R^{23}(u-v)L^{13}(u)L^{12}(v) = L^{12}(v)L^{13}(u)R^{23}(u-v). \tag{1.2}$$

Here L is an operator in the tensor product $\mathbb{C}^N \otimes \mathbb{C}^n$, and all the factors in (1.2) are operators in the tensor product $\mathbb{C}^N \otimes \mathbb{C}^n \otimes \mathbb{C}^n$.

Let $n = 2$ in (1.1); then the most general R-matrix with elliptic dependence of the spectral parameter u corresponds to the famous 8-vertex model (or, equivalently,

---

1991 *Mathematics Subject Classification.* Primary 58F07; Secondary 16S99.

The work of the first author was supported in part by RFBR grant 95-01-00755.

The work of the second author was supported in part by RFBR grant 95-01-01106 and by grant 96-15-96455 for support of scientific schools.

[1] We use the following standard notation: if R is a linear operator acting in the tensor product $\mathbb{C}^n \otimes \mathbb{C}^n$, then R^{ij}, $i, j = 1, 2, 3$, is an operator in the tensor product $\mathbb{C}^n \otimes \mathbb{C}^n \otimes \mathbb{C}^n$ which acts as R in the tensor product of the ith and jth factors of the triple tensor product and as the identity operator on the remaining factor.

to the XYZ magnet):

$$R(u) = \sum_{a=0}^{3} W_a(u+\eta)\sigma_a \otimes \sigma_a. \tag{1.3}$$

Here σ_a are Pauli matrices (σ_0 is the unit matrix), and $W_a(u)$ are functions of u with parameters η and τ:

$$W_a(u) = \frac{\theta_{a+1}(u|\tau)}{\theta_{a+1}(\eta|\tau)} \tag{1.4}$$

(the Jacobi θ-functions are listed in the Appendix).

In [**1, 2**] Sklyanin reformulated the problem of solving equation (1.2) in terms of representations of an algebra with four generators S_0, S_α, $\alpha = 1, 2, 3$, subject to the homogeneous quadratic relations

$$\begin{aligned} [S_0, S_\alpha]_- &= iJ_{\beta\gamma}[S_\beta, S_\gamma]_+, \\ [S_\alpha, S_\beta]_- &= i[S_0, S_\gamma]_+. \end{aligned} \tag{1.5}$$

Here and below $\{\alpha, \beta, \gamma\}$ is any *cyclic* permutation of $\{1, 2, 3\}$, $[A, B]_\pm = AB \pm BA$. The structure constants have the form

$$J_{\alpha\beta} = \frac{J_\beta - J_\alpha}{J_\gamma}, \tag{1.6}$$

where J_α are arbitrary parameters. The algebra generated by the S_a with relations (1.5) and structure constants (1.6) is called a *Sklyanin algebra*. There is a two-parametric family of such algebras. The relations of the Sklyanin algebra imposed on S_a are equivalent to the condition that the L-operator of the form

$$L(u) = \sum_{a=0}^{3} W_a(u)\, S_a \otimes \sigma_a \tag{1.7}$$

(regarded as an operator in $\mathcal{H} \otimes \mathbb{C}^2$, where $\mathcal{H}$ is a module over the algebra) satisfies equation (1.2). Hence, any finite-dimensional representation of the Sklyanin algebra provides a solution to equation (1.2).

As it was shown in [**2**], the operators S_a, $a = 0, \ldots, 3$, admit a realization as second order difference operators in the space of meromorphic functions $F(z)$ of a complex variable z. One of the series of such representations (called the principal analytic series or series a) in [**2**]) is

$$(S_a F)(z) = \frac{(i)^{\delta_{a,2}}\theta_{a+1}(\eta)}{\theta_1(2z)}\left(\theta_{a+1}(2z - 2\ell\eta)F(z+\eta) - \theta_{a+1}(-2z - 2\ell\eta)F(z-\eta)\right) \tag{1.8}$$

(hereafter $\theta(z) \equiv \theta(z|\tau)$). A straightforward but tedious computation shows that the operators (1.8) for any τ, η, ℓ satisfy the commutation relations (1.5) with the following values of the structure constants:

$$J_\alpha = \frac{\theta_{\alpha+1}(0)\theta_{\alpha+1}(2\eta)}{\theta^2_{\alpha+1}(\eta)}. \tag{1.9}$$

Therefore, τ and η parametrize the structure constants, while ℓ characterizes the representation.

In [**3**], a connection between the representation theory of the Sklyanin algebra and the finite-gap theory of soliton equations was found. It was proved that for an integer ℓ the operator S_0 is algebraically integrable and, therefore, is a difference analog of the classical Lamé operator

$$\mathcal{L} = -\frac{d^2}{dx^2} + \ell(\ell+1)\wp(x),$$

which can be obtained from S_0 in the limit as $\eta \to 0$. Finite gap properties of higher Lamé operators (for arbitrary integer values of ℓ) were established in [**4**]. Algebraic integrability of S_0 implies, in particular, some extremely unusual spectral properties of this operator. Putting $F_n = F(n\eta + z_0)$, we assign to (1.8) the difference Schrödinger operators

$$S_a F_n = A_n^a F_{n+1} + B_n^a F_{n-1} \tag{1.10}$$

with quasiperiodic coefficients. The spectrum of a generic operator of this form in the space $l^2(\mathbb{Z})$ (square integrable sequences F_n) has a structure of Cantor set type. If η is a rational number, $\eta = P/Q$, the operators (1.10) have Q-periodic coefficients. In general, Q-periodic difference Schrödinger operators have Q unstable bands in the spectrum.

It was shown in [**3**] that the operator S_0 given by equation (1.8) for positive integer values of ℓ and *arbitrary* η has 2ℓ unstable bands in the spectrum. Its Bloch functions $\psi(z)$ are parametrized by points of a hyperelliptic curve of genus 2ℓ defined by the equation

$$y^2 = P(\varepsilon) = \prod_{i=1}^{2\ell+1} (\varepsilon^2 - \varepsilon_i^2). \tag{1.11}$$

Regarded as a function on the curve, ψ is the Baker–Akhiezer function. Moreover, Bloch eigenfunctions $\psi(z, \pm\varepsilon_i)$ of the operator S_0 at the edges of bands span an invariant functional subspace for all the operators S_a. The corresponding $(4\ell+2)$-dimensional representation of the Sklyanin algebra is a direct sum of two equivalent $(2\ell+1)$-dimensional irreducible representations of the Sklyanin algebra.

It is well-known from the early days of finite-gap theory that the ring of operators commuting with a finite-gap operator is isomorphic to the ring of meromorphic functions on the corresponding spectral curve with poles at "infinite points". For difference operators this was proved in [**5, 6**]. Therefore, the ring of operators commuting with S_0 is generated by S_0 and an operator D such that

$$D^2 = P(S_0) = \prod_{i=1}^{2\ell+1} (S_0^2 - \varepsilon_i^2).$$

In [**7**], for any algebraic curve Γ of genus g with two punctures, a special basis in the ring $\mathcal{M}$ of meromorphic functions A_i with poles at the punctures was introduced. These functions define the *almost graded structure* in $\mathcal{M}$; i.e., the product of the two basis functions has the form

$$A_i A_j = \sum_{k=-g/2}^{g/2} c_{i,j}^k A_{i+j+k}.$$

Therefore, for any finite-gap difference operator S there exist operators M_i commuting with S such that $M_i\psi = A_i\psi$, where ψ is the Baker–Akhiezer eigenfunction of S. The ring generated by the operators M_i has the same almost graded structure as the ring $\mathcal{M}$. A particular case of the last result corresponding to the ring of operators commuting with the Sklyanin operator S_0 was recently rediscovered in [**8**].

The connection of the Sklyanin operator S_0 with the algebro-geometric theory of soliton equations is a part of the theory of integrable multi-dimensional differential or difference linear operators with elliptic coefficients. It turns out that the spectral theory of such operators is isomorphic to the theory of finite-dimensional integrable systems. Among them are spin generalizations of the Calogero–Moser system, the Ruijesenaars–Schneider systems and nested Bethe ansatz equations (see [**9**], [**3**], and [**10**], respectively).

In [**3**], we suggested a relatively simple way to derive the realization (1.8) of the Sklyanin algebra by difference operators. Our approach clarifies their origin as difference operators acting on the *vacuum curve* of the L-operator (1.7). This notion, introduced by one of the authors [**11**], proved to be useful in analyzing the Yang–Baxter equation by the methods of algebraic geometry. The construction of the vacuum curve and vacuum vectors is a suitable generalization of a key property of the elementary R-matrix (1.3), which Baxter used in his famous solution of the 8-vertex model and called "pair-propagation through a vertex" [**12**]. In that particular case, the vacuum curve is an elliptic curve.

In this paper we present a more detailed analysis of the vacuum curve of the higher spin L-operator (1.7). Our main result is a new realization of the Sklyanin algebra by difference operators acting in *two* variables rather than one (Sections 5, 6). Remarkably, the "finite-gap" operator S_0 in this new realization preserves its form, while the form of the other three generators changes drastically, with both variables entering them in a nontrivial way. This realization naturally follows from the explicit construction of the vacuum curve of the elliptic L-operator and corresponding vacuum vectors.

Let us recall the main definitions. Consider an *arbitrary* L-operator L with two-dimensional auxiliary space $\mathbb{C}^2$, i.e., an arbitrary $2N \times 2N$ matrix represented as a 2×2 matrix whose matrix elements are $N \times N$ matrices L_{11}, L_{12}, L_{21}, L_{22}:

$$L = \begin{pmatrix} L_{11} & L_{12} \\ L_{21} & L_{22} \end{pmatrix}. \tag{1.12}$$

They act in the linear space $\mathcal{H} \cong \mathbb{C}^N$, which is called the *quantum space* of the L-operator.

Let $X \in \mathcal{H}$ and $|U\rangle = \begin{pmatrix} U_1 \\ U_2 \end{pmatrix} \in \mathbb{C}^2$ be two vectors such that

$$L(X \otimes |U\rangle) = Y \otimes |V\rangle, \tag{1.13}$$

where $Y \in \mathcal{H}$ and $|V\rangle \in \mathbb{C}^2$ are vectors. Suppose (1.13) holds; then the vector X is called the *vacuum vector* of the L-operator. Multiplying (1.13) from the left by the covector ${}^{\perp}\langle V| = (V_2, -V_1)$ orthogonal to $|V\rangle$, we obtain a necessary and sufficient condition for the existence of the vacuum vectors:

$${}^{\perp}\langle V|L|U\rangle X = 0. \tag{1.14}$$

Here ${}^{\perp}\langle V|L|U\rangle$ is an operator in $\mathcal{H}$. The *vacuum curve* is defined by the equation

$$\det({}^{\perp}\langle V|L|U\rangle) = 0. \tag{1.15}$$

By construction, it is embedded into $\mathbb{C}P^1 \times \mathbb{C}P^1$.

The relation (1.13) (in the particular case $\mathcal{H} \cong \mathbb{C}^2$) was the starting point for Baxter in his solution of the 8-vertex and XYZ models [**12**]. In the context of the quantum inverse scattering method [**13**], the equivalent condition (1.14) is more customary. It defines the local vacua of the (gauge-transformed) L-operator. A generalization of that solution to the higher spin XYZ model was given by Takebe in [**14**], where, in particular, generalized vacuum vectors were constructed. However, the vacuum curve itself was implicit in that work.

In Section 2 we recollect the main formulas related to the Sklyanin algebra and its representations. As we shall see in Section 3, the vacuum curves corresponding to finite-dimensional representations of the principal analytic series [**2**] of the Sklyanin algebra are completely reducible; i.e., their equations have the form

$$P(U,V) = \prod_{k=1}^{M} P_k(U,V) = 0$$

with some polynomials P_k. If the dimension of $\mathcal{H}$ is even, then $2M = \dim \mathcal{H}$ and all the of the irreducible components are elliptic curves isomorphic to the vacuum curve of the R-matrix R (1.3) found by Baxter. If the dimension of $\mathcal{H}$ is odd, then $2M = \dim \mathcal{H} + 1$ and all but one of the irreducible components are isomorphic to elliptic curves, while the last component is rational. We would like to mention, in passing, that one of the variables in the new realization of the Sklyanin algebra arises precisely as the index k marking irreducible components of the vacuum curve. The other one is, as in [**3**], a uniformization parameter on any one of the elliptic components.

The origin of this realization can be traced back to the structure of the vacuum curve of the operator Λ defined by the left-hand (or right-hand) side of equation (1.2). By definition, Λ is an operator in the tensor product $\mathcal{H} \otimes \mathbb{C}^2 \otimes \mathbb{C}^2$. Let us regard it as an operator in $\widehat{\mathcal{H}} \otimes \mathbb{C}^2$, where $\widehat{\mathcal{H}} = \mathcal{H} \otimes \mathbb{C}^2$ is the product of the first two factors. Proceeding in the way explained above, we may assign a vacuum curve to this operator. As was shown in [**11**], this curve is a result of a "composition" of vacuum curves of the operators L and R. This curve is completely reducible, too. Besides, all components but one of this latter curve are *two-fold degenerate*; i.e., the equation of the curve has the form

$$\widehat{P}(U,V) = \widehat{P}_0(U,V)\left[\prod_{i=1}^{N} \widehat{P}_i(U,V)\right]^2 = 0.$$

The multiplicity of components of the composed vacuum curve "mixes" different components of the vacuum curve of the L-operator and, in terms of action to the vacuum vectors, leads to shifts in the index k, which appear in the two-variable realization of the Sklyanin generators. A detailed study of the action of $L(u)$ on the vacuum vectors is given in Section 4.

Finally, in Section 7 we make some remarks on the trigonometric degeneration of the construction presented in Sections 4–6. This is related to representations of the quantum algebra $U_q(sl(2))$. In particular, our approach provides a new type of representation of this algebra, and these representations are q-analogs of representations of the $sl(2)$ algebra by vector fields on the two-dimensional sphere.

2. The elliptic L-operator and the Sklyanin algebra

In this section we collect main formulas related to the Sklyanin algebra and its representations.

Consider the elliptic L-operator (1.7):

(2.1)
$$L(u)=\sum_{a=0}^{3}W_a(u)S_a\otimes\sigma_a=\begin{pmatrix}W_0(u)S_0+W_3(u)S_3 & W_1(u)S_1-iW_2(u)S_2\\ W_1(u)S_1+iW_2(u)S_2 & W_0(u)S_0-W_3(u)S_3\end{pmatrix}$$

with $W_a(u)$ given by (1.4). In the sequel we write $\theta_a(x)\equiv\theta_a(x|\tau)$, $\theta_a(x|\tau/2)\equiv\bar{\theta}_a(x)$ for brevity. The operators S_a obey the Sklyanin algebra relations (1.5) with the structure constants

$$J_{\alpha\beta}=-(-1)^{\alpha-\beta}\frac{\theta_1^2(\eta)\theta_{\gamma+1}^2(\eta)}{\theta_{\alpha+1}^2(\eta)\theta_{\beta+1}^2(\eta)}. \tag{2.2}$$

We recall that $\{\alpha,\beta,\gamma\}$ is any cyclic permutation of $\{1,2,3\}$. Note that the coefficients W_a in (2.1) satisfy the algebraic relations

$$(W_\alpha^2-W_\beta^2)=J_{\alpha\beta}(W_\gamma^2-W_0^2). \tag{2.3}$$

Any two of them are independent and define an elliptic curve $\mathcal{E}_0\subset\mathbb{C}P^3$ as the intersection of two quadrics. The spectral parameter u uniformizes this curve; i.e., (2.3) is identically satisfied under the substitutions (1.4), (2.2). Another useful form of the structure constants is

$$J_{\alpha\beta}=\frac{J_\beta-J_\alpha}{J_\gamma},\qquad J_\alpha=\frac{\theta_{\alpha+1}(2\eta)\theta_{\alpha+1}(0)}{\theta_{\alpha+1}^2(\eta)} \tag{2.4}$$

(see (1.6)). The algebra has two independent central elements:

$$\Omega_0=\sum_{a=0}^{3}S_a^2,\qquad \Omega_1=\sum_{\alpha=1}^{3}J_\alpha S_\alpha^2. \tag{2.5}$$

In some formulas below it is more convenient to deal with the renormalized generators:

$$S_a=(i)^{\delta_{a,2}}\theta_{a+1}(\eta)\mathcal{S}_a. \tag{2.6}$$

The relations (1.5) can be rewritten in the form

$$\begin{aligned}(-1)^{\alpha+1}I_{\alpha 0}\mathcal{S}_\alpha\mathcal{S}_0&=I_{\beta\gamma}\mathcal{S}_\beta\mathcal{S}_\gamma-I_{\gamma\beta}\mathcal{S}_\gamma\mathcal{S}_\beta,\\ (-1)^{\alpha+1}I_{\alpha 0}\mathcal{S}_0\mathcal{S}_\alpha&=I_{\gamma\beta}\mathcal{S}_\beta\mathcal{S}_\gamma-I_{\beta\gamma}\mathcal{S}_\gamma\mathcal{S}_\beta,\end{aligned} \tag{2.7}$$

where $I_{ab}=\theta_{a+1}(0)\theta_{b+1}(2\eta)$. The central elements in the renormalized generators read

$$\Omega_0=\theta_1^2(\eta)\mathcal{S}_0^2+\sum_{\alpha=1}^{3}(-1)^{\alpha+1}\theta_{\alpha+1}^2(\eta)\mathcal{S}_\alpha^2,\qquad \Omega_1=\sum_{\alpha=1}^{3}(-1)^{\alpha+1}I_{\alpha\alpha}\mathcal{S}_\alpha^2. \tag{2.8}$$

Sklyanin's realization [**2**] of the algebra by difference operators has the form (1.8). The parameter ℓ is the "spin" of the representation. When $\ell\in\frac{1}{2}\mathbb{Z}_+$, these operators have a finite-dimensional invariant subspace $\mathcal{T}_{4\ell}^+$ of *even* θ-functions of order 4ℓ, i.e., the space of entire functions $F(z)$, $z\in\mathbb{C}$, such that $F(-z)=F(z)$ and

$$F(z+1)=F(z),\qquad F(z+\tau)=\exp(-4\ell\pi i\tau-8\ell\pi iz)F(z). \tag{2.9}$$

This is the representation space of a $(2\ell+1)$-dimensional irreducible representation (of series a)) of the Sklyanin algebra.[2] In the spin-ℓ representation of series a), the central elements take the values

$$\Omega_0 = 4\theta_1^2((2\ell+1)\eta), \qquad \Omega_1 = 4\theta_1(2\ell\eta)\theta_1(2(\ell+1)\eta). \tag{2.10}$$

In terms of the shift operators $T_\pm \equiv \exp(\pm\eta\partial_z)$, equation (1.8) can be rewritten for the renormalized generators (2.6) in a slightly simpler form:

$$\mathcal{S}_a = \frac{\theta_{a+1}(2z-2\ell\eta)}{\theta_1(2z)}T_+ - \frac{\theta_{a+1}(-2z-2\ell\eta)}{\theta_1(2z)}T_-. \tag{2.11}$$

Plugging the difference operators (1.8) into (2.1), one can represent the L-operator in a "factored" form [**16**], which is especially convenient for computations. Introduce the matrix

$$\widehat{\Phi}(u;z) = \begin{pmatrix} \bar{\theta}_4(z-\ell\eta+u/2) & \bar{\theta}_3(z-\ell\eta+u/2) \\ \bar{\theta}_4(z+\ell\eta-u/2) & \bar{\theta}_3(z+\ell\eta-u/2) \end{pmatrix}. \tag{2.12}$$

Then we have

$$L(u)F(z) = 2\theta_1(u+2\ell\eta)\widehat{\Phi}^{-1}(u+4\ell\eta;z)\begin{pmatrix} F(z+\eta) & 0 \\ 0 & F(z-\eta) \end{pmatrix}\widehat{\Phi}(u;z), \tag{2.13}$$

where the elements of the L-operator are assumed to act on the function $F(z)$ according to (1.8).

3. The vacuum curve

Our goal in this section is to find an explicit parametrization of the vacuum curve and vacuum vectors of the elliptic L-operator (2.1) in a finite-dimensional representation of the Sklyanin algebra.

In practical computations, it will be more convenient to write 2-dimensional vectors like $|U\rangle$ from the left of the L-operator rather than from the right, so the basic equation (1.13) acquires the form

$$\langle U|L(u)\,X = \langle V|\,Y. \tag{3.1}$$

Here $\langle U| = (U_1, U_2)$, $\langle V| = (V_1, V_2)$ are *covectors.* It is easy to see that this leads to a definition of the vacuum curve, $\det(\langle U|L|V^\perp\rangle) = 0$, equivalent to the one given in the Introduction.

Let the operators S_a in (2.1) be realized as in (1.8) and let them act in the finite-dimensional subspace $\mathcal{T}_{4\ell}^+$ (2.9). Then X and Y in (3.1) are functions of z belonging to the space $\mathcal{T}_{4\ell}^+$. In this section it is convenient to normalize $\langle U|$ and $\langle V|$ by the condition that the second components are equal to 1: $\langle U| = (U,1)$, $\langle V| = (V,1)$.

In this notation the equation of the vacuum curve (1.15) acquires the form

$$\det K(U,V) = 0, \tag{3.2}$$

where

(3.3)
$$K(U,V) = (U-V)W_0S_0 + (1-UV)W_1S_1 + i(1+UV)W_2S_2 + (U+V)W_3S_3.$$

[2] There are three other series of irreducible representations [**2, 15, 3**] that we do not discuss here.

In other words, we must find a relation on U and V under which the operator $K(U,V)$ has an eigenvector (belonging to $\mathcal{T}_{4\ell}^{+}$) with zero eigenvalue (a "zero mode"). One can write $K(U,V)$ in the equivalent form

$$K(U,V) = \sum_{a=0}^{3} \zeta_a W_a S_a,$$

where the new variables ζ_a are subject to the constraint

$$\zeta_1^2 + \zeta_2^2 + \zeta_3^2 = \zeta_0^2. \tag{3.4}$$

THEOREM 3.1. *For $\ell \in \mathbb{Z}_+ + 1/2$ the vacuum curve of the L-operator* (2.1) *splits into the union of $\ell + 1/2$ elliptic curves isomorphic to $\mathcal{E}_0$* (2.3), *and for $\ell \in \mathbb{Z}_+$ the vacuum curve splits into ℓ components isomorphic to $\mathcal{E}_0$ and one rational component. The equation defining the vacuum curve in the coordinates U, V has the form*

$$P_\ell(U,V) = 0, \qquad \ell \in \mathbb{Z}_+ + 1/2, \tag{3.5}$$

$$(U-V)^{-1} P_\ell(U,V) = 0, \qquad \ell \in \mathbb{Z}_+, \tag{3.6}$$

where

$$P_\ell(U,V) = \prod_{n=0}^{[\ell]} (U^2 + V^2 - 2\Delta_{\ell-n} UV + \Gamma_{\ell-n}(1 + U^2V^2))$$

($[\ell]$ denotes the integer part of ℓ), $\Gamma_0 = 0$, $\Delta_0 = 1$ and the u-independent constants Γ_k, Δ_k for $k \geqslant 1$ are defined below in (3.24), (3.25).

REMARK. Note that for integer ℓ the polynomial $P_\ell(U,V)$ is divisible by $(U-V)^2$, so the left-hand side of (3.6) is a polynomial.

PROOF. First let us consider a linear combination of the generators S_a with arbitrary coefficients:

$$K = \sum_{a=0}^{3} y_a S_a. \tag{3.7}$$

According to (1.8), the equation $KX = 0$ is equivalent to

$$\frac{s(-z-\ell\eta)}{s(z-\ell\eta)} = \frac{X(z+\eta)}{X(z-\eta)}, \tag{3.8}$$

where

$$s(z) = \sum_{a=0}^{3} (i)^{\delta_{a,2}} y_a \theta_{a+1}(\eta)\theta_{a+1}(2z) \tag{3.9}$$

belongs to the space $\mathcal{T}_4$ of θ-functions of order 4 with the following monodromy properties: $s(z+1) = s(z)$, $s(z+\tau) = \exp(-4\pi i\tau - 8\pi i z)s(z)$. Note that $\dim \mathcal{T}_4 = 4$ and $s(z)$ has 4 zeros in the fundamental domain of the lattice formed by 1 and τ. The function $s(z)$ can be parametrized by its zeros:

$$s(z) = \prod_{i=1}^{4} \theta_1(z - a_i), \qquad \sum_{i=1}^{4} a_i = 0 \tag{3.10}$$

(up to an inessential factor). The functions $\theta_{a+1}(2z)$ form a basis in the space $\mathcal{T}_4$. Expanding (3.10) with respect to this basis, we obtain

$$s(z) = \frac{1}{2}\sum_{a=0}^{3}(-1)^a\theta_{a+1}(2z)\prod_{i=2}^{4}\theta_{a+1}(-a_1 - a_i). \tag{3.11}$$

Let us begin by considering the case $\ell \geqslant 1$. Since $X(z)$ has 4ℓ zeros in the fundamental domain, (3.8) is possible only if the functions $X(z+\eta)$ and $X(z-\eta)$ have $4\ell - 4$ common zeros. Such a cancellation of zeros of the numerator and the denominator in the right-hand side of (3.8) takes place if the zeros of $X(z)$ are arranged into "strings", i.e.,

$$X(z) = \prod_{i=1}^{4}\prod_{j=0}^{m_i-1}\theta_1(z - z_i - 2j\eta) \tag{3.12}$$

with the condition

$$\sum_{i=1}^{4} m_i = 4\ell, \qquad m_i \geqslant 0. \tag{3.13}$$

At this stage we do not impose any other restrictions; in particular, it is not implied that $X(z)$ is even. The number m_i is called the length of the string. So, if all the m_i are positive, there are four strings in (3.12) with total length 4ℓ. If some of m_i equal zero, the number of strings is less than four. Let $X(z)$ be given by (3.12); then

$$\frac{X(z+\eta)}{X(z-\eta)} = \prod_{i=1}^{4}\frac{\theta_1(z - z_i + \eta)}{\theta_1(z - z_i - (2m_i - 1)\eta)}. \tag{3.14}$$

Identifying zeros of the left-hand side of (3.8) with zeros of (3.14), we get

$$z_i + a_i = -(\ell - 1)\eta, \qquad i = 1, \ldots, 4. \tag{3.15}$$

Taking these relations into account, we then identify poles of (3.8) and (3.14) and conclude that the (unordered) sets of points $(z_i + (2m_i - 1)\eta)$ and $(-z_i + \eta)$, $i = 1, \ldots, 4$, must coincide. In order to describe all the possibilities, denote by P any permutation of the indices (1234) and consider the systems of linear equations

$$z_i + z_{P(i)} = -2(m_i - 1)\eta, \qquad i = 1, \ldots, 4, \tag{3.16}$$

for each P. The solutions to these systems consistent with the condition

$$\sum_{i=1}^{4} z_i = -4(\ell - 1)\eta, \tag{3.17}$$

which follows from (3.10) and (3.15), yield all possible values of z_i. Depending on the choice of P, the rank of the linear system (3.16) may equal 4, 3, or 2; i.e., the number of free parameters in the solutions may be respectively 0, 1, or 2. Since the coefficients of the function $s(z)$ with respect to the basis $\theta_a(2z)$ are already constrained by one relation (3.4) (describing the embedding $\mathbb{C}P^1 \times \mathbb{C}P^1 \subset \mathbb{C}P^3$), the vacuum curve corresponds to the case of minimal rank. It is easy to see that system (3.16) has rank 2 for the following three permutations of (1234): (2143), (3412), (4321); otherwise its rank is greater than 2. Solving (3.16) for these three permutations, we arrive at the following result.

The function $s(z)$ has the form

(3.18)
$$\begin{aligned} s(z) &\equiv s^{(m)}(z) \\ &= \theta_1(z-\mu_1)\theta_1(z-\mu_2)\theta_1(z+\mu_1+2(\ell-m)\eta)\theta_1(z+\mu_2-2(\ell-m)\eta), \end{aligned}$$

where $m = 0, 1, \dots, [\ell]$ and μ_1, μ_2 are free parameters. The corresponding "zero mode" of K is given by the formula[3]

(3.19)
$$\begin{aligned} X^{(m)}(z) = &\prod_{j=1}^{m} \theta_1(z+\mu_1+(\ell+1-2j)\eta)\theta_1(z-\mu_1-(\ell+1-2j)\eta) \\ &\times \prod_{j=1}^{2\ell-m} \theta_1(z+\mu_2+(\ell+1-2j)\eta)\theta_1(z-\mu_2-(\ell+1-2j)\eta). \end{aligned}$$

Note that $X^{(m)}(z)$ is an even function.

Let us set $k = \ell - m$, $k = 0, 1, \dots, \ell$ for $\ell \in \mathbb{Z}_+$ and $k = 1/2, 3/2, \dots, \ell$ for $\ell \in \mathbb{Z}_+ + 1/2$, and identify $\mu_1 + \mu_2 = u$ (the spectral parameter), $\mu_2 - \mu_1 = 2\zeta + 2k\eta$, where ζ is the uniformization parameter on the vacuum curve. It follows from (1.4), (3.9), (3.11), and (3.17) that the coefficients ζ_a in (3.3) are given by the formula

(3.20)
$$\zeta_a = \frac{2(i)^{\delta_{a,2}}\theta_{a+1}(2k\eta)\theta_{a+1}(2\zeta)}{\bar{\theta}_3(\zeta+k\eta)\bar{\theta}_3(\zeta-k\eta)}.$$

They satisfy the homogeneous quadratic equations

(3.21)
$$\zeta_1^2 + \zeta_2^2 + \zeta_3^2 = \zeta_0^2, \qquad \sum_{\alpha=1}^{3} \zeta_\alpha^2 \frac{\theta_{\alpha+1}^2(0)}{\theta_{\alpha+1}^2(2k\eta)} = 0,$$

which provide a purely algebraic description of the vacuum curve. Rescaling the coordinates, $\zeta_a = \xi_a \theta_{a+1}(2k\eta)$ (for $k \neq 0$), one may represent (3.21) in a form independent of k:

(3.22)
$$\sum_{\alpha=1}^{3} \xi_\alpha^2 \theta_{\alpha+1}^2(0) = 0, \qquad \sum_{\alpha=1}^{3} \xi_{3-\alpha}^2 \theta_{\alpha+1}^2(0) = 0.$$

It is straightforward to see that the equations (2.3) defining the elliptic curve $\mathcal{E}_0$ can be transformed to the same form. This means that the vacuum curve is reducible: it splits into the union of the components isomorphic to $\mathcal{E}_0$ (corresponding to nonzero values of k).

In the coordinates U, V, system (3.21) is equivalent to a single equation of degree 4:

(3.23)
$$U^2 + V^2 - 2\Delta_k UV + \Gamma_k(1 + U^2V^2) = 0,$$

where

(3.24)
$$\Gamma_k = \frac{\theta_1^2(2k\eta)\theta_4^2(2k\eta)}{\theta_2^2(2k\eta)\theta_3^2(2k\eta)} = \left(\frac{\bar{\theta}_1(2k\eta)}{\bar{\theta}_2(2k\eta)}\right)^2,$$

(3.25)
$$\Delta_k = \frac{\theta_3^2(0)}{\theta_4^2(0)}\left(\frac{\theta_4^2(2k\eta)}{\theta_3^2(2k\eta)} + \frac{\theta_1^2(2k\eta)}{\theta_2^2(2k\eta)}\right) = \frac{\bar{\theta}_2^2(0)\bar{\theta}_3(2k\eta)\bar{\theta}_4(2k\eta)}{\bar{\theta}_2^2(2k\eta)\bar{\theta}_3(0)\bar{\theta}_4(0)}.$$

[3]These functions coincide with the "intertwining vectors" from the paper [**14**], found there by a different method.

For $k = 1/2$ this equation, after trivial redefinitions, coincides with the equation of the vacuum curve for spin 1/2 first obtained by Baxter [**12**]. So formulas (3.21)–(3.25) are valid for $\ell = 1/2$ as well (in this case k takes only one value, namely 1/2).

The case $k = 0$ (4 strings of length ℓ each) requires separate study. In this case $s(z)$ is an even function, so $\zeta_0 = 0$ and the corresponding component of the vacuum curve is a rational curve (the cone). Therefore, for integer ℓ the vacuum curve has a rational component (corresponding to $k = 0$) given by the equation $U = V$. □

REMARK. For $\tau = 0$ the elliptic L-operator (2.1) degenerates into a trigonometric one, corresponding to the higher spin XXZ model. Its vacuum curve was studied in [**17**]. This curve can be obtained from our formulas (3.23)–(3.25) in the limit as $\tau \to 0$ and $\eta \to 0$, provided $\eta' \equiv \eta/\tau$ is finite: $\Gamma_k = 0$, $\Delta_k = \cos(4\pi k\eta')$. Equation (3.23) turns into

$$(U - e^{4\pi i k\eta'}V)(U - e^{-4\pi i k\eta'}V) = 0,$$

so each elliptic component (of degree 4) splits into two rational components (of degree 2 each). The equation for the whole vacuum curve acquires the form

$$\prod_{n=-[\ell]}^{[\ell]} (U - e^{4\pi i n\eta'}V) = 0.$$

Other types of rational degenerations are also possible [**18**]. Their detailed classification is not discussed here.

We call the "copy" of the elliptic curve $\mathcal{E}_0$ defined by equation (3.23) with $k = \ell$ the *highest component* of the vacuum curve. The results of [**3**] suggest that it plays a distinguished role in representations of the Sklyanin algebra.

Let us summarize the results of this section and prepare some formulas which will be extensively used in the sequel. The vacuum curve of the elliptic spin-ℓ L-operator consists of $[\ell + 1/2]$ components marked by $k = 0, 1, \dots, \ell$, for integer ℓ and $k = 1/2, 3/2, \dots, \ell$ for half-integer ℓ. Each elliptic component is uniformized by the variable ζ:

$$U = \frac{\bar\theta_4(\zeta + k\eta)}{\bar\theta_3(\zeta + k\eta)}, \qquad V = \frac{\bar\theta_4(\zeta - k\eta)}{\bar\theta_3(\zeta - k\eta)}$$

(at $k = 0$ it degenerates into a rational component). In the sequel it will be more convenient to pass to homogeneous coordinates and work with the following two-dimensional covectors:

$$\langle\zeta| \equiv (\bar\theta_4(\zeta), \bar\theta_3(\zeta)), \qquad \langle -\zeta| = \langle\zeta|. \tag{3.26}$$

The vacuum vectors are given by (3.19), which we rewrite in terms of ζ, k as

$$\begin{aligned} X_k^\ell(z,\zeta) = \prod_{j=1}^{\ell-k} &\theta_1\Big(z - \zeta + \frac{u}{2} + (\ell - k + 1 - 2j)\eta\Big) \\ &\times \theta_1\Big(z + \zeta - \frac{u}{2} - (\ell - k + 1 - 2j)\eta\Big) \\ \times \prod_{j=1}^{\ell+k} &\theta_1\Big(z + \zeta + \frac{u}{2} + (\ell + k + 1 - 2j)\eta\Big) \\ &\times \theta_1\Big(z - \zeta - \frac{u}{2} - (\ell + k + 1 - 2j)\eta\Big). \end{aligned} \tag{3.27}$$

We call X^ℓ_ℓ the *highest vacuum vector.* Formula (3.27) still has a meaning for negative values of k, too, extending $X^\ell_k(z,\zeta)$ to $-\ell \leqslant k \leqslant \ell$. From now on we assume that k varies in this region. Let us point out the following simple properties of vacuum vectors:

$$X^\ell_k(-z,\zeta) = X^\ell_k(z,\zeta), \qquad X^\ell_k(z,-\zeta) = X^\ell_{-k}(z,\zeta),$$
$$X^\ell_k(z+\eta,\zeta) = X^\ell_k(z,\zeta+\eta)\,\frac{\theta_1(z-\zeta+u/2-(k-\ell)\eta)\theta_1(z-\zeta-u/2+(k+\ell)\eta)}{\theta_1(z-\zeta+u/2+(k-\ell)\eta)\theta_1(z-\zeta-u/2-(k+\ell)\eta)},$$
$$X^\ell_k(z,\zeta+\eta) = X^\ell_{k+1}(z,\zeta)\,\frac{\theta_1(z-\zeta+u/2-(\ell-k)\eta)\theta_1(z+\zeta-u/2+(\ell-k)\eta)}{\theta_1(z+\zeta+u/2-(\ell+k)\eta)\theta_1(z-\zeta-u/2+(\ell+k)\eta)}.$$

In the next section we study how $L(u)$ acts on the vacuum vectors.

4. Action of $L(u)$ to the vacuum vectors

By a straightforward computation we obtain

$$\langle\zeta+k\eta|\,L(u)X^\ell_k(z,\zeta) = 2\theta_1(u-2\ell\eta)\langle\zeta-k\eta|\,Y^\ell_k(z,\zeta+\eta), \tag{4.1}$$

where

$$\begin{aligned} Y^\ell_k(z,\zeta) = \prod_{j=0}^{\ell-k-1} & \theta_1\Big(z-\zeta+\frac{u}{2}+(\ell-k+1-2j)\eta\Big) \\ & \times\theta_1\Big(z+\zeta-\frac{u}{2}-(\ell-k+1-2j)\eta\Big) \\ \times\prod_{j=1}^{\ell+k} & \theta_1\Big(z+\zeta+\frac{u}{2}+(\ell+k+1-2j)\eta\Big) \\ & \times\theta_1\Big(z-\zeta-\frac{u}{2}-(\ell+k+1-2j)\eta\Big). \end{aligned} \tag{4.2}$$

It is easy to see that

$$Y^\ell_k(z,-\zeta) = Y^\ell_{-k}(z,\zeta+2\eta).$$

Indicating explicitly the dependence of the vacuum vector on the spectral parameter u, for any $-\ell \leqslant k \leqslant \ell$ we write

$$Y^\ell_k(z,\zeta+\eta,u) = X^\ell_k(z,\zeta,u+2\eta), \qquad X^\ell_{-k}(z,\zeta,u) = X^\ell_k(z,\zeta,-u). \tag{4.3}$$

Below we suppress the u-dependence of vacuum vectors if it is not misleading.

For the highest vacuum vectors, we have

$$Y^\ell_\ell(z,\zeta) = X^\ell_\ell(z,\zeta).$$

Therefore, at $k=\ell$ one can rewrite equation (4.1) in the form

$$\langle\zeta\pm\ell\eta|\,L(u)X^\ell_{\pm\ell}(z,\zeta) = 2\theta_1(u-2\ell\eta)\langle\zeta\mp\ell\eta|\,X^\ell_{\pm\ell}(z,\zeta\pm\eta). \tag{4.4}$$

The equation with the lower sign is obtained by changing $\zeta\to-\zeta$ and using the above listed properties of the vacuum vectors. From (3.27) we see that

$$X^\ell_\ell(z,\zeta-u/2,u) = X^\ell_{-\ell}(z,\zeta+u/2,u). \tag{4.5}$$

This property allows us to convert (4.4) into a closed system of equations for the vector X^ℓ_ℓ only. Indeed, let us substitute $\zeta\to\zeta\mp u/2$ in the first (second) equality in (4.4) and then make use of (4.5). In this way we get the following system of equations:

$$\begin{aligned}\langle\zeta - u/2 + \ell\eta|\, L(u)X_\ell^\ell(z,\zeta - u/2)\\ = 2\theta_1(u - 2\ell\eta)\langle\zeta - u/2 - \ell\eta|\, X_\ell^\ell(z,\zeta - u/2 + \eta)\\ \langle\zeta + u/2 - \ell\eta|\, L(u)X_\ell^\ell(z,\zeta - u/2)\\ = 2\theta_1(u - 2\ell\eta)\langle\zeta + u/2 + \ell\eta|\, X_\ell^\ell(z,\zeta - u/2 - \eta).\end{aligned} \tag{4.6}$$

In our paper [**3**] it is shown that the representation (1.8) in the variable ζ follows from the solution to this system. Therefore, we can say that these representations are realized in the space of functions on the highest component of the vacuum curve.

Now let us turn to other components of the vacuum curve. It is possible to express action of the L-operator in terms of the vectors $X_k^\ell(z,\zeta)$ only. A straightforward computation leads to the following result:

$$\begin{aligned}&\langle\zeta \pm w + k\eta|\, L(u - 2w)X_k^\ell(z,\zeta,u)\\ &= 2\,\frac{\theta_1(2(\zeta\mp\ell\eta))\theta_1(u - 2w \mp 2k\eta)}{\theta_1(2(\zeta - k\eta))}\langle\zeta \pm w - k\eta|\, X_k^\ell(z,\zeta\pm\eta,u)\\ &\pm 2\,\frac{\theta_1(2(k\mp\ell)\eta)\theta_1(2\zeta\mp u \pm 2w)}{\theta_1(2(\zeta - k\eta))}\langle\zeta \mp w - k\eta|\, X_{k\pm1}^\ell(z,\zeta,u).\end{aligned} \tag{4.7}$$

Here w is an arbitrary parameter. The formulas with upper and lower signs are connected by the transformation $\zeta\to-\zeta$ and $k\to-k$. Equalities of this type are sometimes called "intertwining relations" or "vertex-face correspondence". In the particular case $\ell = 1/2$ they were suggested by Baxter [**12**] and recently generalized to arbitrary half-integer values of ℓ by Takebe [**14**].

Putting w equal to 0 and comparing these formulas with (4.1), we get the following linear relations between vectors Y_k^ℓ and X_k^ℓ:

$$\begin{aligned}Y_k^\ell(z,\zeta+\eta,u) &= \frac{\theta_1(2(\zeta\mp\ell\eta))\theta_1(u\mp 2k\eta)}{\theta_1(2(\zeta - k\eta))\theta_1(u - 2\ell\eta)}\,X_k^\ell(z,\zeta\pm\eta,u)\\ &\pm\frac{\theta_1(2(k\mp\ell)\eta)\theta_1(2\zeta\mp u)}{\theta_1(2(\zeta - k\eta))\theta_1(u - 2\ell\eta)}\,X_{k\pm1}^\ell(z,\zeta,u).\end{aligned} \tag{4.8}$$

The two choices of signs in the right-hand side provide a 4-term linear relation between vectors X_k^ℓ. Moreover, taking into account property (4.3), one can rewrite (4.8) in a more symmetric form:

$$\begin{aligned}&\theta_1(2(\zeta - k\eta))\theta_1(u - 2\ell\eta)X_k^\ell(z,\zeta,u + 2\eta)\\ &= \theta_1(2(\zeta - \ell\eta))\theta_1(u - 2k\eta)X_k^\ell(z,\zeta+\eta,u)\\ &\quad + \theta_1(2(k-\ell)\eta)\theta_1(2\zeta - u)X_{k+1}^\ell(z,\zeta,u)\\ &= \theta_1(2(\zeta + \ell\eta))\theta_1(u + 2k\eta)X_k^\ell(z,\zeta-\eta,u)\\ &\quad - \theta_1(2(k+\ell)\eta)\theta_1(2\zeta + u)X_{k-1}^\ell(z,\zeta,u).\end{aligned} \tag{4.9}$$

Note that the three variables ζ, k, and u, after an appropriate rescaling, enter symmetrically in spite of their very different nature.

5. Difference operators on the vacuum curve

Formulas (4.7) give rise to some distinguished difference operators in *two* variables, ζ and k, related to representations of the Sklyanin algebra. Consider a particular case of the first equation (4.7) (with upper sign) at $u = 0$, $w = -u/2$. From (3.27) it follows that $X_k^\ell(z, \zeta, 0)$ is an *even function of* ζ. Note that this automatically implies $X_{-k}^\ell(z, \zeta, 0) = X_k^\ell(z, \zeta, 0)$. Therefore, the substitution $\zeta \to -\zeta$ provides us with another equation for the same vector $X_k^\ell(z, \zeta, 0)$. Together with the first one they form a closed system:

$$
\begin{aligned}
&\langle \zeta \mp u/2 \pm k\eta |\, L(u) X_k^\ell(z, \zeta, 0) \\
&\qquad = 2\, \frac{\theta_1(2(\zeta \mp \ell\eta))\theta_1(u - 2k\eta)}{\theta_1(2(\zeta \mp k\eta))} \langle \zeta \mp u/2 \mp k\eta |\, X_k^\ell(z, \zeta \pm \eta, 0) \\
&\qquad + 2\, \frac{\theta_1(2(k - \ell)\eta)\theta_1(2\zeta \mp u)}{\theta_1(2(\zeta \mp k\eta))} \langle \zeta \pm u/2 \mp k\eta |\, X_{k+1}^\ell(z, \zeta, 0),
\end{aligned} \tag{5.1}
$$

which is an extension of the simpler system (4.6) on the highest component to the whole vacuum curve.

Let X_i, ${}^i\bar{X}$, $i = 1, 2, \ldots, 2\ell + 1$, be dual bases in the space $\mathcal{T}_{4\ell}^+$ of theta-functions and its dual $\mathcal{T}_{4\ell}^{+,*}$, respectively, i.e., $({}^i\bar{X}, X_j) = \delta_{ij}$, where $(\ ,\ )$ denotes the pairing of the spaces $\mathcal{T}_{4\ell}^{+,*}$ and $\mathcal{T}_{4\ell}^+$. The linear space $\mathcal{F}_{\zeta,k}$ of functions spanned by the functions

$$
{}^iX_k^\ell(\zeta) \equiv ({}^i\bar{X}, X_k^\ell(z, \zeta, 0)),
$$

which depend on ζ, k, but not on z, plays a central role in what follows. Let $\mathcal{A}$ be any difference operator in z. One can translate its action to the space $\mathcal{F}_{\zeta,k}$ according to the definition

$$
(\mathcal{A} \circ {}^iX_k^\ell)(\zeta) = ({}^i\bar{X}, \mathcal{A}X_k^\ell(z, \zeta, 0)). \tag{5.2}
$$

This action can be extended to the whole space $\mathcal{F}_{\zeta,k}$ by linearity.

Note the composition rule

$$
\mathcal{A} \circ (\mathcal{A}' \circ F) = (\mathcal{A}'\mathcal{A}) \circ F \tag{5.3}
$$

for any $F \in \mathcal{F}_{\zeta,k}$, i.e., the order of the operators $\mathcal{A}$, $\mathcal{A}'$ with respect to the action $\circ$ must be reversed. Indeed, let us write

$$
X_k^\ell(z, \zeta) = \sum_i {}^iX_k^\ell(\zeta) X_i(z)
$$

and define the matrix elements of an operator $\mathcal{A}$ to be

$$
(\mathcal{A})_i^j = ({}^i\bar{X},\, \mathcal{A}X_i(z)).
$$

Then we have

$$
\begin{gathered}
\mathcal{A} \circ {}^jX_k^\ell = \sum_i (\mathcal{A})_i^j\, {}^iX_k^\ell, \qquad \mathcal{A}' \circ {}^jX_k^\ell = \sum_i (\mathcal{A}')_i^j\, {}^iX_k^\ell, \\
(\mathcal{A}'\mathcal{A}) \circ {}^jX_k^\ell = ({}^j\bar{X}, \mathcal{A}'\mathcal{A}X_k^\ell) = \sum_{i,l} (\mathcal{A}')_l^j (\mathcal{A})_i^l\, {}^iX_k^\ell = \mathcal{A} \circ (\mathcal{A}' \circ {}^jX_k^\ell).
\end{gathered}
$$

After these preliminaries, we can take the convolution of system (5.1) with the basis covectors ${}^i\bar{X}$ and rewrite it in terms of the action $\circ$ as a relation between functions in the space $\mathcal{F}_{\zeta,k}$:

$$
\begin{aligned}
&\langle \zeta \mp u/2 \pm k\eta |\, L(u) \circ X_k^\ell(\zeta) \\
&\quad = 2\,\frac{\theta_1(2(\zeta \mp \ell\eta))\theta_1(u-2k\eta)}{\theta_1(2(\zeta \mp k\eta))} \langle \zeta \mp u/2 \mp k\eta |\, X_k^\ell(\zeta \pm \eta) \\
&\quad + 2\,\frac{\theta_1(2(k-\ell)\eta)\theta_1(2\zeta \mp u)}{\theta_1(2(\zeta \mp k\eta))} \langle \zeta \pm u/2 \mp \eta |\, X_{k+1}^\ell(\zeta)
\end{aligned} \tag{5.4}
$$

(here and below $X_k^\ell(\zeta) \in \mathcal{F}_{\zeta,k}$). These relations form a system of four linear equations for the four functions $(S_a \circ X_k^\ell)(\zeta)$ appearing in the left-hand side. To make this clear, it is useful to rewrite the system in more explicit form:

$$
\begin{aligned}
&\begin{pmatrix} \bar\theta_4(\zeta - u/2 + k\eta) & \bar\theta_3(z - u/2 + k\eta) \\ \bar\theta_4(z + u/2 - k\eta) & \bar\theta_3(z + u/2 - k\eta) \end{pmatrix} \begin{pmatrix} L_{11}(u) \circ X_k^\ell(\zeta) & L_{12}(u) \circ X_k^\ell(\zeta) \\ L_{21}(u) \circ X_k^\ell(\zeta) & L_{22}(u) \circ X_k^\ell(\zeta) \end{pmatrix} \\
&\quad = 2 \begin{pmatrix} W_{11}\bar\theta_4(\zeta - u/2 - k\eta) X_k^\ell(\zeta+\eta) & W_{11}\bar\theta_3(z - u/2 - k\eta) X_k^\ell(\zeta+\eta) \\ W_{22}\bar\theta_4(z + u/2 + k\eta) X_k^\ell(\zeta-\eta) & W_{22}\bar\theta_3(z + u/2 + k\eta) X_k^\ell(\zeta-\eta) \end{pmatrix} \\
&\quad + 2 \begin{pmatrix} W_{12}\bar\theta_4(\zeta + u/2 - k\eta) & W_{12}\bar\theta_3(z + u/2 - k\eta) \\ W_{21}\bar\theta_4(z - u/2 + k\eta) & W_{21}\bar\theta_3(z - u/2 + k\eta) \end{pmatrix} X_{k+1}^\ell(\zeta),
\end{aligned}
$$

where W_{ij} are elements of the matrix

$$
W = \begin{pmatrix} \dfrac{\theta_1(2(\zeta - \ell\eta))\theta_1(u - 2k\eta)}{\theta_1(2(\zeta - k\eta))} & \dfrac{\theta_1(2(k-\ell)\eta)\theta_1(2\zeta - u)}{\theta_1(2(\zeta - k\eta))} \\ \dfrac{\theta_1(2(k-\ell)\eta)\theta_1(2\zeta + u)}{\theta_1(2(\zeta + k\eta))} & \dfrac{\theta_1(2(\zeta + \ell\eta))\theta_1(u - 2k\eta)}{\theta_1(2(\zeta + k\eta))} \end{pmatrix}.
$$

Solving this system, we obtain the following action rules of the generators S_a to functions on two variables:

$$
(S_0 \circ X_k^\ell)(\zeta) = \frac{\theta_1(\eta)}{\theta_1(2\zeta)} \left(\theta_1(2\zeta - 2\ell\eta) X_k^\ell(\zeta+\eta) + \theta_1(2\zeta + 2\ell\eta) X_k^\ell(\zeta - \eta)\right), \tag{5.5}
$$

(5.6)

$$
\begin{aligned}
(S_\alpha \circ X_k^\ell)(\zeta) = -(i)^{\delta_{a,2}} \frac{\theta_{\alpha+1}(\eta)}{\theta_1(2\zeta)} &\left(\frac{\theta_1(2\zeta - 2\ell\eta)\theta_{\alpha+1}(2\zeta - 2k\eta)}{\theta_1(2\zeta - 2k\eta)} X_k^\ell(\zeta+\eta) \right. \\
&\left. - \frac{\theta_1(2\zeta + 2\ell\eta)\theta_{\alpha+1}(2\zeta + 2k\eta)}{\theta_1(2\zeta + 2k\eta)} X_k^\ell(\zeta - \eta) \right) \\
+ (i)^{\delta_{a,2}} \frac{2\theta_{\alpha+1}(\eta)\theta_{\alpha+1}(2k\eta)\theta_{\beta+1}(2\zeta)\theta_{\gamma+1}(2\zeta)\theta_1(2(k-\ell)\eta)}{\theta_{\beta+1}(0)\theta_{\gamma+1}(0)\theta_1(2\zeta - 2k\eta)\theta_1(2\zeta + 2k\eta)} &X_{k+1}^\ell(\zeta).
\end{aligned}
$$

Note that the formula for S_0 remains the same as in Sklyanin's realization, while the other generators are substantially different. They are nontrivial operators in two variables rather than one.

REMARK. At $k = \ell$ the last term in the right-hand side of (5.6) disappears and we come back to Sklyanin's formulas in the variable ζ. However, the last three generators differ from the ones in (1.8) by sign. This sign can be explained if we recall that the action $\circ$ has a "contravariant" composition rule (5.3) that amounts to the formal transposition of the commutation relations. The commutation relations of the Sklyanin algebra imply that the transposition of all generators is equivalent (up to an automorphism of the algebra) to changing the signs of S_1, S_2, and S_3. Hence formulas (5.5), (5.6) at $k = \ell$ are indeed equivalent to (1.8).

Do the difference operators (5.5), (5.6) obey the (transposed) Sklyanin algebra? The answer is *no*, since, as it is clear from (5.6), the highest shifts in k do not cancel in the commutation relations (2.7). The matter is that functions $X_k^\ell(\zeta)$ are not arbitrary functions of two variables. By construction, they belong to the space $\mathcal{F}_{\zeta,k}$. This implies an additional condition that follows from (4.8) at $u=0$ after convolution with the basis covectors ${}^i\bar{X}$:

$$\begin{aligned}\frac{\theta_1(2\zeta-2\ell\eta)}{\theta_1(2\zeta)}X_k^\ell(\zeta+\eta)+\frac{\theta_1(2\zeta+2\ell\eta)}{\theta_1(2\zeta)}X_k^\ell(\zeta-\eta)\\ =\frac{\theta_1(2k\eta-2\ell\eta)}{\theta_1(2k\eta)}X_{k+1}^\ell(\zeta)+\frac{\theta_1(2k\eta+2\ell\eta)}{\theta_1(2k\eta)}X_{k-1}^\ell(\zeta).\end{aligned}\tag{5.7}$$

This equality means that S_0 has a "dual" realization as a difference operator in the variable k which has the same form.

Investigating the commutation properties of the difference operators appearing in (5.5) and (5.6), it is convenient to modify them in two respects. First, let us change signs of S_α, thus coming back to the "covariant" action (see the remark above). Second, it is natural to disregard the origin of the k, ℓ as (half) integer numbers and allow them to take arbitrary complex values. Namely, let us set $x=k\eta$, $K_\pm=\exp(\pm\eta\partial_x)$, $T_\pm=\exp(\pm\eta\partial_\zeta)$. In this notation, our difference operators in two variables ζ, x are

$$\mathcal{D}_0=\frac{\theta_1(2\zeta-2\ell\eta)}{\theta_1(2\zeta)}T_++\frac{\theta_1(2\zeta+2\ell\eta)}{\theta_1(2\zeta)}T_-,\tag{5.8}$$

$$\begin{aligned}\mathcal{D}_\alpha=\frac{\theta_1(2\zeta-2\ell\eta)\theta_{\alpha+1}(2\zeta-2x)}{\theta_1(2\zeta)\theta_1(2\zeta-2x)}T_+-\frac{\theta_1(2\zeta+2\ell\eta)\theta_{\alpha+1}(2\zeta+2x)}{\theta_1(2\zeta)\theta_1(2\zeta+2x)}T_-\\ -\frac{2\theta_{\alpha+1}(2x)\theta_{\beta+1}(2\zeta)\theta_{\gamma+1}(2\zeta)\theta_1(2x-2\ell\eta)}{\theta_{\beta+1}(0)\theta_{\gamma+1}(0)\theta_1(2\zeta-2x)\theta_1(2\zeta+2x)}K_+.\end{aligned}\tag{5.9}$$

The last line can be transformed as follows:

$$\begin{aligned}\mathcal{D}_\alpha=\frac{\theta_1(2\zeta-2\ell\eta)}{\theta_1(2\zeta)}\frac{\theta_{\alpha+1}(2\zeta-2x)}{\theta_1(2\zeta-2x)}T_+-\frac{\theta_1(2\zeta+2\ell\eta)}{\theta_1(2\zeta)}\frac{\theta_{\alpha+1}(2\zeta+2x)}{\theta_1(2\zeta+2x)}T_-\\ -\frac{\theta_1(2x-2\ell\eta)}{\theta_1(2x)}\left(\frac{\theta_{\alpha+1}(2\zeta-2x)}{\theta_1(2\zeta-2x)}-\frac{\theta_{\alpha+1}(2\zeta+2x)}{\theta_1(2\zeta+2x)}\right)K_+.\end{aligned}\tag{5.10}$$

In the next section we show that these operators do form a nonstandard realization of the Sklyanin algebra on a subspace of functions of two variables.

6. Representations of the Sklyanin algebra

Let us consider the operator

$$\nabla=\frac{\theta_1(2\zeta-2\ell\eta)}{\theta_1(2\zeta)}T_++\frac{\theta_1(2\zeta+2\ell\eta)}{\theta_1(2\zeta)}T_--\frac{\theta_1(2x-2\ell\eta)}{\theta_1(2x)}K_+-\frac{\theta_1(2x+2\ell\eta)}{\theta_1(2x)}K_-.\tag{6.1}$$

The condition (5.7) is then $\nabla X_k^\ell(\zeta)=0$. The main statement of this section is

THEOREM 6.1. *For any complex parameter ℓ the operators $\mathcal{D}_a$ (5.8), (5.9) form a representation of the Sklyanin algebra (2.7) in the invariant subspace of functions* *of two variables $X = X(\zeta, x)$ such that $\nabla X = 0$. The values of the central elements (2.5) in this representation are*

$$\Omega_0 = 4\theta_1^2((2\ell+1)\eta), \qquad \Omega_1 = 4\theta_1(2\ell\eta)\theta_1(2(\ell+1)\eta). \tag{6.2}$$

First of all let us show that the space of solutions to the equation $\nabla X = 0$ is invariant under action of the operators $\mathcal{D}_a$.

LEMMA 6.2. *The following commutation relations hold:*

$$\nabla \mathcal{D}_a = \mathcal{D}'_a \nabla, \tag{6.3}$$

where $\mathcal{D}'_0 = \mathcal{D}_0$ and

$$\begin{aligned} \mathcal{D}'_\alpha = {} & \frac{\theta_1(2\zeta - 2\ell\eta)}{\theta_1(2\zeta)} T_+ \frac{\theta_{\alpha+1}(2\zeta - 2x)}{\theta_1(2\zeta - 2x)} - \frac{\theta_1(2\zeta + 2\ell\eta)}{\theta_1(2\zeta)} T_- \frac{\theta_{\alpha+1}(2\zeta + 2x)}{\theta_1(2\zeta + 2x)} \\ & - \frac{\theta_1(2x - 2\ell\eta)}{\theta_1(2x)} K_+ \left(\frac{\theta_{\alpha+1}(2\zeta - 2x)}{\theta_1(2\zeta - 2x)} - \frac{\theta_{\alpha+1}(2\zeta + 2x)}{\theta_1(2\zeta + 2x)} \right). \end{aligned} \tag{6.4}$$

COROLLARY 6.1. *The condition $\nabla X = 0$ is invariant under the action of the operators $\mathcal{D}_a$.*

The lemma can be proved by straightforward though quite long computations. Let us show how to reduce them to a reasonable amount. Our strategy is to begin with the case $\ell = 0$. For brevity, we use the notation

$$b_\alpha(\zeta) = \frac{\theta_{\alpha+1}(2\zeta)}{\theta_1(2\zeta)}. \tag{6.5}$$

At $\ell = 0$ the operators ∇, $\mathcal{D}_a$, $\mathcal{D}'_a$ take the form

$$\nabla = T_+ + T_- - K_+ - K_-, \tag{6.6}$$

$$\mathcal{D}_0 = \mathcal{D}'_0 = T_+ + T_-, \tag{6.7}$$

$$\mathcal{D}_\alpha = b_\alpha(\zeta - x)T_+ - b_\alpha(\zeta + x)T_- - (b_\alpha(\zeta - x) - b_\alpha(\zeta + x))K_+, \tag{6.8}$$

$$\mathcal{D}'_\alpha = T_+ b_\alpha(\zeta - x) - T_- b_\alpha(\zeta + x) - K_+(b_\alpha(\zeta - x) - b_\alpha(\zeta + x)). \tag{6.9}$$

It is not too difficult to verify the relations (6.3) by the direct substitution of these operators. Note that in this case the specific form of the function $b_\alpha(\zeta)$ given by (6.5) is irrelevant. This proves the lemma for $\ell = 0$.

The relation $\nabla \mathcal{D}_0 = \mathcal{D}'_0 \nabla$ for general values of ℓ is obvious. To prove the other ones in the case $\ell \neq 0$, we modify the shift operators to the form

$$\widetilde{T}_\pm = c_\ell(\pm\zeta) T_\pm, \qquad \widetilde{K}_\pm = c_\ell(\pm x) K_\pm, \tag{6.10}$$

where

$$c_\ell(\zeta) = \frac{\theta_1(2\zeta - 2\ell\eta)}{\theta_1(2\zeta)}. \tag{6.11}$$

The operators $\widetilde{T}_\pm$, $\widetilde{K}_\pm$ possess the same commutation relations as $T_\pm$, $K_\pm$ *except* for the properties $T_+T_- = T_-T_+ = 1$ (and similarly for $K_\pm$). The latter are replaced by

$$\widetilde{T}_\pm \widetilde{T}_\mp = \rho_\ell(\pm\zeta), \qquad \widetilde{K}_\pm \widetilde{K}_\mp = \rho_\ell(\pm x), \tag{6.12}$$

where

$$\rho_\ell(\zeta) \equiv c_\ell(\zeta) c_\ell(-\zeta - \eta) = \frac{\theta_1(2\zeta - 2\ell\eta)\theta_1(2\zeta + 2(\ell+1)\eta)}{\theta_1(2\zeta)\theta_1(2\zeta + 2\eta)}. \tag{6.13}$$

Clearly, the operators (6.6)–(6.9) with the substitutions $T_\pm \to \widetilde{T}_\pm$, $K_\pm \to \widetilde{K}_\pm$ convert into the corresponding operators for $\ell \neq 0$. Moreover, it is easy to see that the same is true for the operator parts of both sides of equation (6.3). Only the c-number contributions are modified. Collecting them together, we come to the relation

$$\nabla \mathcal{D}_\alpha = \mathcal{D}'_\alpha \nabla + f_\alpha(\zeta, x),$$

where the function f_α is given by

$$f_\alpha = (\rho_\ell(x) - \rho_\ell(\zeta)) b_\alpha(\zeta + x + \eta) + (\rho_\ell(-x) - \rho_\ell(\zeta)) b_\alpha(\zeta - x + \eta) + (\zeta \to -\zeta).$$

It is easy to verify that this function has no singularities and has the same monodromy properties in ζ as $b_\alpha(\zeta)$. (This time the explicit form of the functions b_α is of course crucial.) Therefore $f_\alpha(\zeta, x) = 0$, and the assertion is proved.

Now let us turn to the commutation relations of the Sklyanin algebra. It is more convenient to deal with the relations in the form (2.7). We set

$$\begin{aligned} G_\alpha &\equiv I_{\beta\gamma}\mathcal{D}_\beta\mathcal{D}_\gamma - I_{\gamma\beta}\mathcal{D}_\gamma\mathcal{D}_\beta + (-1)^\alpha I_{\alpha 0}\mathcal{D}_\alpha\mathcal{D}_0, \\ \overline{G}_\alpha &\equiv I_{\gamma\beta}\mathcal{D}_\beta\mathcal{D}_\gamma - I_{\beta\gamma}\mathcal{D}_\gamma\mathcal{D}_\beta + (-1)^\alpha I_{\alpha 0}\mathcal{D}_0\mathcal{D}_\alpha, \end{aligned} \tag{6.14}$$

$$\widetilde{\Omega}_0 = \theta_1^2(\eta)\mathcal{D}_0^2 + \sum_{\alpha=1}^{3}(-1)^{\alpha+1}\theta_{\alpha+1}^2(\eta)\mathcal{D}_\alpha^2, \qquad \widetilde{\Omega}_1 = \sum_{\alpha=1}^{3}(-1)^{\alpha+1} I_{\alpha\alpha}\mathcal{D}_\alpha^2. \tag{6.15}$$

Relations (2.7) and (2.8) imply that if $\mathcal{D}_\alpha$ were the generators of the Sklyanin algebra, then one would have $G_\alpha = \overline{G}_\alpha = 0$, $\widetilde{\Omega}_0 = \Omega_0$, $\widetilde{\Omega}_1 = \Omega_1$.

LEMMA 6.3. *For any $\ell \in \mathbb{C}$ we have*

$$G_\alpha = \lambda_\alpha(\zeta, x) c_\ell(x) K_+ \nabla, \qquad \overline{G}_\alpha = -\lambda_\alpha(\zeta, -x-\eta) c_\ell(x) K_+ \nabla, \tag{6.16}$$

$$\begin{aligned} \widetilde{\Omega}_0 &= 4\theta_1^2((2\ell+1)\eta) + \lambda'(\zeta, x) c_\ell(x) K_+ \nabla, \\ \widetilde{\Omega}_1 &= 4\theta_1(2\ell\eta)\theta_1(2(\ell+1)\eta) + \lambda_0(\zeta, x) c_\ell(x) K_+ \nabla, \end{aligned} \tag{6.17}$$

where

(6.18)

$$\lambda_a(\zeta, x) = -2(-1)^a b_a(\zeta) b_a(x) \frac{\theta_1(2x)\theta_1(2x+2\eta)\theta_1(2\zeta)\theta_1(2\zeta+2\eta)}{\theta_1(2\zeta-2x)\theta_1(2\zeta+2x+2\eta)} + (\zeta \to -\zeta),$$

(6.19)

$$\lambda'(\zeta, x) = -2\,\frac{\theta_1(2x)\theta_1(2x+2\eta)\theta_1^2(2\zeta+\eta)}{\theta_1(2\zeta-2x)\theta_1(2\zeta+2x+2\eta)} + (\zeta \to -\zeta).$$

Again, the computations can be essentially simplified by dealing with the case $\ell = 0$ first and performing the substitutions (6.10) after that. Then only the c-number contributions need additional attention. A few identities used in the computation are given in the Appendix.

It follows from (6.16) that in the space of solutions to the equation $\nabla X = 0$ we have $G_\alpha = \overline{G}_\alpha = 0$, which proves the theorem.

Let us mention the relation

$$\mathcal{D}'_\alpha = -\Xi\, \check{\mathcal{D}}_\alpha^\dagger\, \Xi^{-1}, \tag{6.20}$$

where † means transposition of the operator,

$$\begin{aligned} \check{\mathcal{D}}_\alpha = b_\alpha(\zeta - x) c_\ell(-\zeta-\eta) T_+ &- b_\alpha(\zeta + x) c_\ell(\zeta - \eta) T_- \\ &- (b_\alpha(\zeta - x) - b_\alpha(\zeta + x)) c_\ell(-x-\eta) K_+, \end{aligned}$$

and Ξ is the operator changing the sign of x: $\Xi f(x) = f(-x)$, $\Xi^2 = 1$. Note that the operators $\check{\mathcal{D}}_\alpha$ differ from $\mathcal{D}_\alpha$ by a "gauge" transformation of the form $U(\zeta, x)(\cdots)U^{-1}(\zeta, x)$ with some function $U(\zeta, x)$. This transformation acts as follows:

$$c_\ell(\mp\zeta - \eta)T_\pm \longrightarrow c_\ell(\pm\zeta)T_\pm, \qquad c_\ell(\mp x - \eta)K_\pm \longrightarrow c_\ell(\pm x)K_\pm,$$

and takes $\nabla^\dagger$, $\mathcal{D}_0^\dagger$ to ∇, $\mathcal{D}_0$ respectively. This fact allows us to reduce the computation of $\overline{G}_\alpha$ to that of G_α.

REMARK. The theorem remains true for the operators $\widetilde{\mathcal{D}}_a = \mathcal{D}_a + g_a(\zeta, x)\nabla$, where the functions g_a are arbitrary. Using this remark, one can write out many other equivalent realizations of the Sklyanin algebra in two variables. In particular, it is possible to "symmetrize" the $\mathcal{D}_a$, i.e., to include all four shift operators $T_\pm$, $K_\pm$ in a symmetric fashion in the realization. We do not know whether it is possible to choose the functions g_a in such a way that the right-hand side of (6.16) would vanish.

7. Remarks on the trigonometric limit

The construction of this paper admits several trigonometric degenerations. Let us outline the simplest one, arising when the Sklyanin algebra degenerates into the quantum algebra $U_q(sl(2))$ **[19]**–**[22]**. In this section a *multiplicative* parametrization is more convenient than the *additive* one used in the previous sections. To ease the comparison with the previous formulas, we denote variables such as u, z, etc. by the same letters, having in mind, however, that shifts are now "multiplicative":

$$T_\pm f(\zeta) = f(q^{\pm 1}\zeta), \qquad K_\pm f(x) = f(q^{\pm 1}x).$$

The L-operator reads

$$L(u) = \begin{pmatrix} uA - u^{-1}D & (q - q^{-1})C \\ (q - q^{-1})B & uD - u^{-1}A \end{pmatrix}, \tag{7.1}$$

where A, B, C, D are generators of the $U_q(sl(2))$:

$$\begin{gathered} AB = qBA, \quad BD = qDB, \quad DC = qCD, \quad CA = qAC, \\ AD = 1, \quad BC - CB = \frac{A^2 - D^2}{q - q^{-1}}. \end{gathered} \tag{7.2}$$

The central element is

$$\Omega = \frac{q^{-1}A^2 + qD^2}{(q - q^{-1})^2} + BC. \tag{7.3}$$

The standard realization of this algebra by difference operators has the form

$$\begin{aligned} (AF)(z) &= F(qz), \qquad (DF)(z) = F(q^{-1}z), \\ (BF)(z) &= \frac{z}{q - q^{-1}}(q^\ell F(q^{-1}z) - q^{-\ell}F(qz)), \\ (CF)(z) &= \frac{z^{-1}}{q - q^{-1}}(q^\ell F(qz) - q^{-\ell}F(q^{-1}z)). \end{aligned} \tag{7.4}$$

The invariant subspace of the spin-ℓ representation is spanned by $z^{-\ell}, \ldots, z^\ell$. The analog of (2.13) has the form

$$L(u)F(z) = (q^\ell u - q^{-\ell}u^{-1})\widehat{\Phi}^{-1}(uq^{2\ell}; z) \begin{pmatrix} q^{-\ell}F(qz) & 0 \\ 0 & q^\ell F(q^{-1}z) \end{pmatrix} \widehat{\Phi}(u; z) \tag{7.5}$$

with the matrix

$$\widehat{\Phi}(u;z)=\begin{pmatrix}1 & \bar{q}^{\ell}u^{-1}z^{-1}\\ 1 & q^{-\ell}uz^{-1}\end{pmatrix}.$$

The vacuum vectors are (cf. (3.27))

$$X_k^{\ell}(z,\zeta)=z^{-\ell}\prod_{j=1}^{\ell-k}(z-\zeta u^{-1}q^{\ell-k+1-2j})\prod_{j=1}^{\ell+k}(z-\zeta uq^{\ell+k+1-2j}),\tag{7.6}$$

where ζ parametrizes the (rational) vacuum curve.

Skipping all the intermediate steps (which are similar to the elliptic case), we turn directly to the representation of the $U_q(sl(2))$ obtained in this way. One arrives at the following realization of the quantum algebra by difference operators in the two variables ζ, x:

$$\begin{aligned}A&=q^{-\ell}T_+, \qquad B=\frac{\zeta}{q-q^{-1}}(q^{\ell}xT_- - q^{-\ell}x^{-1}T_+ - (q^{-\ell}x-q^{\ell}x^{-1})K_+),\\ C&=\frac{\zeta^{-1}}{q-q^{-1}}(q^{-\ell}xT_+ - q^{\ell}x^{-1}T_- - (q^{-\ell}x-q^{\ell}x^{-1})K_+), \qquad D=q^{\ell}T_-.\end{aligned}\tag{7.7}$$

This is to be compared with (5.8), (5.9). An easy computation shows that these operators obey the algebra (7.2) *without any additional conditions*! However, in our present set-up there *is*, also, an analog of the operator ∇:

$$\nabla=q^{-\ell}T_+ + q^{\ell}T_- - \frac{q^{-\ell}x-q^{\ell}x^{-1}}{x-x^{-1}}K_+ - \frac{q^{\ell}x-q^{-\ell}x^{-1}}{x-x^{-1}}K_-,\tag{7.8}$$

and we have the condition $\nabla X=0$. Now its role is to restrict the function space to the representation space of spin ℓ. Indeed, computing the central element (2.8), we obtain

$$\Omega=\frac{q^{2\ell+1}+q^{-2\ell-1}}{(q-q^{-1})^2}-\frac{(q^{-\ell}x-q^{\ell}x^{-1})(qx-q^{-1}x^{-1})}{(q-q^{-1})^2}K_+\nabla.\tag{7.9}$$

The invariance of the space of solutions to the equation $\nabla X=0$ is then obvious.

Finally, let us discuss the continuum limit ($q\to 1$) of our formulas. In this limit we arrive at representations of the algebra $sl(2)$. In fact there are several different limits as $q\to 1$. The most interesting one reads

$$\begin{aligned}s_0&=\zeta\partial_\zeta-\ell, \qquad s_+=\frac{1}{2}\zeta((x+x^{-1})(2\ell-\zeta\partial_\zeta)-(x^2-1)\partial_x),\\ s_-&=\frac{1}{2}\zeta^{-1}((x+x^{-1})\zeta\partial_\zeta-(x^2-1)\partial_x),\end{aligned}\tag{7.10}$$

where s_0, $s_\pm$ are the standard generators of $sl(2)$. After the change of variables $\zeta=e^{i\varphi}$, $x=i\cot(\theta/2)$ equations (7.10) acquire the form

$$s_0=-i\partial_\varphi-\ell,\quad s_+=e^{i\varphi}(i\partial_\theta-\cot\theta\partial_\varphi+2i\ell\cot\theta),\quad s_-=e^{-i\varphi}(i\partial_\theta+\cot\theta\partial_\varphi).\tag{7.11}$$

For $\ell=0$ this is the well-known realization of sl_2 by vector fields on the two-dimensional sphere. The representation (7.7) for $\ell=0$ is its q-deformation, with the q-deformed variables being $e^{i\varphi}$ and $\cot(\theta/2)$. Another q-deformed version of (7.11) (for $\ell=0$), where $e^{i\varphi}$ and $e^{i\theta}$ are "discretized", has been suggested in [**23**].

Acknowledgements

We thank S. Khoroshkin and M. Olshanetsky for useful discussions.

Appendix

We use the following definition of the θ-functions:

$$\begin{aligned}
\theta_1(z|\tau) &= \sum_{k\in\mathbb{Z}} \exp\left(\pi i\tau\left(k+\frac{1}{2}\right)^2 + 2\pi i\left(z+\frac{1}{2}\right)\left(k+\frac{1}{2}\right)\right),\\
\theta_2(z|\tau) &= \sum_{k\in\mathbb{Z}} \exp\left(\pi i\tau\left(k+\frac{1}{2}\right)^2 + 2\pi i z\left(k+\frac{1}{2}\right)\right),\\
\theta_3(z|\tau) &= \sum_{k\in\mathbb{Z}} \exp(\pi i\tau k^2 + 2\pi i z k),\\
\theta_4(z|\tau) &= \sum_{k\in\mathbb{Z}} \exp\left(\pi i\tau k^2 + 2\pi i\left(z+\frac{1}{2}\right)k\right).
\end{aligned} \tag{A.1}$$

Throughout the paper we write $\theta(z|\tau) = \theta(z)$, $\theta(z|\tau/2) = \bar\theta(z)$.

The list of identities used in the computations is given below:

$$\begin{aligned}
\bar\theta_4(x)\bar\theta_3(y) + \bar\theta_4(y)\bar\theta_3(x) &= 2\theta_4(x+y)\theta_4(x-y),\\
\bar\theta_4(x)\bar\theta_3(y) - \bar\theta_4(y)\bar\theta_3(x) &= 2\theta_1(x+y)\theta_1(x-y),\\
\bar\theta_3(x)\bar\theta_3(y) + \bar\theta_4(y)\bar\theta_4(x) &= 2\theta_3(x+y)\theta_3(x-y),\\
\bar\theta_3(x)\bar\theta_3(y) - \bar\theta_4(y)\bar\theta_4(x) &= 2\theta_2(x+y)\theta_2(x-y),
\end{aligned} \tag{A.2}$$

$$\begin{aligned}
&\bar\theta_4(z-x+\ell\eta)\theta_1(z+y+x-(k+\ell)\eta)\theta_1(z-y+x+(k-\ell)\eta)\\
&\quad - \bar\theta_4(z+x-\ell\eta)\theta_1(z-y-x+(k+\ell)\eta)\theta_1(z+y-x-(k-\ell)\eta)\\
&\quad = \bar\theta_4(y-k\eta)\theta_1(2z)\theta_1(2x-2\ell\eta)
\end{aligned} \tag{A.3}$$

(and the same with the change $\bar\theta_4 \to \bar\theta_3$),

$$\begin{aligned}
&\theta_{\alpha+1}(2\zeta-u+2x)\theta_1(2\zeta-2x)\theta_1(2\zeta+u)\\
&\quad - \theta_{\alpha+1}(2\zeta+u-2x)\theta_1(2\zeta+2x)\theta_1(2\zeta-u)\\
&\quad = \frac{\theta_1(4\zeta)\theta_{\alpha+1}(2x)}{\theta_{\alpha+1}(2\zeta)}\theta_1(u-2x)\theta_{\alpha+1}(u), \qquad \alpha = 1,2,3,
\end{aligned} \tag{A.4}$$

$$\theta_1(2z)\theta_2(0)\theta_3(0)\theta_4(0) = 2\theta_1(z)\theta_2(z)\theta_3(z)\theta_4(z). \tag{A.5}$$

In the main text we use the notation

$$I_{ab} = \theta_{a+1}(0)\theta_{b+1}(2\eta), \qquad b_\alpha(\zeta) = \frac{\theta_{\alpha+1}(2\zeta)}{\theta_1(2\zeta)},$$

$$c_\ell(\zeta) = \frac{\theta_1(2\zeta-2\ell\eta)}{\theta_1(2\zeta)}, \qquad \rho_\ell(\zeta) = c_\ell(\zeta)c_\ell(-\zeta-\eta).$$

To verify the commutation relations (6.3), (6.16), we need the identities

$$\begin{aligned}
&I_{\gamma\beta}\theta_{\gamma+1}(x)\theta_{\beta+1}(x+2\eta) - I_{\beta\gamma}\theta_{\beta+1}(x)\theta_{\gamma+1}(x+2\eta)\\
&\quad = (-1)^\alpha I_{\alpha 0}\theta_{\alpha+1}(x)\theta_1(x+2\eta),
\end{aligned} \tag{A.6}$$

$$b_\alpha(\zeta-x) - b_\alpha(\zeta+x) = 2\,\frac{\theta_{\alpha+1}(2x)\theta_{\beta+1}(2\zeta)\theta_{\gamma+1}(2\zeta)\theta_1(2x)}{\theta_{\beta+1}(0)\theta_{\gamma+1}(0)\theta_1(2\zeta-2x)\theta_1(2\zeta+2x)}, \tag{A.7}$$

$$\rho_\ell(\zeta) - \rho_\ell(x) = \frac{\theta_1(2\ell\eta)\theta_1(2(\ell+1)\eta)\theta_1(2\zeta-2x)\theta_1(2\zeta+2x+2\eta)}{\theta_1(2x)\theta_1(2x+2\eta)\theta_1(2\zeta)\theta_1(2\zeta+2\eta)}. \tag{A.8}$$

To find the right-hand side of equations (6.17), we need the identities

$$\sum_{\alpha=1}^{3}(-1)^{\alpha}\theta_{\alpha+1}^{2}(\eta)b_{\alpha}(x)b_{\alpha}(x+\eta)=\theta_{1}^{2}(\eta), \tag{A.9}$$

$$\sum_{\alpha=1}^{3}(-1)^{\alpha}\theta_{\alpha+1}(0)\theta_{\alpha+1}(2\eta)b_{\alpha}(x)b_{\alpha}(x+\eta)=0, \tag{A.10}$$

$$\begin{aligned}&\sum_{\alpha=1}^{3}(-1)^{\alpha}\theta_{\alpha+1}(0)\theta_{\alpha+1}(x)\theta_{\alpha+1}(y)\theta_{\alpha+1}(z)\\&\quad=2\theta_{1}\Big(\frac{x+y+z}{2}\Big)\theta_{1}\Big(\frac{x-y+z}{2}\Big)\theta_{1}\Big(\frac{x+y-z}{2}\Big)\theta_{1}\Big(\frac{x-y-z}{2}\Big).\end{aligned} \tag{A.11}$$

References

[1] E. K. Sklyanin, *On some algebraic structures related to the Yang–Baxter equation*, Funktsional. Anal. i Prilozhen. **16** (1982), no. 4, 27–34; English transl. in Functional Anal. Appl. **16**, (1982), no. 4.

[2] ———, *On some algebraic structures related to the Yang–Baxter equation. Representations of the quantum algebra*, Funktsional. Anal. i Prilozhen. **17** (1983), no. 4, 34–48; English transl., Functional Anal. Appl. **17** (1983), no. 4, 273–284.

[3] I. Krichever and A. Zabrodin, *Spin generalization of the Ruijsenaars–Schneider model, non-abelian 2D Toda chain and representations of Sklyanin algebra*, Uspekhi Mat. Nauk **50** (1995), no. 6, 3–56; English transl. in Russian Math. Surveys **50** (1995), no. 6, `hep-th/9505039`.

[4] E. L. Ince, *Further investigations into the periodic Lamé functions*, Proc. Roy. Soc. Edinburgh **60** (1940), 83–99.

[5] D. Mumford, *Algebro-geometric construction of commuting operators and of solutions to the Toda lattice equation, Korteweg-de Vries equation and related non-linear equations*, in: Proceedings Int. Symp. Algebraic Geometry, Kyoto, 1977, Kinokuniya Book Store, Tokyo, 1978, pp. 115–153.

[6] I. M. Krichever, *Algebraic curves and non-linear difference equation*, Uspekhi Mat. Nauk **33** (1978), no. 4, 215–216; English transl., Russian Math. Surveys **33** (1978), no. 4, 255–256.

[7] I. M. Krichever and S. P. Novikov, *Virasoro-type algebras, Riemann surfaces and structures of soliton theory*, Funktsional. Anal. i Prilozhen. **21** (1987), no. 2, 46–63; English transl. in Functional Anal. Appl. **21** (1987), no. 2.

[8] G. Felder and A. Varchenko, *Algebraic Bethe ansatz for the elliptic quantum group $E_{\tau,\eta}(sl_2)$*, Nuclear Phys. B **480** (1996), 485–503; `q-alg/9605024`; *Algebraic integrability of the two-body Ruijsenaars operator*, `q-alg/9610024`; Funktsional. Anal. i Prilozhen. **32** (1998), no. 2, 8–25; English transl. in Functional Anal. Appl. **32** (1998), no. 2.

[9] I. Krichever, O. Babelon, E. Billey, and M. Talon, *Spin generalization of the Calogero–Moser system and the Mamrix KP equation*, Topics in Topology and Mathematical Physics, Amer. Math. Soc. Transl. Ser. 2, Vol. 170, Amer. Math. Soc., Providence, RI, 1995, pp. 83–119.

[10] I. Krichever, O. Lipan, P. Wiegmann, and A. Zabrodin, *Quantum integrable models and discrete classical Hirota equations*, Comm. Math. Phys. **188** (1997), 267–304

[11] I. Krichever, *Baxter's equations and algebraic geometry*, Funktsional. Anal. i Prilozhen. **15** (1981), no. 2, 22–35; English transl. in Functional Anal. Appl. **15** (1981), no. 2.

[12] R. Baxter, *Eight-vertex model in lattice statistics and one-dimensional anisotropic Heisenberg chain. I. Some fundamental eigenvectors*, Ann. Phys. **76** (1973), 1–24; II. *Equivalence to a generalized ice-type lattice model*, Ann. Phys. **76** (1973), 25–47; III. *Eigenvectors of the transfer matrix and hamiltonian*, Ann. Phys. **76** (1973), 48–71.

[13] L. D. Faddeev and L. A. Takhtadzhan, *The quantum method of the inverse problem and the Heisenberg XYZ model*, Uspekhi Mat. Nauk **34** (1979), no. 5, 13–62; English transl. in Russian Math. Surveys **34** (1979), no. 5.

[14] T. Takebe, *Generalized Bethe ansatz with the general spin representation of the Sklyanin algebra*, J. Phys. A **25** (1992), 1071–1083; *Bethe ansatz for higher spin eight vertex models*, J. Phys. A **28** (1995), 6675–6706; q-alg/950427.

[15] S. P. Smith and J. M. Staniszkis, *Irreducible representations of the 4-dimensional Sklyanin algebra at points of infinite order*, J. Algebra **160** (1993), 57–86.

[16] S. M. Sergeev, $Z_N^{\otimes n-1}$ *broken model*, Preprint IHEP-92-7; K. Hasegawa, *On the crossing symmetry of the elliptic solution of the Yang–Baxter equation and a new L-operator for Belavin's solution*, J. Phys. A **26** (1993), 3211–3228; Y. Quano and A. Fujii, *Yang–Baxter equation for* $A_{n-1}^{(1)}$ *broken* $\mathbb{Z}_N$ *models*, Modern Phys. Lett. A **8** (1993), 1585–1597.

[17] I. G. Korepanov, *Vacuum curves of L-operators connected with the six-vertex model*, Algebra i Analiz **6** (1994), no. 2 176–194; English transl., St.-Petersburg Math. J. **6** (1995), 349–364.

[18] A. Gorsky and A. Zabrodin, *Degenerations of Sklyanin algebra and Askey–Wilson polynomials*, J. Phys. A **26** (1993), L635–L639.

[19] P. Kulish and N. Reshetikhin, *Quantum non-linear problem for the sine-Gordon equation and higher representations*, Zap. Nauchn. Sem. Leningrad. Otdel. Mat. Inst. Steklov. (LOMI) **101** (1980), 101–110; English transl. in J. Soviet Math. **23** (1983), no. 4.

[20] V. G. Drinfeld, *Hopf algebras and quantum Yang–Baxter equation*, Dokl. Akad. Nauk SSSR **283** (1985), 1060–1064; English transl. in Soviet Math. Dokl. **32** (1985).

[21] M. Jimbo, *q-Difference analogue of U(g) and the Yang–Baxter equation*, Lett. Math. Phys. **10** (1985), 63–69.

[22] L. D. Faddeev, N. Yu. Reshetikhin, and L. A. Takhtajan, *Quantization of Lie groups and Lie algebras*, Algebra i Analiz **1** (1989), no. 1, 178–206; English transl., Leningrad Math. J. **1** (1990), no. 1, 193–225.

[23] Ya. Granovskiĭ and A. Zhedanov, *Spherical q-functions*, J. Phys. A **26** (1993), 4331–4338.

Columbia University, 2990 Broadway, New York, NY 10027, USA; and Landau Institute for Theoretical Physics, Kosygina str. 2, 117940 Moscow, Russia
E-mail address: krichev@shire.math.columbia.edu

Joint Institute of Chemical Physics, Kosygina str. 4, 117334, Moscow, Russia; and ITEP, 117259, Moscow, Russia
E-mail address: zabrodin@heron.itep.ru

Amer. Math. Soc. Transl.
(2) Vol. **191**, 1999

Hierarchies of Isomonodromic Deformations and Hitchin Systems

A. M. Levin and M. A. Olshanetsky

ABSTRACT. We investigate the classical limit of the Knizhnik–Zamolodchikov–Bernard equations, regarded as a system of nonstationary Schrödinger equations on singular curves, where times are the moduli of curves. It has the form of reduced nonautonomous Hamiltonian systems which include as particular examples the Schlesinger equations, the Painlevé VI equation, and their generalizations. In the general case, they are defined as hierarchies of isomonodromic deformations (HID) with respect to changes in the moduli of underling curves. HID are accompanied by the Whitham hierarchies. The phase space of HID is the space of flat connections of G-bundles with some additional data at the marked points. HID can be derived from some free field theory by Hamiltonian reduction under the action of the gauge symmetries and subsequent factorization with respect to diffeomorphisms of the curve. This approach allows us to define the Lax equations associated with HID and the linear system whose isomonodromic deformations are provided by HID. In addition, it leads to a description of solutions of HID by the projection method. In a certain special limit, HID convert into the Hitchin systems. In particular, for $\mathrm{SL}(N, \mathbb{C})$ bundles over elliptic curves with a marked point, we obtain in this limit the elliptic Calogero N-body system.

1. Introduction

1.1. It has been known for a long time that equations providing isomonodromic deformations possess structures typical for integrable systems, such as the Lax representation [**1, 2**]. In this paper we study isomonodromic deformations in the spirit of Hitchin's approach to integrable systems [**3**]. Hitchin discovered a wide class of Hamiltonian integrable systems on the cotangent bundles to the moduli space of holomorphic vector bundles over Riemann curves. Some important facts about the Hitchin systems became clear only later.

1991 *Mathematics Subject Classification.* Primary 58F07.

The work of the first author was supported in part by grants INTAS 94-4720, Award No. RM1-265 of the Civilian Research & Development Foundation (CRDF) for the Independent States of the Former Soviet Union, and grant 96-15-96455 for support of scientific schools.

The work of the second author was supported in part by grants INTAS 96-538, RFBR-96-02-18046, Award No. RM2-150 of the Civilian Research & Development Foundation (CRDF) for the Independent States of the Former Soviet Union, INTAS, and grant 96-15-96455 for support of scientific schools.

First of all, the Hitchin systems are related to the Knizhnik–Zamolodchikov–Bernard (KZB) equations [**4, 5**] on the critical level. KZB equations on the critical level take the form of second order differential equations, and their solutions are partition functions of the Wess–Zumino–Witten (WZW) theory on the corresponding Riemann curves. It turns out that these operators coincide with quantum second order Hitchin Hamiltonians [**6**]–[**11**]. In the KZB equations for correlators of vertex operators we must include curves with marked points where the vertex operators are located. The KZB Hamiltonians for correlators define very important classes of quantum integrable equations such as the Gaudin equations (for genus zero curves), elliptic Calogero equations (for genus one curves), and their generalizations [**12, 13, 14**]. This means that their classical counterparts are particular cases of the Hitchin systems [**7, 8**].

Hitchin's approach, based on the Hamiltonian reduction of some free field Hamiltonian theory, claims to be universal in describing integrable systems. Essentially, it allows us to present almost exhaustive information, e.g. integrals of motion, Lax pairs, action-angle variables, explicit solutions, via the projective method.

The main goal of this paper is to go beyond the critical level in the classical limit of the KZB equations and repeat the Hitchin program as far as possible. For generic values of the level, the KZB equations have the form of nonstationary Schrödinger equations, where the role of times is played by the coordinates of tangent vectors to the moduli space of curves [**9, 10, 15**]. On the classical level, they correspond to nonautonomous Hamiltonian systems. We shall call them hierarchies of isomonodromic deformations (HID). In this situation the analog of the Hitchin phase space is the moduli space of flat connections $\mathcal{A}$ over a Riemann curve $\Sigma_{g,n}$ of genus g with n marked points. While the flatness is a topological property of bundles, the polarization of the connections $\mathcal{A} = (A, \bar{A})$ depends on the choice of the complex structure on $\Sigma_{g,n}$. We consider a bundle $\mathcal{P}$ over the moduli space $\mathcal{M}_{g,n}$ of curves with flat connections $(A, \bar{A})$ as fibers. The fibers are supplemented by elements of coadjoint orbits $\mathcal{O}_a$ at the marked points x_a. The points of the fibers $(A, \bar{A})$ are analogs of momenta and coordinates, while the base $\mathcal{M}_{g,n}$ serves as a set of "times". There exists a closed degenerate two-form ω on $\mathcal{P}$ which is nondegenerate on the fibers. The connection $\bar{A}$ plays the same role as in the Hitchin construction, while A replaces the Higgs field. Essentially, our construction is local, we work in the neighborhood of some fixed curve $\Sigma_{g,n}$ in $\mathcal{M}_{g,n}$. As we have already mentioned, the coordinates of the tangent vector to $\mathcal{M}_{g,n}$ at $\Sigma_{g,n}$ play the role of times, while the Hitchin times have nothing to do with the moduli space. The Hamiltonians of HID are the same quadratic Hitchin Hamiltonians, but now they are time-dependent.

There is a free parameter κ (the level) in our construction. On the critical level ($\kappa = 0$), after rescaling the times, HID convert into Hitchin systems. As the latter, they can be derived by symplectic reduction from the infinite affine space of the connections and the Beltrami differentials with respect of the gauge action on the connection. In addition, to get to the moduli space, we need the subsequent factorization under the action of the diffeomorphisms of $\Sigma_{g,n}$ that effectively act on the Beltrami differentials only. Apart from the last step, our approach is close to [**16**], where the KZB system is derived as the quantization of a very similar symplectic quotient.[1] Using this derivation, we immediately find the Lax pairs, prove that the equations of motion are consistency conditions of the isomonodromic deformations of the auxiliary linear problem, and, therefore, justify the notion of HID. Moreover, we describe solutions via linear procedures (the projection method).

For genus zero our procedure leads to the Schlesinger equations. This case was discussed earlier [**2**]. We restrict ourselves to the simplest cases in which we consider only simple poles of connections. Therefore, we do not include the Stokes parameters in the phase space. In the rational case this phenomenon was investigated in detail in [**2**]. Isomonodromic deformations on genus one curves were considered in [**20**] and on higher genus curves in [**21**]. Here we consider genus one curves with one marked point and obtain for $SL(2,\mathbb{C})$-bundles a particular family of the Painlevé VI equations. The generalization of this case to arbitrary simple groups leads to the multicomponent Painlevé VI related to these groups. If we introduce several marked points, we come to the elliptic generalization of the Schlesinger equations. In fact, the concrete systems we consider here are the deformations from the critical level of the Hitchin systems described in [**7**].

1.2. Painlevé VI and Calogero equations. A very instructive (but not generic) example of our systems is the Painlevé VI equation. It depends on four free parameters ($PVI_{\alpha,\beta,\gamma,\delta}$) and has the form

$$\begin{aligned}\frac{d^2X}{dt^2} &= \frac{1}{2}\left(\frac{1}{X}+\frac{1}{X-1}+\frac{1}{X-t}\right)\left(\frac{dX}{dt}\right)^2 - \left(\frac{1}{t}+\frac{1}{t-1}+\frac{1}{X-t}\right)\frac{dX}{dt} \\ &+ \frac{X(X-1)(X-t)}{t^2(t-1)^2}\left(\alpha+\beta\frac{t}{X^2}+\gamma\frac{t-1}{(X-1)^2}+\delta\frac{t(t-1)}{(X-t)^2}\right).\end{aligned} \tag{1.1}$$

$PVI_{\alpha,\beta,\gamma,\delta}$ has a lot of different applications (see, for example [**22**]). It is a Hamiltonian system [**23**]. We shall write the symplectic form and the Hamiltonian below in other variables. This equation was first derived by Gambier [**24**], as an equation which has no solutions with mobile singularities. Among the distinguished features of this equation, we are interested in its relationship to the isomonodromic deformations of linear differential equations. This approach was investigated by Fuchs [**25**].

There exists an elliptic form of $PVI_{\alpha,\beta,\gamma,\delta}$ derived by Painlevé himself [**26**]. It was investigated recently by Manin [**27**] in connection with Frobenius varieties. A particular solution of $PVI_{1/8,-1/8,0,1/2}$ defines a potential of the quantum homology of P^2 [**28**]. It sheds light on the relationship of the Painlevé equations with the Hitchin systems, and, thereby, with the KZB equations. We briefly describe this approach.

Let $\wp(u|\tau)$ be the Weierstrass function on the elliptic curve $T^2_\tau = \mathbb{C}/(\mathbb{Z}+\mathbb{Z}\tau)$, and

$$e_i = \wp(T_i/2\,|\,\tau), \qquad (T_0,\dots,T_3) = (0,1,\tau,1+\tau).$$

Consider instead of (X,t) in (1.1) the new variables

$$(u,\tau) \to \left(X = \frac{\wp(u|\tau)-e_1}{e_2-e_1},\ t = \frac{e_3-e_1}{e_2-e_1}\right). \tag{1.2}$$

[1] Quantization of isomonodromic deformations on rational and elliptic curves and their relations to KZB was considered in [**17**]–[**19**].

Then $PVI_{\alpha,\beta,\gamma,\delta}$ takes the form

$$\frac{d^2u}{d\tau^2} = -\partial_u U(u|\tau), \qquad U(u|\tau) = \frac{1}{(2\pi i)^2}\sum_{j=0}^{3}\alpha_j\wp\Big(u+\frac{T_j}{2}\Bigm|\tau\Big), \tag{1.3}$$
$$(\alpha_0,\dots,\alpha_3) = (\alpha,-\beta,\gamma,1/2-\delta).$$

The Hamiltonian form of (1.3) is defined by the standard symplectic form

$$\omega_0 = \delta v\,\delta u, \tag{1.4}$$

and the Hamiltonian

$$H = \frac{v^2}{2} + U(u|\tau). \tag{1.5}$$

The equation of motion (1.3) can be derived from the action S

$$\delta S = v\,\delta u - H\,\delta\tau. \tag{1.6}$$

We can regard (1.4) and (1.5) as nonautonomous Hamiltonian system with time-dependent potential.

To find the symmetries, we consider the two-form

$$\omega = \omega_0 - \delta H\,\delta\tau = \delta v\,\delta u - \delta H\,\delta\tau. \tag{1.7}$$

The semidirect product of $\mathbb{Z}+\mathbb{Z}\tau$ and the modular group acting on the dynamical variables (v,u,τ) are the symmetries of (1.7). We discuss them in detail in Section 7.

Let us introduce the new parameter κ and instead of (1.6) consider

$$\omega = \omega_0 - \frac{1}{\kappa}\,\delta H\,\delta\tau. \tag{1.8}$$

Then (1.5) takes the form

$$\kappa^2\,\frac{d^2u}{d\tau^2} = -\partial_u U(u|\tau). \tag{1.9}$$

This corresponds to the overall rescaling of constants $\alpha_j \to \alpha_j/\kappa^2$. Put $\tau = \tau_0 + \kappa t^H$ and consider the system in the limit as $\kappa \to 0$. We come to the equation

$$\frac{d^2u}{(dt^H)^2} = -\partial_u U(u|\tau_0). \tag{1.10}$$

This is just the rank one elliptic Calogero–Inozemtsev equation [**29**]–[**31**], which we denote $CI_{\alpha,\beta,\gamma,\delta}$. Thus, in this limit we have

$$PVI_{\alpha,\beta,\gamma,\delta} \xrightarrow{\kappa\to 0} CI_{\alpha,\beta,\gamma,\delta}. \tag{1.11}$$

So far we do not know how to manage the general forms of both types of equations for arbitrary values of the constants. Here we consider only the one-parametric family in (1.9) $PVI_{\nu^2/4,-\nu^2/4,\nu^2/4,1/2-\nu^2/4}$

$$\alpha_j = \frac{\nu^2}{4} \qquad \Big(\alpha = \frac{\nu^2}{4},\ \beta = -\frac{\nu^2}{4},\ \gamma = \frac{\nu^2}{4},\ \delta = \frac{1}{2} - \frac{\nu^2}{4}\Big).$$

The potential (1.9) takes the form

$$U(u|\tau) = \frac{1}{(2\pi i)^2}\,\nu^2\wp(2u|\tau). \tag{1.12}$$

We shall prove that (1.4) and (1.5) with the potential (1.12) describe the dynamics of flat connections of SL$(2,\mathbb{C})$ bundles over elliptic curves T_τ with one marked point $\Sigma_{1,1}$. In fact, u lies on the Jacobian of T_τ, the pairs (v,u) are related to the flat connections, and τ defines a point in $\mathcal{M}_{1,1}$. Roughly speaking, the triple (v,u,τ) are the coordinates in the total space of the bundle $\mathcal{P}$ over $\mathcal{M}_{1,1}$, while ω (1.8) is a two-form on $\mathcal{P}$ and ω_0 is the symplectic form on its fibers. This finite-dimensional Hamiltonian system is derived as the quotient of the infinite-dimensional phase space of $(A,\bar{A})$ connections with the Beltrami differentials as times under the action of gauge transforms and diffeomorphisms of $\Sigma_{1,1}$. This approach leads directly to the Lax linear system and allows us to define solutions of the Cauchy problem via the projection procedure. Simultaneously, we describe the auxiliary linear problem whose isomonodromic deformations are governed by this particular family of Painlevé VI. The discrete symmetries of (1.6) are nothing but the remaining gauge symmetries. On the critical level, this equation is just the two-body elliptic Calogero system. The corresponding quantum system is identified with the KZB equation for the one-vertex correlator. In a similar way $PVI_{\nu^2/4,-\nu^2/4,\nu^2/4,1/2-\nu^2/4}$ is the classical limit of the KZB for $\kappa \neq 0$. This example will be analyzed in detail in Section 7.

Solutions of $PVI_{\nu^2/4,-\nu^2/4,\nu^2/4,1/2-\nu^2/4}$ in the particular case $\nu^2 = 1/2$ were found by Hitchin [**32**] in connection with his investigations of anti-self-dual Einstein metrics. They are written out in terms of theta functions. But for generic values of ν, the solutions are genuine Painlevé transcendents [**33**].

1.3. Whitham equations. The previous example is not entirely exhaustive. A special phenomenon occurs if $\dim \mathcal{M}_{g,n} > 1$. We have as many Hamiltonians as $\dim \mathcal{M}_{g,n}$, since each Hamiltonian H_s is attached to the tangent vector t_s to $\mathcal{M}_{g,n}$ at some fixed point $\Sigma_{g,n}$. There are consistency conditions of the equations of motion for the nonautonomous multi-time Hamiltonian systems. They take the form of the classical zero-curvature conditions for the connections

$$\partial_{t_s} + H_s, \tag{1.13}$$

where the commutator is replaced by the Poisson brackets. The latter is the inverse tensor to the symplectic form defined on the fibers. The flatness conditions is the so-called Whitham hierarchy (WH) defined in the same terms by Krichever [**34**], though the starting point of his construction is different and is based on the averaging procedure. Let S be the action for HID ($\delta S = \delta^{-1}\omega$ as in (1.6)). The τ-function of WH is

$$\log \tau = S.$$

It allows us to find the Hamiltonians. For HID in the rational case the τ-functions were investigated in [**2**]. For Painlevé I–VI they were considered in [**35**].

The quantum analog of the Whitham equations was exploited in [**15, 9**] to construct the KZB connections. The quantum version of (1.13) is an operator acting in the space V of sections of the holomorphic line bundle over moduli of flat connections. The quantum WH implies the flatness of the bundle of projective spaces $\mathbf{P}V$ over the Teichmüller space $\mathcal{T}_{g,n}$.

1.4. Outline. We start in Section 2 with a general setup about flat bundles over singular curves. In Section 3 we present the basic facts about the abstract nonautonomous Hamiltonian equations. They are defined by a degenerate closed two-form on the extended phase space; this space is a bundle over the space of times. The Whitham equations occur on this stage. We recall the symplectic reduction technique, which is applicable in the degenerate case as well. Section 4 contains our main result: the derivation of the HID corresponding to the flat bundles over Riemann curves. In Section 5 we discuss two limits: the level zero limit of HID to the Hitchin systems and the classical limit of general KZB equations to HID. Then we analyze in detail two feasible examples: flat bundles over rational curves, leading to the Schlesinger equations (Section 6), and flat bundles over elliptic curves, responsible for the Painlevé VI type equations and the elliptic Schlesinger equations (Section 7). In particular, we apply the projection method to construct perturbative solutions of the Schlesinger equations. In the Appendix we summarize some results about elliptic functions which are used in Section 7.

Acknowledgments. We would like to thank our colleagues V. Fock, A. Losev, and N. Nekrasov for fruitful and illuminating discussions. We are grateful to the Max-Planck-Institut für Mathematik in Bonn, where this work was started, for hospitality. We are grateful to Yu. Manin: his lectures and discussions with him concerning PVI stimulated our interests to these problems.

2. Flat bundles over singular curves

We shall describe the general setup more or less naively. In this section we do not consider the mechanism of symplectic reduction in detail; this will be done in Section 4. We shall consider three cases:

1) smooth proper (compact) algebraic curves;
2) smooth proper algebraic curves with punctures;
3) proper algebraic curves with nodal singularities (double points).

Let G be a semisimple group and V an exact representation of G, e.g. $G = \mathrm{SL}(N, \mathbb{C})$, where V is an N-dimensional vector space with volume form.

2.1. Smooth curves. Let S be a smooth oriented compact surface of genus g. Let us consider the moduli space $\mathrm{FBun}_{S,G}$ of flat V-bundles on S. This space can be regarded as the quotient of the space FConn of flat C^∞ connections on the trivial V-bundle by the action of the gauge group $\mathcal{G}$ of G-valued C^∞-functions on S, or as the result of a Hamiltonian reduction of the space Conn of all connections by the action of the gauge group. The symplectic form on the space Conn is the form

$$\omega = \int_S \langle \delta\mathcal{A}, \delta\mathcal{A}\rangle, \tag{2.1}$$

where $\langle\,\cdot\,,\cdot\,\rangle$ denotes the Killing form and $\delta\mathcal{A}$ is a $\mathrm{Lie}(G)$-valued one-form on S. So, $\langle \delta\mathcal{A}, \delta\mathcal{A}\rangle$ is a two-form, and the integral is well defined. The dual space to the gauge Lie algebra is the space of $\mathrm{Lie}(G)$-valued two-forms on S, and the momentum map assigns to any connection $\mathcal{A}$ its curvature $F_{\mathcal{A}} = d\mathcal{A} + \frac{1}{2}[\mathcal{A}, \mathcal{A}]$. Hence, the preimage of 0 under the momentum map is the space of flat connections.

A complex structure Σ on S is a differential operator $\bar\partial\colon \Omega^0_{C^\infty} \to \Omega^1_{C^\infty}$; its kernel is the space of holomorphic functions. For any connection, we can consider its $\bar\partial$-part; it defines a holomorphic bundle. A section is holomorphic if it belongs to the kernel of this operator. Let us recall that:

1. A connection on the holomorphic bundle $\mathcal{E}$ is an operator of type $\partial + A$.
2. A connection is holomorphic if this operator commutes with the complex structure operator on the bundle.
3. A connection is flat if its square vanishes; for curves this condition is empty.
4. The space of all holomorphic connections on $\mathcal{E}$ is an infinite-dimensional affine space, and the gauge group $\mathcal{G}$ acts on it by affine transformations.

5. The quotient of the set of all stable holomorphic connections by the action of all gauge transformations is the moduli space $\mathrm{Bun}_{\Sigma,G}$ of holomorphic vector bundles on Σ.

Let b be a point in $\mathrm{Bun}_{\Sigma,G}$ and $\mathcal{V}_b$ the corresponding V-bundle on Σ. Denote by ad_b the bundle of endomorphisms of $\mathcal{V}_b$ regarded as a bundle of Lie algebras, and by Flat_b the space of flat holomorphic connections on the holomorphic bundle $\mathcal{V}_b$.

Consider the natural projection map from $\mathrm{FBun}_{S,G}$ onto the moduli space $\mathrm{Bun}_{\Sigma,G}$. The fiber of the projection $\mathrm{FBun}_{S,G} \to \mathrm{Bun}_{\Sigma,G}$ over a point b is naturally isomorphic to Flat_b. These fibers are Lagrangian with respect to the form ω (2.1), since any complex structure on S defines a polarization on $\mathrm{FBun}_{S,G}$.

The tangent space to $\mathrm{Bun}_{\Sigma,G}$ at the point b is canonically isomorphic to the first cohomology group $H^1(\Sigma, \mathrm{ad}_b)$. The space Flat_b of holomorphic connections on $\mathcal{V}_b$ is an affine space over the vector space $H^0(\Sigma, \mathrm{ad}_b \otimes \Omega^1)$ of holomorphic ad_b-valued one-forms, since a difference between any connections is an ad_b-valued differential form. The vector spaces $H^1(\Sigma, \mathrm{ad}_b)$ and $H^0(\Sigma, \mathrm{ad}_b \otimes \Omega^1)$ are dual.

The Dolbeault representation of the cohomology classes corresponds to the description based on Hamiltonian reduction. Indeed, let us decompose $\mathcal{A}$ into its $(1,0)$ and $(0,1)$ parts: $\mathcal{A} = A + \bar{A}$. From the zero curvature condition, we see that the $(1,0)$-part δA must be $\bar{A}$-holomorphic: $\bar{\partial}_A \delta A \equiv (\bar{\partial} + \bar{A})\,\delta A = 0$, and $\delta \bar{A}$ is defined up to the infinitesimal gauge transformations $\delta \bar{A} \to \delta \bar{A} + \bar{\partial}_A h$.

2.2. Curves with punctures. For any two isomorphic representations V and V' of G, denote by $\mathrm{Isom}(V, V')$ the space of G-isomorphisms between them; this space is a principal homogeneous space over G. For a curve Σ_n with n marked points x_j, we replace the moduli space $\mathrm{Bun}_{\Sigma,G}$ by the moduli space $\mathrm{Bun}_{\Sigma,\mathbf{x},G}$ ($\mathbf{x} = (x_1, \dots, x_n)$) of holomorphic V-bundles $\mathcal{V}$ with the trivializations $g_j \colon \mathcal{V}_{x_j} \to V$ of fibers at the marked points. We have the natural "forgetting" projection $\pi\colon \mathrm{Bun}_{\Sigma,\mathbf{x},G} \to \mathrm{Bun}_{\Sigma,G}$. The fiber of this projection is the product $\prod \mathrm{Isom}(\mathcal{V}_{x_j}, V)$ of the spaces of isomorphisms between fibers of the bundle at marked points and V. The projection π can be treated as a reduction by the action of n copies of the group G acting transitively on the fibers

$$\mathrm{Bun}_{\Sigma,G} = \mathrm{Bun}_{\Sigma,\mathbf{x},G} \Big/ \prod G.$$

The bundle of endomorphisms of this data is the bundle $\mathrm{ad}_b(-\mathbf{x})$ of endomorphisms vanishing at the marked points, since nonvanishing endomorphisms change trivializations. Hence, the tangent space to $\mathrm{Bun}_{\Sigma,\mathbf{x},G}$ is isomorphic to $H^1(\Sigma, \mathrm{ad}_b(-\mathbf{x}))$. The dual space to this space is $H^0(\Sigma, \mathrm{ad}_b(\mathbf{x}) \otimes \Omega^1)$. Consequently, in order to obtain a symplectic variety, we must replace the affine space Flat_b of flat holomorphic connections by the affine space $\mathrm{Flat}_b(\log)$ of flat connections with logarithmic singularities. As a result, we get the moduli space $\mathrm{FBun}_{\Sigma,\mathbf{x},G}$ of triples: (holomorphic bundle, trivializations at marked points, holomorphic connection with logarithmic singularities). Again, we have the Lagrangian projection $\mathrm{FBun}_{\Sigma,\mathbf{x},G} \to \mathrm{Bun}_{\Sigma,\mathbf{x},G}$. The fiber of this projection is an affine space over $H^0(\Sigma, \mathrm{ad}_b(\mathbf{x}) \otimes \Omega^1)$.

The product of n copies of G acts on $\mathrm{FBun}_{X,\mathbf{x},G}$ by changing the trivialization of fibers, and this action is Hamiltonian. We can consider the Hamiltonian reduction with respect to this action. As a result, for any collection of G-orbits of the adjoint action, we get the moduli space $\mathrm{FBun}_{\Sigma,\mathbf{x},G,\{\mathcal{O}_j\}}$ of pairs (holomorphic bundle, holomorphic connection with logarithmic singularities having residues in the orbits $\mathcal{O}_j$).

According to the Dolbeault theorem, the tangent space $H^1(\Sigma, \mathrm{ad}_b(-\mathbf{x}))$ can be realized as the space of $(0,1)$ ad_b-valued forms *holomorphically* vanishing at the marked points modulo $\bar{\partial}_A$-coboundaries of ad_b-valued functions *holomorphically* vanishing at marked points. A function (or one-form) is said to be *holomorphically vanishing* at the point x if it has the asymptotics $(z-x)O(1)$ at this point.

Second description: the tangent space $H^1(\Sigma, \mathrm{ad}_b(-\mathbf{x}))$ is isomorphic to the space of all $(0,1)$ ad_b-valued forms modulo $\bar{\partial}_A$-coboundaries of ad_b-valued functions vanishing at marked points. Indeed, we have a natural embedding of the space of holomorphically vanishing forms to the space of all forms, and of the space of holomorphically vanishing functions to the space of vanishing functions. It defines a map ϕ from the space from the first description to the space from the second description. We can assume that $\bar{A}$ vanishes in small neighborhoods of marked points. Then $\bar{\partial}_{\bar{A}}$ equals $\bar{\partial}$ in these neighborhoods. Let h be a function vanishing, but not holomorphically vanishing, at the marked points,

$$h \equiv \sum_{i=1}^{\infty} a_i \overline{(z-x_j)}^{\,i} + (z-x_j)o(1).$$

Then

$$\bar{\partial} h \sim \left(\sum_{i=1}^{\infty} i a_i \overline{(z-x_j)}^{\,(i-1)} + (z-x_j)o(1) \right) d\bar{z}$$

is not holomorphically vanishing. So, the map ϕ is injective. On the other hand, for any form ν with the asymptotics

$$\nu \sim \left(\sum_{i=0}^{\infty} a_i \overline{(z-x_j)}^{\,i} + (z-x_j)o(1) \right) d\bar{z}$$

at the points x_j, we can choose some function ϕ such that $\nu - \bar{\partial}_A \phi$ has the asymptotics $(z-x_j)o(1)$, since for any collection of asymptotics

$$\sum_{i=0}^{\infty} \frac{1}{i+1} a_i \overline{(z-x_j)}^{\,i+1} + (z-x_j)o(1)$$

a function with such asymptotic exists. So, the map ϕ is surjective.

The third equivalent description of this space is defined as the cokernel of the map

$$\Omega^{0,0}_{C^\infty}(\mathrm{ad}_b) \to \Omega^{0,1}_{C^\infty}(\mathrm{ad}_b) + \sum \mathrm{ad}_b|_{x_i}, \qquad h \to (\bar{\partial}_A h, h(x_i)).$$

Indeed, for any element $(\delta\bar{A}, \{a_i\})$ in the cokernel, we can choose its representative with vanishing second part $\{a_i\}$. This choice is unique up to $\bar{\partial}_A h$ for functions h vanishing at the marked points. This description is adapted to the forgetting map π. The local part $\{a_i\}$ corresponds to the tangent space of the fiber, and the global part $\delta\bar{A}$ corresponds to the tangent space of the base.

By "integration" of the action of the gauge Lie algebra ad_b yielding the action of the gauge Lie group $\mathcal{G}$, we get two descriptions of the moduli space.

A. The moduli space $\mathrm{Bun}_{\Sigma,\mathbf{x},G}$ is the quotient of the space of $\bar{\partial}$-connections $\bar{A}$ on the trivial V-bundle by the action of the reduced gauge group of G-valued functions, whose values at the marked points are equal to the neutral element 1 of the group G. The isomorphisms between the fibers and V are identity maps.

B. The moduli space $\mathrm{Bun}_{\Sigma,\mathbf{x},G}$ is the quotient of the space of pairs ($\bar{\partial}$-connection $\overline{A}$ on the trivial V-bundle, collection g_j of elements of G) by the action of the gauge group $\mathcal{G}$. The isomorphisms of the fibers onto V are the elements g_i. The isomorphism of these descriptions can be proved by the following consideration. For any collection $(\overline{A}, g_j)$, we can choose a gauge equivalent collection with $g_j = 1$, and such collections are equivalent up to the action of the reduced gauge group.

In what follows we assume the second description of $\mathrm{Bun}_{X\Sigma,\mathbf{x},G}$.

Let us consider the space of data ($\bar{\partial}$-connection $\overline{A}$ on the trivial V-bundle, a collection g_j of elements of G, ∂-connection A with logarithmic singularities at marked points). This space is symplectic with the form $\mathrm{FBun}_{X,\{x_j\},G}$ equal to the Hamiltonian reduction of this space by action of the gauge group $\mathcal{G}$.

It is worthwhile to note that this construction is essentially based on the complex structure on the surface, since $\overline{A}$ is nonsingular and A has singularities at the marked points. We can consider the moduli space of flat bundles over the noncompact surface $S\setminus\bigcup x_j$ as the G-representations of the fundamental group of $S\setminus\bigcup x_j$. According to Deligne, for any complex structure on S, any stable representation can be realized by connections with logarithmic singularities on a holomorphic bundle. Hence, any connection $\mathcal{A}$ is gauge equivalent to a connection with regular $\bar{\partial}$-part, but the corresponding gauge transformation has singularities at marked points a priori. This fact makes such an approach very complicated.

2.3. Curves with double points. A curve with a double point (a nodal singularity) can be treated as the "limit" of nonsingular curves as some circle is pinched to a point. If this circle is homologous to zero, then the resulting curve is the union of two intersecting smooth curves, and the sum of the genera of these curves is equal to the genus of the nonsingular curves. If this circle is homologically nontrivial, then the singular curve is a smooth curve with two different points glued together. The genus of this curve is less than the genus of the initial curves by one. The normalization of the singular curve ("disgluing" the singularity) is a smooth curve (not connected, in general) with marked points.

Let us fix some notation. Denote by $\mathcal{M}_g$ the moduli space of smooth curves of genus g, and denote by $\mathcal{M}_{g,n}$ the moduli space of smooth curves of genus g with n different marked points ($\mathcal{M}_g = \mathcal{M}_{g,0}$). Let $\overline{\mathcal{M}}_{g,n}$ be the Deligne-Mumford compactification of $\mathcal{M}_{g,n}$. Then the compactification divisor $D_\infty = \overline{\mathcal{M}}_g \setminus \mathcal{M}_g$ is the union of components $D_\infty^{g_1,g_2}$, $g_1 + g_2 = g$, and D_∞^{g-1}. These components are covered by $\overline{\mathcal{M}}_{g_1,1} \times \overline{\mathcal{M}}_{g_2,1}$ and $\overline{\mathcal{M}}_{g-1,2}$ respectively.

Consider the moduli space $\mathrm{Bun}_{g,\Sigma}$ of pairs (smooth curve of genus g, G-bundle on it). Evidently, this space is fibered over $\mathcal{M}_g$ with $\mathrm{Bun}_{X,G}$ as fibers. Unfortunately, we do not know the canonical compactification $\overline{\mathrm{Bun}_{g,X}}$, which is fibered over $\overline{\mathcal{M}}_g$. Let us assume that such a compactification does exist. Then the open part of the fiber over the singular curve may be described as the moduli space of the following data (G-bundle over normalized curve, isomorphisms of fibers over the "glued" points).

Denote by Σ^0 a singular curve with nodal points $y_1, \dots, y_n$, and denote by Σ its normalization; let x_a and x_{n+a} be the preimages of y_a under the normalization map. Then the moduli space $\mathrm{Bun}_{\Sigma^0,G}$ corresponding to Σ^0 is the moduli space of (holomorphic V-bundle on Σ, isomorphisms between fibers of this bundle over x_a and x_{n+a}). This space is the quotient of $\mathrm{Bun}_{\Sigma,\mathbf{x},G}$ by the action of n copies of the group G, with jth G acting on $\mathrm{Isom}(\mathcal{V}_{x_a}, V)$ and $\mathrm{Isom}(\mathcal{V}_{x_{n+a}}, V)$ from the right:

$$\{h_a\}_{1\leqslant a\leqslant n}\colon \{g_a, g_{a+n}\}_{1\leqslant a\leqslant n} \to \{g_a h_a, g_{a+n} h_a\}_{1\leqslant a\leqslant n}.$$

The corresponding symplectic variety $\mathrm{FBun}_{\Sigma^0,G}$ is the result of the symplectic reduction of $\mathrm{FBun}_{\Sigma,\mathbf{x},G}$ by the action described above with zero level of the momentum map. This space is the moduli space of data (holomorphic V-bundle on X, isomorphisms between fibers of this bundle over x_a and x_{n+a}, connection with logarithmic singularities such that the residues at x_a and x_{n+a} are opposite). In the last part of these data, we use the isomorphisms of fibers from the second part of the data.

3. Hamiltonian formalism

We consider nonautonomous Hamiltonian systems. For this type of system it is customary in classical mechanics to deal with a degenerate symplectic form on the extended phase space, which includes, besides the usual coordinates and momenta, the space of times and the corresponding Hamiltonians as the conjugate variables. Then the Hamiltonian equations of motion are defined as variations of dynamical variables along the null leaves of this symplectic form.

3.1. Equations of motion. Let $\mathcal{R}$ be an (infinite-dimensional) phase space endowed with a nondegenerate symplectic structure. For simplicity, we take it in the canonical form

$$\omega_0 = (\delta\mathbf{v}, \delta\mathbf{u}), \qquad \mathbf{v} = (v_1, \dots, v_i, \dots),\ \mathbf{u} = (u_1, \dots, u_i, \dots).$$

Consider the space of "times" $\mathcal{N} = \{\mathbf{t} = (t_1, \dots, t_a, \dots)\}$ and the corresponding dual Hamiltonians $(H^1, \dots, H^a, \dots)$ on $\mathcal{R}$ depending on times as well. We consider the extended phase space, which is a bundle $\mathcal{P}$ over $\mathcal{N}$ with fibers $\mathcal{R}$. Introduce a symplectic form on $\mathcal{P}$,

$$\omega = \omega_0 - \sum_a \delta H^a\, \delta t_a = (\delta\mathbf{v}, \delta\mathbf{u}) - \sum_a \delta H^a \delta t_a. \tag{3.1}$$

This form is closed but degenerate on $\mathcal{P}$. The vector fields

$$\mathcal{V}^a = \sum_i (A_i^a \partial_{v_i} + B_i^a \partial_{u_i}) + \partial_{t_a}$$

lie in the kernel of ω if and only if

$$\begin{aligned} &A_i^a + \frac{\partial H^a}{\partial u_i} = 0, \qquad -B_i^a + \frac{\partial H^a}{\partial v_i} = 0, \\ &-A_i^a \frac{\partial H^b}{\partial v_i} - B_i^a \frac{\partial H^b}{\partial u_i} - \frac{\partial H^a}{\partial t_b} + \frac{\partial H^b}{\partial t_a} = 0. \end{aligned} \tag{3.2}$$

Then the vector field $\mathcal{V}^a \in \operatorname{Ker}\omega$ takes the form

$$\mathcal{V}^a = -\frac{\partial H^a}{\partial u_i}\partial_{v_i} + \frac{\partial H^a}{\partial v_i}\partial_{u_i} + \partial_{t_a}.$$

It can be checked immediately that $[\mathcal{V}^a, \mathcal{V}^b] = [\partial_{t_b}, \partial_{t_a}]$. For any functions $f(V, U, t)$ on $\mathcal{P}$, its evolution is defined as

$$\frac{df(\mathbf{v}, \mathbf{u}, \mathbf{t})}{dt_a} = \mathcal{V}^a f(\mathbf{v}, \mathbf{u}, \mathbf{t}), \qquad \mathcal{V}^a \in \operatorname{Ker}\omega. \tag{3.3}$$

It follows from (3.2) that the Hamiltonians satisfy the classical zero curvature conditions (the *generalized Whitham hierarchy*)

$$\frac{dH^a}{dt_b} - \frac{dH^b}{dt_a} + \{H^b, H^a\}_{\omega_0} = 0, \qquad a, b = 1, \dots . \tag{3.4}$$

Evidently, for time-independent Hamiltonians, (3.3) and (3.4) give the standard approach. Define the action S of the system as $\delta^{-1}\omega = \delta S$, where

$$\delta S = (\mathbf{v}, \delta\mathbf{u}) - \sum_a H_a\,\delta t_a, \tag{3.5}$$

and the τ*-function*,

$$\delta \log \tau = \delta S. \tag{3.6}$$

The equations of motion can be written in Hamilton–Jacobi form,

$$\frac{\partial S}{\partial t_a} = -H_a\left(\frac{\delta S}{\delta \mathbf{u}}, \mathbf{u}, \mathbf{t}\right). \tag{3.7}$$

3.2. Moment map. The symplectic reduction can be applied to the nonautonomous system with symmetries in the standard way. We briefly describe this approach. Let $\mathcal{G}$ be the symmetry group of the system. This means that for any $\epsilon \in \mathrm{Lie}(\mathcal{G})$ there exists a vector field acting on $\mathcal{P}$ such that the Lie derivative $\mathcal{L}_\epsilon$ annihilates ω (see (3.1)),

$$\mathcal{L}_\epsilon \omega = (\delta j_\epsilon + j_\epsilon \delta)\omega = 0.$$

This vector field is said to be a *Hamiltonian vector field* with respect to ω. Since ω is closed, locally we can write $j_\epsilon \omega = \delta F_\epsilon$. If $j_\epsilon\omega$ is exact, then j_ϵ is a *strictly Hamiltonian vector field.* The function $F_\epsilon = F_\epsilon(\mathbf{v}, \mathbf{u}, \mathbf{t})$ is a linear function on $\mathrm{Lie}(\mathcal{G})$,

$$F_\epsilon = \langle \epsilon, \mathcal{J}(\mathbf{v}, \mathbf{u}, \mathbf{t})\rangle,$$

and thereby defines the *moment map*

$$\mathcal{J} : \mathcal{P} \to \mathrm{Lie}^*(\mathcal{G}).$$

Assume that we impose the moment constraint

$$\mathcal{J}(\mathbf{v}, \mathbf{u}, \mathbf{t}) = 0. \tag{3.8}$$

The quotient

$$\mathcal{P}^{\mathrm{red}} = \mathcal{P}//\mathcal{G} := \mathcal{J}^{-1}(0)/\mathcal{G}$$

is a symplectic space with the symplectic form ω^{red} obtained as the reduction of the original form ω from (3.1). It is defined by a two-step procedure.

i) Fixing the gauge. In other words, we must define a surface (in the phase space $\mathcal{P}$) transversal to the orbits of the gauge group $\mathcal{G}$.
ii) Solving the moment constraint equations (3.8) for the dynamical variables, restricted to the gauge fixing surface.

The main difference from autonomous Hamiltonian systems is that we require gauge invariance for the whole degenerate symplectic form (3.1) only, and do not consider the form ω_0 and the Hamiltonians separately.

4. Symplectic reduction and factorization

4.1. Space of "times" $\mathcal{N}'$. Let $\Sigma_{g,n}$ be a Riemann curve of genus g with n marked points. Let us fix the complex structure on $\Sigma_{g,n}$ by defining local coordinates $(z,\bar{z})$ in open charts covering $\Sigma_{g,n}$. Assume that the marked points $(x_1,\dots,x_n)$ are in general position; i.e., there exists a set of their neighborhoods $(\mathcal{U}_1,\dots,\mathcal{U}_n)$, such that $\mathcal{U}_a\cap\mathcal{U}_b=\varnothing$ for $a\neq b$.

The deformations of the basic complex structure are determined by the Beltrami differentials μ, which are smooth $(-1,1)$ differentials on $\Sigma_{g,n}$. We identify this set with the space of times $\mathcal{N}'$. The Beltrami differentials can be defined in the following way. Consider a chiral smooth transformation of $\Sigma_{g,n}$, which in some local map can be represented as

$$w=z-\epsilon(z,\bar{z}),\qquad \bar{w}=\bar{z}. \tag{4.1}$$

Up to the conformal factor $1-\partial\epsilon(z,\bar{z})$, the corresponding one-form dw is equal to

$$dw=dz-\mu\,d\bar{z},\qquad \mu=\frac{\bar{\partial}\epsilon(z,\bar{z})}{1-\partial\epsilon(z,\bar{z})}. \tag{4.2}$$

The Beltrami differential defines a new holomorphic structure, the deformed antiholomorphic operator annihilating dw, while the antiholomorphic structure is unchanged:

$$\partial_{\bar{w}}=\bar{\partial}+\mu\partial,\qquad \partial_w=\partial.$$

In addition, assume that μ vanishes at the marked points:

$$\mu(z,\bar{z})|_{x_a}=0. \tag{4.3}$$

In our construction we consider small deformations of the basic complex structure $(z,\bar{z})$. This allows us to replace (4.2) by

$$\mu=\bar{\partial}\epsilon(z,\bar{z}). \tag{4.4}$$

Nevertheless, in some cases we shall use the exact representation (4.2) as well.

4.2. Fibers $\mathcal{R}'$. Let $\mathcal{E}$ be a principal stable G bundle over a Riemann curve $\Sigma_{g,n}$. Assume that G is a complex simple Lie group. The phase space $\mathcal{R}'$ is constructed from the following data:

i) The affine space $\{\mathcal{A}\}$ of a Lie(G)-valued connection on $\mathcal{E}$. Its componentwise description is as follows:
   a) the C^∞ connection $\{\bar{A}\}$ corresponding to the $d\bar{w}=d\bar{z}$ component of $\mathcal{A}$;
   b) the dual to the previous space, the space $\{A\}$ of dw components of the connection $\mathcal{A}$; a component A can have simple poles at the marked points; moreover, assume that $A\mu$ is a well-defined function.

ii) The cotangent bundles $T^*G_a=\{(p_a,g_a),\,p_a\in \text{Lie}^*(G_a),\,g_a\in G_a\}$ $(a=1,\dots,n)$ at the points $(x_1,\dots,x_n)$.

There is the canonical symplectic form on $\mathcal{R}'$

$$\omega_0=\int_\Sigma\langle\delta A,\delta\bar{A}\rangle+2\pi i\sum_{a=1}^{n}\delta\langle p_a,g_a^{-1}\delta g_a\rangle, \tag{4.5}$$

where $\langle\ ,\ \rangle$ denotes the Killing form on Lie(G).

According to Section 2, we can regard the elements g_a as trivializations of fibers at the marked points and p_a as the residues of holomorphic connections at the same points:

$$g_a \in \mathrm{Isom}(\mathcal{V}_a, V) \quad (V \sim \mathrm{Lie}(G)), \qquad p_a = \mathrm{res}|_{x_a} A.$$

Thus, the symplectic form (4.5) is the generalization of (2.1) to singular curves.

4.3. Extended phase space $\mathcal{P}'$. According to the general prescription, the bundle $\mathcal{P}'$ over $\mathcal{N}'$ with $\mathcal{R}'$ as the fiber plays the role of the extended phase space. Consider the degenerate form on $\mathcal{P}'$

$$\omega = \omega_0 - \frac{1}{\kappa}\int_\Sigma \langle \delta A, A\rangle\, \delta\mu. \tag{4.6}$$

Thus, we deal with the infinite set of Hamiltonians $\langle A, A\rangle(z,\bar z)$, parametrized by points of $\Sigma_{g,n}$ and by the corresponding set of times $\mu(z,\bar z)$. We apply the formalism presented in the previous section to these systems.

4.4. Equations of motion. They take the form (see (3.3), (4.6))

$$\frac{\delta A}{\delta\mu}(z,\bar z)=0, \quad \kappa\frac{\delta \bar A}{\delta\mu}(z,\bar z)=A(z,\bar z), \quad \frac{\delta p_b}{\delta\mu}=0, \quad \frac{\delta g_b}{\delta\mu}=0. \tag{4.7}$$

We can introduce the modified connection

$$\bar A' = \bar A - \frac{1}{\kappa}\,\mu A. \tag{4.8}$$

In terms of it, (4.6) takes the canonical form

$$\omega = \int_\Sigma \langle \delta A, \delta \bar A'\rangle + \sum_{a=1}^{n} \delta\langle p_a, g_a^{-1}\delta g_a\rangle, \tag{4.9}$$

and the equations of motion (4.7) become trivial:

$$\frac{\delta A}{\delta\mu}(z,\bar z)=0, \qquad \frac{\delta \bar A'}{\delta\mu}(z,\bar z)=0. \tag{4.10}$$

In other words, as we pointed out in Section 2, if we forget about the complex structures of curves, then the bundle become trivial: $\mathcal{P}' \sim \mathcal{R}'\times\mathcal{N}'$.

4.5. Symmetries. The form ω (4.6) (or (4.9)) is invariant with respect to the action of the group $\mathcal{G}_0$ of diffeomorphisms of $\Sigma_{g,n}$, which are trivial in the neighborhoods $\mathcal{U}_a$ of marked points:

$$\mathcal{G}_0 = \{z\to N(z,\bar z),\ \bar z\to \bar N(z,\bar z),\ N(z,\bar z)=z+o(|z-x_a|),\ z\in\mathcal{U}_a\}. \tag{4.11}$$

In particular, its action on the Beltrami differentials takes the form of the Möbius transform

$$\mu \to \frac{\partial z/\partial\bar N + \mu\,\partial\bar z/\partial\bar N}{\partial z/\partial N + \mu\,\partial\bar z/\partial N}. \tag{4.12}$$

Another infinite gauge symmetry of the form (4.6) (or (4.9)) is given by the group

$$\mathcal{G}_1 = \{f(z,\bar z)\in C^\infty(\Sigma_g, G)\},$$

which acts on the dynamical fields as follows:

$$
\begin{gathered}
A+\kappa\partial \to f(A+\kappa\partial)f^{-1}, \qquad \bar{A}+\bar{\partial}+\mu\partial \to f(\bar{A}+\bar{\partial}+\mu\partial)f^{-1}, \\
(\bar{A}'+\bar{\partial} \to f(\bar{A}'+\bar{\partial})f^{-1}), \\
p_a \to f_a p_a f_a^{-1}, \quad g_a \to g_a f_a^{-1} \quad \Big(f_a = \lim_{z\to x_a} f(z,\bar{z})\Big), \quad \mu \to \mu.
\end{gathered} \tag{4.13}
$$

In other words, the gauge action of $\mathcal{G}_1$ does not touch the base $\mathcal{N}'$ and transforms only the fibers $\mathcal{R}'$. The whole gauge group is the semidirect product

$$
\mathcal{G}_1 \oslash \mathcal{G}_0. \tag{4.14}
$$

There is an additional finite-dimensional symmetry group $\mathcal{G}_2$, which commutes with (4.14). It acts only on the singular curves in the fibers at the marked points. It is the remnant of $\mathcal{G}_1$ on the desingular curves (see Section 2):

$$
\mathcal{G}_2 = \bigotimes_{a=1}^{n} G_a, \quad p_a \to p_a, \quad g_a \to h_a g_a, \quad h_a \in G_a \quad (a=1,\dots,n). \tag{4.15}
$$

This action commutes with (4.14).

4.6. Symplectic reduction with respect to $\mathcal{G}_1$. The infinitesimal action (4.13) of $\mathcal{G}_1$ generates the vector field $\epsilon_1 \in \mathcal{A}^{(0)}(\Sigma_{g,n}, \mathrm{Lie}(G))$:

$$
\begin{gathered}
j_{\epsilon_1} A = -\kappa\partial\epsilon_1 + [\epsilon_1, A], \qquad j_{\epsilon_1}\bar{A}' = -\bar{\partial}\epsilon_1 + [\epsilon_1, \bar{A}'], \\
j_{\epsilon_1} p_a = [\epsilon_1(z_a,\bar{z}_a), p_a], \qquad j_{\epsilon_1} g_a = -g_a\epsilon_1(z,\bar{z}_a), \qquad j_{\epsilon_1}\mu = 0.
\end{gathered}
$$

Let

$$
F_{A,\bar{A}'} = \bar{\partial}A - \kappa\partial\bar{A}' + [\bar{A}, A] = (\bar{\partial}+\partial\mu)A - \kappa\partial\bar{A} + [\bar{A}, A].
$$

Note that the action of the operator $(\bar{\partial}+\partial\mu)$ on the connection A is well defined.

The corresponding moment map

$$
\mathcal{J}_1 \colon \mathcal{P} \to \mathrm{Lie}^*(\mathcal{G}_1)
$$

takes the form

$$
\mathcal{J}_1 = -F_{A,\bar{A}'}(z,\bar{z}) + 2\pi i \sum_{a=1}^{n} \delta^2(x_a^0) p_a.
$$

Let $\mathcal{J}_1 = 0$. In other words, the moment constraints equation has the form

$$
F_{A,\bar{A}'}(z,\bar{z}) = 2\pi i \sum_{a=1}^{n} \delta^2(x_a^0) p_a. \tag{4.16}
$$

This means that we deal with the flat connection everywhere on $\Sigma_{g,n}$ except at the marked points. The holonomies of $(A, \bar{A})$ around the marked points are conjugate to $\exp 2\pi i p_a$.

Let $(L, \bar{L})$ be the gauge transformed connections

$$
\bar{A} = f\bar{L}f^{-1} + f(\bar{\partial}+\mu\partial)f^{-1}, \tag{4.17}
$$

$$
A = fLf^{-1} + \kappa f\partial f^{-1}. \tag{4.18}
$$

Then (4.16) takes the form

$$
(\bar{\partial}+\partial\mu)L - \kappa\partial\bar{L} + [\bar{L}, L] = 2\pi i \sum_{a=1}^{n} \delta^2(x_a^0) p_a. \tag{4.19}
$$

By fixing the gauge we can choose $\bar{A}$ in a such way that $\partial\bar{L} = 0$. In fact, the antiholomorphity of $f^{-1}(\bar{\partial} + \mu\partial)f + f^{-1}\bar{A}f$ amounts to the classical equations of motion for the Wess–Zumino–Witten functional $S_{WZW}(f, \bar{A})$ in the external field $\bar{A}$, which has extremal points. Then instead of (4.19) we have

$$(\bar{\partial} + \partial\mu)L + [\bar{L}, L] = \sum_{a=1}^{n} \delta^2(x_a)p_a. \tag{4.20}$$

We can rewrite it as

$$\partial_{\bar{w}}L + [\bar{L}, L] = 2\pi i \sum_{a=1}^{n} \delta^2(x_a)p_a. \tag{4.21}$$

By choosing $\bar{L}$ we fix the gauge somehow in the generic case. The last form of the moment constraint (4.21) coincides with that for the Hitchin systems [**7**], which allows us to apply the known solutions.

4.7. Symplectic reduction with respect to $\mathcal{G}_2$. The gauge transforms $h_a \in G_a$ at the points x_a act on T^*G_a. In the case of punctures, this allows us to fix p_a on some coadjoint orbit $p_a = g_a p_a^{(0)} g_a^{-1}$ and obtain the symplectic quotient $\mathcal{O}_a = T^*G_a//G_a$. In fact, the moment corresponding to this action is

$$\mu_a = g_a p_a g_a^{-1} \in \mathrm{Lie}^*(G).$$

Let $\mu_a = J_a$ be some fixed point in $\mathrm{Lie}^*(G)$. Then by gauge fixing we can choose g_a up to the stabilizator of J_a, and

$$p_a = g_a^{-1} J_a g_a \in \mathcal{O}_a.$$

Thus, in (4.19) or (4.20) the elements p_a belong to $\mathcal{O}_a$. The symplectic form on $\mathcal{O}_a$ keeps the same form as on T^*G_a.

Consider the double point case (nodal singularity). Let x_1 and x_2 be the preimages of the nodal point y under the normalization map. Then the symplectic form on the normalization of the singular curve

$$\omega = \delta\langle p_1, g_1^{-1}\delta g_1\rangle + \delta\langle p_2, g_2^{-1}\delta g_2\rangle$$

generates the moment $\mu = g_1 p_1 g_1^{-1} + g_2 p_2 g_2^{-1}$. Put $\mu = 0$. Then $p_1 = -\tilde{g}_2 p_2 \tilde{g}_2^{-1}$ and $\tilde{g}_2 = g_1^{-1} g_2$. The pair $(p_2, \tilde{g}_2)$ is an arbitrary element of T^*G. Therefore, in this case $T^*G_1 \oplus T^*G_2//G = T^*G$.

In what follows we shall concentrate on the case of curves with punctures (without double points). Let $\mathcal{I}_{g,n}$ be the equivalence classes of connections $(A, \bar{A})$ with respect to the gauge action (4.17), (4.18), i.e., the moduli space of stable flat G-bundles over $\Sigma_{g,n}$. It is a smooth finite-dimensional space. If we fix the conjugacy classes of holonomies $(L, \bar{L})$ around the marked points, $\mathcal{I}_{g,n}$ becomes a symplectic manifold. It is extended here by the coadjoint orbits $\mathcal{O}_a$ at the marked points x_a $(a = 1, \dots, n)$ in a consistent way (see (4.19)). Fixing the gauge, we come to the symplectic quotient

$$\mathcal{R} \subset \mathcal{I}_{g,n} \times \prod_{a=1}^{n} \mathcal{O}_a, \qquad \mathcal{R} = \mathcal{R}'//(\mathcal{G}_1 \oplus \mathcal{G}_2) = \mathcal{J}_1^{-1}(0)/(\mathcal{G}_1 \oplus \mathcal{G}_2).$$

It has the dimension

$$\dim(\mathcal{R}) = \begin{cases} \dim(\sum_{a=1}^{n} \mathcal{O}_a//G), & g = 0, \\ 2\,\mathrm{rank}\, G + \dim(\sum_{a=1}^{n} \mathcal{O}_a//H), & g = 1 \\ (2g-2)\dim G + \dim(\sum_{a=1}^{n} \mathcal{O}_a), & g \geqslant 2, \end{cases}$$

where H is the Cartan subgroup $H \subset G$, and $\mathcal{O}_a//G$ and $\mathcal{O}_a//H$ are the symplectic quotients of the symplectic spaces $\mathcal{O}_a$ under the actions of the automorphisms of the bundles in the rational and the elliptic cases respectively. The connections $(L, \bar{L})$ in addition to $\mathbf{p} = (p_1, \dots, p_n)$ depend on a finite even number $2r$ of free parameters:

$$(\mathbf{v}, \mathbf{u}), \quad \mathbf{v} = (v_1, \dots, v_r), \quad \mathbf{u} = (u_1, \dots, u_r), \qquad r = \begin{cases} 0, & g = 0, \\ \mathrm{rank}\, G, & g = 1, \\ (g-1)\dim G, & g \geqslant 2. \end{cases}$$

$\mathcal{R}$ is a symplectic manifold with the nondegenerate symplectic form obtained as the reduction of (4.5):

$$\omega_0 = \int_\Sigma \langle \delta L, \delta \bar{L} \rangle + 2\pi i \sum_{a=1}^{n} \delta \langle p_a, g_a \delta g_a^{-1} \rangle. \tag{4.22}$$

At this stage we come to the bundle $\mathcal{P}''$ with finite-dimensional fiber $\mathcal{R}$ over the infinite-dimensional base $\mathcal{N}'$ with the symplectic form

$$\omega = \int_\Sigma \langle \delta L, \delta \bar{L} \rangle + 2\pi i \sum_{a=1}^{n} \delta \langle p_a, g_a^{-1} \delta g_a \rangle - \kappa \int_\Sigma \langle L, \delta L \rangle\, \delta\mu. \tag{4.23}$$

4.8. Factorization with respect to the diffeomorphisms $\mathcal{G}_0$. We can use the invariance of ω with respect to $\mathcal{G}_0$ and reduce $\mathcal{N}'$ to the finite-dimensional space $\mathcal{N}$, which is isomorphic to the moduli space $\mathcal{M}_{g,n}$. Let ϵ_0 be the vector field generated by the diffeomorphisms (4.11). Consider the action of the Lie derivative $\mathcal{L}_{\epsilon_0}$ on $\mathcal{A} = (A, \bar{A})$,

$$\mathcal{L}_{\epsilon_0}\mathcal{A} = dj_{\epsilon_0}\mathcal{A} + j_{\epsilon_0} d\mathcal{A} = d_{\mathcal{A}}(j_{\epsilon_0}\mathcal{A}) + j_{\epsilon_0} F_{\mathcal{A}} \qquad (F_{\mathcal{A}} = d\mathcal{A} + \tfrac{1}{2}[\mathcal{A}, \mathcal{A}]).$$

Thus for the flat connections $F_{\mathcal{A}} = 0$, the action of diffeomorphisms from $\mathcal{G}_0$ on the connection field is generated by the gauge transforms $j_{\epsilon_0}\mathcal{A} \in \mathcal{G}_1$. But we already have performed the symplectic reduction with respect to $\mathcal{G}_1$. Therefore, j_{ϵ_0} belongs to the kernel of ω (4.23), and we can push it down to the quotient space $\mathcal{P}''/\mathcal{G}_0$. Since $\mathcal{G}_0$ acts only on $\mathcal{N}'$, this can be done by fixing the dependence of μ on the coordinates in the Teichmüller space $\mathcal{T}_{g,n}$. According to (4.4), we can represent μ as

$$\mu = \sum_{s=1}^{l} \mu_s, \quad \mu_s = t_s \mu_s^0, \quad l = \dim \mathcal{T}_{g,n}, \quad \mu_s^0 = \bar{\partial} n_s. \tag{4.24}$$

The Beltrami differential (4.24) defines the tangent vector $\mathbf{t} = (t_1, \dots, t_l)$ to the Teichmüller space $\mathcal{T}_{g,n}$ at any fixed point of $\mathcal{T}_{g,n}$.

We specify the dependence of μ on the positions of the marked points in the following way. Let $\mathcal{U}_a' \supset \mathcal{U}_a$ be two neighborhoods of the marked point x_a such that

$\mathcal{U}'_a \cap \mathcal{U}'_b = \varnothing$ for $a \neq b$. Let $\chi_a(z, \bar{z})$ be the smooth function

$$\chi_a(z,\bar{z}) = \begin{cases} 1, & z \in \mathcal{U}_a, \\ 0, & z \in \Sigma_g \setminus \mathcal{U}'_a. \end{cases} \tag{4.25}$$

Introduce times related to the positions of the marked points: $t_a = x_a - x_a^0$. Then

$$\mu_a^0 = \bar{\partial} n_a(z,\bar{z}), \qquad n_a(z,\bar{z}) = (1 + c_a(z - x_a^0))\chi_a(z,\bar{z}). \tag{4.26}$$

In other words, $n_a(z, \bar{z})$ defines a local vector field deforming the complex coordinates only in $\mathcal{U}'_a$, namely

$$w = z - t_a n_a(z, \bar{z}).$$

The action of $\mathcal{G}_0$ on the phase space $\mathcal{P}''$ reduces the infinite-dimensional component $\mathcal{N}'$ to $\mathcal{T}_{g,n}$. After the reduction, we come to a bundle with base $\mathcal{T}_{g,n}$. Substituting

$$\delta\mu = \sum_{s=1}^{l} \mu_s^0 \, \delta t_s \qquad (\partial_s = \partial_{t_s}) \tag{4.27}$$

into (4.23), we obtain

$$\omega = \omega_0(\mathbf{v}, \mathbf{u}, \mathbf{p}, \mathbf{t}) - \frac{1}{\kappa} \sum_{s=1}^{l} \delta H_s(\mathbf{v}, \mathbf{u}, \mathbf{p}, \mathbf{t}) \, \delta t_s, \tag{4.28}$$

where ω_0 is defined by (4.22), and the functions H_s are the Hamiltonians

$$H_s = \int_{\Sigma} \langle L, L \rangle \, \partial_s \mu. \tag{4.29}$$

In fact, we still have a remaining discrete symmetry, since ω is invariant under the mapping class group $\pi_0(\mathcal{G}_0)$. Finally, we come to the moduli space $\mathcal{M}_{g,n} = \mathcal{T}_{g,n}/\pi_0(\mathcal{G}_0)$.

Summarizing, we have defined the extended phase space (the bundle $\mathcal{P}$) as the result of symplectic reduction with respect to the $\mathcal{G}_1 \oplus \mathcal{G}_2$ action and subsequent factorization under the $\mathcal{G}_0$ action. We can write symbolically

$$\mathcal{P} = (\mathcal{P}''//\mathcal{G}_1 \oplus \mathcal{G}_2)/\mathcal{G}_0.$$

This bundle is endowed with the symplectic form (4.28).

4.9. The hierarchies of isomonodromic deformations. The equations of motion can be extracted from the symplectic form (4.28), as described in Section 3 (see (3.3)). They will be referred as the *hierarchies of isomonodromic deformations* (HID). This terminology will be justified later. In local coordinates, (3.3) takes the form

$$\kappa \partial_s \mathbf{v} = \{H_s, \mathbf{v}\}_{\omega_0}, \quad \kappa \partial_s \mathbf{u} = \{H_s, \mathbf{u}\}_{\omega_0}, \quad \kappa \partial_s \mathbf{p} = \{H_s, \mathbf{p}\}_{\omega_0}. \tag{4.30}$$

The Poisson bracket $\{\,\cdot\,,\cdot\,\}_{\omega_0}$ is the inverse tensor to ω_0. We also have the *Whitham hierarchy* (3.4) accompanying (4.30). It follows from (4.22) that the Hamiltonians H_s (4.29) commute, $\{H_r, H_s\}_{\omega_0} = 0$. Therefore,

$$\partial_s H_r - \partial_r H_s = 0, \tag{4.31}$$

and there exists a one-form on $\mathcal{M}_{g,n}$ defining the *tau function* of the HID:

$$\delta \log \tau = -\frac{1}{\kappa} \sum H_s \, dt_s. \tag{4.32}$$

The following three statements are valid for the hierarchy of isomonodromic deformations (4.30):

PROPOSITION 4.1. *The flatness condition* (4.19) *and the HID* (4.30) *are equivalent to the consistent system of linear equations*

$$(\kappa\partial + L)\Psi = 0, \tag{4.33}$$

$$(\kappa\partial_s + M_s)\Psi = 0 \qquad (s = 1, \dots, l = \dim \mathcal{M}_{g,n}), \tag{4.34}$$

$$(\bar\partial + \mu\partial + \bar L)\Psi = 0, \tag{4.35}$$

where M_s is a solution to the linear equation

$$\partial_{\bar w} M_s - [M_s, \bar L] = \kappa\partial_s \bar L - L\mu^0. \tag{4.36}$$

PROPOSITION 4.2. *The linear conditions* (4.34) *provide isomonodromic deformations of the linear system* (4.33), (4.35) *with respect to the change of "times" on* $\mathcal{M}_{g,n}$.

Therefore, the HID (4.30) are monodromy-preserving conditions for the linear system (4.33), (4.35).

The presence of the derivative with respect to the spectral parameter $w \in \Sigma_{g,n}$ in the linear equation (4.33) is a distinguishing feature of the isomonodromy-preserving equations. It hampers the application of the inverse scattering method to this type of systems. This means solving the Riemann–Hilbert problem, which amounts to reconstructing the pair $(L, \bar L)$ from the monodromy data. In [**36**] this technique was applied to calculate the asymptotics of solutions for bundles over rational curves. But in the general case, we have an explicit form (in a certain sense) of solutions:

PROPOSITION 4.3 (projection method). *The solution of the Cauchy problem for* (4.30) *with initial data* $\mathbf{v}^0$, $\mathbf{u}^0$, $\mathbf{p}^0$ *at the time* $\mathbf{t} = \mathbf{t}^0$ *is defined in terms of the elements* L^0, $\bar L^0$ *as the gauge transform*

$$\bar L(\mathbf{t}) = f^{-1}(L^0(\mu(\mathbf{t}) - \mu(\mathbf{t}^0)) + \bar L^0)f + f^{-1}(\bar\partial + \mu(\mathbf{t})\partial)f, \tag{4.37}$$

$$L(\mathbf{t}) = f^{-1}(\partial + L^0)f, \qquad \mathbf{p}(\mathbf{t}) = f^{-1}(\mathbf{p}^0)f, \tag{4.38}$$

where $f = f(z, \bar z)$ is a smooth G-valued functions on $\Sigma_{g,n}$ fixing the gauge.

This means that solutions of HID are gauge transformations of free motion in the upstairs system. Equations (4.37), (4.38) look like the dressing transform of free motion. To find solutions, one must know the gauge transform f from the upstairs system to a fixed gauge. For example, in case of genus zero, we consider the holomorphic solutions Ψ. Therefore, f should kill the $\bar L$-operator in (4.37) (see below).

PROOFS. To prove the first statement, represent A as (4.18). The first equation in (4.7), $\partial_s A = 0$, means that

$$\kappa\partial_s L - \kappa\partial M_s + [M_s, L] = 0. \tag{4.39}$$

Then (4.39) is the consistency condition for the linear system (4.33), (4.34). At this stage M_s is defined as $M_s = \kappa f^{-1}\partial_s f$. To find the linear equation (4.36) defining M_s, we must substitute the gauge-transformed form of $\bar A$ (4.17) together with (4.18) into the equation of motion, expressed as $\kappa\partial_s\bar{A} = A\partial_s\mu$. This equation is the same as

$$\kappa(\partial_s\mu\partial + \partial_s\bar{L}) - \partial_{\bar{w}}M_s + [M_s, \bar{L}] = \partial_s\mu(\kappa\partial + L).$$

The latter equation coincides with (4.36). Finally,

$$\begin{aligned}
&\text{compatibility of (4.33), (4.34)} \iff \text{Lax equations (4.39)},\\
&\text{compatibility of (4.33), (4.35)} \iff \text{flatness of (4.19)},\\
&\text{compatibility of (4.34), (4.35)} \iff \text{equation (4.36)}.
\end{aligned}$$

This proves the first statement.

To prove the second statement, note that (4.33) and (4.35) are equivalent to

$$(4.40) \qquad (\kappa\partial + A)\Psi^f = 0, \quad (\bar{\partial} + \mu\partial + \bar{A})\Psi^f = 0 \qquad (\Psi^f = f^{-1}\Psi).$$

Due to the equations of motion ($\partial_s A = 0$, $\partial_s\bar{A} = (1/\kappa)A\partial_s\mu$), the monodromies of this system are independent of the moduli, i.e., $\partial_s\Psi^f = 0$. The monodromies of the reduced system (4.33), (4.35) are conjugate to the monodromies of system (4.40). Thus, we come to the second statement.

To derive the expressions for the dynamical variables in the projection method, we lift the initial data L^0, $\bar{L}^0$, $p_1^0, \ldots, p_n^0$ from the reduced phase space $\mathcal{R}$ at the point t^0 to $\mathcal{R}'$ by the trivial gauge transform. Due to the equations of motion (4.7), the evolution in $\mathcal{R}$ is trivial:

$$A(t) = L^0, \qquad \bar{A} = \kappa L^0(\mu(t) - \mu(t^0)) + (\bar{L}^0),$$

and can be pulled back to the reduced phase space $\mathcal{R}$ by the gauge transform to the fixed gauge. This procedure is reflected in the projection method formulas (4.37) and (4.38). □

5. Remarks about Hitchin systems and KZB equations

5.1. Scaling limit. Consider our system in the limit as $\kappa \to 0$. We shall prove that in this limit we obtain the Hitchin systems, which live on the cotangent bundles to the moduli space of holomorphic G-bundles over $\Sigma_{g,n}$. The value $\kappa = 0$ is called *critical* and appears to be singular (see (4.6), (4.28)). To get around this, we rescale the times, putting

$$(5.1) \qquad \mathbf{t} = \mathbf{T} + \kappa\mathbf{t}^H,$$

where $\mathbf{t}^H$ are the fast (Hitchin) times and $\mathbf{T}$ are the slow times. Therefore,

$$\delta\mu(\mathbf{t}) = \kappa\sum_s \mu_s^0\delta t_s^H \qquad (\mu_s^0 = \bar{\partial}n_s).$$

After this rescaling, the forms (4.6), (4.28) become regular. The rescaling procedure means that we blow up a neighborhood of the fixed point (5.1) in $\mathcal{M}_{g,n}$ and the whole dynamics of the Hitchin systems is developed in this neighborhood.[2] This gauge fixing is defined by the complex coordinates

$$(5.2) \qquad w_0 = z - \sum_s T_s n_s(z, \bar{z}), \qquad \bar{w}_0 = \bar{z}.$$

---

[2]We are grateful to A. Losev for elucidating this point.

Instead of (4.5) and (4.6), we now have

$$\omega = \int_\Sigma \langle \delta\varphi, \delta\bar{A}\rangle + 2\pi i \sum_{a=1}^{n} \delta\langle p_a, g_a^{-1}\delta g_a\rangle - \sum_s \int_\Sigma \langle \delta\varphi, \varphi\rangle \partial_s \mu^0 \delta t^H, \tag{5.3}$$

where φ is the Higgs field. It is the one-form $\varphi \in \Omega^{(1,0)}(\Sigma_{g,n}, \mathrm{Lie}(G))$ obtained in the limit as $\kappa \to 0$ of the connection A (see (4.13)).

An important point is that the Hamiltonians now become time-independent. The form (5.3) is the starting point in the derivation of the Hitchin systems from the symplectic reduction [**3, 7**]. Essentially, it is the same procedure as described above. Namely, we obtain the same moment constraint (4.20) and the same gauge fixing (4.17). But now we are at a fixed point $\mu(\mathbf{t}^0)$ of the moduli space $\mathcal{M}_{g,n}$ and do not need the factorization under the action of the diffeomorphisms and, therefore, need not worry about the modular properties of solutions $(L, \bar{L})$ of the moment constraint (4.20). This is the only difference between the solutions $(L, \bar{L})$ and the quadratic Hamiltonians H_s in the Hitchin systems as compared to those in the HID. The symplectic reduction allows us to identify the phase space $\mathcal{R}$ with the cotangent bundle to the moduli of holomorphic stable bundles.

Propositions 4.1 and 4.3 are valid for Hitchin systems in a slightly modified form. According to (5.1), we put $L(\mathbf{t}) = L^{(0)} + O(\kappa)$, $L^{(0)} = L(\mathbf{T})$. We define $M_s^{(0)}$ and $\bar{L}^{(0)}$ in a similar way.

PROPOSITION 5.1. *There exists a consistent system of linear equations*

$$(\lambda + L^{(0)})Y = 0, \qquad \lambda \in \mathbb{C}, \tag{5.4}$$

$$(\partial_s + M_s^{(0)})Y = 0, \quad \partial_s = \frac{\partial}{\partial t_s^H} \qquad (s = 1, \dots, l = \dim \mathcal{M}_{g,n}), \tag{5.5}$$

$$(\partial_{\bar{w}_0} + \bar{L}^{(0)})Y = 0, \tag{5.6}$$

where $M_s^{(0)}$ *is a solution to the linear equation*

$$\bar{\partial} M_s^{(0)} - [M_s^{(0)}, \bar{L}^{(0)}] = \partial_s \bar{L}^{(0)} - L^{(0)} \mu_s^0. \tag{5.7}$$

Here we have

compatibility of (5.4), (5.5) $\Longleftrightarrow$ Lax equation $\partial_{T_s} L^{(0)} = [L^{(0)}, M_s^{(0)}]$,

compatibility of (5.4), (5.6) $\Longleftrightarrow$ moment equation (4.20),

compatibility of (5.5), (5.6) $\Longleftrightarrow$ equation (5.7).

To derive these equations in the general case from (4.33), (4.34), and (4.35), we use the WKB approximation

$$\Psi = \Phi \exp\left(\frac{\mathcal{S}^{(0)}}{\kappa} + \mathcal{S}^{(1)}\right), \tag{5.8}$$

where Φ is a group valued function. There are no terms of order κ^{-1} if

$$\frac{\partial}{\partial \bar{w}_0} \mathcal{S}^{(0)} = 0, \qquad \frac{\partial}{\partial t_s^H} \mathcal{S}^{(0)} = 0.$$

In other words, it follows from the definition of the fixed point in the moduli of complex structures (5.2) that

$$\mathcal{S}^{(0)} = \mathcal{S}^{(0)}\left(T_1, \dots, T_l \,\middle|\, z - \sum_s T_s n_s(z, \bar{z})\right). \tag{5.9}$$

We also put

$$\mathcal{S}^{(1)} = \partial\mathcal{S}^{(0)} \sum_s t_s^H n_s(z, \bar{z}). \tag{5.10}$$

In the quasiclassical approximation, we set

$$\partial\mathcal{S}^{(0)} = \lambda. \tag{5.11}$$

Then (4.33) takes the form (5.4). Substituting this in (4.34), (5.8) and taking into account (5.10), we obtain

$$\partial_{t_s^H}\Phi^{(0)} + \frac{d\mathcal{S}^{(0)}}{dT_s} + \partial_{t_s^H}\mathcal{S}^{(1)}\Phi^{(0)} + M_s^{(0)}\Phi^{(0)} = 0.$$

From (5.10), $\partial_{t_s^H}\mathcal{S}^{(1)} = \partial\mathcal{S}^{(0)} n_s(z, \bar{z})$. But (see (5.9))

$$\frac{d\mathcal{S}^{(0)}}{dT_s} = \partial_{T_s}\mathcal{S}^{(0)} - \partial\mathcal{S}^{(0)} n_s(z, \bar{z})$$

and we come to the equation

$$\left(\partial_{t_s^H} + \frac{d}{dT_s}\mathcal{S}^{(0)} + M_s^{(0)}\right)\Phi^{(0)} = 0.$$

The comparison of this equation to (5.5) suggests the form of Y:

$$Y = \Phi^0 \exp\left\{\sum_s t_s^H \frac{d}{dT_s}\mathcal{S}^{(0)}\right\}. \tag{5.12}$$

It can be directly checked that the function Ψ from (5.8) satisfies the last linear equation (4.35) up to the first order in κ. In the zero order, it yields

$$(\partial_{\bar{w}_0} + \bar{L}^{(0)})\Phi^{(0)} = 0.$$

Thus, it is consistent with the form of Y (5.12). A detailed analysis of the perturbation in the rational case was undertaken in [**37**].

Equation (5.4) allows us to introduce the fixed spectral curve

$$\mathcal{C}: \quad \det(\lambda \cdot \mathrm{Id} + L) = 0, \qquad \mathcal{C} \in \mathbf{P}(T^*\Sigma \oplus 1),$$

where λ is the coordinate in the cotangent space. The Hitchin phase space $\mathcal{R}$ has a "spectral" description, namely the bundle

$$\pi\colon \mathcal{R} \to \mathcal{M}_C$$

over the moduli $\mathcal{M}_C$ of spectral curves with abelian varieties as generic fibers. The map π acts from the pair $(A, \bar{A}) \sim (L, \bar{L})$ (L is now the Higgs field $L \in \Omega^{(1,0)}(\Sigma_{g,n}, \mathrm{Lie}^*(G))$) to the set of coefficients of the characteristic polynomial $\det(\lambda \cdot \mathrm{Id} + L)$. The one-form $\theta = \lambda\, dw^0$, after integration over the corresponding cycles in $\mathcal{C}$, gives rise to the action variables. The angle variables can be also extracted from the spectral curve. All together this defines the symplectic structure on the Hitchin phase space in the spectral picture. This original Hitchin construction works in the singular case as well [**7**]. A symplectic structure of this type connected with hyperelliptic curves was introduced originally in soliton theory by Novikov and Veselov [**38**]. In terms of 4d gauge theories, the form θ is the Seiberg–Witten differential. It follows from (5.11) that $\theta = d\mathcal{S}^0$, and along with (5.10) it determines a first order approximation to the solutions of the linear form (4.33), (4.34), (4.35) of the HID.

It is possible to define the dynamics of the spectral curve beyond the critical level as it was done for the Painlevé equations in [**39**]. It follows from the Lax representation (4.39) that

$$\partial_s \log\det(\lambda\cdot\mathrm{Id}+L)=\mathrm{tr}\,\partial_w M_s(\lambda\cdot\mathrm{Id}+L)^{-1}.$$

Note that this equation defines the motion of $\mathcal{C}$ only within the subset $\mathcal{M}_{g,n}\subset\mathcal{M}_C$.

When L and thereby M can be found explicitly, the simplified form of (4.20) allows us to apply the inverse scattering method to find solutions of the Hitchin hierarchy, in the same way as this was done for the $\mathrm{SL}(N,\mathbb{C})$ holomorphic bundles over $\Sigma_{1,1}$ corresponding to the elliptic Calogero system [**40**]. We present an alternative way of describing the solutions:

PROPOSITION 5.2 (projection method).

$$\bar L(t_s)=f^{-1}(L^0(t_s-t_s^0)\partial_s\mu^0+\bar L^0)f+f^{-1}\bar\partial f,$$
$$L(t)=f^{-1}L^0 f,\qquad p_a(t)=f^{-1}(p_a^0)f.$$

The degenerate version of these expressions has been known for a long time [**41**].

5.2. About the KZB equations. The Hitchin systems are the classical limit of the KZB equations on the critical level [**7, 10**]. The latter have the form of Schrödinger equations obtained as the result of geometric quantization of the moduli of flat G-bundles [**16, 15**]. The conformal blocks of the WZW theory on $\Sigma_{g,n}$ with vertex operators at marked points are the ground state wave functions $\widehat{H}_s F=0$, $s=1,\dots,l$. The classical limit means that one replaces operators by their symbols, and generators of finite-dimensional representations in the vertex operators by the corresponding elements of the coadjoint orbits.

Generically, for the quantum level $\kappa^{\text{quant}}\neq 0$, the KZB equations can be written in the form of nonstationary Schrödinger equations [**15, 10**]:

$$(\kappa^{\text{quant}}\partial_s+\widehat{H}_s)F=0.$$

To pass to the classical limit in this equation, we replace the conformal block by its quasiclassical expression

$$F=\exp S\hbar^{-1},\qquad \hbar^{-1}=\kappa^{\text{quant}},$$

where S is the classical action ($S=\log\tau$ (4.32)). The classical limit $\hbar\to 0$, $\kappa^{\text{quant}}\to\infty$ leads to the Hamilton–Jacobi equations for S, (3.7) with $\kappa=1$, which are equivalent to the HID (4.30).

To summarize, let us arrange these quantum and classical systems in a commutative diagram. The vertical arrows denote passing to the classical limit and mean simultaneous rescaling of the quantum level, while the limit $\kappa^{\text{quant}}\to 0$ ($\kappa\to 0$) on the horizontal arrows also includes rescaling the moduli of complex structures. The examples in the bottom diagram will be considered in the next sections.

$$\begin{array}{ccc}
\left\{\begin{array}{c}\text{KZB eqs., }(\kappa,\mathcal{M}_{g,n},G),\\ (\kappa^{\text{quant}}\partial_{t_a}+\widehat{H}_a)F=0\\ (a=1,\dots,\dim\mathcal{M}_{g,n})\end{array}\right\} & \xrightarrow{\kappa^{\text{quant}}\to 0} & \left\{\begin{array}{c}\text{KZB eqs. on the critical level,}\\ (\mathcal{M}_{g,n},G),\ (\widehat{H}_a)F=0\\ (a=1,\dots,\dim\mathcal{M}_{g,n})\end{array}\right\}\\
\Big\downarrow{\scriptstyle \hbar\to 0} & & \Big\downarrow{\scriptstyle \hbar\to 0}\\
\left\{\begin{array}{c}\text{Hierarchies of Isomonodromic}\\ \text{Deformations on }\mathcal{M}_{g,n}\end{array}\right\} & \xrightarrow{\kappa\to 0} & \{\text{Hitchin systems}\}.
\end{array}$$

EXAMPLES.

$$\left\{\begin{matrix}\text{Schlesinger eqs.}\\ \text{Painlevé type eqs.}\\ \text{Elliptic Schlesinger eqs.}\end{matrix}\right\} \xrightarrow{\kappa\to 0} \left\{\begin{matrix}\text{Classical Gaudin eqs.}\\ \text{Calogero eqs.}\\ \text{Elliptic Gaudin eqs.}\end{matrix}\right\}.$$

6. Genus zero Schlesinger equations

6.1. Derivation of the equations. Consider the projective line $\mathbb{C}P^1$ with n punctures $(x_1, \dots, x_n \mid x_a \neq x_b)$. The Beltrami differential μ is related only to the positions of the marked points. Then from (4.26) we obtain

$$\delta\mu_a = \bar{\partial}(1 + c_a\bar{\partial}(z - x_a)\chi_a(z, \bar{z}))\,\delta t_a, \qquad \delta t_a = \delta x_a. \tag{6.1}$$

On $\mathbb{C}P^1$ the gauge transform (4.17) allows us to choose $\bar{A}$ to be identically zero. After the gauge fixing

$$\bar{A} = f\partial_{\bar{w}}f^{-1}, \qquad A = fLf^{-1} + \kappa f\partial_w f^{-1}, \tag{6.2}$$

the moment equation takes the form

$$\partial_{\bar{w}}L = 2\pi i \sum_{a=1}^{n} \delta^2(x_a)p_a.$$

This allows us to find L, namely

$$L = \sum_{a=1}^{n} \frac{p_a}{w - x_a}. \tag{6.3}$$

Then from (6.1) we have

$$\begin{aligned}\frac{1}{2}\delta \int_{\mathbb{C}P^1} \langle L, L\rangle\,\delta\mu &= \frac{1}{2}\sum_{b,a}\delta \int_{\mathbb{C}P^1} \frac{\langle p_a, p_b\rangle}{(w - x_b)(w - x_a)} \sum_c \bar{\partial}(1 + c_c(w - x_c))\chi_c\,\delta x_c\\ &= \sum_a (\delta H_{a,1} + \delta H_{a,0})\,\delta x_a,\end{aligned}$$

where

$$H_{a,1} = \sum_{b \neq a} \frac{\langle p_a, p_b\rangle}{x_a - x_b} \tag{6.4}$$

and

$$H_{2,a} = c_a\langle p_a, p_a\rangle.$$

The functions $H_{1,a}$ are precisely the Schlesinger Hamiltonians. On the symplectic quotient, the form ω (4.23) can be expressed as

$$\omega = \sum_{a=1}^{n} \delta\langle p_a, g_a^{-1}\delta g_a\rangle - \frac{1}{\kappa}\sum_{b=1}^{n}(\delta H_{b,1} + \delta H_{b,0})\,\delta x_b.$$

Note that we still have gauge freedom with respect to the coordinate-independent G action, since this action does not change our gauge fixing (6.2). The corresponding moment constraint means that the sum of residues of L vanishes:

$$\sum_{a=1}^{n} p_a = 0. \tag{6.5}$$

While the Hamiltonians $H_{2,a}$ are Casimir invariants and lead to trivial equations, the equations of motion for $H_{1,a}$ are the Schlesinger equations

$$\kappa\partial_b p_a = \frac{[p_a, p_b]}{x_a - x_b} \quad (a \neq b), \qquad \kappa\partial_a p_a = -\sum_{b\neq a} \frac{[p_a, p_b]}{x_a - x_b}.$$

As a byproduct, by this procedure we obtain the corresponding linear problem (4.33), (4.34), and (4.35) with $L = (6.3)$, $\bar{L} = 0$, and $M_{a,1} = -p_a/(w - x_a)$ as a solution of (4.36). The tau-function (4.32) for the Schlesinger equations has the form [**2**]

$$\delta \log \tau = -\sum_{c\neq b} \langle p_b, p_c\rangle \, \delta \log(x_c - x_b).$$

6.2. Solutions via the projection method. We shall find the dressing transform defining the evolution on the coadjoint orbits

$$p_a(\mathbf{t}) = f^{-1}(z, \bar{z}, \mathbf{t}) p_a^0 f(z, \bar{z}, \mathbf{t}). \tag{6.6}$$

Recall that the times in the Schlesinger equations are $t_a = x_a - x_a^0$, and assume that $t_a^0 = 0$. From (6.3) we obtain

$$L(\mathbf{t}_0 = 0) = L^0 = \sum_{a=1}^{n} \frac{p_a^0}{z - x_a^0}.$$

It follows from (6.2) that $\bar{L} = 0$. Then the projection method (4.37) in this case yields

$$f^{-1}(z, \bar{z}, \mathbf{t}) L^0 \mu(\mathbf{t}) f(z, \bar{z}, \mathbf{t}) + f^{-1}(z, \bar{z}, \mathbf{t})(\bar{\partial} + \mu(\mathbf{t})\partial) f(z, \bar{z}, \mathbf{t}) = 0.$$

In other words, the gauge transform $f(z, \bar{z}, \mathbf{t})$ defining the evolution of solutions (the dressing transform) can be found from the equation

$$\left[\bar{\partial} + \sum_k t_k \bar{\partial}\chi_k(z, \bar{z})(\partial + L^0)\right] f(z, \bar{z}, \mathbf{t}) = 0. \tag{6.7}$$

We seek smooth solutions to this equation, assuming that the times $t_s = x_s - x_s^0$ are small. To this end, we consider the perturbative series

$$f(z, \bar{z}, \mathbf{t}) = id + \sum_k t_k a_k + \sum_{j\leqslant k} t_j t_k a_{jk} + \sum_{i\leqslant j\leqslant k} t_i t_j t_k a_{ijk} + \dots, \tag{6.8}$$

where $a_k, a_{jk}, a_{ijk}, \dots$ are smooth maps $\mathbb{C}P^1 \to$ (universal enveloping algebra (G)). Then (6.7) leads to the system

$$\begin{aligned} 1) \quad \bar{\partial} a_k &= -L^0 \bar{\partial}\chi_k, \\ 2) \quad \bar{\partial} a_{jk} &= -(\partial + L^0)(\bar{\partial}\chi_j a_k + \bar{\partial}\chi_k a_j), \\ 3) \quad \bar{\partial} a_{ijk} &= -(\partial + L^0)\bar{\partial}\chi_{[j} a_{jk]}, \\ 4) \quad & \dots\dots\dots\dots\dots\dots, \end{aligned}$$

where the subscript $_{[\dots]}$ stands for symmetrization. All equations have the same structure; their solutions depend only on the previous step. Since all $(0,1)$-forms on $\mathbb{C}P^1$ are exact, the equations can be integrated and the solutions found step by step.

In the first order, one has

$$a_k(z,\bar z) = -\sum_{a=1}^{n} \frac{p_a^0}{z-x_a^0}\,\chi_k^a(z,\bar z) \qquad (\chi_k^a(z,\bar z)=\chi_k(z,\bar z)-\chi_k(x_a,\bar x_a)). \tag{6.9}$$

Note that $a_k(z,\bar z)$ is a nonsingular function on $\mathbb{C}P^1$ due to the definition of $\chi_k^a(z,\bar z)$. Now consider the second order approximation

$$\bar\partial a_{jk}(z,\bar z) = \sum_{a=1}^{n} \partial\Big(\frac{p_a^0}{z-x_a^0}\Big)\bar\partial\chi_{[j}(z,\bar z)\chi_{k]}^a(z,\bar z) + \sum_{a=1}^{n}\frac{p_a^0}{z-x_a^0}\,\bar\partial\chi_{[j}(z,\bar z)\partial\chi_{k]}^a(z,\bar z)$$
$$+\sum_{a=1}^{n}\frac{p_a^0}{z-x_a^0}\sum_{b=1}^{n}\frac{p_b^0}{z-x_b^0}\,\bar\partial\chi_{[j}(z,\bar z)\chi_{k]}^b(z,\bar z).$$

The result of integration is

$$\begin{aligned} a_{jk}(z,\bar z) &= \sum_{a=1}^{n} p_a^0\Big(\frac{\psi_j^a(z,\bar z)\delta_{jk}}{z-x_a^0} - \frac{\chi_j^a(z,\bar z)\chi_k^a(z,\bar z)}{(z-x_a^0)^2}\Big) \\ &\quad + \sum_{a=1}^{n}\frac{(p_a^0)^2}{(z-x_a^0)^2}\,\chi_j(z,\bar z)\chi_k^a(z,\bar z) \\ &\quad - \sum_{a\neq b}\frac{p_a^0p_b^0}{z-x_b^0}\,(\chi_j^a(z,\bar z)\chi_k^a(z,\bar z)-\chi_j^a(x_b^0)\chi_k^a(x_b^0)). \end{aligned} \tag{6.10}$$

Here the function $\psi_j^a(z,\bar z)$ is also defined as the result of integration:

$$\bar\partial\psi_j^a(z,\bar z) = 2\chi_j(z,\bar z)\chi_j^a(z,\bar z) \qquad (\psi_j^a(x_a)=0). \tag{6.11}$$

This provides the absence of poles at $z=x_a$ in the first term of the right-hand side of (6.10). By subtracting $\chi_j^a(x_b^0)\chi_k^a(x_b^0)$ in the last term, we kill the pole of $a_{jk}(z,\bar z)$ at $z=x_b$. Thus, in the second order approximation we obtain a regular solution as well. It is defined almost explicitly up to the integration (6.11).

Thus we have defined the dressing transformation of the initial data (6.6) up to the third order (6.8), (6.9), (6.10). The calculations of the higher order corrections reproduce the same procedure as for the second order, and we can repeat them step by step.

7. Genus one elliptic Schlesinger, Painlevé VI, ...

In the genus one case it is still feasible to write out explicit formulas for the Hamiltonians and the equations of motion.

7.1. Deformations of elliptic curves. In addition to the moduli coming from the positions of the marked points, there is the elliptic modulus τ, $\operatorname{Im}\tau>0$, of the curves $\Sigma_{1,n}$. As in (4.24), (4.26), we take the Beltrami differential in the form

$$\mu = \sum_{a=1}^{n}\mu_a + \mu_\tau \qquad (\mu_a = t_a\,\bar\partial n_a),$$

where $n_a(z,\bar z)$ is the same as in (4.26) and

$$n_\tau = (\bar z - z)\Big(1-\sum_{a=1}^{n}\chi_a(z,\bar z)\Big). \tag{7.1}$$

We perform the changes $t_\tau \to t_\tau/\rho$, $t_\tau = \tau - \tau_0$, $\rho = \tau_0 - \bar{\tau}_0$, where τ_0 defines the reference complex structure on the curve

$$T_0^2 = \{0 < x \leqslant 1,\, 0 < y \leqslant 1,\, z = x + \tau_0 y,\, \bar{z} = x + \bar{\tau}_0 y\}.$$

For small t_τ, from (7.1) we have

$$\mu_\tau = \tilde{\mu}_\tau \bar{\partial}(\bar{z} - z)\Big(1 - \sum_{a=1}^{n} \chi'_a(z, \bar{z})\Big) \qquad \Big(\tilde{\mu}_\tau = \frac{t_\tau}{\tau - \tau_0}\Big), \tag{7.2}$$

or

$$\mu_\tau = \frac{t_\tau}{\rho}\bar{\partial}(\bar{z} - z)\Big(1 - \sum_{a=1}^{n} \chi_a(z, \bar{z})\Big). \tag{7.3}$$

As we assumed from the very beginning, μ_τ vanishes at the marked points. It describes not only a small neighborhood of T_0^2 in $\mathcal{M}_{1,1}$, but also the whole Teichmüller space as well. In terms of μ_τ, the Teichmüller space is the unit disk $|\mu_\tau| < 1$ (see (4.12) for the transformation law). In terms of τ, it is the upper half-plane $\operatorname{Im} \tau > 0$. On the other hand, the times $\mathbf{t} = (t_\tau, t_1 = x_1 - x_1^0, \dots, t_n = x_n - x_n^0)$ define deformations

$$T_0^2(x_1^0, \dots, x_n^0) \xrightarrow{\mathbf{t}} T_\tau^2(x_1, \dots, x_n)$$

in a neighborhood of T_0^2 in $\mathcal{M}_{1,1}$. Finally, for small t_τ and t_a from (7.3) we obtain

$$\delta\mu = \delta\tilde{\mu}_\tau \bar{\partial}(\bar{z} - z)\Big(1 - \sum_{a=1}^{n} \chi_a(z, \bar{z})\Big) + \sum_{a=1}^{n} \delta t_a \bar{\partial}(1 + c_a(z - x_a))\chi_a(z, \bar{z}). \tag{7.4}$$

7.2. Flat bundles on a family of elliptic curves. First note that $\bar{A}$ as a $\bar{\partial}$-connection determines a holomorphic G-bundle $\mathcal{E}$ over T_τ^2. For stable bundles, $\bar{A}$ can be gauge-transformed by (4.17) to the Cartan $(z, \bar{z})$-independent form $\bar{L}$,

$$\bar{A} = f\bar{L}f^{-1} + f(\bar{\partial} + \mu\partial)f^{-1},$$

where $\bar{L} \in \mathcal{H}$ is the Cartan subalgebra of Lie(G). Therefore, the stable bundle $\mathcal{E}$ is decomposed into the direct sum of line bundles

$$\mathcal{E} = \bigoplus_{k=1}^{r} \mathcal{L}_k, \qquad r = \operatorname{rank}(G).$$

The set of gauge-equivalent connections represented by $\{\bar{L}\}$ can be identified with the rth power of the Jacobian of T_τ^2, factored by the action of the Weyl group W of G. Put

$$\bar{L} = 2\pi i \frac{1 - \tilde{\mu}_\tau}{\rho}\mathbf{u}, \quad \mathbf{u} \in \mathcal{H} \quad \Big(\frac{1 - \tilde{\mu}_\tau}{\rho} = \frac{1}{\tau - \bar{\tau}_0}\Big). \tag{7.5}$$

This means that

$$\int_{T_\tau^2} \bar{L}\, dw = \mathbf{u}. \tag{7.6}$$

Let $\bar{L} = \bar{\partial}\log\phi$. Then the integral

$$\int_{T_\tau^2} \bar{L}\, dw = \int_{P_0}^{P} \log\phi\, dw$$

defines the Abel map $P \in T_\tau^2 \to \mathbf{u}$. We shall return to this point later.

The flatness condition (the moment constraints (4.21)) for the gauge transformed connections $(L, \bar{L})$ takes the form

$$\partial_{\bar{w}} L + [\bar{L}, L] = 2\pi i \sum_{a=1}^{n} \delta^2(x_a^0) p_a. \tag{7.7}$$

Let $R = \{\alpha\}$ be the root system of $\mathrm{Lie}(G) = \mathcal{G}$, and let $\mathcal{G} = \mathcal{H} \bigoplus_{\alpha \in R} \mathcal{G}_\alpha$ be the root decomposition. Impose the vanishing of the residues in (7.7),

$$\sum_{a=1}^{n} p_a|_{\mathcal{H}} = 0, \tag{7.8}$$

where $p_a|_{\mathcal{H}}$ is the Cartan component of p_a and we have identified $\mathcal{G}$ with its dual space $\mathcal{G}^*$. This condition is similar to (6.5) and has the sense of moment constraints for the remaining gauge action (see subsection 7.3 below).

We shall parametrize the set of solutions of (7.7) by two elements $\mathbf{v}, \mathbf{u} \in \mathcal{H}$. Let $E_1(w)$ be the Eisenstein function of the module τ (A.2), and let $(p_a)_{\mathcal{H}}$ and $(p_a)_\alpha$ be the Cartan component and the root component of $p_a \in \mathcal{O}_a$.

LEMMA 7.1. *Solutions of the moment constraint equation* (7.7) *have the form*

$$L = P + X, \quad P \in \mathcal{H}, \quad X = \sum_{\alpha \in R} X_\alpha, \tag{7.9}$$

$$P = 2\pi i \left(\frac{\mathbf{v}}{1 - \tilde{\mu}_\tau} - \kappa \frac{\mathbf{u}}{\rho} + \sum_{a=1}^{n} (p_a)_{\mathcal{H}} E_1(w - x_a) \right), \tag{7.10}$$

$$\begin{gathered} X_\alpha = \sum_{a=1}^{n} X_\alpha^a, \\ X_\alpha^a = \frac{(p_a)_\alpha}{1 - \tilde{\mu}_\tau} \exp 2\pi i \left\{ \frac{(w - x_a) - (\bar{w} - \bar{x}_a)}{\tau - \bar{\tau}_0} \alpha(\mathbf{u}) \right\} \phi(\alpha(\mathbf{u}), w - x_a), \end{gathered} \tag{7.11}$$

where $\phi(u, w)$ *is defined in* (A.4).

PROOF. First consider the Cartan component. Since $\bar{L} \in \mathcal{H}$, we have

$$\partial_{\bar{w}} P = 2\pi i \sum_{a=1}^{n} \delta^2(x_a) p_a.$$

From (A.28), (A.30) we obtain (7.10). The special choice of the constant part of P will be explained later. Note that $\mathbf{v} \in \mathcal{H}$ is a new parameter.

For the root components, equation (7.7) takes the form

$$\left(\partial_{\bar{w}} + 2\pi i \frac{1 - \tilde{\mu}_\tau}{\rho} \alpha(\mathbf{u}) \right) X_\alpha = 2\pi i \sum_{a=1}^{n} \delta^2(x_a)(p_a)_\alpha.$$

Comparing it with (A.31) and its solution (A.32), we come to (7.11). □

Therefore we have found the flat connections $L = L(\mathbf{v}, \mathbf{u})$, $\bar{L} = \bar{L}(\mathbf{u})$.

7.3. Symmetries. The remaining gauge transforms do not change the gauge fixing and thereby preserve the chosen Cartan subalgebra $\mathcal{H} \subset G$. These transformations are generated by the Weyl subgroup W of G and elements $f(w, \bar{w}) \in \mathrm{Map}(T_\tau^2, \mathrm{Cartan}(G))$. Let Π be the system of simple roots, $R^\vee = \{\alpha^\vee = 2\alpha/(\alpha|\alpha)\}$

the dual root system, and $\mathbf{m} = \sum_{\alpha\in\Pi} m_\alpha \alpha^\vee$ the element from the dual root lattice $\mathbb{Z}R^\vee$. Then the Cartan valued harmonics

$$f_{\mathbf{m},\mathbf{n}} = \exp 2\pi i\left(\mathbf{m}\,\frac{w-\bar{w}}{\tau-\bar{\tau}_0} + \mathbf{n}\,\frac{\tau\bar{w}-\bar{\tau}_0 w}{\tau-\bar{\tau}_0}\right) \qquad (\mathbf{m},\mathbf{n}\in R^\vee) \tag{7.12}$$

generate the basis in the space of gauge transforms. They act as follows:

$$\bar{L}\to\bar{L}+2\pi i\,\frac{\mathbf{m}-\mathbf{n}\tau}{\tau-\bar{\tau}_0},\quad P\to P+2\pi i\kappa\,\frac{-\mathbf{m}+\mathbf{n}\bar{\tau}_0}{\tau-\bar{\tau}_0},\quad X^a_\alpha\to X^a_\alpha\varphi(m_\alpha,n_\alpha),$$

$$\varphi(m_\alpha,n_\alpha)=\exp\frac{4\pi i}{\tau-\bar{\tau}_0}\,[(m_\alpha+n_\alpha\bar{\tau}_0)(w-x_a)-(m_\alpha+n_\alpha\tau)(\bar{w}-\bar{x}_a)]. \tag{7.13}$$

In terms of the new variables $\mathbf{v}$ and $\mathbf{u}$, the transforms have the simple form

$$\mathbf{u}\to\mathbf{u}+\mathbf{m}-\mathbf{n}\tau,\qquad \mathbf{v}\to\mathbf{v}-\kappa\mathbf{n}. \tag{7.14}$$

The whole discrete gauge symmetry is the semidirect product $\widehat{W}$ of the Weyl group W and the lattice $\mathbb{Z}R^\vee\oplus\tau\mathbb{Z}R^\vee$. The latter is the Bernstein–Schvartsman complex crystallographic group [**43**]. The quotient space $\mathcal{H}/\widehat{W}$ is the true space for the coordinates $\mathbf{u}$ that we discussed above (see (7.5) and (7.6)).

The transformations (7.12), in accord with (4.13), act also on $p_a\in\mathcal{O}_a$. This action leads to the symplectic quotient $\mathcal{O}_a//H$ and generates the moment equation (7.8).

The modular group $\mathrm{PSL}_2(\mathbb{Z})$ is a subgroup of the mapping class group for the Teichmüller space $\mathcal{T}_{1,n}$. Here we do not consider the action of the permutation of the marked points on the dynamical variables $(\mathbf{v},\mathbf{u},\mathbf{p},\tau,x_a)$. Due to (4.12) and (7.2), its action on τ takes the standard form

$$\tau\to\gamma\tau=\frac{a\tau+b}{c\tau+d},\qquad \gamma\in\mathrm{PSL}_2(\mathbb{Z}).$$

We summarize the action of the Bernstein–Schvartsman group and the modular group on the dynamical variables in the following table.

	$W=\{s\}$	$\mathbb{Z}R^\vee\oplus\tau\mathbb{Z}R^\vee$	$\mathrm{PSL}_2(\mathbb{Z})$
$\mathbf{v}$	$s\mathbf{v}$	$\mathbf{v}+\kappa\mathbf{n}$	$\mathbf{v}(c\tau+d)-\kappa c\mathbf{u}$
$\mathbf{u}$	$s\mathbf{u}$	$\mathbf{u}-\mathbf{m}+\mathbf{n}\tau$	$\mathbf{u}(c\tau+d)^{-1}$
$(p_a)_{\mathcal{H}}$	$s(p_a)_{\mathcal{H}}$	p_a	p_a
$(p_a)_\alpha$	$(p_a)_{s\alpha}$	$\varphi(m_\alpha,n_\alpha)(p_a)_\alpha$	$(p_a)_\alpha$
τ	τ	τ	$(a\tau+b)(c\tau+d)^{-1}$
x_a	x_a	x_a	$x_a(c\tau+d)^{-1}$

Here $\varphi(m_\alpha,n_\alpha)$ is defined by (7.13).

7.4. The symplectic form. The set

$$(\mathbf{v},\mathbf{u})\in\mathcal{H},\qquad \mathbf{p}=(p_1,\dots,p_n)\in\bigoplus_{a=1}^n\mathcal{O}_a,$$

of dynamical variables along with the times $\mathbf{t} = (t_\tau, t_1, \dots, t_n)$ describe local coordinates in the total space of the bundle $\mathcal{P}$. In accordance with the general prescription, we can define the Hamiltonian system on this set. The main statement, formulated in terms of theta-functions and Eisenstein functions (see the Appendix), is as follows.

PROPOSITION 7.1. *The symplectic form* ω (4.28) *on* $\mathcal{P}$ *is*

(7.15)
$$\frac{1}{4\pi^2}\,\omega = (\delta\mathbf{v}, \delta\mathbf{u}) + \sum_{a=1}^{n} \delta\langle p_a, g_a^{-1}\delta g_a\rangle - \frac{1}{\kappa}\Big(\sum_{a=1}^{n} \delta H_{2,a} + \delta H_{1,a}\Big)\delta t_a - \frac{1}{\kappa}\,\delta H_\tau \delta\tau,$$

with the Hamiltonians

(7.16)
$$H_{2,a} = c_a\langle p_a, p_a\rangle,$$

(7.17)
$$\begin{aligned} H_{1,a} &= 2\Big(\frac{\mathbf{v}}{1-\tilde{\mu}_\tau} - \kappa\,\frac{\mathbf{u}}{\rho}, p_a|_{\mathcal{H}}\Big) + \sum_{b\neq a}(p_a|_{\mathcal{H}}, p_b|_{\mathcal{H}})E_1(x_a - x_b) \\ &\quad + \sum_{b\neq a}\sum_{\alpha}(p_a|_\alpha, p_b|_{-\alpha})\phi(\alpha(\mathbf{u}), x_a - x_b), \end{aligned}$$

(7.18)
$$\begin{aligned} H_\tau = \frac{(\mathbf{v},\mathbf{v})}{2} &- \frac{1}{4\pi^2}\sum_{a=1}^{n}\sum_{\alpha}(p_a|_\alpha, p_a|_{-\alpha})E_2(\alpha(\mathbf{u})) \\ &+ \sum_{a\neq b}^{n}(p_a|_{\mathcal{H}}, p_b|_{\mathcal{H}})(E_2(x_a - x_b) - E_1^2(x_a - x_b)) \\ &- \frac{1}{4\pi^2}\sum_{a\neq b}\sum_{\alpha}(p_a|_\alpha, p_b|_{-\alpha})\phi(-\alpha(\mathbf{u}), x_a - x_b)(E_1(\alpha(\mathbf{u})) \\ &- E_1(x_b - x_a + \alpha(\mathbf{u}))), \end{aligned}$$

$$\phi(\alpha(\mathbf{u}), x_a - x_b) = \frac{\theta(\alpha(\mathbf{u}) + x_a - x_b)\theta'(0)}{\theta(\alpha(\mathbf{u}))\theta(x_a - x_b)}.$$

PROOF. The form we must calculate is (4.23):

$$\omega = \int_\Sigma \langle \delta L, \delta\bar{L}\rangle + 2\pi i\sum_{a=1}^{n}\delta\langle p_a, g_a^{-1}\delta g_a\rangle - \frac{1}{\kappa}\int_\Sigma \langle L, \delta L\rangle\,\delta\mu$$

with $\delta\mu$ (7.4). First, we gauge-transform $(L, \bar{L})$ by

$$f(w,\bar{w}) = \prod_{a=1}^{n}\exp\Big(2\pi i\,\frac{w-\bar{w}}{\tau-\tau_0}\,\chi_a'(w,\bar{w})\mathbf{u}\Big),$$

where we choose $\chi_a'(w,\bar{w})$ in a such way that

(7.19) $\operatorname{supp}\chi_a'(w,\bar{w}) \subset \operatorname{supp}\chi_a(w,\bar{w}), \qquad \operatorname{supp}\chi_a'(w,\bar{w}) \cap \operatorname{supp}\bar{\partial}\chi_a(w,\bar{w}) = \varnothing,$

where $\chi_a(w,\bar{w})$ is related to the moduli curves by (7.1). In fact, the first condition follows from the second.

As we know, the gauge transformations do not change ω, but do change the relations between its summands. Instead of (7.5), (7.10), and (7.11), we obtain

$$\bar{L} = 2\pi i \frac{1-\tilde{\mu}_\tau}{\rho}\, \mathbf{u}\partial_{\bar{w}}(\bar{w}-w)\Big(1-\sum_{a=1}^{n}\chi'_a(w,\bar{w})\Big), \tag{7.20}$$

$$\begin{aligned} P = 2\pi i \Big(&\frac{\mathbf{v}}{1-\tilde{\mu}_\tau} - \kappa\frac{\mathbf{u}}{\rho} + \sum_{a=1}^{n}(p_a)_{\mathcal{H}}E_1(w-x_a)\Big) \\ &- \frac{1-\tilde{\mu}_\tau}{\rho}\,\mathbf{u}\partial_w(\bar{w}-w)\sum_{a=1}^{n}\chi'_a(w,\bar{w}), \end{aligned} \tag{7.21}$$

$$X_\alpha = \frac{1}{1-\tilde{\mu}_\tau}\exp\Big\{\frac{w-\bar{w}}{\tau-\bar{\tau}_0}\alpha(\mathbf{u})\Big\}\sum_{a=1}^{n}(p_a)_\alpha\phi(\alpha(\mathbf{u}),w-x_a). \tag{7.22}$$

Taking into account the explicit form (7.5) of $\bar{L}$, we obtain

$$\langle \delta L, \delta\bar{L}\rangle = \frac{(\delta\mathbf{v},\delta\mathbf{u})}{\rho} + S(\delta\tau,\delta\mathbf{t}), \tag{7.23}$$

where $S(\delta\tau,\delta\mathbf{t})$ is a sum of terms linearly depending on the "time" differentials. It is compensated by terms coming from

$$-\frac{1}{4\pi^2\kappa}\int_{T^2_\tau}\langle L,\delta L\rangle\,\delta\mu.$$

Let us calculate the Hamiltonians

(7.24)

$$\begin{aligned} -\frac{1}{4\pi^2}\langle L,L\rangle = &\Big(\frac{\mathbf{v}}{1-\tilde{\mu}_\tau} - \kappa\frac{\mathbf{u}}{\rho} + \sum_{a=1}^{n}p_a|_{\mathcal{H}}E_1(w-x_a)\Big)^2 \\ &- \frac{4\pi^2}{1-\tilde{\mu}_\tau}\sum_{a,b}\sum_{\alpha\in R}(p_a|_\alpha, p_b|_{-\alpha})\phi(-\alpha(\mathbf{u}),w-x_a)\phi(\alpha(\mathbf{u}),w-x_b) \\ &- \frac{1}{2\pi i}\Big(\Big(\frac{\mathbf{v}}{1-\tilde{\mu}_\tau} - \kappa\frac{\mathbf{u}}{\rho} + \sum_{a=1}^{n}(p_a)_{\mathcal{H}}E_1(w-x_a)\Big), \frac{1-\tilde{\mu}_\tau}{\rho}\mathbf{u}\Big) \\ &\times \partial_w(\bar{w}-w)\sum_{a=1}^{n}\chi'_a(w,\bar{w}). \end{aligned}$$

To take the integral over T_0^2, we must couple this two-form with $\delta\mu$ (7.4):

$$\begin{aligned} -\frac{1}{4\pi^2}\int_{T^2_\tau}\langle L,L\rangle\,\delta\mu = -\frac{1}{4\pi^2}\int_{T^2_\tau}&\langle L,L\rangle\,\delta\tilde{\mu}_\tau\bar{\partial}(\bar{w}-w)\Big(1-\sum_{a=1}^{n}\chi_a(w,\bar{w})\Big) \\ &+ \sum_{a=1}^{n}\delta t_a\partial_{\bar{w}}(1+c_a(w-x_a))\chi_a(w,\bar{w}). \end{aligned}$$

Due to our choice (7.19) of $\chi'_a(w,\bar{w})$, the last line of (7.24) does not contribute to the integral. Therefore, we are left with the holomorphic double periodic part of the two-form $\langle L, L\rangle$. Using (A.10), we rewrite (7.24) as

$$-\frac{1}{4\pi^2}\langle L, L\rangle = \left(\frac{\mathbf{v}}{1-\tilde{\mu}_\tau} - \kappa\frac{\mathbf{u}}{\rho}, \frac{\mathbf{v}}{1-\tilde{\mu}_\tau} - \kappa\frac{\mathbf{u}}{\rho}\right)$$
$$+ 2\sum_{a=1}^{n}\left(\frac{\mathbf{v}}{1-\tilde{\mu}_\tau} - \kappa\frac{\mathbf{u}}{\rho}, p_a|_{\mathcal{H}}\right)E_1(w - x_a)$$
$$+ \sum_{a\neq b}^{n}(p_a|_{\mathcal{H}}, p_b|_{\mathcal{H}})E_1(w - x_a)E_1(w - x_b)$$
$$- \frac{4\pi^2}{1-\tilde{\mu}_\tau}\sum_{a\neq b}\sum_{\alpha\in R}(p_a|_\alpha, p_b|_{-\alpha})\phi(-\alpha(\mathbf{u}), w - x_a)\phi(\alpha(\mathbf{u}), w - x_b)$$
$$- \frac{4\pi^2}{1-\tilde{\mu}_\tau}\sum_{a}\sum_{\alpha\in R}(E_2(w - x_a) - E_2(\alpha(\mathbf{u}))).$$

Since L has only the first order poles (7.21), (7.22), we expand it on the deformed torus according to (A.34):

$$-\frac{1}{4\pi^2}\langle L, L\rangle = \left(\sum_{a=1}^{n}H_{2,a}E_2(w - x_a) + H_{1,a}E_1(w - x_a)\right) + h_0. \tag{7.25}$$

Due to (A.34), (A.39), and (A.40),

$$-\frac{1}{4\pi^2}\int_{T_\tau^2}\langle L, L\rangle\,\delta\mu = \sum_{a=1}^{n}\left(\int_{T_\tau^2}(H_{2,a}E_2(w - x_a)\partial_{\bar{w}}(w - x_a)\chi_a(w,\bar{w}) + H_{1,a}E_1(w - x_a))\partial_{\bar{w}}\chi_a(w,\bar{w})\right)\delta t_a$$
$$+ h_0\delta\tilde{\mu}_\tau\bar{\partial}(\bar{w} - w)\left(1 - \sum_{a=1}^{n}\chi_a(w,\bar{w})\right).$$

Taking into account (A.35), (A.36), we find that

$$H_{2,a} = \mathrm{res}|_{x_a}\langle L, L\rangle(w - x_a) = c_a\langle p_a, p_a\rangle + \mathrm{const},$$
$$H_{1,a} = \mathrm{res}|_{x_a}\langle L, L\rangle = (7.17).$$

The constant term in (7.25) h_0 is defined by (A.37). To find it, we use (A.10) and (A.38). Then $H_\tau = h_0\partial_\tau\tilde{\mu}_\tau$. After some calculations, we obtain (7.18). □

7.5. Example 1. $PVI_{\nu^2/4,-\nu^2/4,\nu^2/4,1/2-\nu^2/4}$. Consider the $\mathrm{SL}(2,\mathbb{C})$ bundles over the family $\Sigma_{1,1}$. Then (7.5) takes the form

$$\bar{L} = 2\pi i\frac{1-\tilde{\mu}_\tau}{\rho}\,\mathrm{diag}(u, -u). \tag{7.26}$$

In this case the position of the marked point is no longer a modulus, and we put $x_1 = 0$. We have from (7.2)

$$w = z - \frac{\tau - \tau_0}{\rho}(\bar{z} - z), \quad \bar{w} = \bar{z}, \qquad \partial_{\bar{w}} = \bar{\partial} + \frac{\tau - \tau_0}{\tau - \bar{\tau}_0}\partial.$$

Since $\dim\mathcal{O} = 2$, the orbit degrees of freedom can be gauged away by the Hamiltonian action of the diagonal group. We assume that

$$p = \nu[(1,1)^T \otimes (1,1) - \mathrm{Id}].$$

Then from (7.9), (7.10), (7.11) we obtain

$$L = \begin{pmatrix} 2\pi i\Big(\dfrac{v}{1-\tilde{\mu}_\tau} - \kappa\,\dfrac{u}{\rho}\Big) & x(2u,w,\bar{w}) \\ x(-2u,w,\bar{w}) & 2\pi i\Big(-\dfrac{v}{1-\tilde{\mu}_\tau} + \kappa\,\dfrac{u}{\rho}\Big) \end{pmatrix}, \tag{7.27}$$

$$x(u,w,\bar{w}) = \frac{\nu}{2\pi i(1-\tilde{\mu})_\tau}\exp 2\pi i\Big\{(w-\bar{w})u\,\frac{1-\tilde{\mu}_\tau}{\rho}\Big\}\phi(u,w).$$

The symplectic form (7.15) becomes

$$-\frac{1}{8\pi^2}\,\omega = (\delta v, \delta u) - \frac{1}{\kappa}\,\delta H_\tau\,\delta\tau,$$

and

$$H_\tau = \frac{v^2}{2} + U(u|\tau), \qquad U(u|\tau) = -\Big(\frac{\nu}{2\pi i}\Big)^2 E_2(2u|\tau) - \frac{\nu^2}{2\pi i}.$$

Then the equations of motion are

$$\kappa\,\frac{\partial u}{\partial\tau} = v, \tag{7.28}$$

$$\kappa\,\frac{\partial v}{\partial\tau} = \frac{\nu^2}{(2\pi i)^2}\,\frac{\partial}{\partial u}\,E_2(2u|\tau). \tag{7.29}$$

From (A.10) we obtain

$$\kappa^2\,\frac{\partial^2 u}{\partial\tau^2} = \frac{\nu^2}{(2\pi i)^2}\,\frac{\partial}{\partial u}\,\wp(2u|\tau),$$

which coincides with (1.12). This equation provides the isomonodromic deformation for the linear system (4.33), (4.35) with L (7.27) and $\bar{L}$ (7.26) with respect to the change in the modulus τ. The Lax pair is given by L (7.27) and M_τ,

$$M_\tau = \begin{pmatrix} 0 & y(2u,w,\bar{w}) \\ y(-2u,w,\bar{w}) & 0 \end{pmatrix},$$

where $y(u,w,\bar{w})$ is defined by the equation (see (4.36))

$$\Big(\partial_{\bar{w}} + \frac{2\pi i}{\tau-\bar{\tau}_0}\,u\Big)y(u,w,\bar{w}) = -\frac{\rho}{\kappa(\tau-\bar{\tau}_0)^2}\,x(u,w,\bar{w}). \tag{7.30}$$

Using the representation (A.33) for $x(u,w,\bar{w}) = g_2(u,w,\bar{w})$, we find that

$$y(u,w,\bar{w}) = \frac{\rho}{2\pi i\kappa(\tau-\bar{\tau}_0)}\,\partial_u x(u,w,\bar{w}). \tag{7.31}$$

The equivalence of the Lax equation

$$\partial_\tau L - \kappa\partial_w M + [M,L] = 0$$

to the equations of motion (7.28), (7.29) can be checked by replacing L (7.27) and M by $y(u,w,\bar{w})$ (7.31). Details of this procedure will be presented for $\mathrm{SL}(N,\mathbb{C})$ bundles in the next example.

The projection method determines solutions of (7.28), (7.29) as the result of diagonalizing L (7.27) by the gauge transform on the deformed curve T^2_τ:

$$\begin{aligned}\frac{1}{\tau-\bar{\tau}_0}&\operatorname{diag}(u(\tau),-u(\tau))\\ &=f^{-1}(z,\bar{z},\tau)\left(\frac{2\pi i(\tau-\tau_0)}{\kappa\rho}\begin{pmatrix} v^0-\kappa u^0/\rho & x(2u^0,z,\bar{z})\\ x(-2u^0,z,\bar{z}) & -v^0+\kappa u^0/\rho\end{pmatrix}\right.\\ &\quad\left.+\frac{1}{\rho}\operatorname{diag}(u^0,-u^0)\right)f(z,\bar{z},\tau)+f^{-1}(z,\bar{z},\tau)\left(\bar{\partial}+\frac{\tau-\tau_0}{\tau-\bar{\tau}_0}\partial\right)f(z,\bar{z},\tau).\end{aligned}$$

On the critical level ($\kappa=0$), we come to the two-body elliptic Calogero system.

7.6. Example 2. For flat G-bundles over $\Sigma_{1,1}$, we obtain PVI-type equations related to arbitrary root systems. They are described by a system of second order differential equations for the variables $\mathbf{u}=(u_1,\dots,u_r)$, where $r=\operatorname{rank}G$. In addition, there are the orbit variables $\mathbf{p}\in\mathcal{O}(G)$ satisfying the Euler top equations.

Consider the $\mathrm{SL}(N,\mathbb{C})$ case with the most degenerate orbits $\mathcal{O}=T^*\mathbb{C}P^{N-1}$ in detail. They have dimension $2N-2$. The orbit variables can be gauged away by diagonal gauge transforms, and we are left with the coupling constant ν. We already have the $\bar{L}$ and L matrices (see (7.5) and (7.9)),

$$L=P+X,\qquad P=2\pi i\left(\frac{\mathbf{v}}{1-\tilde{\mu}_\tau}-\kappa\frac{\mathbf{u}}{\rho}\right),$$

$$X=\{x_\alpha\}=(\tau-\bar{\tau}_0)\nu\exp 2\pi i\left\{\frac{w-\bar{w}}{\tau-\bar{\tau}_0}\alpha(\mathbf{u})\right\}\phi(\alpha(\mathbf{u}),w).$$

Here $\alpha=e_j-e_k$, $\alpha(\mathbf{u})=u_j-u_k$ $(j\neq k)$, and

$$\mathbf{u}=\operatorname{diag}(u_1,\dots,u_N),\qquad \mathbf{v}=\operatorname{diag}(v_1,\dots,v_N).$$

The equations of motion take the form

$$\kappa\frac{du_j}{d\tau}=v_j,\tag{7.32}$$

$$\kappa\frac{dv_j}{d\tau}=-\partial_{u_j}U(\mathbf{u}|\tau),\qquad U(\mathbf{u}|\tau)=\frac{1}{(2\pi i)^2}\sum_{j\neq k}p_{j,k}p_{k,j}E_2(u_j-u_k|\tau).\tag{7.33}$$

On the critical level, this Painlevé type system degenerates into the N-body elliptic Calogero system.

Let us check the consistency of the equations of motion with the Lax representation

$$\partial_\tau L-\kappa\partial_w M+[M,L]=0.\tag{7.34}$$

Put the M-operator in the form $M=-D+Y$, where $Y=\{y(u_j-u_k)\}=\{y_\alpha\}$ and D is a diagonal matrix. We have already found y_α (7.31) using (4.36). But the diagonal part D is not fixed by (4.36) and must be found from the consistency of the Lax equation with the equations of motion. We take D in the same form as in the Calogero system [**42**],

$$D=\operatorname{diag}(d_1,\dots,d_N),\qquad d_j=\sum_{i\neq j}^{N}s(u_j-u_i).$$

First, we shall prove a few facts concerning the matrix elements of L and M. We shall prove that they satisfy the following functional equation:

$$x(u,z,\bar{z})y(v,z,\bar{z}) - x(v,z,\bar{z})y(u,z,\bar{z}) = (s(v)-s(u))x(u+v,z,\bar{z}). \tag{7.35}$$

In particular, its solution $s(u,w)$ is independent of w,

$$s(u) = \frac{1}{\kappa}\,\wp(u) + \text{const}\,. \tag{7.36}$$

The relation (7.35) is the so-called Calogero functional equation. Using (7.31), we can put in it the derivatives of x instead of y. The exponential factor in the expression of x cancels and (7.35) takes the form of the addition formula (A.26). Simultaneously, we obtain (7.36). We also need the following identity:

$$\left(\partial_\tau - \frac{\rho}{2\pi i}\,\partial_z\partial_u + \frac{u}{\tau-\bar{\tau}_0}\,\partial_u\right)x(u,z,\bar{z}) = 0. \tag{7.37}$$

It can be derived from the representation (A.33) for $x(u,z,\bar{z}) = g_2(u,z,\bar{z})$.

Now consider the Lax equation (7.34). Separation of the diagonal and nondiagonal terms leads to the system

$$\frac{d}{d\tau}\,P + \kappa\partial_w D + \sum_\alpha (y_\alpha x_{-\alpha} - y_{-\alpha}x_\alpha) = 0, \tag{7.38}$$

$$\frac{d}{d\tau}\,x_\alpha - \kappa\partial_w y_\alpha - \alpha(D)x_\alpha + [Y,X]_\alpha = 0. \tag{7.39}$$

Using (7.36) and (7.32), the first equation (7.38) can be rewritten as

$$\begin{aligned}2\pi i\,\frac{d}{d\tau}\,v_k &= \frac{(1-\tilde{\mu}_\tau)^2}{2\pi i\kappa}\sum_{j\neq k}(x(u_j-u_k)x'(u_k-u_j) - x(u_k-u_j)x'(u_j-u_k))\\ &= \sum_{j\neq k}[x(u_j-u_k)x(u_k-u_j)]'.\end{aligned}$$

But

$$x(u)x(-u) = \frac{\nu^2}{(1-\mu)^2}\,\phi(u,w)\phi(-u,w) = \frac{\nu^2}{(1-\mu)^2}\,(E_2(w)-E_2(u))$$

(see (A.10)). Therefore we obtain (7.33).

Now let us check the off-diagonal part of (7.39). It is convenient to return from

$$w = \left(1+\frac{\tau-\tau_0}{\rho}\right)z - \frac{\tau-\tau_0}{\rho}\,\bar{z}, \qquad \bar{w} = \bar{z},$$

to the $(z,\bar{z})$ variables. Then (7.39) takes the form

$$\begin{aligned}\frac{d}{d\tau}\,x(u_j-u_k) - \kappa\partial_z y(u_j-u_k) - 2\pi i\left(\frac{v_j-v_k}{1-\tilde{\mu}_\tau} - \kappa\,\frac{u_j-u_k}{\rho}\right)y(u_j-u_k)\\ - x(u_j-u_k)\left[\sum_{i\neq j}^N s(u_j-u_i) - \sum_{i\neq k}^N s(u_k-u_i)\right]\\ + \sum_{i=1}^N y(u_j-u_i)x(u_i-u_k) - y(u_i-u_k)x(u_j-u_i) = 0.\end{aligned}$$

It follows from (7.31), (7.32), and (7.37) that

$$\frac{d}{d\tau}x(u_j-u_k)-\kappa\partial_z y(u_j-u_k)-2\pi i\left(\frac{v_j-v_k}{1-\tilde{\mu}_\tau}-\kappa\frac{u_j-u_k}{\rho}\right)y(u_j-u_k)=0.$$

On the other hand, the equation

$$x(u_j-u_k)(s(u_j-u_i)-s(u_k-u_i))+y(u_j-u_i)x(u_i-u_k)-y(u_i-u_k)x(u_j-u_i)=0$$

is the addition formula (7.35). This concludes the proof of the equivalence of the Lax equation and the equations of motion.

The new ingredients of this construction as compared to its original form [**42**] are the dependence of the L-matrix on the spectral parameter,[3] the presence of the derivative $\partial_z M$ in the Lax equation, and the replacement of the external time t by the modular parameter τ. Nevertheless, the form of the Lax matrices is defined as for the Calogero system by the same functional equation (7.35). It turns out that its solutions satisfy the additional differential equations (7.36), (7.37), which allows us to apply them in the isomonodromic situation as well. Again, the solutions of the equations of motion (7.32), (7.33) are obtained by diagonalization:

$$2\pi i\,\frac{1-\tilde{\mu}_\tau}{\rho}\operatorname{diag}(u_1,\dots,u_N)=\tilde{\mu}_\tau f^{-1}(L(\mathbf{v}^0,\mathbf{u}^0,\tau^0))f+f^{-1}(\bar{\partial}+\tilde{\mu}_\tau\partial)f.$$

Appendix

We summarize the main formulas for elliptic functions, mainly borrowed from [**44**]. We assume that $q=\exp 2\pi i\tau$, and the curve T^2_τ is a $\mathbb{C}/\mathbb{Z}+\tau\mathbb{Z}$ factor of $\mathbb{C}$ under the shifts generated by $(1,\tau)$.

The basic element is the theta function:

(A.1)
$$\begin{aligned}\theta(z|\tau)&=q^{1/8}\sum_{n\in\mathbb{Z}}(-1)^n e^{\pi i(n(n+1)\tau+2nz)}\\&=q^{1/8}e^{-i\pi/4}(e^{i\pi z}-e^{-i\pi z})\prod_{n=1}^{\infty}(1-q^n)(1-q^n e^{2i\pi z})(1-q^n e^{-2i\pi z}).\end{aligned}$$

The Eisenstein functions.

(A.2) $$E_1(z|\tau)=\partial_z\log\theta(z|\tau),\qquad E_1(z|\tau)\sim\frac{1}{z}+\dots,$$

(A.3) $$E_2(z|\tau)=-\partial_z E_1(z|\tau)=\partial_z^2\log\theta(z|\tau),\qquad E_2(z|\tau)\sim\frac{1}{z^2}+\dots.$$

The next important function is

(A.4) $$\phi(u,z)=\frac{\theta(u+z)\theta'(0)}{\theta(u)\theta(z)}.$$

It has a pole at $z=0$ and

(A.5) $$\phi(u,z)=\frac{1}{z}+E_1(u)+\cdots.$$

[3]The dependence on the spectral parameter was first introduced in [**40**] for the Calogero system in a slightly different form.

Relationship with the Weierstrass functions.

$$\zeta(z|\tau) = E_1(z|\tau) + 2\eta_1(\tau)z, \tag{A.6}$$

$$\wp(z|\tau) = E_2(z|\tau) - 2\eta_1(\tau), \tag{A.7}$$

where

$$\eta_1(\tau) = \zeta\left(\frac{1}{2}\right) = \frac{3}{\pi^2}\sum_{m=-\infty}^{\infty}\sum_{n=-\infty}^{\infty}\frac{1}{(m\tau+n)^2} = \frac{24}{2\pi i}\frac{\eta'(\tau)}{\eta(\tau)} \tag{A.8}$$

and

$$\eta(\tau) = q^{1/24}\prod_{n>0}(1-q^n)$$

is the Dedekind function.

Further,

$$\phi(u,z) = \exp(-2\eta_1 uz)\frac{\sigma(u+z)}{\sigma(u)\sigma(z)}, \tag{A.9}$$

$$\phi(u,z)\phi(-u,z) = \wp(z) - \wp(u) = E_2(z) - E_2(u). \tag{A.10}$$

Series representations.

$$\begin{aligned} E_1(z|\tau) &= -2\pi i\left(\frac{1}{2} + \sum_{n\neq 0}\frac{e^{2\pi iz}}{1-q^n}\right) \\ &= -2\pi i\left(\sum_{n<0}\frac{1}{1-q^n e^{2\pi iz}} + \sum_{n\geqslant 0}\frac{q^n e^{2\pi iz}}{1-q^n e^{2\pi iz}} + \frac{1}{2}\right), \end{aligned} \tag{A.11}$$

$$E_2(z|\tau) = -4\pi^2\sum_{n\in\mathbb{Z}}\frac{q^n e^{2\pi iz}}{(1-q^n e^{2\pi iz})^2}, \tag{A.12}$$

$$\phi(u,z) = 2\pi i\sum_{n\in\mathbb{Z}}\frac{e^{-2\pi inz}}{1-q^n e^{-2\pi iu}}. \tag{A.13}$$

Parity.

$$\theta(-z) = -\theta(z), \tag{A.14}$$

$$E_1(-z) = -E_1(z), \tag{A.15}$$

$$E_2(-z) = E_2(z), \tag{A.16}$$

$$\phi(u,z) = \phi(z,u) = -\phi(-u,-z). \tag{A.17}$$

Behavior on the lattice.

$$\theta(z+1) = -\theta(z), \qquad \theta(z+\tau) = -q^{-1/2}e^{-2\pi iz}\theta(z), \tag{A.18}$$

$$E_1(z+1) = E_1(z), \qquad E_1(z+\tau) = E_1(z) - 2\pi i, \tag{A.19}$$

$$E_2(z+1) = E_2(z), \qquad E_2(z+\tau) = E_2(z), \tag{A.20}$$

$$\phi(u+1,z) = \phi(u,z), \qquad \phi(u+\tau,z) = e^{-2\pi iz}\phi(u,z). \tag{A.21}$$

Modular properties.

$$\theta\left(\frac{z}{c\tau+d}\,\middle|\,\frac{a\tau+b}{c\tau+d}\right) = \epsilon e^{\pi i/4}(c\tau+d)^{1/2}\exp\left(\frac{i\pi c z^2}{c\tau+d}\right)\theta(z|\tau) \quad (\epsilon^8=1), \tag{A.22}$$

$$E_1\left(\frac{z}{c\tau+d}\,\middle|\,\frac{a\tau+b}{c\tau+d}\right) = (c\tau+d)E_1(z|\tau) + 2\pi i z, \tag{A.23}$$

$$E_2\left(\frac{z}{c\tau+d}\,\middle|\,\frac{a\tau+b}{c\tau+d}\right) = (c\tau+d)^2E_2(z|\tau) + 2\pi i(c\tau+d). \tag{A.24}$$

Addition formula.

$$\phi(u,z)\partial_v\phi(v,z) - \phi(v,z)\partial_u\phi(u,z) = (E_2(v)-E_2(u))\phi(u+v,z), \tag{A.25}$$

or

$$\phi(u,z)\partial_v\phi(v,z) - \phi(v,z)\partial_u\phi(u,z) = (\wp(v)-\wp(u))\phi(u+v,z). \tag{A.26}$$

The proof of (A.25) is based on (A.5), (A.17), and (A.21).

In fact, $\phi(u,z)$ satisfies a more general relation, which follows from the Fay three-section formula

$$\begin{aligned}\phi(u_1,z_1)\phi(u_2,z_2) - \phi(u_1+u_2,z_1)\phi(u_2,z_2-z_1)\\ - \phi(u_1+u_2,z_2)\phi(u_1,z_1-z_2) = 0.\end{aligned} \tag{A.27}$$

Green functions. The Green functions are $(1,0)$-forms on T^2_τ. The first function $g_1(z)$ is defined by the equation

$$\bar\partial g_1(z) = 2\pi i\sum_a p_a\delta^2(x_a) \qquad \left(\sum_a p_a = 0\right), \tag{A.28}$$

where

$$\delta^2(x_a) = \sum_{m,n\in\mathbb{Z}} f_{m,n}(z-x_a,\bar z-\bar x_a), \tag{A.29}$$

$$f_{m,n}(z,\bar z) = \exp\frac{2\pi i}{\rho}\{m(z-\bar z)+n(\tau\bar z-\bar\tau z)\} \qquad (\rho=\tau-\bar\tau),$$

$$g_1(z) = \sum_a p_aE_1(z-x_a) + \text{const}. \tag{A.30}$$

For the equation

$$\left(\bar\partial + \frac{2\pi i}{\rho}u\right)g_2(u,z) = 2\pi i\delta^2(0), \tag{A.31}$$

the Green function is

$$g_2(u,z) = \frac{1}{\rho}e^{2\pi i(z-\bar z)u/\rho}\phi(u,z), \tag{A.32}$$

$$g_2(u,z) = \rho\sum_{m,n\in\mathbb{Z}}\frac{f_{m,n}(z,\bar z)}{u-m+n\tau}. \tag{A.33}$$

Expansion of elliptic functions. Let M_2 be the space of meromorphic elliptic functions on T_0^2 with poles of order two or less. Then any $f(z) \in M_2$ can be decomposed into the sum of Eisenstein functions

$$f(z) = \sum_a (c_{2,a} E_2(z - x_a|\tau) + c_{1,a} E_1(z - x_a|\tau)) + c_0, \tag{A.34}$$

where

$$\sum_a c_{1,a} = 0, \tag{A.35}$$

$$c_{1,a} = \mathrm{res}|_{x_a} f(z), \qquad c_{2,a} = \mathrm{res}|_{x_a} (z - x_a) f(z), \tag{A.36}$$

and

$$c_0 = \text{const. part}\left[f(z) - \sum_a (c_{2,a} E_2(z - x_a|\tau) + c_{1,a} E_1(z - x_a|\tau)\right]. \tag{A.37}$$

In particular, for $a \neq b$

$$\begin{aligned} &\text{const. part}\,[\phi(u, z - x_a)\phi(-u, z - x_b)] \\ &\quad = \phi(-u, x_a - x_b)[E_1(u) - E_1(u + x_b - x_a)]. \end{aligned} \tag{A.38}$$

According to (A.34), let us denote by

$$e_{2,a} = E_2(z - x_a), \quad e_{1,a} = E_1(z - x_a), \quad e_0 = 1 \tag{A.39}$$

the basis in M_2. The dual basis with respect to the integration on T_0^2 is

$$\begin{aligned} f_{2,a} &= \bar{\partial}(z - x_a)\chi_a(z, \bar{z}), \qquad f_{1,a} = \bar{\partial}\chi_a(z, \bar{z}), \\ f_0 &= \bar{\partial}(\bar{z} - z)\Big(1 - \sum_{a=1}^{n} \chi_a(z, \bar{z})\Big), \end{aligned} \tag{A.40}$$

where χ_a is the characteristic function of the neighborhood $\mathcal{U}_a$ of x_a (see (4.25)).

Integrals.

$$\sum_a c_{1,a} \int_{T_\tau^2} E_1(z - x_a|\tau) = \sum_a c_{1,a}(x_a - \bar{x}_a), \tag{A.41}$$

$$\int_{T_\tau^2} E_2(z - x_a|\tau) = -2\pi i. \tag{A.42}$$

References

[1] H. Flashka and A. Newell, *Monodromy and spectrum preserving deformations*, I, Comm. Math. Phys. **76** (1980), 65–116.

[2] M. Jimbo, T. Miwa, and K. Ueno, *Monodromy preserving deformations of linear ordinary differential equations*, I, II, Physica D **2** (1981), 306–352; 407–448

[3] N. Hitchin, *Stable bundles and integrable systems*, Duke Math. J. **54** (1987), 91–114.

[4] V. Knizhnik and A. Zamolodchikov, *Current algebra and Wess–Zumino model in two dimensions*, Nuclear Phys. B **247** (1984), 83–103.

[5] D. Bernard, *On the Wess–Zumino–Witten models on the torus*, Nuclear Phys. B **303** (1988), 77–93; *On the Wess–Zumino–Witten models on the Riemann surfaces*, Nuclear Phys. B **309** (1988), 145–174.

[6] D. Ivanov, *Knizhnik–Zamolodchikov–Bernard equations on Riemann surfaces*, Internat. J. Modern Phys. A **10** (1995), 2507–2536.

[7] N. Nekrasov, *Holomorphic bundles and many-body systems*, Comm. Math. Phys. **180** (1996), 587–604; `hep-th/9503157`.

[8] B. Enriques and V. Rubtsov, *Hitchin systems, higher Gaudin operators and r-matrices*, Math. Res. Lett. **3** (1996), 343–357.

[9] G. Felder, *The KZB equations on Riemann surfaces*, hep-th/9609153. Symétries Quantiques (Les Houches, 1995), North Holland, Amsterdam, 1998, pp. 687–725.
[10] D. Ivanov, *KZB eqs. as a quantization of nonstationary Hitchin systems*, hep-th/9610207.
[11] A. Beilinson and V. Drinfeld, *Quantization of Hitchin's fibration and Langlands program*, Preprint, 1994; Algebraic and Geometric Methods in Mathematical Physics (Kaciveli, 1993; A. Boutet de Monvel and V. Marchenko, eds.), Math. Phys. Stud., vol. 19, Reidel, Dordrecht, 1996, pp. 3–7.
[12] F. Falceto and K. Gawedzky, *Elliptic Wess–Zumino–Witten model from elliptic Chern–Simons theory*, hep-th/9502161; Lett. Math. Phys. **38** (1996), 155–175.
[13] P. Etingof and A. Kirillov, *Representations of affine Lie algebras, parabolic differential equations and Lamé functions*, Duke Math. J. **74** (1994), 585–614.
[14] G. Felder and C. Wieczerkowski, *Conformal blocks on elliptic curves and Knizhnik–Zamolodchikov–Bernard equations*, hep-th/9411004; Comm. Math. Phys. **176** (1996), 133–161.
[15] N. Hitchin, *Flat connections and geometric quantization*, Comm. Math. Phys. **131** (1990), 347–380.
[16] S. Axelrod, S. Della Pietra, and E. Witten, *Geometric quantization of the Chern–Simons gauge theory*, J. Differential Geom. **33** (1991), 787–902.
[17] N. Reshetikhin, *The Knizhnik–Zamolodchikov system as a deformation of the isomonodromic problem*, Lett. Math. Phys. **26** (1992), 167–177.
[18] J. Harnad, *Quantum isomonodromic deformations and the Knizhnik–Zamolodchikov equations*, hep-th/9406078; Symmetries and Integrability of Differential Equations (Estérel, 1994; D. Levi et al., eds.), CRM Proc. & Lecture Notes, vol. 9, Amer. Math. Soc., Providence, RI, 1996, pp. 155–161.
[19] D. Korotkin and J. Samtleben, *On the quantization of isomonodromic deformations on the torus*, hep-th/9511087; Internat. J. Modern Phys. A **12** (1997), 2013–2030.
[20] K. Okamoto, *Déformation d'une équation différentielle linéaire avec une singularité irrégulière sur un tore*, J. Fac. Sci. Univ. Tokyo Sect. IA Math. **26** (1979), 501–518.
[21] K. Iwasaki, *Fuchsian moduli on a Riemann surface—its Poisson structure and Poincaré–Lefschetz duality*, Pacific J. Math. **155** (1992), 319–340.
[22] D. Levi and P. Winternitz (eds.), *Workshop on Painlevé Transcedents, Their Asymptotics and Physical Applications*, NATO ASI Ser. B: Physics Vol. 278, Plenum Press, New York, 1992.
[23] K. Okamoto, *Isomonodromic deformation and the Painlevé equations and the Garnier system*, J. Fac. Sci. Univ. Tokyo Sect. IA Math. **33** (1986), 575–618.
[24] B. Gambier, *Sur les équations différentielles du second ordre et du premier degré dont l'intégrale générale a ses points critiques fixes*, C. R. Acad. Sci. Paris **142** (1906), 266–269.
[25] R. Fuchs, *Über lineare homogene Differentialgleichungen zweiter Ordnung mit im endlich gelegne wesentlich singulären Stellen*, Math. Annalen **63** (1907), 301–323.
[26] Painlevé, *Sur les équations différentielles du second ordre à points critiques fixes*, C. R. Acad. Sci. **143** (1906), 1111–1117.
[27] Yu. I. Manin, *Sixth Painlevé equation, universal elliptic curve, and mirror of* $\mathbf{P}^2$, alg-geom/9605010, Geometry of Differential Equations, A. Khovanskii, A. Varchenko, V. Vassiliev (eds.), Amer. Math. Soc. Transl. (2) **186** (1998), 131–151.
[28] B. Dubrovin, *Integrable systems in topological field theories*, Nuclear Phys. B **379** (1992), 627–689.
[29] F. Calogero, *Exactly solvable one-dimensional many-body problems*, Lett. Nuovo Cimento **13** (1976), 411–417.
[30] V. Inozemtsev, *Lax representation with spectral parameter on a torus for integrable particle systems*, Lett. Math. Phys. **17** (1989), 11–17.
[31] A. Treibich and J.-L. Verdier, *Revêtements tangentiels et sommes de 4 nombres triangulaires*, C. R. Acad. Sci. Paris Sér. I Math. **311** (1990), 51–54.
[32] N. Hitchin, *Twistors spaces, Einstein metrics and isomonodromic deformations*, J. Diff. Geom. **3** (1995), 52–134.
[33] K. Okamoto, *Studies on the Painlevé equations* I, *Sixth Painlevé*, Ann. Mat. Pura Appl. **146** (1987), 337–381.
[34] I. Krichever, *The tau-function of the universal Whitham hierarchy, matrix models and topological field theories*, Comm. Pure Appl. Math. **47** (1994), 437–475.

[35] K. Okamoto, *On the tau-function of the Painlevé equations*, Phys. D **2** (1981), 525–535.

[36] A. Its and V. Novokshenov, *The isomonodromic deformations method in the theory of Painlevé equations*, Lecture Notes Math., Vol. 1191, Springer-Verlag, Berlin–New York, 1986.

[37] K. Takasaki, *Spectral curves and Whitham equations in the isomonodromic problems of Schlesinger type*, solv-int/9704004.

[38] S. Novikov and A. Veselov, *Poisson brackets that are compatible with algebraic geometry and Korteweg–de Vries dynamics on the space of finite-zone potentials*, Dokl. Akad. Nauk SSSR **266** (1982), no. 3, 533–537; English transl., Soviet Math. Dokl. **26** (1982), 357–362.

[39] V. Vereshchagin, *Nonlinear quasiclassics and Painlevé equations*, hep-th/9605092.

[40] I. Krichever, *Elliptic solutions of the Kadomtsev–Petviashvili equation and many-body problems*, Funktsional. Anal. i Prilozhen. **14** (1980), no. 4, 45–54; English transl., Functional Anal. Appl. **14** (1980) no. 4, 282–290.

[41] M. Olshanetsky and A. Perelomov, *Explicit solutions of the Calogero models in the classical case and geodesic flows on symmetric spaces of zero curvature*, Lett. Nuovo Cimento **16** (1976), 333–339; *Explicit solutions of some completely integrable systems*, Lett. Nuovo Cimento **17** (1976), 97–101.

[42] F. Calogero, *On a functional equation connected with integrable many-body problems*, Lett. Nuovo Cimento **16** (1976), 77–80; Lectures at Spring school, Trieste (1990).

[43] J. Bernstein and O. Schvartsman, *Chevalley's theorem for complex crystallographic Coxeter groups*, Funktsional. Anal. i Prilozhen. **12** (1978), no. 4, 79–80; English transl., Functional Anal. Appl. **12** (1978), no. 4, 308–310.

[44] A. Weil, *Elliptic functions according to Eisenstein and Kronecker*, Springer-Verlag, Berlin–Heidelberg, 1976.

Institute of Oceanology, Moscow, Russia
E-mail address: andrl@landau.ac.ru

Institute of Theoretical and Experimental Physics, Moscow, Russia
E-mail address: olshanet@heron.itep.ru

Amer. Math. Soc. Transl.
(2) Vol. **191**, 1999

Infinite-Dimensional Algebras, Many-Body Systems and Gauge Theories

N. Nekrasov

ABSTRACT. In this survey we present several constructions of Hamiltonian reductions, leading to integrable many-body systems. The original phase spaces are naturally constructed with the help of infinite-dimensional current algebras. Along these lines we construct integrable systems of Calogero type, their relativistic and spin generalizations. The degenerations of Hitchin systems are particular cases of these systems. We discuss various dualities relating integrable systems and show that these dualities are easily obtained in the framework of Hamiltonian reductions. We also discuss the applications to solutions of gauge theories in various dimensions.

1. Introduction

1.1. Motivations. Tractable (integrable, solvable, exactly solvable, quasi-exactly-solvable, ...) many-body systems are interesting for several reasons:

1. They inform us about the behavior of real many-particle systems and may serve as an interesting approximation to them.
2. The wavefunctions of the many-body systems are spherical functions on various group-like objects: Lie algebras, homogeneous spaces, Lie groups, affine groups, double-loop groups, gauge groups, quantum groups, etc.
3. The coordinates of particles in the many-body system behave like eigenvalues of random matrices in a broad sense. This principle suggests connections to gauge theories in various dimensions. In particular, the large N limit in gauge theory corresponds to the thermodynamic limit in the particle system.
4. The many-body systems encode information about the exact low-energy effective actions of supersymmetric gauge theories; mathematically speaking, the geometry of the phase spaces of the integrable system is the way to describe the intersection theory on the moduli spaces of curves, stable maps, holomorphic bundles and Higgs systems.

1991 *Mathematics Subject Classification.* Primary 17B65, 58F07; Secondary 81T13.
The research was supported by the Harvard Society of Fellows, partially by the NSF under grant PHY-92-18167, and partially by the RFBR under grant 96-02-18046 and by grant 96-15-96455 for scientific schools.

1.2. Notation. Throughout the paper the following notations are used:

$\underline{\mathbf{g}}$ a compact simple finite-dimensional Lie algebra;
f_{ab}^{c} its structure constants in some basis T_a;
$\underline{\mathbf{t}}$ a Cartan subalgebra of $\underline{\mathbf{g}}$;
$\underline{\mathbf{g}}_{\mathbb{C}}$ the complexified Lie algebra $\underline{\mathbf{g}}$;
$\underline{\mathbf{t}}_{\mathbb{C}}$ the complexified Cartan subalgebra;
$\mathbf{a}$ a Lie algebra (perhaps infinite-dimensional);
$\mathbf{A}$ the corresponding Lie group;
W the Weyl group of $\underline{\mathbf{g}}$;
Δ the set of roots in $\underline{\mathbf{g}}_{\mathbb{C}}$;
Δ_+ the set of positive roots;
G the compact simple Lie group whose Lie algebra is $\underline{\mathbf{g}}$;
$\mathbf{T}$ the maximal torus;
$G_{\mathbb{C}}$ the complex simple Lie group whose compact real form is G;
$G_{\mathbb{R}}$ the real algebraic group, noncompact real form of $G_{\mathbb{C}}$;
Λ the weight lattice, $\Lambda \subset \underline{\mathbf{t}}$;
α a root, $\alpha \in \Delta$, also an element of $\underline{\mathbf{t}}^*$;
e_α the element of the Chevalley basis of $\underline{\mathbf{g}}$ corresponding to α;
h_i a Chevalley base vector in $\underline{\mathbf{t}}$;
$\langle\ ,\ \rangle$ the positive definite Killing form; for $G = SU(N)$ this form is normalized in such a way that $\langle X, Y\rangle = -\operatorname{Tr} XY$, where Tr is taken in the basic N-dimensional representation; the dual algebra $\underline{\mathbf{g}}^*$ is identified by means of $\langle\ ,\ \rangle$ with $\underline{\mathbf{g}}$ once and for all;
(P, Ω) a symplectic manifold P with symplectic form Ω;
I the interval $[0, 1]$, including the endpoints;

Finally, (H, P, Ω) denotes the Hamiltonian system on the symplectic manifold (P, Ω), generated by the Hamiltonian H.

2. The cast

The main characters of our play are the integrable systems describing the interaction of the collections of indistinguishable particles. The phase space of the many-body system has the form

$$P = X/W,$$

where X is the phase space of the system of particles (which are labelled) and W is the group of allowed permutations.

Basic example. $X = Y \times \cdots \times Y$ (N factors) and W is the symmetric group S_N. The space Y is the fiber bundle over the space Z with fiber F:

$$\begin{array}{ccc} F & \to & Y \\ & & \downarrow \\ & & Z \end{array}$$

One may distinguish the following cases:

	Nonrelativistic	Relativistic
rational	$Y = T^*Z,\ Z = \mathbb{R}^1, \mathbb{C}^1$	$F = S^1, \mathbb{C}^*,\ Z = \mathbb{R}^1, \mathbb{C}^1$
trigonometric	$Y = T^*Z,\ Z = S^1, \mathbb{C}^*$	$F = S^1, \mathbb{C}^*,\ Z = S^1, \mathbb{C}^*$
elliptic	$Y = T^*Z$, Z an elliptic curve	$F = S^1, \mathbb{C}^*$, Z an elliptic curve

One fixes the symplectic form $\omega = dp \wedge dq$ on Y which determines the symplectic form on X/W:

$$\Omega = \pi_* \sum_{i=1}^{N} p_i^* \omega, \tag{2.1}$$

where $\pi\colon Y^{\times N} \to X/W$ is the projection-factorization, π_* is the corresponding pushforward map and $p_i\colon Y^{\times N} \to Y$ is the projection onto the ith factor.

The typical system has a basic Hamiltonian H, which gives rise to the time evolution of the model. One has also the higher Hamiltonians, which Poisson-commute with H and give rise to the higher flows. These Hamiltonians provide a complete set of Poisson-commuting integrals of motion.

The basic Hamiltonian H in the nonrelativistic case has the form of the sum of kinetic T and potential U energies. The kinetic energy is the quadratic form in momenta

$$T = \sum_{i=1}^{N} \frac{1}{2} p_i^2 \tag{2.2}$$

while the potential energy is the sum of the pairwise potentials

$$U = \sum_{i,j} U(q_i, q_j), \qquad q_i \in Z,\ i = 1, \dots, N. \tag{2.3}$$

The potentials $U(q_i, q_j)$ can be roughly associated to the root systems and/or Coxeter groups.

EXAMPLE 2.1. For the root system of A_N type, the basic Hamiltonian in the elliptic case has the form

$$H = \sum_{i=1}^{N} \frac{1}{2} p_i^2 + \sum_{i \neq j} g^2 \wp(q_i - q_j \mid \omega_1, \omega_2), \tag{2.4}$$

where g^2 is the coupling constant (a complex number). The Weierstrass function $\wp(x \mid \omega_1, \omega_2)$ is the following series:

$$\begin{aligned} \wp(x \mid \omega_1, \omega_2) &= \frac{1}{\omega_1^2} \wp\left(\frac{x}{\omega_1} \,\middle|\, \frac{\omega_2}{\omega_1} \right), \\ \wp(x; \tau) &= \frac{1}{x^2} + \sum_{(m,n) \in \mathbb{Z}^2 - (0,0)} \frac{1}{(m\tau + n + x)^2} - \frac{1}{(m\tau + n)^2}. \end{aligned} \tag{2.5}$$

In the limit as $\operatorname{Im} \tau \to \infty$ the function $\wp(x; \tau)$ reduces to $\pi^2 \sinh^{-2}(\pi x) - \pi^2/3$. The model defined by the Hamiltonian (2.4) is known as the *elliptic Calogero–Moser system.*

The relativistic Hamiltonians are trigonometric functions in the momenta p. They do not split as the sum of kinetic and potential energies. Rather, the dependence on the momenta is mixed in an intricate way with the dependence on the coordinates.

EXAMPLE 2.2. Ruijsenaars–Schneider model:

$$H = \sum_{i=1}^{N} \cosh \beta p_i \prod_{j \neq i} \sqrt{1 - g^2 \wp(q_{ij})}. \tag{2.6}$$

3. The philosophy

In this section we explain the mechanism of the appearence of the integrable systems described in the previous section. In the subsequent sections we show that all of them can be obtained as a result of the reduction of simple "free motion" systems defined on somewhat bigger (infinite-dimensional) phase spaces. The reduction involved is the Hamiltonian reduction with respect to the action of a certain natural symmetry group.

DEFINITION 3.1. The *"big" phase space* P is the cotangent bundle to an $\mathbf{A}$-manifold $M_{\mathbf{A}}$, where $\mathbf{A}$ is a (possibly, infinite-dimensional) Lie group: $P = T^*M_{\mathbf{A}}$.

The group $\mathbf{A}$ acts on P and gives rise to the moment map

$$\mu \colon P \to \mathbf{a}^*.$$

Explicitly, if $V_a \in \mathrm{Vect}(M_{\mathbf{A}})$ is the vector field on $M_{\mathbf{A}}$ corresponding to the element $\xi_a \in \mathbf{a}$ of the Lie algebra of $\mathbf{A}$, then

$$\mu_a(x,p) = \langle p, V_a(x) \rangle,$$

where $p \in T_x^*M_{\mathbf{A}}$ and $\langle\ ,\ \rangle$ is the canonical pairing between $T_xM_{\mathbf{A}}$ and $T_x^*M_{\mathbf{A}}$. We proceed by imposing the moment map equations. We choose a certain coadjoint orbit $\mathcal{O} \in \mathbf{a}^*$ and consider its preimage $\mu^{-1}(\mathcal{O}) \subset P$. This preimage is $\mathbf{A}$-invariant.

DEFINITION 3.2. The *reduced phase space* $P_{\mathcal{O}}$ is the quotient

$$P_{\mathcal{O}} = \mu^{-1}(\mathcal{O})/A.$$

It carries a canonically induced symplectic form. In the problems that we are going to consider, the quotients may have some mild singularities (they are orbifolds).

There exists a parallel theory in which the big phase space, the group $\mathbf{A}$, etc. are complex. The quotients in the complex setting must be defined in a more intricate way. One approach is to define the reduced phase spaces as hyper-Kähler quotients [**1, 2**]. In the present paper, we almost never use them, except when the complex setup is forced on us, for example in the study of elliptic systems.

PROPOSITION 3.1. *A Hamiltonian system on P gives rise to a system on $P_{\mathcal{O}}$ if the restriction of the Hamiltonian to $\mu^{-1}(\mathcal{O})$ is $\mathbf{A}$-invariant. It is called the* reduced *system.*

PROPOSITION 3.2. *Suppose that $l = \dim(\mu^{-1}(\mathcal{O})/A)$ Poisson-commuting functions $F_1, \ldots, F_l$ are defined on $P \times \mathcal{O}^\vee$, where $\mathcal{O}^\vee$ denotes the orbit $\mathcal{O}$ with the symplectic form which is equal to minus the standard Kirillov symplectic form on $\mathcal{O}$. Let $\mu \colon P \times \mathcal{O}^\vee \to \mathbf{a}^*$ be the moment map. Suppose that $dF_1 \wedge \cdots \wedge dF_l \neq 0$ on $\mu^{-1}(0)$. Let H be one of the functions $F_1, \ldots, F_l$. Suppose that $\{F_k, \mu^a\} = 0$ on $\mu^{-1}(0)$ for any k, a. Then $(P_{\mathcal{O}}, H, \Omega_{\mathcal{O}})$ is an integrable system.*

In the next sections we study the rational, trigonometric and elliptic cases of nonrelativistic systems.

4. Nonrelativistic systems: rational case

The structure of the big phase space encodes the nature of the reduced system one gets as the output. For example, the rational models correspond to the case in which the phase space is the cotangent bundle $P = T^*\underline{\mathbf{g}}$ to the Lie algebra $\underline{\mathbf{g}}$. The natural symmetry group in question is the Lie group G acting on $\underline{\mathbf{g}}$ via the adjoint representation. Let $\mathbf{T}$ be the maximal torus $\mathbf{T} \subset G$. The symplectic quotient $T^*\underline{\mathbf{g}}//G$ depends on the level of the moment map. In fact, the reduction data contains the choice of the coadjoint orbit $\mathcal{O} \subset \underline{\mathbf{g}}^*$. The set of orbits $\underline{\mathbf{g}}^*/G$ can be identified with $\underline{\mathbf{t}}^*/W$, where $\underline{\mathbf{t}}$ is the Cartan subalgebra of $\underline{\mathbf{g}}$ and W is the Weyl group. For the simplest choice of trivial orbit, the quotient is $X = T^*\underline{\mathbf{t}}/W$. Let $\pi\colon T^*\underline{\mathbf{t}}/W \to \underline{\mathbf{t}}/W$ be the projection.

EXAMPLE 4.1. For $G = SU(N)$ this space is the set of N unordered pairs (p_i, q_i) with the restriction

$$\sum_{i=1}^{N} q_i = \sum_{i=1}^{N} p_i = 0.$$

DEFINITION 4.2. The *discriminant* $\Sigma \subset \underline{\mathbf{t}}/W$ is the set of irregular orbits. We denote by $\widetilde{\Sigma}$ its preimage under the projection π: $\widetilde{\Sigma} = \pi^{-1}(\Sigma)$.

THEOREM 4.3. *Let G be a simple compact group Lie. Let $\mathcal{O} \subset \underline{\mathbf{g}}^*$ be any coadjoint orbit of G. The reduced phase space $P_{\mathcal{O}}$ contains an open dense subset $\widetilde{P}_{\mathcal{O}}$ which is a fiber bundle over $\widetilde{X} = X \setminus \widetilde{\Sigma}$. The fibers are isomorphic to the symplectic quotient $\mathcal{O}//T$ of the orbit $\mathcal{O}$ by the action of the torus $\mathbf{T}$ at the zero level of the moment map.*

PROOF. The moment map $\mu\colon T^*\underline{\mathbf{g}} \to \underline{\mathbf{g}}^*$ sends the pair (P, Q) to $\mathrm{ad}^*_Q(P) \in \underline{\mathbf{g}}^*$. Using the identifications $\underline{\mathbf{g}} \approx \underline{\mathbf{g}}^*$ provided by the Killing form, one may write the moment equation as

$$[P, Q] = -J, \tag{4.1}$$

where J is the moment map $\mathcal{O} \to \underline{\mathbf{g}}^*$ (which is actually the tautological embedding). Let $q \in \underline{\mathbf{t}}$ be a representative of the conjugacy class of Q. The element q is defined uniquely up to the action of W. The complexified (co)algebra decomposes as the sum of the root spaces with respect to q:

$$\underline{\mathbf{g}}_{\mathbb{C}} = \underline{\mathbf{t}}_{\mathbb{C}} \oplus \bigoplus_{\alpha} \mathbb{C} \cdot e_\alpha. \tag{4.2}$$

The elements P and j decompose accordingly:

$$P = p \oplus \bigoplus_{\alpha} P_\alpha e_\alpha, \qquad J = j \oplus \bigoplus_{\alpha} J_\alpha e_\alpha,$$

where $p, j \in \underline{\mathbf{t}}$, $P_\alpha, J_\alpha \in \mathbb{C}$, $P_{-\alpha} = P^*_\alpha$, $J_{-\alpha} = J^*_\alpha$. The equation (4.1) is equivalent to

$$\langle q, \alpha\rangle P_\alpha = -J_\alpha, \qquad j = 0. \tag{4.3}$$

Thus P_α is uniquely reconstructed once J_α is known provided $\langle q, \alpha\rangle \neq 0$, which is the condition that $x = (p, q) \in \widetilde{X}$. The quotient $\mu^{-1}(\mathcal{O})/G$ is isomorphic to the quotient of the space $\{(p, q, J_\alpha)\}$ by the action of the normalizator of q, i.e., the semidirect product $\mathbf{T} \ltimes W$. Forgetting the J_α's maps the quotient $P_{\mathcal{O}}$ to the space

$\{p,q\}/W$, which is isomorphic to $T^*\underline{\mathbf{t}}/W$. The fiber is the quotient of the set of J_α's by the action of $\mathbf{T}$. It remains to prove that this quotient is isomorphic to $\mathcal{O}//\mathbf{T}$. Indeed, it is easy to show that the moment map is nothing but j. Setting it equal to zero and factoring with respect to $\mathbf{T}$ gives the desired result. □

REMARK 4.4. For certain $\mathcal{O}$ there is a stronger statement: $\widetilde{P}_{\mathcal{O}} = P_{\mathcal{O}}$. An example is $G = SU(N)$, $\mathcal{O} = \mathbb{CP}^{N-1}$.

DEFINITION 4.5. The orbit $\mathcal{O}$ is called *nice* if for any $J \in \mathcal{O}$ and for any $\alpha \in \Delta$ such that $J \perp \underline{\mathbf{t}}$ we have $\langle J, e_\alpha\rangle \neq 0$.

BASIC EXAMPLE. Here $G = SU(N)$ and $\mathcal{O} = \mathbb{CP}^{N-1}$, which is the conjugacy class of the element

$$j = \nu(\mathrm{Id} - e \otimes e^\dagger), \quad \text{where } e \in \mathbb{C}^N,\ (e^\dagger, e) = N. \tag{4.4}$$

PROPOSITION 4.1. *The orbit $\mathcal{O}$ from the basic example is nice.*

The reduced symplectic form. From the proof of the theorem, it is clear that the reduced symplectic form splits as

$$\omega_{P_{\mathcal{O}}} = \langle \delta p \wedge \delta q\rangle + \omega_{\mathcal{O}}, \tag{4.5}$$

where $\omega_{\mathcal{O}}$ is the symplectic form on $\mathcal{O}//\mathbf{T}$.

Hamiltonians. The obvious set of Poisson-commuting Hamiltonians is the set of functions on the reduced phase space that descend from the functions on $\underline{\mathbf{g}}^*$. But not all the functions on $\underline{\mathbf{g}}^*$ are G-invariant, even when they are restricted to $\mu^{-1}(0)$. Of course, the Casimir functions $u_k \in \mathrm{Fun}(\underline{\mathbf{g}}^*)^G$, $k = 1, \ldots, \mathrm{rk}\, G$, are invariant. The quadratic Casimir $H_2 = \frac{1}{2}\langle P, P\rangle$, $P \in \underline{\mathbf{g}}^*$, always gives rise to an invariant function. Its expression in the coordinates (p, q, J_α) on the reduced phase space $P_{\mathcal{O}}$, if these coordinates are compatible with its fiber bundle structure, is

$$H_2(p,q,J_\alpha) = \frac{1}{2}\langle p, p\rangle + \sum_{\alpha\in\Delta_+} \frac{g_\alpha^2 |J_\alpha|^2}{\langle q, \alpha\rangle^2}, \tag{4.6}$$

where $g_\alpha^2 = \langle e_\alpha, e_{-\alpha}\rangle$. In the basic example, the numerator in (4.6) is equal to ν^2.

THEOREM 4.6. *The Hamiltonian system $(H_2, \widetilde{P}_{\mathcal{O}}, \Omega_{P_{\mathcal{O}}})$ is Liouville-integrable for any $\mathcal{O}$.*

PROOF. We present the complete set of Poisson-commuting functions on $\widetilde{P}_{\mathcal{O}}$ containing H_2. Their functional independence is left to the reader as an exercise. Introduce the notation $\phi(z) = P - J/z \in \underline{\mathbf{g}}^*$, $z \in \mathbb{C}^*$. It is clear that

$$\{\phi_a(z), \phi_b(w)\} = f_{ab}^c \frac{\phi_c(z) - \phi_c(w)}{z - w}. \tag{4.7}$$

Let $u_1, \ldots, u_r$, $r = \mathrm{rk}\, G$, be the homogeneous multiplicative basis in the space of G-invariant polynomials on $\underline{\mathbf{g}}^*$. The degree d_l of u_l equals $m_l + 1$, where m_l is the lth exponent of the Weyl group of G. Obviously, the functions $u_l(\phi(z))$ and $u_{l'}(\phi(w))$ Poisson-commute for any z, w, l, l' and descend to $P_{\mathcal{O}}$. The function $I_l(z) = u_l(\phi(z))$ is the polynomial in z^{-1} of degree d_l. The space of such polynomials is of dimension $d_l + 1$. We get Poisson-commuting functions $I_{l,m}$, $0 \leqslant m \leqslant d_l + 1$, defined as

$$I_l(z) = \sum_{m=0}^{d_l+1} \frac{I_{l,m}}{z^m}.$$

The leading coefficient I_{l,d_l} in the z^{-1} expansion of $I_l(z)$ is fixed by the choice of $\mathcal{O}$. Also notice that

$$\phi(z) = \mathrm{Ad}^*_{e^{Q/z}} P \mod \frac{1}{z^2}$$

when restricted to $\mu^{-1}(0)$. Hence $I_{l,1} = 0$. We obtain the total number

$$\sum_{l=1}^{r} (d_l - 1) = \frac{1}{2} \dim(G/\mathbf{T}) \tag{4.8}$$

of Poisson-commuting functions, since for any simple Lie group (see **[3]** for various explanations of this identity) we have

$$\dim G = \sum_{l=1}^{r} (2d_l - 1).$$

In the case of a generic orbit $\mathcal{O}$ this concludes the proof, since $\dim P_{\mathcal{O}} = \dim \mathcal{O} = \dim(G/\mathbf{T})$. For the degenerate orbits, the number of functionally independent functions drops in the same way the dimension $\mathcal{O}$ does. □

5. Nonrelativistic systems: trigonometric case

In the trigonometric case there are two approaches.

DEFINITION 5.1. The *trigonometric big phase space* is T^*G.

This approach has its own advantages **[4]**, but in order to keep our presentation more logical, we start with a more recent point of view, originating in **[5, 6]**.

5.1. Affine algebra approach.

Let $\mathbf{a}_I = \mathrm{Map}(I, \underline{\mathbf{g}})$ be the Lie algebra of the group of smooth (C^1 will be sufficient) maps of the interval to the Lie group G. This algebra has no nontrivial central extensions, but its subalgebra $L\underline{\mathbf{g}}$ consisting of maps $\phi(t)$ such that $\phi(0) = \phi(1)$ and $\partial_t\phi|_{t=0} = \partial_t\phi|_{t=1}$ does. The extended algebra $L\underline{\mathbf{g}} \oplus \mathbb{R}$ is denoted by $\hat{\underline{\mathbf{g}}}$. Its adjoint representation can be extended to the action of the group LG on $\mathbf{a} = \mathbf{a}_I \oplus \mathbb{R}$ defined by the formula

$$g \cdot (\phi \oplus c) = \left(g\phi g^{-1} \oplus c + \int_I dt \, \langle \phi, g^{-1}\partial_t g \rangle \right). \tag{5.1}$$

DEFINITION 5.2. The *trigonometric big phase space* P is the cotangent bundle to the Lie algebra $\mathbf{a}$.

The dual space $\mathbf{a}^*$ can be identified with the first order operators $\nabla = k\partial + A$ acting on the sections of the trivial G-bundle over I **[7]**. Here k is a number.

The symmetry group is the loop group $\mathbf{A} = LG$. Its action on $\mathbf{a}^*$ is induced from (5.1),

$$A \to g^{-1}Ag + kg^{-1}\partial g, \qquad k \to k. \tag{5.2}$$

The big phase space has a canonical LG-invariant symplectic structure

$$i\Omega = \delta c \wedge \delta k + \int_I \langle \delta\phi \wedge \delta A \rangle. \tag{5.3}$$

The corresponding moment map sends (∇, ϕ, c) to

$$\mu_{\mathbf{a}} = [\nabla, \phi] = k\partial\phi + [A, \phi] - k\delta(t)(\phi(1) - \phi(0)). \tag{5.4}$$

The expression (5.4) should be understood as taking values in $\underline{\mathbf{g}}^*$-valued functions on the circle. Since $\phi(0)$ need not be equal to $\phi(1)$, the derivative $\partial_t\phi$ may have delta-function singularities, which are canceled by the compensating term with delta-function.

More generally, the loop group acts in the space $T^*\mathbf{a}_S$, where $\mathbf{a}_S$ is the Lie algebra of the group of smooth maps of any subset $S \subset S^1 = I/(\{0\} \sim \{1\})$ extended by $\mathbb{R}$. In the sequel we use the notation $S = S^1 \setminus \{t_1, \dots, t_K\}$. We shall denote the elements of $\mathbf{a}$ by $\phi(t) \oplus c$ and think of $\phi(t)$ as a function on the circle with possible jumps at $t_1, \dots, t_K$.

DEFINITION 5.3. The *evaluation orbit* $\mathcal{O}$ of LG is the pair $(\vec{t}, \vec{\mathcal{O}})$, where $\vec{t}$ is a collection of points $t_1, \dots, t_K \in S^1$ and $\vec{\mathcal{O}}$ is a collection of coadjoint orbits of the finite-dimensional group G: $\mathcal{O}_1, \dots, \mathcal{O}_K \subset \underline{\mathbf{g}}^*$.

The loop group acts on $\mathcal{O}$ via evaluation at the points t_l, i.e., $g(t_l)\colon \mathcal{O}_l \to \mathcal{O}_l$.

PROPOSITION 5.1. *The moment map for the action of LG on $\mathcal{O}$ is a sum of delta-functions supported at $t_1, \dots, t_K$,*

$$\mu_{\mathcal{O}} = \sum_{l=1}^{K} J_l \delta(t - t_l)\, dt. \tag{5.5}$$

The coefficient in front of $\delta(t - t_l)\, dt$ is the moment map of the action of G on $\mathcal{O}_l$ (which is simply the tautological map $J_l\colon \mathcal{O}_l \to \underline{\mathbf{g}}^*$).

The moment equation $\mu = -\mu_{\mathcal{O}}$ can be solved in the particular gauge

$$A = a \in \underline{\mathbf{t}}. \tag{5.6}$$

The element a is defined uniquely up to the action of the affine Weyl group. There are two kinds of gauge transformations preserving (5.6): the gauge transformations

$$g(t) = \exp 2\pi i t\lambda, \qquad \lambda \in \Lambda \subset \underline{\mathbf{t}}, \tag{5.7}$$

where Λ is the weight lattice (these shift a by $(ik/R)\lambda$), and the action of the Weyl group W. The semidirect product $\mathbf{T}_\Lambda \ltimes W$ is the affine Weyl group $\widehat{W}$ of G. It acts on a as follows:

$$(\lambda, w)\colon a \mapsto w \cdot a + \lambda. \tag{5.8}$$

DEFINITION 5.4. The *trigonometric discriminant* $\Sigma_{\mathbf{T}}$ is the image in $\underline{\mathbf{t}}/\widehat{W}$ of the union of all mirrors $\bigcup_\alpha \{q \in \underline{\mathbf{t}} \mid \langle q, \alpha \rangle = 0\}$. The variety $\widetilde{\Sigma}_{\mathbf{T}}$ is, by definition, the pullback $\pi^{-1}(\Sigma_{\mathbf{T}})$, where $\pi\colon T^*\underline{\mathbf{t}}/\widehat{W} \to \underline{\mathbf{t}}/W$ is the natural projection.

DEFINITION 5.5. A collection of orbits $\vec{\mathcal{O}}$ is called *nice* if for any vector $\vec{J} = (J_1, \dots, J_K) \in \mathcal{O}_1 \times \dots \mathcal{O}_K$ and for any $\alpha \in \Delta$ such that $\sum_l J_l \perp \underline{\mathbf{t}}$ we have $\langle \sum_l J_l, e_\alpha \rangle \neq 0$.

PROPOSITION 5.2. *The reduced phase space $P_{\mathcal{O}}$ is a fiber bundle over $\widetilde{X}_{\mathbf{T}} = (T^*\underline{\mathbf{t}}/\widehat{W})\backslash\widetilde{\Sigma}_{\mathbf{T}}$. The fibers are isomorphic to the symplectic quotient $\mathcal{O}_1\times\cdots\times\mathcal{O}_P//\mathbf{T}$ at the zero level of the moment map. If $\vec{\mathcal{O}}$ is not nice, then the proposition holds with $P_{\mathcal{O}}$ replaced by its dense open subset $\widetilde{P}_{\mathcal{O}}$.*

PROOF. The moment equation reads

$$k\partial\phi + [A,\phi] = -\sum_{l=1}^{K} J_l\delta(t-t_l). \tag{5.9}$$

Let $H(t)$ be the piecewise constant function $H(t) = [t]$. It is a function on the universal covering of S^1. It may be viewed as a section of a nontrivial $\mathbb{Z}$-bundle over S^1. We assume that $0 < t_1 \leqslant t_2 \leqslant \ldots \leqslant t_K < 1$. Choose the branch of $H(t)$ that takes the values 0 and 1 on the interval $[0,1)$. In the gauge (5.6), equation (5.9) is solved immediately:

$$\begin{gathered}\phi = p - \sum_{l=1}^{K}\sum_{i}\langle J_l, h_i\rangle H(t-t_l)h_i + \sum_{l=1}^{K}\sum_{\alpha\in\Delta}\langle J_l, e_{-\alpha}\rangle h\Big(t-t_l, \frac{\langle a,\alpha\rangle}{k}\Big),\\ h(x,y) = e^{-xy}\Big(\frac{1}{1-e^y} - H(x)\Big)e_\alpha.\end{gathered} \tag{5.10}$$

Here $p\in\underline{\mathbf{t}}$ and $(p,a)\in T^*\underline{\mathbf{t}}$. Equation (5.10) defines a function on the circle S^1 if and only if for any i we have

$$\sum_{l=1}^{K}\langle J_l, h_i\rangle = 0. \tag{5.11}$$

Condition (5.11) is the moment equation on the J_l's. Formula (5.10) makes perfect sense as long as there is no $\alpha\in\Delta$ such that $\langle a,\alpha\rangle\in 2\pi ik\mathbb{Z}$. This condition is equivalent to the requirement $(p,a)\in\widetilde{X}_{\mathbf{T}}$. The proposition is proved. □

EXAMPLE 5.6. $G = SU(N)$, $K=1$, $\mathcal{O}_1 = \mathbb{CP}^{N-1}$, $J_1 = i\nu(\mathrm{Id}-e\otimes e^\dagger)$, $e\in\mathbb{C}^N$, $(e^\dagger, e) = N$, $e\sim\lambda e$, for $|\lambda| = 1$. The equation (5.9) can be written explicitly in matrix notation:

$$k\partial\phi_{ii} = \nu(1-|e_i|^2)\delta(t), \qquad k\partial\phi_{ij} + (a_i-a_j)\phi_{ij} = -\nu e_i^* e_j\delta(t), \quad i\neq j. \tag{5.12}$$

Here $a = \mathrm{diag}(a_1,\ldots,a_N)$, $\sum a_i = 0$. The equation is solved by the following ansatz:

$$\begin{gathered}\phi_{ii} = p_i = \mathrm{const}_t, \qquad |e_i| = 1,\\ \phi_{ij}(t) = \exp\Big(-\frac{t}{k}(a_i-a_j)\Big)\phi_{ij}(0), \qquad \phi_{ij}(1)-\phi_{ij}(0) = \frac{\nu}{k}e_i^*e_j.\end{gathered} \tag{5.13}$$

The last equation follows from the relation

$$\partial f = l\delta(t) \implies f(+0) - f(-0) = l$$

and from the periodicity of $\phi_{ij}(t)$. The equation $|e_i| = 1$ allows us to set all the e_i's equal to 1 by the constant diagonal gauge transformation. Finally, we get

$$\phi_{ij}(t) = \frac{(\nu/k)e^{-ta_{ij}/k}}{e^{-a_{ij}/k}-1},$$

where $a_{ij} = a_i - a_j$.

COROLLARY 5.1. *Under the conditions of Proposition* 5.2, *the symplectic form on the reduced phase space equals*

$$\Omega_{P_{\mathcal{O}}} = -i\delta k \wedge \delta c - i\,\langle \delta a \wedge \delta p\rangle + \omega_{\mathcal{O}}, \tag{5.14}$$

where $\omega_{\mathcal{O}}$ *is the symplectic form on the symplectic quotient* $(\mathcal{O}_1 \times \cdots \times \mathcal{O}_P)//\mathbf{T}$.

PROOF. Direct computation. □

REMARK 5.7. In our basic example, the form $\omega_{\mathcal{O}}$ is empty since the orbit is reduced to a point.

The variable canonically conjugate to p is

$$q = ia.$$

Hamiltonians. The Hamiltonian H_2 comes from the quadratic Casimir on the big phase space:

$$H_2 = \frac{1}{2}\int_{S^1} dt\langle \phi, \phi\rangle; \tag{5.15}$$

this Hamiltonian is equal to

$$\begin{aligned} H_2 = {} & \frac{1}{2}\,\langle p, p\rangle - \frac{1}{2}\sum_{l=1}^{K}(1-t_l)\langle p - J_l, J_l\rangle \\ & + \sum_{K\geqslant l>l'\geqslant 1}(1-t_l)\langle \mathrm{Ad}_{e^{t_l q/k}} J_l, \mathrm{Ad}_{e^{t_{l'} q/k}} J_{l'}\rangle \\ & + \sum_{\alpha\in\Delta}\sum_{K\geqslant l>l'\geqslant 1} J_{l,\alpha}J_{l',-\alpha}\frac{e^{(t_l - t_{l'})\langle q,\alpha\rangle/k}}{4\sin^2(\langle q,\alpha\rangle/(2k))} \\ & \times [(e^{i\langle q,\alpha\rangle/k} - 1)(1-t_l) - (1 - e^{-i\langle q,\alpha\rangle/k})(1 - t_{l'}) + 1]. \end{aligned} \tag{5.16}$$

In the basic example, this expression simplifies and becomes the Sutherland Hamiltonian:

$$H_2 = \sum_{i=1}^{N}\frac{1}{2}p_i^2 + \sum_{i<j}\frac{\nu^2}{4k^2\sin^2(q_{ij}/(2k))}. \tag{5.17}$$

One can notice that the role of the level k is to set the period of the potential equal to $2\pi k$. One can rewrite the final potential in the form of an infinite sum of rational potentials:

$$U(x) = \sum_{n\in\mathbb{Z}}\frac{\pi\nu^2}{(x+2\pi k)^2}.$$

In the limit as $k \to \infty$ it reduces to the rational potential. In the rest of this section the level k is set equal to 1 (we hope that the reader will not confuse it with the index k).

THEOREM 5.8. *The reduced trigonometric system* $(H_2, \widetilde{P}_{\mathcal{O}}, \Omega_{P_{\mathcal{O}}})$ *is Liouville-integrable for any* $\mathcal{O}$.

PROOF. Fix the basic polynomials u_k as in the rational case. Consider the gauge invariant functions $I_{k,l}(z) = u_k(\phi_l(z))$, where $\phi_l(z) = \phi(t_l + 0) + J_l/z$. Expand in z^{-1}:

$$I_{k,l}(z) = \sum_{m=0}^{d_k} \frac{I_{k,l,m}}{z^m}. \tag{5.18}$$

As in the rational case, it is easy to show that all the $I_{k,l,m}$ Poisson-commute. Again the coefficients I_{k,l,d_k} are fixed by the choice of $\mathcal{O}$. Also, on the set of solutions to the moment map equations (5.9) we have the identity

$$\sum_{m=0}^{d_k} I_{k,l+1,m} = I_{k,l,0},$$

which follows from the equations

$$u_k(\phi(t_l - 0)) = u_k(\phi(t_l + 0) + J_l), \qquad u_k(\phi(t_{l+1} - 0)) = u_k(\phi(t_l + 0)).$$

Thus we have a total of $K \sum_{l=1}^{r}(d_l - 1)$ Poisson-commuting functions, which is enough to ensure integrability in the case of generic $\mathcal{O}$. What remains to prove is that H_2 can be expressed in terms of $I_{k,l,m}$ for suitable (k, l, m). In fact, an easy computation shows that

$$H_2 = \sum_{l=1}^{K} (t_{l+1} - t_l) I_{1,l,0}, \tag{5.19}$$

where $t_{K+1} := 1 + t_1$.

In the case $K = 1$, the model we get is a spin generalization of the Calogero–Moser system (see [**8, 9, 10**]). □

5.2. Models associated with graphs.

DEFINITION 5.9. An *oriented graph* Γ is a pair of finite sets E_Γ and V_Γ with two maps $(v_1, v_2)\colon E_\Gamma \to V_\Gamma \times V_\Gamma$. The elements of V_Γ are called *vertices*, the elements of E_Γ are called *edges*. The number of pairs $F = (v, e)$ such that $v_1(e) = v$ or $v_2(e) = v$ for a given v is denoted by $\mathrm{val}(v)$.

We shall also denote by Γ the one-dimensional CW-complex obtained from $E_\Gamma \times I$ by identifying the common endpoints, i.e., $(e, 0) \sim (e', 0)$ if and only if $v_1(e) = v_1(e')$, $(e, 0) \sim (e', 1)$ if and only if $v_1(e) = v_2(e')$, and $(e, 1) \sim (e', 1)$ if and only if $v_2(e) = v_2(e')$. We denote by I_e the interval $(e, I) \subset \Gamma$. It is supplied with the parameter t for which $t = 0$ corresponds to $v_1(e)$ and $v_2(e)$ corresponds to $t = 1$.

DEFINITION 5.10. The *graph gauge group* $\mathcal{G}_\Gamma$ is the subgroup in

$$\mathop{\times}_{e \in E_\Gamma} \mathrm{Map}(I_e, G)$$

consisting of the continuous maps of the CW-complex Γ to G that are smooth on $\Gamma \setminus V_\Gamma$. The Lie algebra $\underline{\mathbf{g}}_\Gamma$ is defined analogously.

PROPOSITION 5.3. *Let Γ be the graph with metric. The graph gauge group $\mathcal{G}_\Gamma$ has a universal central extension $\widetilde{\mathcal{G}_\Gamma}$ by the torus $H^1(\Gamma, \mathbb{R}/2\pi i\mathbb{Z})$:*

$$1 \to H^1(\Gamma, \mathbb{R}/2\pi i\mathbb{Z}) \to \widetilde{\mathcal{G}_\Gamma} \to \mathcal{G}_\Gamma \to 1. \tag{5.20}$$

The corresponding $H^1(\Gamma,\mathbb{R})$-valued Lie algebraic cocycle is given by the formula

$$(c(\phi_1,\phi_2),e)=\int_e \langle\phi_1,d\phi_2\rangle-\langle\phi_2,d\phi_1\rangle, \tag{5.21}$$

where e is any chain.

The proof is an easy adaptation of the constructions of [**11, 12, 7**].

DEFINITION 5.11. The *evaluation orbit* $\mathcal{O}$ of the group $\mathcal{G}_\Gamma$ is the collection $\mathcal{O}_v$, $v\in V_\Gamma$, of coadjoint orbits of G. An element $g\in\mathcal{G}_\Gamma$ acts on $\mathcal{O}_v$ via evaluation of g at the vertex v, which is well defined.

We also denote by $\mathcal{O}_\Gamma$ the product

$$\mathcal{O}_\Gamma=\underset{v\in V_\Gamma}{\times}\ \mathcal{O}_v.$$

DEFINITION 5.12. $\mathbf{a}_\Gamma=\times_{e\in E_\Gamma}\operatorname{Map}(I_e,\underline{\mathbf{g}})$.

Clearly, $\mathbf{a}_\Gamma$ is a $\underline{\mathbf{g}}_\Gamma$-module. The difference between $\underline{\mathbf{g}}_\Gamma$ and $\mathbf{a}_\Gamma$ is the possibility of having discontinuities at the vertices of the graph. The vector space $\mathbf{a}_\Gamma\oplus\operatorname{Map}(E_\Gamma,\mathbb{R})$ is acted on by the graph gauge group as follows:

$$g\colon(\phi,f)\mapsto(g\phi g^{-1},g\cdot f), \tag{5.22}$$

where for any $e\in E_\Gamma$ we have

$$g\cdot f(e)=f(e)+\int_e\langle\phi(t),g^{-1}\partial_t g(t)\rangle\,dt. \tag{5.23}$$

This action descends in an obvious way to the space $K_\Gamma=\mathbf{a}_\Gamma\oplus H^1(\Gamma,\mathbb{R})$.

Take an evaluation orbit $\mathcal{O}$ and take the product $P_\Gamma=T^*K_\Gamma\times\mathcal{O}$ as the big phase space. The symmetry group is the graph gauge group. It acts on P_Γ and preserves its canonical symplectic form. The action on K_Γ^* is the action given by the following explicit formulas. An element of K_Γ^* is a collection of pairs (A_e,k_e), where A_e is the 1-form on the unit interval representing the edge e and k_e is a real number such that $e\mapsto k_e$ represents the element k of $H^1(\Gamma,\mathbb{R})$. The gauge transformation g maps (A_e,k_e) to $(g^{-1}A_eg+k_eg^{-1}\partial_tg,k_e)$.

PROPOSITION 5.4. *Let $\varepsilon\in\underline{\mathbf{g}}_\Gamma$. The corresponding Hamiltonian H_ε is equal to*

$$H_\varepsilon=\langle\mu_{P_\Gamma},\varepsilon\rangle=\sum_{e\in E_\Gamma}\int_e\langle\phi,k_e\partial_t\varepsilon+[A,\varepsilon]\rangle+\sum_{v\in V_\Gamma}\langle J_v,\varepsilon(v)\rangle. \tag{5.24}$$

Here J_v is the tautological map $J_v\colon\mathcal{O}_v\to\underline{\mathbf{g}}^*$.

To each edge e let us assign a group element g_e and a pair of elements of $\underline{\mathbf{g}}$, $\phi_{1,e}$ and $\phi_{2,e}$, with the properties

$$\sum_{e:v_1(e)=v}k_e\phi_{1,e}-\sum_{e:v_2(e)=v}k_e\phi_{2,e}=-J_v,\qquad J_v\in\mathcal{O}_v. \tag{5.25}$$

Let us denote the space of solutions to (5.25) by $M_{\Gamma,\mathcal{O}}$. This space is acted on by the group $G_\Gamma=\operatorname{Map}(V_\Gamma,G)$.

THEOREM 5.13. *The quotient $\mathcal{M}_{\Gamma,\mathcal{O}}=M_{\Gamma,\mathcal{O}}/G_\Gamma$ is isomorphic to the quotient of P_Γ at the zero level of the moment map $\mu_{P_\mathcal{O}}$. The dimension of $\mathcal{M}_{\Gamma,\mathcal{O}}$ is equal to*

$$\dim\mathcal{M}_{\Gamma,\mathcal{O}}=\dim\mathcal{O}-2\chi(\Gamma)\dim G,$$

where $\chi(\Gamma) = \#V_\Gamma - \#E_\Gamma$ *is the Euler characteristic of* Γ.

PROOF. Choose the gauge $A_e = \text{const}$ for every edge e. Set $g_e = \exp A_e$. The moment equation for ϕ on the edge has a unique solution once the values $\phi(0)$ and $\phi(1)$ are given. They are provided by $\phi_{1,e}$ and $\phi_{2,e}$. The conditions (5.25) easily follow from (5.24). □

COROLLARY 5.2. *The symplectic form* $\Omega_{\Gamma,\mathcal{O}}$ *can be computed from the symplectic form on* $\mathrm{Map}(E_\Gamma, T^*G) \times \mathcal{O}$:

$$\Omega = \sum_{e\in E_\Gamma} k_e \langle \delta\phi_{1,e} \wedge \delta g_e g_e^{-1}\rangle + \langle \phi_{1,e}, \delta g_e g_e^{-1} \wedge \delta g_e g_e^{-1}\rangle + \sum_{v\in V_\Gamma} \omega_{\mathcal{O}_v}. \tag{5.26}$$

DEFINITION 5.14. A *metric* on the graph Γ is nonnegative function $\ell\colon E_\Gamma \to \mathbb{R}_+$. Fix a metric ℓ and define a function H_2 on $\mathcal{M}_{\Gamma,\mathcal{O}}$ as follows:

$$\begin{aligned} H_2 &= \frac{1}{2} \sum_{e\in E_\Gamma} \ell(e) \int_0^1 dt\, \langle \phi_e, \phi_e\rangle \\ &= \frac{1}{2} \sum_{e\in E_\Gamma} \ell(e) \langle \phi_{1,e}, \phi_{1,e}\rangle = \frac{1}{2} \sum_{e\in E_\Gamma} \ell(e) \langle \phi_{2,e}, \phi_{2,e}\rangle. \end{aligned} \tag{5.27}$$

THEOREM 5.15. *For generic* $\mathcal{O}_v$, *the system* $(H_2, \mathcal{M}_{\Gamma,\mathcal{O}}, \Omega_{\mathcal{M}_{\Gamma,\mathcal{O}}})$ *is integrable.*

PROOF. To every vertex $v \in V_\Gamma$ we associate a complex line $\mathbb{P}^1$ with parameter z and $\mathrm{val}(v)+1$ marked points on it. We denote them by $z_e^\pm$ and ∞. The points z_e^+ correspond to the edges e with $v_1(e) = v$ (outgoing edges), while z_e^- correspond to the edges e with $v_2(e) = v$. Define a $\underline{\mathbf{g}}_{\mathbb{C}}$-valued meromorphic 1-differential on $\mathbb{P}^1$:

$$\phi_v(z) = \sum_{e\colon v_1(e)=v} \frac{k_e \phi_{1,e}\, dz}{z - z_e^+} - \sum_{e\colon v_2(e)=v} \frac{k_e \phi_{2,e}\, dz}{z - z_e^+}. \tag{5.28}$$

Its residue at infinity is equal to J_v due to (5.25). The polynomials $u_k(\phi_v(z))$ are gauge invariant and descend to some functions on $\mathcal{M}_{\Gamma,\mathcal{O}}$. Using (5.26) it is easy to show that (4.7) holds. Therefore the functions $u_k(\phi_v(z))$ Poisson-commute for all k, z. It remains to compute the total number of independent functions and show that it is equal to $\frac{1}{2}\dim\mathcal{O} - \chi(\Gamma)\dim G$. The total number of meromorphic d_k-differentials on $\mathbb{P}^1$ with poles of order d_k at the $\mathrm{val}(v)+1$ points ($z_e^\pm$ and ∞) is equal to $d_k(\mathrm{val}(v)-1)+1$. The leading coefficient in the expansion of $\phi_v(z)$ near infinity is fixed by the choice of the orbit $\mathcal{O}_v$. Thus we have

$$\sum_{v\in V_\Gamma} \sum_{k=1}^{r} d_k(\mathrm{val}(v)-1) = \frac{1}{2}(\dim G + r)(2\#E_\Gamma - \#V_\Gamma)$$

parameters. The condition $u_k(\phi_{1,e}) = u_k(\phi_{2,e})$ reduces the total number of parameters by $r\cdot\#E_\Gamma$. Thus the net number of independent integrals of motion is equal to

$$(\#E_\Gamma - \tfrac{1}{2}\#V_\Gamma)\dim G - \tfrac{1}{2}\#V_\Gamma r = \#V_\Gamma \dim(G/\mathbf{T}) + (\#E_\Gamma - \#V_\Gamma)\dim G, \tag{5.29}$$

which is enough for generic orbits $\mathcal{O}_v$. The degenerate case can be treated by a more thorough inspection of the behavior of $L_v(z)$ near infinity. □

THEOREM 5.16. *Let all the orbits* $\mathcal{O}_v$ *be trivial. Let* Γ *be connected and* $\#E_\Gamma - \#V_\Gamma > 0$. *Then* $(H_2, \mathcal{M}_{\Gamma,\mathcal{O}}, \Omega_{\mathcal{M}_{\Gamma,\mathcal{O}}})$ *is an integrable system.*

PROOF. First of all we explain a certain homotopy equivalence. Suppose that some vertex v has $\mathrm{val}(v) = 1$; then there is only one edge e which contains this v. Let v' be the other end of e. The moment equation implies that $\phi_{1,e} = \phi_{2,e} = 0$. Also, the group element g_e can be gauged away by the action of a gauge transformation g_v. The result is that one may remove the vertex v and the edge e, leaving the system $(H_2, \mathcal{M}_{\Gamma,\mathcal{O}}, \Omega_{\mathcal{M}_{\Gamma,\mathcal{O}}})$ unchanged.

Now suppose that some vertex has $\mathrm{val}(v) = 2$. To be definite, assume that there are edges e, e', such that $v_2(e) = v$, $v_1(e') = v$. First assume that $v_1(e) \neq v$. The moment equation implies that $\phi_{2,e} = \phi_{1,e'}$ and $\phi_{2,e'} = \mathrm{Ad}_{g_{e'}} \phi_{1,e} = -\mathrm{Ad}_{g_{e'} g_e} \phi_{2,e}$. By a gauge transformation we can set g_e equal to 1 (since $v_1(e) \neq v$). The result is that the map $(g_e, g_{e'}) \mapsto g_{e'} g_e$, $(\phi_{1,e}, \phi_{2,e}, \phi_{1,e'}, \phi_{2,e'}) \mapsto (\phi_{1,e}, \phi_{2,e'})$ establishes an isomorphism between $(\mathcal{M}_\Gamma, \Omega_{\mathcal{M}_\Gamma})$ and $(\mathcal{M}_{\Gamma'}, \Omega_{\mathcal{M}_{\Gamma'}})$, where Γ' is obtained from Γ by deleting the vertex v and replacing the edges e, e' by a single edge which goes from $v_1(e)$ to $v_2(e')$.

If the graph Γ has a vertex v with $\mathrm{val}(v) = 2$, and an edge e with $v_2(e) = v$ also has $v_1(e) = v$, then the graph is a loop and has $\#E_\Gamma = \#V_\Gamma$, which contradicts the assumptions of the theorem.

Hence, by using these simple transformations, we can get rid of all vertices with valency less than 3. We can now assume that all the vertices of the graph have valency greater or equal to 3. Counting valencies as in the previous proof, we obtain

$$\sum_{k=1}^{r} d_k(\mathrm{val}(v) - 2) + 1 \tag{5.30}$$

functions for such a vertex. The sum over v of such quantities is reduced by $r\#E_\Gamma$ due to the constraints $u_k(\phi_{1,e}) = u_k(\phi_{2,e})$. Thus we have

$$\left(\sum_{k=1}^{r} \sum_{v \in V_\Gamma} d_k(\mathrm{val}(v) - 2) + 1\right) - r\#E_\Gamma = (\#E_\Gamma - \#V_\Gamma) \dim G. \tag{5.31}$$

□

REMARK 5.17. By complexifying the space and the symmetry group, one may notice the similarity of the phase space obtained (by treating the parameter z seriously) with the degeneration of the Hitchin system. We shall come back to this point later.

5.3. Models associated with quivers. Let Γ be an oriented graph with a function $d\colon V_\Gamma \to \mathbb{N}$. To each vertex $v \in V_\Gamma$ we assign a Hermitian vector space $L_v = \mathbb{C}^{d(v)}$. Consider the big phase space

$$P_\Gamma = \bigoplus_{e \in E_\Gamma} \mathrm{Hom}(L_{v_1(e)}, L_{v_2(e)}). \tag{5.32}$$

It is acted on by the gauge group

$$G_\Gamma = \mathop{\times}_{v \in V_\Gamma} U(d(v)), \tag{5.33}$$

and by its various subgroups $G_{\Gamma,S}$, $S \subset V_\Gamma$:

$$G_{\Gamma,S} = \mathop{\times}_{v \in S} U(d(v)). \tag{5.34}$$

This action preserves the canonical symplectic form:

$$\Omega_\Gamma = \frac{1}{2i} \sum_{e \in E_\Gamma} \operatorname{Tr} \delta B_{v_1(e),v_2(e)} \wedge \delta B^\dagger_{v_1(e),v_2(e)} \tag{5.35}$$

for $B_{v,v'} \in \operatorname{Hom}(L_v, L_{v'})$. Let $\underline{\mathbf{g}}_S$ be the Lie algebra of $G_{\Gamma,S}$. The group $G_{\Gamma,S}$ has the center $U(1)^S$. Let $\zeta\colon S \to \mathbb{R}$ be a function. The moment map $\mu\colon P_\Gamma \to \underline{\mathbf{g}}^*_S$ is given by the explicit formulas

$$\begin{gathered} \mu = \bigoplus_{v \in S} \mu_v, \\ i\mu_v = \sum_{e\colon v_1(e)=v} B^\dagger_{v_1(e),v_2(e)} B_{v_1(e),v_2(e)} - \sum_{e\colon v_2(e)=v} B^\dagger_{v_1(e),v_2(e)} B_{v_1(e),v_2(e)}. \end{gathered} \tag{5.36}$$

DEFINITION 5.18. The reduced phase space $\mathcal{M}_{\Gamma,S,\zeta}$ is the quotient of the space of solutions to the equations $\mu_v = \zeta(v) \cdot \operatorname{Id}_{L_v}$ by the group $G_{\Gamma,S}$.

PROPOSITION 5.5. *For generic ζ, the space $\mathcal{M}_{\Gamma,S,\zeta}$ is a smooth symplectic manifold of dimension*

$$2\Big(\sum_{e \in E_\Gamma} d(v_1(e))\, d(v_2(e)) - \sum_{v \in S} d(v)^2 \Big).$$

REMARK 5.19. The analysis of the previous section extrapolated to the case $G = U(N)$ (the subtlety is that it is not a simple group) for the graph $\widetilde{\Gamma}$ is a particular case of the new construction. To this end one takes as the graph Γ the union of $\widetilde{\Gamma}$ and $\#V_\Gamma$ copies of the A_N Dynkin diagram,[1] glued at the vertices of $\widetilde{\Gamma}$. The dimensions d are the following: on $V_{\widetilde{\Gamma}}$, $d = N$, and on every Dynkin diagram, $d(i) = N - i$, $i = 0, \ldots, N-1$.

There exists a set of Poisson-commuting functions on the quotient $\mathcal{M}_{\Gamma,S,\zeta}$ defined as in the construction in (5.28), where

$$\phi_{1,e} = B_{v_1(e)v_2(e)} B^\dagger_{v_1(e)v_2(e)}, \qquad \phi_{2,e} = B^\dagger_{v_1(e)v_2(e)} B_{v_1(e)v_2(e)}. \tag{5.37}$$

PROBLEM 5.20. To describe the conditions on S and d under which the reduced phase space is an integrable system.

Let Γ be an oriented graph. Let $H_\Gamma = E_\Gamma \amalg E_\Gamma$. Let $\sigma\colon H_\Gamma \to H_\Gamma$ be the involution which acts by permuting the summands E_Γ. Let $v_1, v_2\colon H \to V_\Gamma$ be defined as follows. On the first summand they act as before, while the action on the second is deduced from the formula $v_{1,2} \circ \sigma = v_{2,1}$. The elements of H_Γ are denoted by h. The elements of the first E_Γ in H_Γ are still denoted by e.

DEFINITION 5.21. The structure $Q = (\Gamma, H, \sigma, v_{1,2})$ is called a *quiver*. Let $d\colon V_\Gamma \to \mathbb{N}$ be a function on the vertices with values in positive integers. The *big phase space P_Q associated to the quiver Q* is the product over all edges $h \in H_\Gamma$ of the vector spaces

$$P_Q = \bigoplus_{h \in H_\Gamma} \operatorname{Hom}(\mathbb{C}^{d(v_1(h))}, \mathbb{C}^{d(v_2(h))}). \tag{5.38}$$

[1] $E_\Gamma = \mathbb{Z}_{N-1}$, $V_\Gamma = \mathbb{Z}_N$, $v_1(i) = i$, $v_2(i) = i+1$, $i = 0, \ldots, N-2$. There is a distinguished vertex 0 which is to be identified with v.

The elements of $\mathrm{Hom}(\mathbb{C}^{d(v_1(h))}, \mathbb{C}^{d(v_2(h))})$ are denoted by B_h. The full symmetry group is

$$G_Q = \mathop{\times}_{v \in V_\Gamma} \mathrm{GL}(d(v), \mathbb{C}). \tag{5.39}$$

For $S \subset V_\Gamma$ we define

$$U_S = \mathop{\times}_{v \in S} \mathrm{U}(d(v)). \tag{5.40}$$

Let $\vec{\zeta}\colon S \to \mathbb{R}^3$ be a function and let $\mathcal{M}_{\Gamma,H,S,\vec{\zeta}} = \vec{\mu}_S^{-1}(\vec{\zeta})/U_S$, where

$$\vec{\mu}_S = \bigoplus_{v \in S} (\mu_{v,\mathbb{R}}, \mu_{v,\mathbb{C}}, \bar{\mu}_{v,\mathbb{C}})\colon P_Q \to \underline{\mathbf{g}}_S \otimes \mathbb{R}^3$$

is given by

$$\begin{aligned} \mu_{v,\mathbb{R}} &= \sum_{h\colon v_1(h)=v} B_h^\dagger B_h - \sum_{h\colon v_2(h)=v} B_h^\dagger B_h, \\ \mu_{s,\mathbb{C}} &= \sum_{h\colon v_1(h)=v} B_{\sigma(h)} B_h - \sum_{h\colon v_2(h)=v} B_{\sigma(h)} B_h. \end{aligned} \tag{5.41}$$

PROPOSITION 5.6. *The manifold $\mathcal{M}_{\Gamma,H,S,\vec{\zeta}}$ is a hyper-Kähler smooth manifold if it is not empty.*

PROBLEM 5.22. When is $\mathcal{M}_{\Gamma,H,S,\vec{\zeta}}$ an integrable holomorphic system in one (and then in any!) of its complex structures?

6. Nonrelativistic systems: elliptic case

In accordance with the idea that the structure of the reduced phase space and the Hamiltonians must be encoded in the original "big" phase space, we proceed with a further enhancement of the "big" phase space. Namely, we take as the space P roughly the cotangent bundle to the central extension of the double loop current algebra $\hat{\underline{\mathbf{g}}}_E$. The precise definition follows.

Let E be an elliptic curve.

DEFINITION 6.1. The algebra $\hat{\underline{\mathbf{g}}}_E$ is the space of pairs $\phi \oplus c$, where ϕ is a map of a two-torus E into the Lie algebra $\underline{\mathbf{g}}$ and c is a number. The Lie structure on $\hat{\underline{\mathbf{g}}}_E$ is defined as follows:

$$[\phi_1(t) \oplus c_1, \phi_2(t) \oplus c_2] = [\phi_1, \phi_2](t) \oplus \int_E dt\, d\bar{t}\, \langle \phi_1, \bar{\partial}_t \phi_2 \rangle. \tag{6.1}$$

This central extension depends explicitly on the complex structure of E; in particular, t is a holomorphic coordinate on E. We identify E with the quotient $\mathbb{C}/(\mathbb{Z}+\tau\mathbb{Z})$, $\mathrm{Im}\,\tau > 0$.

A more refined definition replaces c by an element of the one-dimensional complex vector space $F = H^0(E, \Omega^1)^* \equiv H^1(E, \mathcal{O})$. Then the last formula in (6.1) can be reinterpreted as defining an element of F by the formula

$$c_{\phi_1,\phi_2}(\eta) = \int_E \eta \wedge \mathrm{Tr}(\phi_1 \bar{\partial}_t \phi_2), \qquad \eta \in H^0(E, \Omega^1).$$

The co-algebra $\hat{\underline{\mathbf{g}}}_E^*$ is the space of $(0,1)$-connections $\nabla = k\bar{\partial}_t + \bar{A}$. The rest of the construction is very similar to the trigonometric case. The group $\mathbf{A}$ is the group of maps of E to the group G. The moment map is $\mu = k\bar{\partial}_t \phi + [\bar{A}, \phi]$.

DEFINITION 6.2. The *evaluation orbit* $\mathcal{O}$ of the group $\mathbf{A}$ is the structure $(\vec{\mathcal{O}}, \vec{t})$, $\vec{\mathcal{O}} = (\mathcal{O}_1, \dots, \mathcal{O}_K)$, $\vec{t} = (t_1, \dots, t_K)$, where $\mathcal{O}_l$ are the coadjoint orbits of the complex Lie group $G_{\mathbb{C}}$ and $t_l \in E$. The group $\mathbf{A}$ acts on $\mathcal{O}_l$ via evaluation $g(t_l)$ at the point t_l.

DEFINITION 6.3. The *elliptic big phase space* is the product $T^*\widetilde{\mathbf{a}} \times \mathcal{O}$, where an evaluation orbit $\mathcal{O}$ of $\mathbf{a}$ is chosen and the Lie algebra $\widetilde{\mathbf{a}}$ consists of all the maps of $E - \{t_1, \dots, t_K\}$ to $\underline{\mathbf{g}}_{\mathbb{C}}$.

DEFINITION 6.4. The *reduced phase space* is the quotient of P by the action of $\mathbf{A}$ at the zero level of the moment map.

THEOREM 6.5. *The reduced phase space is an integrable system whose phase space is a fibration over the space* $(T^*\underline{\mathbf{t}}_{\mathbb{C}})/W$ *with generic fiber isomorphic to*

$$(\mathcal{O}_1 \times \cdots \times \mathcal{O}_K)//\mathbf{T}_{\mathbb{C}}.$$

PROOF. The moment equation has the form

$$k\bar{\partial}_t\phi + [\bar{A}, \phi] = -\sum_{l=1}^{K} J_l \delta^{(2)}(t - t_l). \tag{6.2}$$

We begin solving this equation by imposing the gauge

$$A = a = \text{const} \in \underline{\mathbf{t}}. \tag{6.3}$$

There are two kinds of gauge transformations that preserve (6.3): the Cartan transformations

$$g(t) = \exp \frac{\pi}{\operatorname{Im}\tau} n(\bar{t} - t) + m(\tau\bar{t} - \bar{\tau}t), \qquad n, m \in \Lambda, \tag{6.4}$$

which shift a by $i\pi k(\operatorname{Im}\tau)^{-1}(n + \tau m)$, and the action of W on the a's. Together they form the elliptic affine Weyl group $\widehat{W_E} = W \ltimes \Lambda \oplus \tau\Lambda$ of G. Introduce the following functions on $\mathbb{C} \times \mathcal{H} \ni x, y \times \tau$:

$$\begin{gathered} h(x,y) = e^{y(x-\bar{x})/(\operatorname{Im}\tau)} H(x,y), \\ H(x,y) = \frac{\theta_{11}(x+y)\theta'_{11}(0)}{\theta_{11}(x)\theta_{11}(y)}, \qquad H(x,y) = \frac{1}{x} + o(1), \quad x \to 0, \\ H(x+1,y) = H(x,y), \qquad H(x+\tau,y) = e^{-2\pi i y} H(x,y) \end{gathered} \tag{6.5}$$

with

$$\theta_{11}(x) = \sum_{n \in \mathbb{Z}+1/2} e^{i\pi\tau n^2 + 2\pi i n(z+1/2)}.$$

Using (6.5), we can write out the expression for ϕ in the generic case; i.e., in the case $\langle a, \alpha\rangle \neq 0$ for any α:

$$\phi = p - \sum_{l=1}^{K}\sum_{i} \langle J_l, h_i\rangle \frac{\theta'_{11}(t-t_l)}{\theta_{11}(t-t_l)} h_i + \sum_{l=1}^{K}\sum_{\alpha\in\Delta} \langle J_l, e_{-\alpha}\rangle h\Big(t - t_l, \frac{\langle a, \alpha\rangle}{k}\Big). \tag{6.6}$$

The condition of single-valuedness of ϕ implies that for any $i = 1, \dots, r$

$$\sum_{l=1}^{K} \langle J_l, h_i\rangle = 0, \tag{6.7}$$

which is nothing but the moment equation for the action of $\mathbf{T}_{\mathbb{C}}$ on $\mathcal{O}_1 \times \cdots \times \mathcal{O}_K$. Taking the quotient with respect to $\mathbf{T}_{\mathbb{C}}$ is part of the reduction with respect to

$\mathbf{A}$, since the torus is what is left after the gauge-fixing (6.3). Factoring along the action of the Weyl group W on the pair (p, a) projects it to a point of $T^*\underline{\mathbf{t}}_{\mathbb{C}}/W$, as promised. $\square$

EXAMPLE 6.6. $\mathbf{G}_{\mathbb{C}} = SL_N(\mathbb{C})$. For the coadjoint orbit $\mathcal{O}$ we take the orbit of the element

$$J = \nu(\mathrm{Id} - v \otimes u), \tag{6.8}$$

where $(v, u) = N$ and v, u are defined up to the $\mathbb{C}^*$ action:

$$\begin{pmatrix} v \\ u \end{pmatrix} \mapsto \begin{pmatrix} \lambda^{-1} v \\ \lambda u \end{pmatrix} \quad \text{for } \lambda \in \mathbb{C}^*,$$

$$\begin{aligned} k\bar{\partial}_t \phi_{ii} &= \nu(1 - v_i u_i)\delta^2(t, \bar{t}), \\ k\bar{\partial}_t \phi_{ij} + (a_i - a_j)\phi_{ij} &= -\nu v_i u_j \delta^2(t, \bar{t}), \qquad i \neq j. \end{aligned} \tag{6.9}$$

The equation is solved by the following ansatz:

$$\begin{gathered} \phi_{ii} = p_i = \mathrm{const}_t, \qquad v_i u_i = 1, \qquad \phi_{ij}(t, \bar{t}) = \exp\Big(\frac{t - \bar{t}}{k} a_{ij}\Big)\psi_{ij}(t), \\ \psi_{ij}(t + 1) = \psi_{ij}(t), \qquad \psi_{ij}(t + \tau) = \exp\Big(-2i\frac{\operatorname{Im}\tau}{k} a_{ij}\Big)\psi_{ij}(t), \\ \psi_{ij}(t) = -\frac{\nu}{2k\pi i t} v_i u_j + o(1), \qquad t \to 0. \end{gathered} \tag{6.10}$$

The last equation follows from the relation $\bar{\partial}_t f = l\delta^2(t) \implies f(t) \sim l/(2\pi i t)$ and from the periodicity of $\phi_{ij}(t)$. The equation $v_i u_i = 1$ allows us to set all the v_i's and u_i's equal to 1 by the constant diagonal gauge transformation. Introduce the variables

$$q_i = \frac{\operatorname{Im}\tau}{k} a_i.$$

They are defined up to additions of $i\pi(n_i + m_i\tau)$ and permutations.

Finally, we get

$$\phi_{ij}(t, \bar{t}) = -\frac{\nu}{2\pi i k} h\Big(t, \frac{q_{ij}}{\pi k}\Big), \tag{6.11}$$

where the Hamiltonian H_2 descends from the quadratic Casimir on the big phase space

$$H_2 = \frac{1}{\operatorname{Im}\tau}\int_E dt\, d\bar{t}\, \operatorname{Tr}\phi^2,$$

which is equal to

$$H_2 = \sum_{i=1}^{N} \frac{1}{2} p_i^2 - \sum_{i<j} \frac{\nu^2}{4\pi^2 k^2} \wp\Big(\frac{q_{ij}}{\pi k}; \tau\Big) - \mathrm{const}$$

on the solutions of the moment equation. The symplectic form equals

$$\Omega = -i\delta k \wedge \delta c - i\sum_{i=1}^{N} \delta q_i \wedge \delta p_i.$$

One can notice that the role of the "level" k is to set the periods of the potential equal to πk and $\tau\pi k$. One can again rewrite the final potential in the form of infinite sum of rational potentials. In the limit as $k \to \infty$, it reduces to the rational potential. In the limit as $\operatorname{Im}\tau \to \infty$, k fixed, it reduces to the trigonometric potential.

The generalization to the case of a simple Lie algebra $\underline{\mathbf{g}}$ and of the collection of orbits $\mathcal{O}_l$, located at the points t_l, $l = 1, \dots, K$, is straightforward. One can also present a hyper-Kähler structure on the reduced phase space, simultaneously giving a precise definition of how the quotient in Definition 6.4 with respect to the complex gauge group is defined. In addition to $\bar{A}$ and ϕ, introduce their Hermitian conjugates $\bar{\phi}$, A. The complex coadjoint orbits $\mathcal{O}_l$ are hyper-Kähler manifolds. Let $\vec{J}_l$ be the corresponding hyper-Kähler maps. Then the quotient is defined as a quotient with respect to the compact gauge group of the space of solutions to the equations (we set $k = 1$ for brevity):

$$\begin{aligned} F_A + [\phi, \bar{\phi}] &= -\sum_l J_{l,\mathbb{R}} \delta^{(2)}(t - t_l), \\ \bar{\partial}\phi + [\bar{A}, \phi] &= -\sum_{l=1}^{K} J_{l,\mathbb{C}} \delta^{(2)}(t - t_l), \\ \partial\bar{\phi} + [A, \bar{\phi}] &= -\sum_{l=1}^{K} \bar{J}_{l,\mathbb{C}} \delta^{(2)}(t - t_l). \end{aligned} \tag{6.12}$$

One can also elaborate further on the models associated to graphs and quivers to include the elliptic systems into the game. See [**13**] for explicit examples.

7. Relativistic case

The approach discussed in the previous sections has a number of generalizations. One of them yields relativistic models as the result of reduction. Another one yields systems whose configuration spaces involve higher-genus Riemann surfaces.

The systems described above had one thing in common—they were all obtained as a result of the Hamiltonian reduction of a linear, perhaps infinite-dimensional, symplectic space. Apparently, replacing the linear space by a curved one may give rise to more interesting dynamics of the reduced model. On the other hand, the curved space must have a tractable Hamiltonian system to begin with.

We "curve" each of the cases considered so far. More concretely, every cotangent bundle $T^*\mathbf{a}$, where $\mathbf{a}$ is the Lie algebra of a Lie group $\mathbf{A}$, will be replaced by the cotangent bundle to this group, $T^*\mathbf{a} \to T^*\mathbf{A}$. All the rest of the construction is going to be unchanged, i.e., the group action, the level of the moment map, and the Hamiltonians of the original system can be copied from the corresponding equations above.

7.1. Rational case: $T^*\underline{\mathbf{g}} \to T^*G$. Let (p, g) be the coordinates on T^*G, $g \in G$, $p \in T_1^*G$. The moment map for the adjoint action of G is

$$\mu = \mathrm{Ad}_g^*(p) - p = g^{-1}pg - p \tag{7.1}$$

under obvious identifications. The moment equation is solved as usual, with the diagonalization of g first and then solving for p, with the result

$$p = \mathrm{diag}(p_1, \dots, p_N) + \left\| \frac{\nu g_i}{g_i - g_j} \right\|, \tag{7.2}$$

where the g_i's are the eigenvalues of g. The reduced symplectic form is equal to

$$\Omega = \sum_i \delta p_i \wedge \delta \log g_i$$

and the reduced Hamiltonian $\mathrm{Tr}\, p^2$ coincides with the Sutherland Hamiltonian (5.17). If, on the other hand, one solves the moment equation by diagonalizing p first, and then solving for g, then the answer turns out to be different:

$$g = -\frac{e^{i\theta_i}}{p_j + \nu - p_i}\sqrt{\frac{P(p_i - \nu)P(p_j + \nu)}{-\nu^2 P'(p_i)P'(p_j)}}, \tag{7.3}$$

where

$$P(x) = \prod_{i=1}^{N}(x - p_i). \tag{7.4}$$

Now it makes sense to choose as Hamiltonians the G-invariant functions of g, i.e., $\mathrm{Tr}\, g^k$, $k \in \mathbb{Z}$. In particular, $H_{\mathrm{rel}} = \mathrm{Tr}(g + g^{-1})$ gives rise to

$$H_{\mathrm{rel}} = \sum_i \cos\beta\theta_i \prod_{j\neq i}\sqrt{1 - \frac{(\beta\nu)^2}{(p_i - p_j)^2}}. \tag{7.5}$$

This is the *rational* limit of the Ruijsenaars–Schneider model [**14**].

Since the first system turned out to be equivalent to the one we obtained when we used affine Lie algebras, one may wonder whether the same is true for the second one. The answer is positive. Indeed, by choosing the gauge where ϕ is diagonal (and necessarily constant along the circle except possibly at the point $t = 0$), one can prove that the same answer for

$$g = P\exp\int_0^{2\pi}\frac{1}{k}A_t\, dt$$

can be obtained, with the identification $\beta \sim 1/k$.

7.2. Trigonometric case: $T^*\hat{\underline{\mathbf{g}}} \to T^*\widehat{LG}$**.** Here we replace the affine Lie algebra $\hat{\underline{\mathbf{g}}}$ by the corresponding group, i.e., the central extension of the loop group of G. For brevity, we only sketch the moment map equations and their solution

$$\mu = kg^{-1}\partial g + g^{-1}Ag - A. \tag{7.6}$$

There are (as in the previous case) two ways to parametrize the solutions. The first one is to diagonalize A [**6**]:

$$A = \mathrm{diag}(\varphi_1, \ldots, \varphi_N).$$

In the second approach, the gauge for which g is diagonal is chosen. It turns out that both lead to the same system. This is a manifestation of the self-duality of the *trigonometric* Ruijsenaars–Schneider model (see Section 9).

The reduced model is shown in [**6**] to have the Hamiltonian

$$H_{\mathrm{rel}} = \int_{S^1} dt\, \mathrm{Tr}(g + g^{-1}),$$

which can be computed to be

$$H_{\mathrm{rel}} = \sum_i \cos\beta\theta_i \prod_{j\neq i}\sqrt{1 - \frac{\sin^2(\beta\nu/2)}{\sin^2(\varphi_i - \varphi_j)/2}}, \tag{7.7}$$

where the $i\varphi_i$'s are the eigenvalues of A. The proof that θ_i and φ_i are canonically conjugate is easy in the gauge for which g is diagonal.

7.3. Elliptic case: $T^*\hat{\underline{\mathbf{g}}}_E \to T^*\widehat{G}_E$. This generalization produces the elliptic version of Ruijsenaars–Schneider model and unifies all the systems described so far. The phase space one starts with is the cotangent bundle to the central extension of the double loop group, the group of maps of the two-torus E to the complex semisimple Lie group G (which is taken to be $G = SL_N(\mathbb{C})$). This is a fairly straightforward generalization of [**15**], and it had been conjectured to yield the elliptic Ruijsenaars–Schneider model already at the end of 1993 in [**15, 6**]. The problem with the actual proof of this conjecture is in the singular nature of the moment equations. The problem was successfully attacked recently in [**16, 17, 18**].

8. Higher genera

The next line of generalizations starts with the observation of V. Fock that our construction of elliptic Calogero–Moser systems closely resembles the construction of Hitchin's systems.

8.1. Hitchin's setup. In [**3**] Hitchin introduced a collection of beautiful integrable systems on the partial resolution and compactification of $T^*\mathcal{M}$, where $\mathcal{M}$ is the moduli space of (semi-)stable holomorphic G-bundles over an algebraic curve Σ. For our purposes we do not need the exact definition of his resolution. The main point of his construction is to start with the infinite-dimensional affine symplectic space (he works in the complex category, so his space is a complex space with holomorphic $(2,0)$ symplectic form) with the Hamiltonian action of the infinite-dimensional group. The space is the cotangent bundle to the space $\mathcal{A}$ of $(0,1)$ forms with values in $\underline{\mathbf{g}}$, a complex Lie algebra (which we take to be $gl_N(\mathbb{C})$). The generic $(0,1)$ $\underline{\mathbf{g}}$-valued form $\bar{A}$ defines a holomorphic structure in the (topologically) trivial vector bundle F over a curve Σ. Let us denote the corresponding holomorphic bundle by the script letter $\mathcal{F}$. The local holomorphic sections of this bundle are solutions of the equation

$$\bar{\partial}_A\psi = \bar{\partial}\psi + \bar{A}\psi = 0. \tag{8.1}$$

The cotangent space to the space of those forms is the space of $\underline{\mathbf{g}}$-valued $(1,0)$ differentials ϕ. Sometimes the field ϕ is called a *Higgs field*. The symmetry group in question is the gauge group $\mathcal{G}$ of the maps $\Sigma \to G$. It acts on the differential operator $\bar{\partial}_A = \bar{\partial} + \bar{A}$ via conjugation:

$$\bar{\partial}_A \to g^{-1}\bar{\partial}_A g, \qquad \bar{A} \to g^{-1}\bar{A}g + g^{-1}\bar{\partial}g, \tag{8.2}$$

and on ϕ in a similar manner: $\phi \to g^{-1}\phi g$. The action preserves the symplectic form

$$\Omega = \int_\Sigma \operatorname{Tr} \delta\phi \wedge \delta\bar{A} \tag{8.3}$$

and gives rise to the moment map

$$\mu = \bar{\partial}\phi + [\bar{A}, \phi]. \tag{8.4}$$

Reduction at the zero level of the moment map produces the finite dimensional space, which can be (formally) identified with the cotangent bundle to the quotient

$\mathcal{M} = \mathcal{A}/\mathcal{G}$. Its dimension equals

$$\dim \mathcal{M} = \dim \mathrm{Ad}(G)(g-1) + \dim Z(G)g = (N^2-1)(g-1) + g. \tag{8.5}$$

One can find an integrable system on $T^*\mathcal{M}$ by looking at the Poisson-commuting gauge-invariant functions on $T^*\mathcal{A}$. A nice (and the only possible!) set of functions is obtained as follows. Pick on the Lie algebra $\underline{\mathbf{g}}$ a set of independent invariant polynomials $u_k(\phi)$, where k denotes the degree of the polynomial. For $G = GL_N(\mathbb{C})$ these are the traces:

$$P_k(\phi) = \mathrm{Tr}\, \phi^k, \qquad k = 1, \ldots, N.$$

For ϕ obeying $\mu = 0$ with μ given by (8.4), one has

$$\bar{\partial} u_k(\phi) = 0;$$

i.e., $u_k(\phi)$ is a holomorphic $(k, 0)$ differential on Σ. The space $H^0(\Sigma, \omega_\Sigma^k)$ of such differentials is finite dimensional, and its dimension is equal to $(2k-1)(g-1)$ for $k > 1$ and g for $k = 1$. Therefore, one gets a total of

$$g + \sum_{k=2}^{N} (2k-1)(g-1) = g + (N^2-1)(g-1) = \frac{1}{2} \dim T^*\mathcal{M} \tag{8.6}$$

holomorphic Poisson-commuting functions on $T^*\mathcal{M}$. This is precisely the number needed for integrability. We have therefore represented $T^*\mathcal{M}$ as a fibration over the linear space:

$$p\colon T^*\mathcal{M} \to \mathcal{H} = \bigoplus_{i=1}^{N} H^0(\Sigma, \omega_\Sigma^i), \qquad p(\bar{\partial}_A, \phi) = (\mathrm{Tr}\, \phi^2, \ldots, \mathrm{Tr}\, \phi^N). \tag{8.7}$$

The next step in Hitchin's construction is to introduce the spectral curve C. It is the N-sheeted cover of Σ, defined as a divisor of the N-differential $\det(\phi(z) - \lambda)$ in $\mathbb{P}T^*\Sigma$. Hitchin proves that it is generically a smooth curve of genus

$$g(C) = N^2(g-1) + 1. \tag{8.8}$$

Here $\lambda \in T_z^*\Sigma$, $z \in \Sigma$. The secret of the spectral curve is that it "abelianizes" the description of the holomorphic bundle $\mathcal{F}$ with the Higgs field ϕ. Indeed, one has a natural projection

$$\pi\colon C \to \Sigma \tag{8.9}$$

which enables one to pull back the holomorphic bundle $\mathcal{F}$ from Σ to C: $\widetilde{\mathcal{F}} = \pi^*\mathcal{F}$. The bundle $\widetilde{\mathcal{F}}$ has a line subbundle $\mathcal{L}$, whose fiber at any generic point (λ, z) is the eigenspace of $\phi(z)$ corresponding to the eigenvalue λ. At a generic branch point (let us call it $z = 0$) the local picture is the following (we have chosen here the relevant 2×2 piece of ϕ):

$$\phi(z) \sim \begin{pmatrix} l + \sqrt{z} & 0 \\ 0 & l - \sqrt{z} \end{pmatrix} \sim \begin{pmatrix} l & 1 \\ z & l \end{pmatrix}; \tag{8.10}$$

i.e., at the point $z = 0$ the matrix ϕ degenerates to a Jordan block with a well-defined one-dimensional eigenspace (the degenerations to the semisimple elements of $gl_N(\mathbb{C})$ occur at the higher codimension locus in $T^*\mathcal{M}$ and do not affect the argument). Conversely, given a line bundle $\mathcal{L}$ over C, one forms the direct image sheaf $\pi_*\mathcal{L}$ over Σ, which eventually turns out to be the vector bundle $\mathcal{F}$.

The importance of the spectral curve for integrable systems is that the Hamiltonian flows generated by the Hamiltonians (8.7) linearize on the Jacobian Jac(C) of C. This was shown by Hitchin in a very nice way in [**3**]. Below we present a direct computational proof which is useful for further applications.

8.2. Action-angle variables. There exist local coordinates on the phase space in which the dynamics of the integrable Hamiltonian system is the simplest. More precisely, the action-angle variables (I_i, φ^i) are canonically conjugate variables for which the Hamiltonians are functions of the I_i's only.

In Hitchin's setup, the action-angle variables can be identified rather directly. Let us choose a gauge for which ϕ is diagonal,

$$\phi(z) = \mathrm{diag}(\lambda_1(z)\, dz, \ldots, \lambda_N(z)\, dz).$$

This gauge cannot be chosen globally on Σ. The trick is to cut out the small discs D_ϵ around the branch points and to choose such a gauge on $\Sigma - \bigcup D_\epsilon$. There the equation $\mu = 0$ implies that $\bar{A}$ is also diagonal,

$$\bar{A} = \mathrm{diag}(\bar{a}_1\, d\bar{z}, \ldots, \bar{a}_N\, d\bar{z}).$$

The "eigen"forms $\bar{a}_i$ do not need to be holomorphic or antiholomorphic. The gauge transformations which leave ϕ unchanged are the diagonal matrices $g = \mathrm{diag}(g_1, \ldots, g_N)$, $g_i \in \mathbb{C}^*$, $\prod_i g_i = 1$. They shift the a_i's as follows:

$$\bar{a}_i\, d\bar{z} \longrightarrow \bar{a}_i\, d\bar{z} + \bar{\partial} \log g_i.$$

One can choose a gauge for which $\partial \bar{a}_i = 0$ (we leave aside the fact that the curve Σ with the discs deleted is not compact and this gauge might not be sufficient, because the final assertion can be justified further). A more precise treatment shows that the $\lambda_i dz$'s and $\bar{a}_i$'s are well defined on $C - \pi^{-1}(\bigcup D_\epsilon)$. The symplectic form can be computed as follows:

$$\Omega = \lim_{\epsilon \downarrow 0} \int_{\Sigma \setminus (\bigcup D_\epsilon)} \sum_{i=1}^{N} \delta\lambda_i \wedge \delta\bar{a}_i\, d^2 z = \lim_{\epsilon \downarrow 0} \int_{C \setminus \pi^{-1}(\bigcup D_\epsilon)} \delta\omega \wedge \delta\bar{a}, \tag{8.11}$$

where ω is the canonical holomorphic $(1,0)$ differential on $T^*\Sigma$, i.e., $\omega = \lambda\, dz$, and $\bar{a}$ is the $(0,1)$-connection on C determining the holomorphic structure on $\mathcal{L}$. Now, since $\bar{\partial}\omega = \partial\bar{a} = 0$, one can rewrite the last integral in (8.11) as follows:

$$\Omega = \sum_{a=1}^{g(C)} \delta I_a \wedge \delta\varphi^a, \tag{8.12}$$

where I_a are the periods of ω and φ^a are the linear coordinates on Jac(C) conjugate to I_a:

$$I_a = \frac{1}{2\pi} \oint_{A^a} \lambda\, dz, \qquad \varphi^a = \int_{B_a} \bar{a}, \qquad \oint_{A^b} \omega^a = \delta^a_b. \tag{8.13}$$

Here we have used the following fact: if α is a holomorphic 1-differential on a compact curve C and $\bar{\beta}$ is an antiholomorphic one, then

$$\int \alpha \wedge \bar{\beta} = \sum_{a,b} \eta^{ab} \int_{C^a} \alpha \int_{C^b} \bar{\beta},$$

where C^a runs through the base in $H_1(C;\mathbb{Z})$ and η^{ab} is the intersection form.

In the context of integrable systems, the Higgs field ϕ is usually referred to as the Lax operator, depending on the spectral parameter z. The importance of this derivation is that it is applicable to degenerate cases, considered below.

Relation to other constructions. It is amusing to note that Krichever in 1978–80 [**19**] presented essentially the same construction of the Lax operator and the action-angle-like variables for the elliptic Calogero–Moser system. He discovered that the evolution of the model is part of the evolution of solutions to the Kadomtsev–Petviashvili (KP) hierarchy. The knowledge of certain finite-gap solutions and their algebro-geometric constructions enabled him to find the structures we described above. Later on, in [**20**], a more abstract presentation of the elliptic solitons in the KP hierarchy was given. What was lacking was the description of the elliptic Calogero–Moser system as a reduced system. The presentation of the next sections fills this gap and at the same time suggests a generalization of the Calogero–Moser systems.

Indeed, what is the place of elliptic Calogero–Moser systems in the framework of the Hitchin systems?

Claim. The construction of elliptic Calogero–Moser system is a particular case of the generalization of Hitchin's approach, which can be carried out for degenerate curves.

In the sequel we denote by $\underline{\mathbf{g}}$ a complex simple Lie algebra and by G the corresponding complex Lie group.

8.3. Degenerate curves and bundles. To elaborate on this question, it is useful to keep in mind the following paradigm of two-dimensional conformal field theories (CFT). CFT's are formulated on compact Riemann surfaces or on Riemann surfaces with punctures. The punctures correspond to the insertions of vertex operators. When the Riemann surface Σ degenerates, by developing a node or a double point, the result can be represented as a CFT on the normalization $\widetilde{\Sigma}$ with the insertion of a complete set of operators at the resolved double point or a node.

One can show that a similar phenomenon happens in the context of Hitchin systems. Indeed, we have described the systems corresponding to the smooth curve Σ. What happens when the curve degenerates? This question was analyzed in [**13**] and the following answer obtained. Let $\widetilde{\Sigma}$ denote the normalization of the stable curve Σ. It has several components, which we label as Σ_v. The normalization $\widetilde{\Sigma}$ projects down to Σ, say $\rho\colon \widetilde{\Sigma} \to \Sigma$. Each component Σ_v has a set of punctures x_v^α. The nodal points on Σ correspond to the pairs (x_v^α, x_w^β) such that $v \neq w$ and $\rho(x_v^\alpha) = \rho(x_w^\beta)$. The double points correspond to the pairs (x_v^α, x_v^β) such that $\alpha \neq \beta$ and $\rho(x_v^\alpha) = \rho(x_v^\beta)$.

DEFINITION 8.1. A *holomorphic bundle* $\mathcal{E}$ *over the degenerate curve* Σ is a collection of holomorphic bundles $\mathcal{E}_v$ over the components Σ_v with the isomorphisms g_v^α and $g_w^\beta = (g_v^\alpha)^{-1}$ of the fibers over the normalizations x_v^α, x_w^β of the singular points.

DEFINITION 8.2. A *holomorphic subbundle* $\mathcal{E}'$ *of the holomorphic bundle* $\mathcal{E}$ *over the degenerate curve* Σ is a collection of the subbundles $\mathcal{E}'_v \subset \mathcal{E}_v$ over Σ_v with the condition $g_v^\alpha(\mathcal{E}'|_{x_v^\alpha}) = \mathcal{E}'|_{x_w^\beta}$.

DEFINITION 8.3. A holomorphic bundle $\mathcal{E}$ over the degenerate curve Σ is called *stable* (*semistable*) if for every subbundle $\mathcal{E}' \subset \mathcal{E}$ the slope $\mu(\mathcal{E}') = c_1(\mathcal{E}')/\operatorname{rk}\mathcal{E}'$ obeys the inequalities

$$\mu(\mathcal{E}') < \mu(\mathcal{E}) \qquad (\mu(\mathcal{E}') \leqslant \mu(\mathcal{E})). \tag{8.14}$$

DEFINITION 8.4. Let R be an irreducible representation of $\mathrm{GL}_N(\mathbb{C})$. The *holomorphic Higgs field* ϕ with values in the holomorphic bundle $R(\mathcal{E})$ associated to the holomorphic bundle $\mathcal{E}$ over the degenerate curve Σ is the collection of meromorphic sections ϕ_v of the bundles $R(\mathcal{E}_v) \otimes \omega_{\Sigma_v}$, whose poles are only at x_v^α's and which obey the conditions:

$$T_R(g_v^\alpha) \operatorname*{Res}_{x_v^\alpha} \phi_v = - \operatorname*{Res}_{x_w^\beta} \phi_w \tag{8.15}$$

for any pair (x_v^α, x_w^β) corresponding to a nodal point.

In the sequel we consider only the adjoint representation $R = \mathrm{Ad}$.

Graph. A simple way to visualize the construction is to build a graph Γ whose vertices correspond to the components Σ_α and whose edges are the punctures. More precisely, let us introduce the notion of flag. A *flag* F is a pair: (vertex v, edge e, containing v). If $F = (v, e)$ and $v = v_1(e)$, then define $\bar{F} = (v_2(e), e)$, and vice versa. The set of flags is denoted by F_Γ. The punctures x_v^e are in one-to-one correspondence with the flags F, and the singular points on Σ correspond to the edges. Each vertex v has two genera: an internal genus $g(v) \equiv \operatorname{genus}(\Sigma_v)$ and $\bar{g}(v) = g(v) + \#\{\text{double points on } \Sigma_v\}$. The condition of stability of the degenerate curve states that for each vertex v we have $2g(v) + \operatorname{val}(v) > 2$, where $\operatorname{val}(v)$ is the valency of the vertex.

One can assign the following data to the graph Γ and Σ. First of all, to each vertex v assign the space $\mathcal{A}_v$ of $\underline{\mathbf{g}}$-valued $(0,1)$ forms $\bar{A}_v$ on Σ_v and the corresponding cotangent space of $\underline{\mathbf{g}}$-valued $(1,0)$ forms ϕ_v, with possible singularities at the punctures x_v^e, such that the pairing

$$\int_{\Sigma_v} \operatorname{Tr} \bar{A}_v \phi_v$$

is well defined. The additional data consists of the space

$$\mathcal{M}_\Gamma = (T^*G)^{\#F_\Gamma - \#E_\Gamma}.$$

We represent this space as the cotangent space to the space $\mathcal{F}_G = \operatorname{Map}(F_\Gamma, G)^{\mathbb{Z}_2}$ of G-valued functions g_F on F_Γ which obey the condition that for two flags F and $\bar{F}$, the corresponding elements g_F and $g_{\bar{F}}$ are mutually inverse, $g_F g_{\bar{F}} = 1$. The cotangent space can be described by putting an element p_F of $\underline{\mathbf{g}}^*$ on every flag F so as to satisfy the condition

$$p_F = - \operatorname{Ad}^*_{g_F} p_{\bar{F}}.$$

There is a gauge group $\mathcal{G}_\Gamma$ defined as the product over all vertices v of the gauge groups $\mathcal{G}_v$ on Σ_v. The group $\mathcal{G}_v$ acts in a standard way on $T^*\mathcal{A}_v$. It also acts on $\mathcal{M}_\Gamma$ as follows.

Let $g(x_v) \in \mathcal{G}_v$ be the gauge transformation, i.e., a function on Σ_v. Its evaluation $g(x_v^e)$ at the puncture acts on p_F and g_F, for $F = (v, e)$, in a natural way:

$$g_F \to g_F g(x_v^e)^{-1}, \qquad p_F \to \operatorname{Ad}^*_{g(x_v^e)^{-1}} p_F.$$

The moment map for this action is the direct sum of the moments μ_v for each component Σ_v:

$$\mu_v = \bar{\partial}\phi_v + [\bar{A}_v, \phi_v] = \sum_{F=(v,e)} p_F \delta^2(x - x_v^e). \tag{8.16}$$

Analysis for special components. The curve Σ may have special components Σ_v which have genus $g(v)$ equal to zero or one. Their special feature is the existence of the automorphism groups. In the presence of punctures, the automorphism groups are eaten up ("Higgsed" in the physical slang) by the data p_F, g_F. Let $\mathcal{E}_v$ be the holomorphic bundle over Σ_v determined by $\bar{\partial}_{A_v}$. Generically this bundle has the automorphism group $\mathrm{Aut}(\mathcal{E}_v)$ presented in the table:

$g(v)$	$\mathrm{Aut}(\mathcal{E}_v)$
0	G
1	$\mathbf{T}$
$\geqslant 2$	$\mathbf{1}$

Here $\mathbf{T}$ is the maximal torus of G.

DEFINITION 8.5. The *local phase space* $\mathcal{P}_v$ associated to a vertex v is the symplectic quotient of the space of $(\bar{\partial}_{A_v}, \phi_v)$ multiplied by the coadjoint orbits $\mathcal{O}_e$ for all e such that $v_1(e) = v$ and $v_2(e) \neq v$, and $\mathcal{O}_e^\vee$ for all e such that $v_2(e) = v$ and $v_1(e) \neq v$, by the action of the gauge group $\mathcal{G}_v$.

In formulas, this is the quotient of the space of solutions of the equation

$$\bar{\partial}_{A_v}\phi_v = \sum_{e:\, v_1(e)=v} J_e \delta^{(2)}(z - x_v^e) - \sum_{e:\, v_2(e)=v} J_e \delta^{(2)}(z - x_v^e) \tag{8.17}$$

by the action of the gauge group. Now let us consider the following fairly simple cases, in which one can write explicit formulas for the systems on $\mathcal{P}_v$.

1. *Genus zero component without double points*, i.e., the corresponding vertex v has no edges that start and end at v (no loops). In this case, omitting the label v and taking the gauge for which $\bar{A} = 0$ (this is always possible for stable bundles over $\mathbb{P}^1$, since they are trivial), instead of (8.16) we get the equation

$$\bar{\partial}\phi = \sum_\alpha p^\alpha \delta^{(2)}(z - x^\alpha), \tag{8.18}$$

whose solution exists if and only if $\sum_\alpha p^\alpha = 0$ and is given by

$$i\pi\phi(z) = \sum_\alpha \frac{p^\alpha}{z - x^\alpha}. \tag{8.19}$$

As a result one gets the Schlesinger–Gaudin model.

2. *Genus zero with double points.* This case differs from the previous one by the conditions imposed on the residues of ϕ at the normalizations of the double points. Let us denote the set of punctures by B, $\alpha \in B$, the set of double points by S, and its normalizations by $S^\pm \subset B$. The phase space is the symplectic quotient of the product of coadjoint orbits $\mathcal{O}_e$ corresponding to the punctures in $B \setminus S^+ \amalg S^-$ times the product over $\sigma \in S$ of copies of T^*G. The residues of the Higgs field ϕ at the normalizations of the double point z must be opposite to each other up to a conjugation. More precisely if $\sigma \in S^+$, $\sigma' \in S^-$ are two normalizations of a single double point, then $p^\sigma = -g^{-1} p^{\sigma'} g$, for $g \in G$. The pairs (p^σ, g) span a copy of T^*G corresponding to the given double point.The symplectic quotient is taken

with respect to G, which acts on g by conjugation. The moment map is the same as (8.18), where now α runs over the whole of B.

For example, if there is one double point and one single puncture, then the integrable model one gets as the result of the reduction is the spin generalization of the Calogero–Moser system.

3. *Genus one without double points.* In this case, in addition to the product of orbits $\mathcal{O}_e$ one gets a copy of the infinite-dimensional space of pairs $(\bar{\partial}_A, \phi)$ which live on the elliptic curve. The moment map is given by

$$\bar{\partial}\phi + [\bar{A}, \phi] = \sum_{\alpha} p^{\alpha}\delta^{(2)}(z - x^{\alpha}); \tag{8.20}$$

one may choose a gauge $\bar{A} = \mathrm{diag}(a_1, \ldots, a_N)$ and solve (8.20) by (6.6) with the substitution $p^{\alpha} \to J_l$.

4. *Genus one with double points.* This case is treated by simply merging cases 3 and 2, so we shall not discuss it.

9. Duality

The reduction of one phase space may give rise to several integrable systems defined on the same reduced phase space. The presentation of this section follows [**21**] closely.

9.1. General philosophy.

DEFINITION 9.1. Two integrable systems (H_1, P, ω) and (H_2, P, ω) defined on the same phase space are called *weakly dual.* Dual integrable systems are called *strongly dual* in the domain $\mathcal{U}$ if the integrals of motion of the first and the second systems form local coordinates on $\mathcal{U}$. If there exists a symplectomorphism s of (P, ω) which maps (H_1, P, ω) to (H_2, P, ω), then the system is called *self-dual.*

REMARK 9.2. Let (H_1, P, ω) and (H_2, P, ω) be strongly dual in $\mathcal{U}$. There are two (noncanonical) distinguished sets of Darboux coordinates on $\mathcal{U}$: the action-angle variables for the first and for the second system. The passage from one system of coordinates to another one is an interesting canonical transformation associated to the pair of dual systems. It is defined noncanonically (like the coordinates themselves).

EXAMPLE 9.3. The rational Calogero–Moser system is self-dual. The Sutherland model is dual to the rational Ruijsenaars model. The trigonometric Ruijsenaars system is self-dual.

REMARK 9.4. The importance of dual systems and the canonical transformations associated with them becomes apparent in quantum theory. The integrable systems $(H_1, \ldots)$ and $(H_2, \ldots)$ become maximal collections $\widehat{H}_1^l$ and $\widehat{H}_2^l$ of commuting operators acting in the Hilbert space $\mathcal{H}$. Diagonalizing $\widehat{H}_1^l$, we identify $\mathcal{H}$ with $\mathrm{Fun}(\mathcal{G}_1)$, where $\mathcal{Q}_1 = \mathrm{Spec}\{\widehat{H}_1^l\}$. Similarly we identify $\mathcal{H}$ with $\mathrm{Fun}(\mathcal{G}_2)$, where $\mathcal{Q}_2 = \mathrm{Spec}\{\widehat{H}_2^l\}$. The identity map $\mathcal{H} \to \mathcal{H}$, viewed as an element of $\mathrm{Map}(\mathrm{Fun}(\mathcal{G}_1) \to \mathrm{Fun}(\mathcal{G}_2))$, may be represented with the help of the kernel

$$\mathcal{K}(q, \lambda), \qquad q \in \mathcal{Q}_1,\ \lambda \in \mathcal{Q}_2. \tag{9.1}$$

In the applications the q's are called *coordinates* and the λ's are the *spectral parameters.* Both q and λ may be continuous or discrete.

In the next subsections we work out a few examples of dual systems explicitly.

9.2. Classical systems. The two-particle systems that we are going to consider reduce (after excluding the center of mass motion) to a one-dimensional problem. The action-angle variables can be written explicitly, and the dual system emerges immediately after the natural Hamiltonians are chosen. The problem is the following. Suppose the phase space is (locally) coordinatized by (p, q). The dual Hamiltonian (in the sense of AC duality) must be a certain function $H_D(I, \varphi) = H_D(q)$. In all the cases below there a natural choice of $H_D(q)$.

Calogero oscillator. Consider the model with the Hamiltonian in the center of mass frame:

$$H(p, q) = \frac{p^2}{2} + \frac{\omega^2 q^2}{2} + \frac{\nu^2}{2q^2}, \tag{9.2}$$

where ω and ν are the parameters. In the limiting cases $\nu = 0$ and $\omega = 0$, one gets the ordinary oscillator and the rational Calogero–Moser system respectively. The action-angle variables I, φ can be found by the standard procedure:

$$I = \frac{1}{2\pi} \oint p\, dq = \frac{1}{2\pi} \oint \sqrt{2\left(E - \frac{\omega^2 q^2}{2} - \frac{\nu^2}{2q^2}\right)}\, dq, \qquad d\varphi = \frac{dq}{p} \left(\frac{\partial I}{\partial E}\right)_I^{-1}, \tag{9.3}$$

with the result

$$\begin{aligned} I &= \frac{E - \omega\nu}{2\omega} = \frac{1}{4\omega}\left[p^2 + \left(\omega q - \frac{\nu}{q}\right)^2\right], \\ H_D(I, \varphi) &= \frac{q^2}{2} = \frac{I}{\omega}\left[1 + \frac{\nu}{2I} + \sqrt{1 + \frac{\nu}{I}} \cos\varphi\right]. \end{aligned} \tag{9.4}$$

The limit as $\nu \to 0$ is straightforward, yet tricky. We must rescale $\varphi \to 2\varphi$, since the period of motion jumps as ν approaches zero. We obtain

$$H_D(I, \varphi) = q = 2\sqrt{I} \cos\varphi. \tag{9.5}$$

The limit $\omega \to 0$ is more subtle, since the classical motion become infinite. For the system with the Hamiltonian

$$H(p, q) = \frac{p^2}{2} + \frac{\nu^2}{2q^2} \tag{9.6}$$

the action variable can be defined as the asymptotic value of the momentum: $I = \sqrt{2E}$. This choice give rise to the evolution, linear in the "angle"-like variable,

$$\varphi = \sqrt{q^2 - \frac{\nu^2}{2E}}, \qquad H_D(I, \varphi) = \frac{q^2}{2} = \frac{\varphi^2}{2} + \frac{\nu^2}{2I^2}. \tag{9.7}$$

Sutherland model. The Hamiltonian (again, in the center of mass frame) is

$$H(p, q) = \frac{1}{2} p^2 + \frac{\nu^2}{2 \sin^2(q)}. \tag{9.8}$$

The action variable I in this case turns out to be the asymptotic value of the momentum, shifted by ν:

$$I = \sqrt{2E} - \nu. \tag{9.9}$$

To prove this, one can pass to the coordinate $t = \cos q$ and compute the integral $(2\pi)^{-1} \oint p\, dq$ using residues. It is interesting to note that if one defines the action variable as the integral

$$\int_0^{2\pi} \frac{p\, dq}{2\pi},$$

then the answer will be $\tilde{I} = \sqrt{2E}$. In some formulas below it is more convenient to use $\tilde{I}$ instead of I. Since they differ by a constant, this does not change the dynamics. The angle variable φ can be determined from the condition $dp \wedge dq = d\tilde{I} \wedge d\varphi$. We get

$$d\varphi = \frac{\tilde{I}\, dq}{\sqrt{\tilde{I}^2 - 2\nu^2 \sin^{-2} q}}, \tag{9.10}$$

$$H_D(I, \varphi) = \cos q = \cos\varphi \sqrt{1 - \frac{2\nu^2}{\tilde{I}^2}}. \tag{9.11}$$

Notice that $\cos q = \operatorname{Tr} g$ coincides with the Hamiltonian of the rational Ruijsenaars model (see below).

Elliptic Calogero–Moser system. The Hamiltonian is

$$H(p, q) = \frac{p^2}{2} + \nu^2 \wp_r(q). \tag{9.12}$$

Here $\wp_r(q)$ is the Weierstrass function on the elliptic curve E_τ,

$$\wp_r(q) = \frac{1}{\pi^2} \wp\left(\frac{q}{\pi}\right). \tag{9.13}$$

Let us introduce Weierstrass notation: $x = \wp_r(q)$, $y = \wp_r(q)'$. We have the following equation defining the curve E_τ:

$$y^2 = 4x^3 - g_2(\tau)x - g_3(\tau) = 4\prod_{i=1}^{3}(x - e_i), \qquad \sum_{i=1}^{3} e_i = 0. \tag{9.14}$$

The holomorphic differential dq on E_τ equals $dq = dx/y$. Introduce the variable $e_0 = 2E/\nu^2$. The action variable is one of the periods of the differential $(2\pi)^{-1} p\, dq$ on the curve $E = H(p, q)$ (this is specific to the two-particle case, but there exists a general prescription):

$$I = \frac{1}{2\pi} \oint_A \sqrt{2(E - \nu^2 \wp_r(q))} = \frac{1}{4\pi i} \oint_A \frac{dx \sqrt{x - e_0}}{\sqrt{(x - e_1)(x - e_2)(x - e_3)}}. \tag{9.15}$$

The angle variable can be determined from the condition $dp \wedge dq = dI \wedge d\varphi$:

$$d\varphi = \frac{1}{2iT(E)} \frac{dx}{\sqrt{\prod_{i=0}^{3}(x - e_i)}}, \tag{9.16}$$

where $T(E)$ normalizes $d\varphi$ in such a way that the A period of $d\varphi$ is equal to 2π:

$$T(E) = \frac{1}{4\pi i} \oint_A \frac{dx}{\sqrt{\prod_{i=0}^{3}(x - e_i)}}. \tag{9.17}$$

Thus

$$2iT(E)\,d\varphi = \frac{dx}{\sqrt{4\prod_{i=0}^{3}(x-e_i)}}, \qquad \omega\,d\varphi = \frac{dt}{\sqrt{4\prod_{i=1}^{3}(t-t_i)}}, \tag{9.18}$$

where

$$\begin{gathered} \omega = -2i\sqrt{e_{01}e_{02}e_{03}}\,T(E) = \frac{1}{2\pi}\oint_A \frac{dt}{\sqrt{4\prod_{i=1}^{3}(t-t_i)}}, \\ t = \frac{1}{x-e_0} + \frac{1}{3}\sum_{i=1}^{3}\frac{1}{e_{0i}}, \qquad t_i = \frac{1}{3}\sum_{j=1}^{3}\frac{e_{ji}}{e_{0i}e_{0j}}, \qquad e_{ij} = e_i - e_j. \end{gathered} \tag{9.19}$$

Introduce a meromorphic function on E_τ as follows:

$$\widehat{cn}_\tau(z) = \sqrt{\frac{x-e_1}{x-e_3}}, \tag{9.20}$$

where z has periods 2π and $2\pi\tau$. This function is the elliptic analog of the cosine (in fact, up to a rescaling of z, it coincides with the Jacobi elliptic cosine). Then we have

$$H_D(I,\varphi) = \widehat{cn}_\tau(z) = \widehat{cn}_{\tau_E}(\varphi)\sqrt{1-\frac{\nu^2 e_{13}}{2E-\nu^2 e_3}}, \tag{9.21}$$

where τ_E is the modular parameter of the relevant spectral curve $v^2 = 4\prod_{i=1}^{3}(t-t_i)$ given by the formula

$$\tau_E = \left(\oint_B \frac{dt}{\sqrt{4\prod_{i=1}^{3}(t-t_i)}}\right) \Big/ \left(\oint_A \frac{dt}{\sqrt{4\prod_{i=1}^{3}(t-t_i)}}\right). \tag{9.22}$$

For large I, $2E(I) \sim I^2$.

Elliptic Ruijsenaars model. The Hamiltonian is

$$H(p,q) = \cos(\beta p)\sqrt{1-2(\beta\nu)^2\wp_\tau(q)}. \tag{9.23}$$

As the curve E_τ degenerates, one moves to the trigonometric ($\wp_r(q) \to 1/\sin^2 q$) or rational ($\wp_r(q) \to 1/q^2$) Ruijsenaars system. The spectral curve $H(p,q) = E$ helps to define the action variable I:

$$I = \frac{1}{2\pi}\oint_A p\,dq, \tag{9.24}$$

which is defined up to the transformations $I \to I^D$ ($A \to B$) and $I \to I + 2\pi n/\beta$, $n \in \mathbb{Z}$ (this fact was used in [**22**]). We can write an explicit formula for the following quantity, which is better defined,

$$\frac{\partial I}{\partial E} = \frac{1}{2\sqrt{2}\pi\beta^2\nu}\oint_A \frac{dx}{\sqrt{\prod_{i=0}^{3}(x-e_i)}}, \tag{9.25}$$

where now $e_0 = (1-E^2)/(2(\beta\nu)^2)$. Under the transformation $I \to I^D$, $\partial I/\partial E$ is multiplied by τ_E, where τ_E is defined as in (9.22). Quite similarly to (9.18), we obtain

$$d\varphi = \frac{1}{T(E)}\frac{dx}{\sqrt{\prod_{i=0}^{3}(x-e_i)}} \tag{9.26}$$

with

$$T(E) = \frac{1}{2\pi} \oint_A \frac{dx}{\sqrt{\prod_{i=0}^{3}(x - e_i)}}. \tag{9.27}$$

Finally, for H_D given by (9.20) we can write

$$H_D(I, \varphi) = \widehat{cn}_\tau(z) = \widehat{cn}_{\tau_E}(\varphi)\sqrt{1 + \frac{2(\beta\nu)^2 e_{13}}{E(I)^2 - 1 - 2(\beta\nu)^2 e_3}}. \tag{9.28}$$

Asymptotically, for large I, $E(I) \sim \cos(\beta I)$.

9.3. Quantum systems. Here we work out a few examples of quantum dual systems.

Harmonic oscillator. The Hamiltonian (9.2) in the limit $\nu = 0$ quantizes to

$$\widehat{H} = -\frac{1}{2}\frac{\partial^2}{\partial q^2} + \frac{\omega^2 q^2}{2}. \tag{9.29}$$

Its normalized eigenfunctions are [**23**]

$$\widehat{H}\psi_n = \omega\left(n + \frac{1}{2}\right)\psi_n, \qquad \psi_n(q) = \left(\frac{\omega}{\pi}\right)^{1/4} \frac{e^{-\omega q^2/2}}{2^{n/2}\sqrt{n!}} H_n(q\sqrt{\omega}), \tag{9.30}$$

where $H_n(\xi)$ is the Hermite polynomial: $H_n(\xi) = e^{\xi^2}(-\partial_\xi)^n e^{-\xi^2}$. Using this representation of the wave function, one can easily obtain the following recurrence relation (details are in the Appendix):

$$\sqrt{n+1}\,\psi_{n+1}(q) + \sqrt{n}\,\psi_{n-1}(x) = \sqrt{2\omega}\, q\psi_n(x). \tag{9.31}$$

This means that $\psi_n(q)$ is an eigenfunction of the following difference operator:

$$\widehat{H}_D = T_+\sqrt{n} + \sqrt{n}\,T_-, \qquad T_\pm = e^{\pm\partial/\partial n}, \tag{9.32}$$

acting on the subscript n. It is easy to recognize in (9.32) the quantized version of (9.5).

Sutherland model. Here we deal with the Hamiltonian

$$\widehat{H} = -\frac{1}{2}\frac{\partial^2}{\partial q^2} + \frac{\nu(\nu-1)}{2\sin^2 q}. \tag{9.33}$$

Its normalized eigenfunctions are [**23**]

$$\begin{gathered} \widehat{H}\psi_n = \frac{n^2}{2}\psi_n, \qquad \psi_n(q) = \sin^\nu(q)\sqrt{n\frac{(n-\nu)!}{(n+\nu-1)!}}\,\Pi_{n-1/2}^{\nu-1/2}(\cos q), \\ \Pi_l^m(x) = \frac{1}{l!}\,\partial_x^{l+m}\left(\frac{x^2-1}{2}\right)^l. \end{gathered} \tag{9.34}$$

For simplicity we take ν and n to be half-integers. One can change $\nu \to -\nu - 1$ to get another eigenfunction with the same eigenvalue. Using the fact that the generating function for the Π_l^0's is

$$Z(y, x) = \sum_{l=0}^{\infty} y^l \Pi_l^0 = \frac{1}{\sqrt{1 - 2xy + y^2}}, \tag{9.35}$$

one derives the recurrence relations (details are in the Appendix)

$$
(9.36)\qquad
\begin{aligned}
x\Pi_l^m &= \frac{l+1-m}{2l+1}\,\Pi_{l+1}^m + \frac{l+m}{2l+1}\,\Pi_{l-1}^m,\\
\cos(q)\psi_n &= \frac{1}{2}\left(\sqrt{\frac{1-\nu(\nu-1)}{n(n+1)}}\,\psi_{n+1} + \sqrt{\frac{1-\nu(\nu-1)}{n(n-1)}}\,\psi_{n-1}\right);
\end{aligned}
$$

this means that ψ_n is an eigenfunction of the finite-difference operator acting on the subscript n:

$$
(9.37)\quad \widehat{H}_D\psi(q) = \cos(q)\psi(q), \qquad \widehat{H}_D = T_+\sqrt{\frac{1-\nu(\nu-1)}{n(n-1)}} + \sqrt{\frac{1-\nu(\nu-1)}{n(n-1)}}\,T_-,
$$

which is a quantum version of (9.11).

Moral of the story. The moral of the previous discussion is that the polynomial dependence on momenta of the Hamiltonian is traded with the rational potential of the dual system. The trigonometric potential is mapped to the trigonometric (= relativistic) dependence on momenta of the dual system. The elliptic potential gives rise to an elliptic (="doubly-relativistic") dependence on the momentum of the dual Hamiltonian system. When the system with trigonometric dependence on momentum is quantized, its Hamiltonian becomes a finite-difference operator. The wave functions become functions of discrete variables. The origin of this is in the Bohr–Sommerfeld quantization condition. Indeed, since the trigonometric dependence of momenta implies that the leaves of the polarization are compact and moreover non-simply-connected, the covariantly constant sections of the prequantization connection along the polarization fiber generically ceases to exist. It is only for special "quantized" values of the action variables that the section exists. In the elliptic case, the quantum dual Hamiltonian will be a difference operator of infinite order.

Appendix. To derive the recurrence relation for the oscillator wave-functions, we use the creation operator representation:

$$
\psi_n = \frac{1}{\sqrt{2n}}\,(-\partial_\xi + \xi)\psi_n.
$$

Applying this relation twice and using the fact that ψ_n is an eigenfunction of $\widehat{H}$, one arrives at (9.31). For the Sutherland model, we use two obvious relations:

$$
(9.38)\qquad (x-y)\partial_x Z = y\partial_y Z,
$$

$$
(9.39)\qquad (1-2xy+y^2)\partial_y Z = (x-y)Z.
$$

Next, (9.38) implies

$$
(9.40)\qquad (y\partial_y - m)\partial_x^m Z = (x-y)\partial_x^{m+1} Z
$$

and (9.39) yields

$$
(9.41)\qquad ((1-2xy+y^2)\partial_y + y - x)\partial_x^m Z = m(1+2y\partial_y)\partial_x^{m-1} Z.
$$

Combining those two relations, we obtain (9.36).

10. Gauge theories at last

It is the time to fulfill the promise of the title and to explain the relationship of the constructions we have presented so far to gauge theories.

10.1. What is gauge theory? Gauge theory in space-time X of dimension d is the study of the space of gauge fields, i.e., connections A in some principal G-bundle P over X and associated geometrical objects, such as sections ψ of the bundles E over Σ associated to P via a certain representation R of G. The dynamics is governed by the gauge-invariant Lagrangian $L(A, \psi)$. Quantum mechanical expectation values are given by the path integral

$$\langle \mathcal{O}_1 \dots \mathcal{O}_p \rangle = \int DAD\psi e^{-L} \mathcal{O}_1(A, \psi) \dots \mathcal{O}_p(A, \psi) \tag{10.1}$$

over a suitable space of fields.

There are several ways the gauge theory can realize an integrable system within it. The most straightforward one is the case in which the dynamics of gauge invariant degrees of freedom is integrable by itself. We shall see that this is the case for certain gauge theories for $d = 2$ and $d = 3$. The less transparent way is realized in supersymmetric gauge theories for $d = 3$ and $d = 4$. In this case the low energy effective theory is expressed in terms of the sigma model on the phase space of the integrable system ($d = 3$) or the base of the Liouville fibration ($d = 4$).

If the space-time manifold X is the product $Y \times T^k$, then the effective theory is macroscopically $(d - k)$-dimensional. The geometry of the space of vacua of the effective theory depends on the geometry of the torus T^k. Therefore it is natural to expect that the integrable system which governs the effective dynamics of the gauge theory on Y can be included into a family of integrable systems.

An example of such a family is provided by relativistic integrable systems of Calogero type. The parameter β interpolates between the nonrelativistic case $\beta = 0$ and the ultrarelativistic case $\beta = \infty$.

The rational-trigonometric-elliptic series also has an interpretation in gauge theory terms.

10.2. Low dimensional theories. Let us consider the two-dimensional Yang–Mills theory. The space-time manifold is some Riemann surface $X = \Sigma$, the fields are just gauge fields, and the Lagrangian equals

$$L(A) = \frac{1}{2e^2} \int_\Sigma \langle F_A, \star F_A \rangle, \tag{10.2}$$

where $F_A = dA + A^2$ is the curvature of the connection A. The path integral (10.1) can be put into the first order form by the standard trick of introducing the electric field ϕ:

$$\int DAD\phi \exp \left(\int_\Sigma i \langle \phi, F \rangle - \frac{e^2}{2} \langle \phi, \phi \rangle \right). \tag{10.3}$$

The gauge-invariant observables $\mathcal{O}_k$, which can be constructed from A and ϕ only, are of two kinds. The local operators are simply our invariant polynomials $u_k(\phi(x))$ evaluated at some point $x \in \Sigma$. There are also interesting nonlocal operators, namely Wilson loops $W_{C,k} = u_k(P \exp \oint_C A)$. Now let us study the canonical quantization of the theory. To this end we take $\Sigma = S \times \mathbb{R}$, where S is some one-dimensional PL-manifold, for example S^1 or even a graph Γ. Let A_0 be the component of A along $\mathbb{R}$, and A the component of A along S. The action in (10.3) can be written in components as

$$\int_\mathbb{R} \int_S \langle A_0, \partial_t \phi + [A, \phi] \rangle + \langle \phi, \partial_0 A \rangle - \frac{e^2}{2} \langle \phi, \phi \rangle. \tag{10.4}$$

We see that A_0 enters linearly and can be integrated out, giving rise to the constraint we encountered above in Section 6. The modification of the action (10.4) appears upon including Wilson loops. One can show [**5**] that in this case the constraints involving the orbits $\mathcal{O}_v$ occur.

In particular, by taking $S = S^1$ and one Wilson line corresponding to an orbit $\mathcal{O}$, one obtains the spin Sutherland model as an equivalent quantum mechanical system.

A deformation of the two-dimensional Yang–Mills theory on Σ is the three-dimensional Chern–Simons theory on $\Sigma \times S^1$. In the appropriate scaling limit in which the circle shrinks to a point, the three-dimensional theory becomes equivalent to the two-dimensional one. The canonical quantization of Chern–Simons theory should involve the study of the gauge fields on $S \times S^1$. The analog of the constraint coming from integrating out A_0 is the flatness condition. By passing to the fields A and g, where A is the gauge field on S and g is the holonomy around the circle S^1, we get the constraints mentioned above. More details can be found in [**6**].

10.3. Supersymmetric gauge theories. We shall consider a specific example of a supersymmetric gauge theory in four dimensions. The fields of the theory are: the gauge field A, the scalars ϕ and $\bar{\phi}$, and the fermions, namely the 1-form ψ, the scalar η and the self-dual 2-form χ. All fields take values in the adjoint bundle $\mathrm{ad}(P)$ over the four-dimensional Riemannian manifold X. The Lagrangian is given by the expression

(10.5)
$$L = \int_X \frac{1}{e^2} \langle F, \star F \rangle + \frac{i\theta}{8\pi^2} \langle F \wedge F \rangle + \langle \chi \oplus \eta, (d_A^+ \oplus d_A^*)\psi \rangle + \langle d_A\phi, \star d_A\bar{\phi} \rangle \ldots,$$

where $\ldots$ denote cubic and quartic terms. The Lagrangian (10.5) describes various fields, massless and massive. Upon integrating out massive fields, one arrives at the low-energy effective action, which is of similar form except for the fact that the coupling

$$\tau = \frac{\theta}{2\pi} + \frac{4\pi i}{e^2}$$

may depend on ϕ. It can be shown on supersymmetry grounds that the low-energy action has the form

$$\tau_{ij} F_-^i \wedge F_-^j - \bar{\tau}_{ij} F_+^i \wedge F_+^j + \mathrm{Im}\,\tau_{ij}\, da^i \wedge \star d\bar{a}^j + \ldots, \tag{10.6}$$

where the a^i's are the components of the field ϕ, which is bound to take values in $\underline{\mathfrak{t}}$, and F^i are the corresponding components of the curvature of the gauge field.

A striking result which was obtained first in the particular case of $G = SU(2)$ in [**24**] and then subsequently extended to the general case of $G = SU(N)$ in [**25, 26, 27**] is that the effective coupling matrix τ_{ij} can be identified with the period matrix of the Liouville torus of the holomorphic integrable system whose space of integrals of motion is identified with the moduli space of vacua of the supersymmetric gauge theory. Then this relation was derived on general grounds in [**28**], essentially from the special geometry of vector multiplets, and in particular the integrable system describing the $SU(N)$ supersymmetric gauge theory with adjoint matter was presented. It turned out to be precisely the elliptic Calogero–Moser system. As for the gauge theories with classical gauge groups other than $SU(n)$, no substantial progress has been made so far [**29**]. Since there are a number of interesting theories in four dimensions, one may wonder about the theories which are compactified versions of higher dimensional supersymmetry. The point is that the theory (10.5) corresponds to the $\mathcal{N}=2$ super-Poincaré algebra in four dimensions, which is a contraction of the $\mathcal{N}=1$ super-Poincaré algebra in five dimensions. Integrable systems which govern five-dimensional theories have been proposed in [**22**] and subsequently checked to some extent in [**30**]. See also [**31**] for the latest developments.

11. Conclusions and prospects for the future

11.1. Beyond integrability. The condition of integrability is very restrictive. Sometimes the Hamiltonian system may be solvable without being integrable in the Liouville sense.

EXAMPLE 11.1. Let $M=T^*G$. Consider the geodesic motion on G. It is generated by the Hamiltonian $\frac{1}{2}\operatorname{Tr} p^2$. There are $\frac{1}{2}(\dim G+\operatorname{rk} G)$ Poisson-commuting Hamiltonians on T^*G. This is not enough in general for the system to be integrable. Nevertheless the solution can be effectively found:

$$G^t(p,g)=(p,\exp(tp)g).$$

DEFINITION 11.2. A Hamiltonian system H on the $2m$-dimensional phase space (M^{2m},ω) is called *Lie integrable* if there are m almost-everywhere functionally independent functions H_a, $a=1,\dots,m$, whose Poisson brackets form a Lie algebra of the semisimple Lie group G:

$$\{H_a,H_b\}=f^c_{ab}H_c. \tag{11.1}$$

In other words, the Hamiltonian flow generated by H must be included in the group action on M.

A warning. The reader should keep in mind that the notion of Lie integrability is actually very different from the notion of Liouville integrability. Of course, Lie integrability with the group $\mathbf{T}$ implies Liouville integrability, but Liouville integrability in general does not imply Lie integrability.

Solution. Given a Lie integrable system, one may easily solve the equations of motion by means of the exponential map. The application of this principle to the reduced system is the following. Suppose the Hamiltonian system on $M_{\mathbf{A}}$ is Lie integrable. Suppose that the group A is a subgroup of the group G in the definition of Lie integrability. Then the solution of the reduced system is achieved by using the so-called projection method [**32**].

11.2. Duality for general elliptic models. One can address the following question: What would be the most general integrable system in the holomorphic sense with compact phase space of dimension 2? Since the phase space M must carry a holomorphic 2-form, nowhere degenerate, one has $c_1(M)=0$. Thus, M is either T^4 or $K3$. Since we are looking for an integrable system, the manifold M must be a fibration over a one-dimensional base B, the fiber being a one-dimensional abelian variety (perhaps degenerate). Since the topological Euler characteristic of a degenerate elliptic curve is greater than or equal to one, then either there are no degenerate fibers and $\chi(M)=0, M=T^4$, or there are degenerate fibers, and $M=K3$. Since $\chi(K3)=24$, generically one needs 24 fibers.

Elliptic $K3$ is a fibration over $B = \mathbb{P}^1$: there exists a holomorphic map $p\colon M \to B$. In the formula (9.14) the coefficients g_2 and g_3 become sections of the line bundles $\mathcal{O}(4n)$ and $\mathcal{O}(6n)$ respectively (over B). The elliptic curve degenerates over the divisor of its discriminant,

$$\Delta = g_2^3 - 27g_3^2, \tag{11.2}$$

which is a section of $\mathcal{O}(12n)$. The latter has generically $6n$ zeroes. As we said, we need 24 singular fibers, which fixes $n = 2$. The Hamiltonian of the integrable system we consider is any function on B. It gives rise to a meromorphic vector field on M, which linearizes along the elliptic fibers. The symplectic form reads

$$\omega = \frac{dx \wedge dz}{y}, \tag{11.3}$$

where z is the projective coordinate on B (when one changes $z \to 1/z$, $y \to -y/z^6$, $x \to x/z^4$, the form (11.3) is unchanged and the equation (9.14) is multiplied by $1/z^{12}$). Now, in order to get a dual system, we must impose a condition on M. Namely, it must be a double elliptic fibration: there should be two bases B_1 and B_2 and two projections $p_{1,2}\colon K3 \to B_{1,2}$, whose fibers generically are the elliptic curves.

Acknowledgements

I am grateful to H. Braden, R. Donagi, B. Enriquez, P. Etingof, L. Faddeev, B. Feigin, V. Fock, I. Frenkel, V. Ginzburg, A. Gorsky, A. Kirillov, Jr., I. Krichever, D. Lebedev, E. Martinec, M. Olshanetsky, A. Polychronakos, A. Rosly, V. Rubtsov, S. Ruijsenaars, S. Shatashvili, F. Smirnov, A. Stoyanovsky, and C. Vafa for inspiring discussions.

References

[1] N. Hitchin, A. Karlhede, U. Lindström, and M. Rocek, *Hyper-Kähler metrics and supersymmetry*, Comm. Math. Phys. **108** (1987), 535–589.

[2] P. Kronheimer, *The construction of ALE spaces as hyper-Kähler quotients*, J. Differential Geom. **28** (1989), 665–683.

[3] N. Hitchin, *Stable bundles and integrable systems*, Duke Math. J. **54** (1987), 91–114.

[4] D. Kazhdan, B. Kostant, and S. Sternberg, *Hamiltonian group actions and dynamical systems of Calogero type*, Comm. Pure Appl. Math. **31** (1978), 481–507.

[5] A. Gorsky and N. Nekrasov, *Hamiltonian systems of Calogero type and two-dimensional Yang–Mills theory*, Nuclear Phys. B **414** (1994), 213–238.

[6] ———, *Relativistic Calogero–Moser model as gauged WZW theory*, Nuclear Phys. B **436** (1995), 582–608.

[7] A. Pressley and G. Segal, *Loop groups*, Clarendon Press, Oxford, 1986.

[8] J. Gibbons and T. Hermsen, *A generalization of the Calogero–Moser system*, Physica D **11** (1984), 337–348.

[9] S. Wojciechowski, *On the integrability of the Calogero-Moser system in an external quartic potential and other many-body systems*, Phys. Lett. A **102** (1984), 85–88.

[10] I. Krichever, O. Babelon, E. Billey, and M. Talon, *Spin generalization of the Calogero–Moser system and the matrix KP equation*, `hep-th/9411160`; Topics in Topology and Mathematical Physics (S. P. Novikov, ed.), Amer. Math. Soc. Transl. (2) **170** (1995), 83–119.

[11] J. Mickelsson, *Chiral anomalies in even and odd dimensions*, Comm. Math. Phys. **97** (1985), 361–370.

[12] ———, *Kac–Moody groups, topology of the Dirac determinant bundle and fermionization*, Comm. Math. Phys. **110** (1987), 173–183.

[13] N. Nekrasov, *Holomorphic bundles and many-body systems*, `hep-th/9503157`; Comm. Math. Phys. **180** (1996), 587–603.

[14] S. N. M. Ruijsenaars and H. Schneider, *A new class of integrable systems and its relation to solitons*, Ann. Physics **170** (1986), 370–405.

[15] A. Gorsky and N. Nekrasov, *Elliptic Calogero–Moser system from two dimensional current algebra*, hep-th/9401021.

[16] G. E. Arutyunov, S. A. Frolov, and P. B. Medvedev, *Elliptic Ruijsenaars–Schneider model via the Poisson reduction of the affine Heisenberg double*, hep-th/9607170; J. Phys. A **30** (1997), 5051–5063.

[17] G. E. Arutyunov, S. A. Frolov, and P. B. Medvedev, *Elliptic Ruijsenaars–Schneider model from the cotangent bundle over the two-dimensional current group*, hep-th/9608013; J. Math. Phys. **38** (1997), 5682–5689.

[18] G. E. Arutyunov, L. O. Chekhov, and S. A. Frolov, *R-matrix quantization of the elliptic Ruijsenaars–Schneider model*, q-alg/9612032; Comm. Math. Phys. **192** (1998), 405–432; G. E. Arutyunov and S. A. Frolov, *Quantum dynamical R-matrices and quantum Frobenius group*, q-alg/9610009; Comm. Math. Phys. **191** (1998), 15–29.

[19] I. Krichever, *Rational solutions of the Kadomtsev–Petviashvili equations and the integrable systems of N particles on the line*, Funktsional. Anal. i Prilozhen. **12** (1978), no. 1, 76–78; English transl., Functional Anal. Appl. **12** (1978), no. 1, 59–61; *Elliptic solutions of the Kadomtsev–Petviashvili equations, and integrable systems of particles*, Funktsional. Anal. i Prilozhen. **14** (1980) no. 4, 45–54; English transl., Functional Anal. Appl. **14** (1980), 282–290.

[20] A. Treibich and J.-L. Verdier, *Solitons elliptiques*, Grothendieck Festschrift, III, Birkhauser, Boston, 1990, pp. 437–480.

[21] V. Fock, A. Gorsky, N. Nekrasov, and V. Rubtsov, *Duality in many-body systems and gauge theories*, in preparation.

[22] N. Nekrasov, *Five dimensional gauge theories and relativistic integrable systems*, hep-th/9609219.

[23] L. Landau and E. Lifshitz, *Quantum mechanics (non-relativistic theory)*, Pergamon Press, 1977.

[24] N. Seiberg and E. Witten, *Electric-magnetic duality, monopole condensation, and confinement in $N = 2$ supersymmetric Yang–Mills theory*, Nuclear Phys. B **426** (1994), 19–52 (erratum: Ibid. **430** (1994), 485–486); *Monopoles, duality and chiral symmetry breaking in $N = 2$ supersymmetric QCD*, hep-th/9408099; Nuclear Phys. B **431** (1994), 484–550.

[25] P. C. Argyres and A. E. Farragi, *The vacuum structure and spectrum of $N = 2$ supersymmetric $SU(n)$ gauge theory*, hep-th/9411057; A. Klemm, W. Lerche, S. Theisen, and S. Yankielowicz, *Simple singularities and $N = 2$ supersymmetric Yang–Mills theory*, hep-th/9411048; Phys. Lett. B. **344** (1995), 169–175.

[26] A. Gorsky, I. Krichever, A. Marshakov, A. Morozov, and A. Mironov, *Integrablity and Seiberg–Witten exact solution*, hep-th/9505035, Phys. Lett. B **355** (1995), 466–474.

[27] E. Martinec and N. Warner, *Integrable systems and supersymmetric gauge theory*, hep-th/9509161; Nuclear Phys. B **459** (1996), 97–112.

[28] R. Donagi and E. Witten, *Supersymmetric Yang–Mills theory and integrable systems*, hep-th/9510101; Nuclear Phys. B **460** (1996), 299–334.

[29] R. Y. Donagi, *Seiberg–Witten integrable systems*, alg-geom/9705010; Algebraic Geometry—Santa Cruz, 1995 (J. Kollár et al., eds.), Proc. Sympos. Pure Math., vol. 62, part 2, Amer. Math. Soc., Providence, RI, 1997, pp. 3–43.

[30] A. Lawrence and N. Nekrasov, *Instanton sums and five dimensional gauge theories*, hep-th/9706025; Nuclear Phys. B **513** (1998), 239–265.

[31] A. Marshakov and A. Mironov, *5d and 6d supersymmetric gauge theories: prepotentials from integrable systems*, hep-th/9711156, Nuclear Phys. B **518** (1998), 59–91.

[32] M. Olshanetsky and A. Perelomov, *Completely integrable Hamiltonian systems connected with semisimple Lie algebras*, Invent. Math. **37** (1976), 93–108; Lett. Math. Phys. **1** (1975/76), 187–193.

INSTITUTE OF THEORETICAL AND EXPERIMENTAL PHYSICS, 117259, MOSCOW, RUSSIA; AND LYMAN LABORATORY OF PHYSICS, HARVARD UNIVERSITY, CAMBRIDGE, MA 02138, USA

E-mail address: nikita@string.harvard.edu

Selected Titles in This Subseries

(Continued from the front of this publication)

15 **A. T. Fomenko, Editor,** Minimal surfaces, 1993
14 **Yu. S. Il′yashenko, Editor,** Nonlinear Stokes phenomena, 1992
13 **V. P. Maslov and S. N. Samborskiĭ, Editors,** Idempotent analysis, 1992
12 **R. Z. Khasminskiĭ, Editor,** Topics in nonparametric estimation, 1992
11 **B. Ya. Levin, Editor,** Entire and subharmonic functions, 1992
10 **A. V. Babin and M. I. Vishik, Editors,** Properties of global attractors of partial differential equations, 1992
9 **A. M. Vershik, Editor,** Representation theory and dynamical systems, 1992
8 **E. B. Vinberg, Editor,** Lie groups, their discrete subgroups, and invariant theory, 1992
7 **M. Sh. Birman, Editor,** Estimates and asymptotics for discrete spectra of integral and differential equations, 1991
6 **A. T. Fomenko, Editor,** Topological classification of integrable systems, 1991
5 **R. A. Minlos, Editor,** Many-particle Hamiltonians: spectra and scattering, 1991
4 **A. A. Suslin, Editor,** Algebraic K-theory, 1991
3 **Ya. G. Sinaĭ, Editor,** Dynamical systems and statistical mechanics, 1991
2 **A. A. Kirillov, Editor,** Topics in representation theory, 1991
1 **V. I. Arnold, Editor,** Theory of singularities and its applications, 1990